Senescence in Animals

Senescence in Animals

Edited by **Brandon Chesser**

Callisto Reference

New York

Published by Callisto Reference,
106 Park Avenue, Suite 200,
New York, NY 10016, USA
www.callistoreference.com

Senescence in Animals
Edited by Brandon Chesser

International Standard Book Number: 978-1-63239-556-6 (Hardback)

Printed in China.

Contents

Preface

This book explains the phenomena related to ageing in animals. It covers various aspects relating to ageing and discusses the factors associated with it. It provides elucidative information regarding ageing processes and senescent changes that living creatures such as animals face. Renowned experts and researchers in the field of senescence have made their valuable contributions to this book. It aims to benefit students and researchers interested in this field.

This book is a comprehensive compilation of works of different researchers from varied parts of the world. It includes valuable experiences of the researchers with the sole objective of providing the readers (learners) with a proper knowledge of the concerned field. This book will be beneficial in evoking inspiration and enhancing the knowledge of the interested readers.

In the end, I would like to extend my heartiest thanks to the authors who worked with great determination on their chapters. I also appreciate the publisher's support in the course of the book. I would also like to deeply acknowledge my family who stood by me as a source of inspiration during the project.

Editor

Animal

1

Senescence in Animals: Why Evolutionary Theories Matter

Thiago Monaco, Daniel Silvestre and Paulo S.P. Silveira
Department of Pathology, School of Medicine at the University of Sao Paulo, Sao Paulo Brazil

1. Introduction

Senescence, considered from the individual viewpoint can be characterized as a "progressive loss of fertility and increasing probability of death with increasing age"(Kirkwood & Austad, 2000). This phenomenon can also be considered from the populational perspective: senescent populations present increasingly higher death rates with increasing age (Masoro & Austad, 2006).

This is a clearly deleterious process, which seems difficult to conciliate with natural selection, which predicts evolution towards increasing fitness. Historically, the first evolutionary explanation able to conciliate these two processes is known as the mutation accumulation theory (Medawar, 1952). According to this theory, in age-structured populations the force of selection decreases with increasing age, allowing the accumulation of deleterious genes with age-specific effects on mortality rate (Hamilton, 1966). Under population genetics mechanisms, senescence is not necessarily deleterious: the original Medawar's proposition implies that the postponement of age-specific effects of harmful genes is equivalent to their elimination in such a way that they become effectively neutral. Hence, such postponement is beneficial and senescence can be regarded as a side effect of the process.

Medawar was convinced that these genes could only account for senescent manifestations encountered in protected populations after they reached ages not achievable in the wild and, therefore, that further explanation involving pleiotropy and linkage would be required to account for a gradual process of organic degeneration, but he did not elaborate on them. This was noted by Williams, who explained the maintenance of beneficial and deleterious traits together, giving rise to the antagonistic pleiotropy theory.

Essentially, the antagonistic pleiotropy theory relies on the existence of genes of a special kind, which are capable of increasing and decreasing fitness depending on the somatic environment and/or age. It is not necessary that the beneficial effects precede the deleterious effect as commonly believed e. g. (Futuyma, 1998; Masoro & Austad, 2006). Instead, Williams' original proposition only required an influx of pleiotropic alleles that may fixate in the population due to their overall beneficial effect. In this scenario, the observed senescence is understood as the composition of deleterious components from all present pleiotropic genes (Williams, 1957).

Instead of basing his arguments on genetics, a somewhat different view was offered by Kirkwood, elaborating on the error catastrophe of Orgel (Orgel, 1963). He approached senescence from an ecological argument in which energy resources may be allocated either to somatic cell maintenance or to reproduction, thus generating some sort of soma-germ conflict. Called disposable soma theory, it ultimately relies on the existence of specific genes controlling the accuracy of the transcription/translation machinery in an age-dependent manner. Kirkwood himself regards his theory as a specialization of the antagonistic pleiotropy of Williams (Kirkwood, 1977; Kirkwood & Holliday, 1975a;b; 1979). The difference is that Williams invokes the existence of genes responsible for beneficial and deleterious effects, but Kirkwood's theory, while not denying the existence of these genes, does not require them. The conflicting destination of energy either to body or reproduction maintenance would suffice for the evolution of senescence.

Senescence is a process that causes animals to become progressively less fertile (Medawar, 1946) and more vulnerable (Comfort, 1956) with age. It has long been noticed that senescence-associated frailty causes population death rates to rise exponentially with age (Gompertz, 1825).

Although a number of evidences have since been collected in support of each of these theories, in the last decades some phenomena have challenged all of them. This includes the effect of caloric restriction on longevity, the late-life mortality deceleration and the longevity pathways controlled by either a single or a few genes, such as the insulin pathway and the effect of sirtuins on longevity.

2. Measuring senescence

Although generally considered together, it is useful to take some time to consider the effects of senescence on individuals' survival and fertility (physiological senescence) or on populational survival curves (demographic senescence) separetly. By not doing so, the researcher may unwittingly take the risk of assuming demographic senescence to stem directly from physiological senescence. Although it might well be the case, there is no theoretical reason why it must be so.

The fact is that the genetic architecture of senescence, i.e., which genes are related to which measurable effects that we call senescence and how they relate to each other, will dictate the relationship between physiological and demographic senescence(s).

2.1 What is the genetic architecture of senescence?

Genetic architecture refers to the genetic basis of a phenotypic trait. Beyond comprehending the map of the genes linked to a given trait, genetic architecture considers all phenomena through which such genetic map produces the phenotype Masoro & Austad (2006).

The most common definition of the senescent phenotype combines individual effects (decrease in functional and reproductive abilities) with an effect which is measurable only in a population (age-dependent increase in mortality). This often leads us to conclude that it is exactly the same phenomenon that makes us individually more fragile and at greater risk of dying as we age.

Figure 1 shows that this is only one of the possible relationships between physiological senescence (progressive fall on functional capacity and fertility) and demographic senescence (increased mortality accompaining chronological aging) (Promislow et al., 2006).

While it is not necessarily clear what the relationship between the physiological and demographic components of senescence is, most "aging genes" described in the literature are simply genes whose variations influence the longevity of the studied species regardless of their physiological effect, and few genes were shown to affect the Mortality Rate Doubling Time (MRDT) of populations of mutants for such genes, and, therefore, to affect the speed of senescence (de Magalhães et al., 2005). Additionally, when strains carrying alleles for many of the so called *longevity genes* are mixed with wild populations, generally the "beneficial" mutation is lost over a few generations, indicating that although such variants increase longevity, they may exert a deleterious effect for fitness (Promislow et al., 2006).

For these reasons, the first decision before staring to seek for "aging genes" should be which model of senescence to assume. Otherwise, we might not know how to interpret the findings in a coherent way: suppose that human carriers of a given mutation have an increase of 5% in their annual mortality from 30 years of age – are them carriers of a genetic disease or of a deleterious mutation in a senescence pathway?

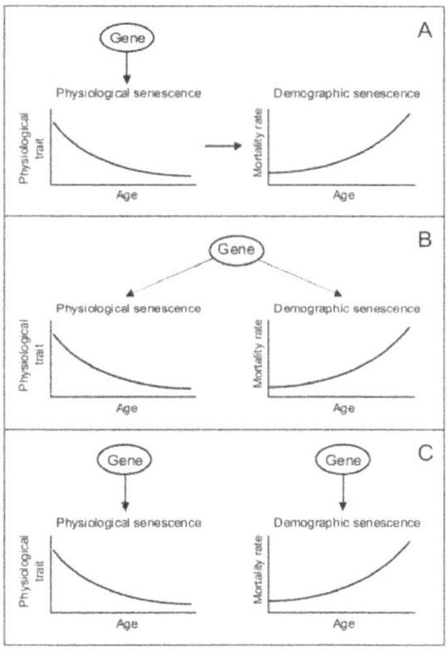

Fig. 1. Three different models for the relationship between physiological and demographic senescence on the genetic architecture of senescence. (a) Genes negatively influence physiological processes, which, then, lead to increasing effects on age-specific mortality. (b) The same genes that lead to physiological senescence independently lead to increasing age-dependent death rates, which are demografically measurable. (c) Different genes operate over physiological and demographic processes that are linked with senescence. Extracted from *Promislow, D. E. L. et al. Evolutionary Biology of Aging: Future Directions. In: Masoro EJ, Austad SN (Ed.). Handbook of the Biology of Aging. 6th. ed. San Diego: Academic Press, 2006. 217-242.*

If we suppose that senescence is a unique genetic phenomenon whose physiological effects lead to its demographic aspects (Figure 1 (a)), then "genes of senescence" should exert age-dependent deleterious effects in the physiology of organisms, and because more frail individuals are more prone to dye from a given insult, such genes would also increase mortality from their ages of onset.

On the other hand, genes that determine effects on demographic senescence may exert independent effects on the physiology of organisms. Such effects might not be linked to the demographic effects of the same genes (Figure 1 (b)).

Finally, physiological senescence could be genetically independent from demographic senescence, so that there would be a "genetic modularity" between the two phenomena, in which different groups of genes participate in each process (Figure 1 (c)).

This text assumes when necessary that genes linked to physiological senescence may impact probability of death (1 (a)). It does so relying on the fact that there is little evidence that there may be a genetic variability to the age-dependent physiological decline without its influencing on demographic senescence (Wessells et al., 2004).

Once delimited the senescent phenotype, we review some genetic phenomena that may have importance for the genetic architecture of senescence. Such phenomena include:

- Epistasis, when the expression of a gene negatively influences the expression of one another;
- Polygyny, where multiple genes contribute to a phenotypic trait;
- Pleiotropy, when multiple phenotypic characteristics are influenced by a single gene;
- Quasi-continuity, while a variation in a gene affects minimally a phenotype;
- Plasticity, when a single genotype can produce more than one distinct phenotype, such phenotypic diversity may occur among individuals of the same genotype, by action of different environmental influences on the same individual or *in the same individual at different ages;*
- Evolvability, when genotypic variations of a phenotype exist in a population and can lead to different degrees of adaptability, so that environmental changes will lead to readaptations.

Epistasis could function similarly to what is predicted on antagonistic pleiotropy theory: assuming two genes with positive effects for fitness, in which the first gene exerts a negative effect on the expression of the second gene, the first gene would have positive and negative effects on fitness The effect under selection, however, would be the average effect.

It is believed, since the formulation of the theory of mutation accumulation by Medawar, that senescence is a polygenic phenotype (Medawar, 1952). Indeed, recent decades have seen the description of "hundreds of aging genes" (Promislow et al., 2006). Summed to the fact that senescence is an early onset and gradually progressive phenotype in almost all of the species that has been described, it points to a polygenic inheritance with almost-continuity in organic response to genes that determine senescence.

3. The evolutionary theories of senescence

3.1 Introduction

It has always been difficult to conciliate senescence with natural selection, a biological mechanism generally expected to increase population fitness. Although acknowledged by Darwin (1872), the first tentative explanation for the evolution of senescence was offered by August Weismann in 1881. For Weismann, senescence had evolved for the good of species, in that the removal of older, weaker and less fertile individuals from a population would enhance the survival of younger individuals and overall reproduction of the species (Weismann, 1889).

Realizing his argument was circular (since it depended on older individuals being weaker and less fertile for senescence to evolve) Weismann withdrew his theory (Weismann, 1892).

More than half a century later, Medawar proposed the mutation accumulation theory of senescence (Medawar, 1952). He realized that even in an imaginary non senescent population, older individuals would be very rare simply because the cumulative incidence of death is necessarily dependent of age. This means that late acting mutations will affect population fitness only in the proportion of surviving individuals after such late ages. In other words, the force of natural selection decreases with age and deleterious mutations with effects that are late enough are in fact neutral mutations, which could randomly accumulate.

For Medawar, this would explain the existence of deleterious mutations fixed in ages to which individuals of a given species are not expected to survive in nature. Senescence evolved through such a process would only be observable in protected species, such as laboratory animals or our own species. Medawar, nonetheless, believed that animals did senesce in nature and, therefore felt the need for an early benefit to explain how a not so late acting deleterious mutation could evolve to fixation.

This was developed into the antagonistic pleiotropy theory (Williams, 1957). In short, Williams proposed that the fitness associated to mutations with more than one effect is the average fitness. Therefore, a mutation with earlier beneficial effects and later deleterious ones could be fixed by natural selection if the overall fitness be positive. Deleterious effects early enough to impact mortality in nature could be compensated by beneficial effects. Williams' theory depended on the existence of such special, pleiotropic genes, in numbers sufficient to explain the observed progressive increase in the effects of senescence.

In 1977, elaborating on the mechanistic error catastrophe theory of Orgel (1963), Kirkwood proposed an ecological argument for the evolution of senescence (Kirkwood, 1977). Since evolution is centered on reproduction and not directly in survival, the energetic and metabolic cost of maintenance and repair could affect reproduction negatively if taken to perfection. Therefore, the level of body maintenance and repair that can evolve is the minimum to assure reproduction. Any deleterious mutation that do not decrease reproduction can not only be neutral, but can enhance fitness if it results in more reproductive resources. This is the disposable soma theory of senescence.

These three theories are not mutually excluding, and can explain different aspects of the evolution of senescence (Kirkwood & Austad, 2000).

3.2 The mutation accumulation theory of senescence

According to Haldane, a deleterious mutation with effect only on later ages may escape natural selection, because either most individuals will be dead or will have reproduced at such ages. For Haldane, this implies in the fall of the force of natural selection with advancing ages (Haldane, 1941). Nevertheless, he failed to turn this observation into a theory of senescence.

It was Medawar who would do so. The gap between Haldane's observation of the falling force of natural selection and an evolutionary theory of senescence relies on the requirement of an age structure on populations for the fell in the force of selection. It is natural to suppose an age structure with many young individuals and rare older ones if senescence exists. In such a population, Haldane's explanation for Huntington's Disease works well, but such a model, which already pressuposes senescence, cannot acount for its evolutionary origins.

Medawar postulated that age-independent environmental hazards such as hunger, predation, accidents, etc. were a sufficient condition for the establishment of populational age structures. Older individuals, Medawar claimed, were rare because they have been exposed to such risks (termed extrinsic mortality) longer than young individuals. In other words, the existence of age structures in wild populations is a function of environmental hazards and not of senescence. Even a non senescent population would have an age structure (Medawar, 1952).

The importance of Medawar's reasoning is that older individuals in age structured populations, being necessarily rarer than younger ones, not only do not compete for environmental resources: they also don't contribute much offspring to newer generations. This is the key for Medawar's mutation accumulation theory of senescence – and also a basis for the next two hypotheses, antagonistic pleiotropy and disposable soma theories of senescence.

Figure 3 (age structured population)

Deleterious mutations, provided that it effects happen in sufficiently late ages, could accumulate in the genome According to mechanisms of population genetics, senescence is not necessarily harmful: Medawar's original proposition implies that the postponement of the effects of age-specific deleterious genes for late ages is equivalent to their elimination. Thus, these genes become effectively neutral (Medawar, 1952).

According to this theory, such evolutionary mechanism could only explain the manifestations found in senescent populations protected after individuals reach ages above those found in nature, since in nature, the age of accumulation would coincide with the maximum age of living individuals.

Medawar accepted that the effects of senescence also occurred at ages commonly found in nature. For this reason, he became convinced that another mechanism, involving either pleiotropy or linkage, would be necessary to explain the process of early and gradual degeneration which is charachteristic of senescence (Medawar, 1952). Medawar, however, didn't advance more details on this hypothesis.

3.3 The antagonistic pleiotropy theory

Although the accumulation of mutations justifies the existence of deleterious genes with late expression (and thus the already established senescent state), it didn't seem to explain the slow onset of senescence (Williams, 1957).

Seeking to understand how deleterious genetic effects expressed in relatively early ages could escape selection, Williams grounded his theory on four assumptions: the existence of a somatic cell line, i.e., non-transferable in whole or part by sexual or asexual reproduction, the natural selection of different alleles at a population, a decreasing probability of reproduction with increasing ages, the existence of pleiotropic genes with different effects on fitness at distinct ages (antagonistic pleiotropy). According to this idea, the evolutionary fundamental process to the establishment of senescence is a selective action on the inheritance of a gene with antagonistic effects on its carrier's fitness (Williams, 1957).

It is to note that the very existence of pleiotropic genes with antagonistic effects was not postulated by Williams. In a previous article, Sewall Wright describes an equation for calculating the impact of a pleiotropic gene on fitness:

$$W = (1 + S_1)(1 + S_2)...(1 + S_n) , \tag{1}$$

where W is the fitness of a gene and S_1, S_2 ... S_n are the separate selective coefficients for each age-specific effect of such gene on fitness (Wright, 1956).

Williams' merit was to note the implication of Wright's equation for the evolution of senescence in age structured populations. He applied to the Equation 1 the same reasoning applied by Medawar in relation to the age structure of populations: the magnitude of the effect of S_n of a gene may be reduced if it only starts at advanced ages. In a gene capable of expressing different effects on different ages, later effects will less be subject to natural selection than earlier effects.

By considering the effects of a given gene in distinct ages of expression, however, Williams proposed that the measure of the magnitude of each effect in question (advantages or disadvantages) is given by

$$S_n = m_n p_n , \tag{2}$$

where S_n is the effect under consideration, m_n its magnitude or impact on fitness and p_n is the proportion of a population's reproductive probability that is *relevant* to the age of manifestation of the effect S_n.

This allowed him to rewrite the Equation 1 considering the effect of age structure in the final selective coefficient of a pleiotropic gene:

$$W = (1 + m_1 p_1)(1 + m_2 p_2)...(1 + m_n p_n) . \tag{3}$$

From this equation we may extract the simplest case: the one of a pleiotropic gene with a late deleterious effect and a very early beneficial effect. To demonstrate the formula we need to know the values of p for each age.

Let us imagine a population (structured for simplicity on a human age-scale) with constant birth and death. This is necessary not to create an *ad hoc* argument, by starting with a previously non-senescent population and therefore with no age differences in mortality.

Let's say that this population has a constant mortality of 0.25 each 4 years, ie, 0.0625 per year, and that each 4-years extract is composed of 1000 individuals. This population age distribution is represented in Figure 2. The familiarity of this age distribution with any wild population is noticiable, as it is with any high-mortality human population (as an example, the Figure 3 represents the age distribution of the population of Afghanistan in 2008).

In this non-senescent population, organisms do not lose fertility with the progression of age and all individuals have the same reproductive probability. Therefore, p_x, i.e., the proportion of the reproductive probability associated with each age will be the proportion of remaining individuals with ages equal or superior to that in the population, since the effect of S remain after activated, i.e., be constant from its manifestation.

Williams could thus formulate an evolutionary hypothesis for the permanence (regardless of natural selection) of detrimental effects whose expression was sufficiently early to be influential in wild populations. It is important to notice that neither the equation nor any

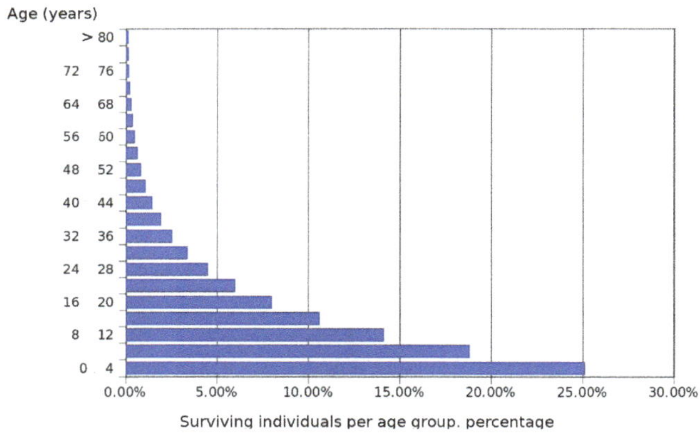

Fig. 2. Age distribution of an alleged non-senescent population exposed to a mortality of 6.25% per year.

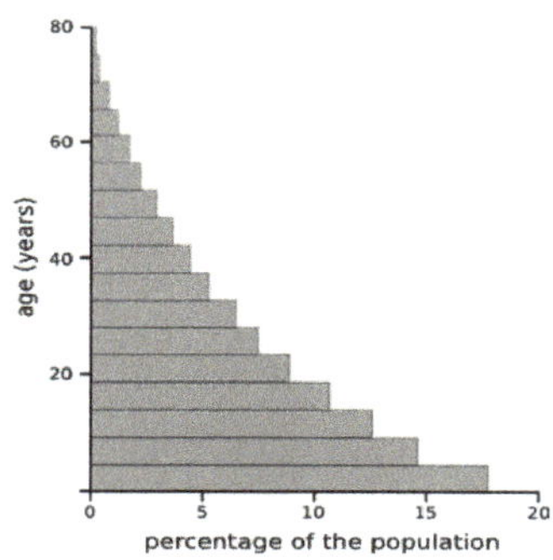

Fig. 3. Age distribution of the Afghan population in 2008 (data from U.S. Census, International Database, available at http://www.census.gov/).

comment on Williams' classic article (Williams, 1957) infer what the number of effects of a pleiotropic gene should be, nor which effect, beneficial or deleterious should happen earlier. While it is clear that the simplest case of antagonistic pleiotropy would be of a gene with an early beneficial effect and a late deleterious one, Equation 3 does not imply that this is the only possibility.

The literature, however, did not follow this conclusion. Williams' antagonistic pleiotropy came to be regarded precisely as the case in which a mutation with two actions, an early advantage and a later disadvantage, it is positively selected. It is worth quoting verbatim the concept of antagonistic pleiotropy found in the *Handbook of Biological Aging*, of 2006, in a chapter written by the editor of the book (Masoro, 2006):

"Another genetic mechanism, proposed by Williams (1957), is referred to as antagonistic pleiotropy. It proposes that those genes that increase evolutionary fitness in early life will be selected for, even if they have catastrophic deleterious effects in late life. Again, the deleterious effects of these genes will be evident only in subjects in protected environments that enable a long life. "

While Medawar discussed the age of onset of a characteristic deleterious and Williams called attention to the magnitude of genetic effects, no theory explicited discussed the influence of the magnitude of extrinsic mortality and random genetic drift on the selection of deleterious genes.

3.4 The disposable soma theory

From the 1970s, a new theory for the evolution of senescence was proposed by Thomas Kirkwood (Kirkwood & Holliday, 1975a;b), reasoning initially not about evolutionary mechanisms as maladaptive as Medawar or on pleiotropic genes as Williams, but on a strong ecological basis. Elaborating on the theory Orgell's "error catastrophe " (Orgel, 1963), Kirkwood gave us a somewhat different view from what had previously been formulated on the evolution of senescence.

Kirkwood addressed the issue of senescence under an ecological constraint in which energy resources available to individuals could be allocated either for maintenance somatic cells or for reproduction, generating a soma-germ conflict. Called disposable soma theory, this ultimately depends on the existence of specific genes that either influence or control the precision of the genetic replication / transcription and translation machineries in an age-dependent fashion; Kirkwood himself considered his theory as a specialization of the antagonistic pleiotropy theory of Williams (Kirkwood, 1977; Kirkwood & Holliday, 1979). The difference is that Williams posits the existence of genes for beneficial and deleterious effects, but the theory of Kirkwood, despite not denying the existence of these genes, do not need them. If, under Williams theory, natural selection would act on the average effects of a selected gene's mutant alleles, under Kirkwood's assumption a single genetic effect, determining the distribution of resources between reproduction and body maintenance would be sufficient for the evolution of senescence. Senescence would be the inevitable result of selection for an "'ideal" energy allocation between reproduction and body maintenance.

This is the first theory to propose that the evolution of organisms can optimize the allocation of metabolic resources between the maintenance of the somatic lineage (the individual itself) and the effort of reproduction (investment in next generation). Under it, the physiological mechanisms that postpone senescence consume metabolic resources, which become less available for reproduction, and vice versa. As reproduction of the species by natural selection is prioritized, the body is "disposable" after its reproductive function has been sufficiently fulfilled, and the aging process may appear, without sufficient opposition from natural selection.

There is some experimental support for this theory. When fruit flies are selected for a longer life expectancy, there is decrease in fertility. Conversely, exposure of females to earlier reproduction was correlated with a decline in their lifetimes as compared to virgin females (Sgro & Partridge, 1999).

The disposable soma theory changes the fundamental question about the evolution of senescence: instead of questioning why we senesce, we could wonder why we, humans, live as much as we do (Kirkwood & Rose, 1991).

4. The evolutionary theories of senescence today

4.1 Is there senescence in the wild indeed?

Does anyone die of old age? The existence of senescence in wild populations in their habitats, which, as mentioned, led Medawar and Williams to think of antagonistic pleiotropy, was harshly questioned in the literature by influential researchers like Hayflick and Comfort (Comfort, 1956; Hayflick, 2000).

In a large study, however, Promislow described significant evidence of senescence in 26 species of mammals *in natura* (Promislow, 1991). In fact, in recent years, several studies have demonstrated demographic senescence (Austad, 1993; Bronikowski et al., 2002; Ericsson et al., 2001; Orell & Belda, 2002) and reproductive senescence in mammals and birds in their habitats (Austad, 1993; Broussard et al., 2003; Ericsson et al., 2001; Reid et al., 2003; Saino et al., 2003).

Finally, a strong indication that senescence does exist in wild populations comes from our species: if the data collected by Gompertz and those who followed him can be extrapolated to primitive humans, then our mortality progresses under a Gompertzian regime from around 12 years of age. Senescence starting in such an early age certainly would have impacted on mortality of early human populations (Gompertz, 1825).

Evidence is thus, that death due to senescence is actually happening in the wild. We must change the question from someone dies of old to " someone dies because of senescence '?"And the answer, backed by extensive literature is a resounding yes (Carey & Judge, 2000).

4.2 Findings supporting each evolutionary theory

Evidence for the genetic basis of senescence accumulate in the literature. With respect to the specific theories on the evolution of senescence, experimental evidence supports each of the three theories mentioned above (Hughes & Reynolds, 2005).

In Drosophila, it is possible to obtain two distinct lineages in relation the speed of installation of its senescence by systematically separating over generations, the first (beginning of reproductive life) or the last oviposition (immediately prior to reproductive senescence). The flies of the the second group senesce more slowly and live up to 50% longer than the first group flies (Baret & Lints, 1993; Fukui et al., 1995; Luckinbill & Clare, 1985; Rose & Charlesworth, 1980; 1981).

According to Rose, while the flies in the control group focus their greater efficiency in the early reproductive life, the group submitted to selective pressure for older reproduction requires the of the opposite strategy. In both groups, according to the evolutionary concept of senescence, mutations of late manifestation accumulate and propagate to the new generations. These changes will not affect the flies of the first group, but the second group will largely benefit if such manifestations are delayed. Mutations that make deleterious effects to occur later will improve fitness of individuals in the second group, but not in the first group. This mechanism, over generations, makes the senescent manifestations in the second group to become even later, so that the flies of this group evolve longevity. Interestingly, even with the suspension of the selective pressure, the difference persists, and the strains of flies arising from these experiments remain more long-lived than wild flies (Rose, 1991).

Hughes and others have found evidence in favor of mutation accumulation in the experimental evolution of accelerated senescence (decrease in MRDT) in fruit flies as the predominant phenomenon (Hughes et al., 2002), argument which is sustained by Cortopassi in relation to human senescence (Cortopassi, 2002). Physiological senescence in our species, however, is easily noticeable between the fourth and fifth decades of life. This is not incompatible with Medawar's observation (Medawar, 1952) that the accumulation of mutations explain only senescence in artificially protected populations: modern man is an excellent example of a protected population.

Pleiotropic mechanisms have currently been described. The gene for the juvenile hormone (JH) found in specimens of wild Drosophila is expressed early in its life cycle, and takes to an increase in fertility, early sexual maturation and augmented vitellogenesis. On the other hand, it undermines resistance to stress factors, reduces immunity and the maximum life expectancy (Flatt et al., 2005). In geese, artificial selection for early sexual maturation causes, as adverse effects, a faster reproductive senescence; quantitative analysis revealed a genetic correlation between these two features (Charmantier et al., 2006). Finally, several programmed cell death mechanisms known in mammals also perform "important vital functions such as energy production, metabolism differentiation or the cell cycle" (Ameisen, 2004; 2005). It was recently suggested that Alzheimer's disease, which appears to be specific to humans, could be an example of antagonistic pleiotropy (Bufill & Blesa, 2006).

Holliday argues that the mammals' life cycle strongly illustrates the idea behind the disposable soma theory: there is in mammals an inverse relationship between maximum reproductive potential and maximum longevity; small mammals are very short-lived and fertile, the great mammals are less fertile and very long-lived (Holliday, 1997; 2005). This contrast between longevity and reproduction also appears in a historical cohort analysis of demographic data of the aristocracy of Britain in which with female longevity correlated with a lower number of children (Westendorp & Kirkwood, 1998).

It is noteworthy that the evolutionary theories of senescence (mutation accumulation, antagonistic pleiotropy, and disposable soma) are not mutually exclusive. Although the three currently accepted evolutionary processes for the evolution of senescence can coexist, a current problem of the evolutionary research on senescence is to know how much each of the processes have contributed to the emergence of this phenomenon (Gavrilov & Gavrilova, 2006).

4.3 The evolutionary theories of senescence and the increasing knowledge on evolution

Charlesworth's cumulative effect model (Charlesworth, 2001). The potential effect of genetic drift and natural selection on deleterious and pleiotropic mutations: effective population size on age-structured populations(Charlesworth, 1980; Felsenstein, 1971), infinite sites model of molecular mutation (Kimura & CROW, 1964), infinite alleles model of polymorphism and the neutral theory of evolution (Kimura and Ohta).

4.3.1 Randon genetic drift

A greater effect of random genetic drift is expected in populations with age structure (where the effective population size, N_e is smaller than the real population size, N), than in populations without age structure (and where $N_e \approx N$). Suppose that a genetic effect in a population is expressed at an advanced age (arbitrarily defined as significantly higher than the age of reproductive maturation of a species). We can use this to divide this population age into two subsets: that of individuals who have not expressed the genetic trait and those who have already expressed.

Let us consider what happens with the genetic trait in question under natural selection on these two subpopulations. Obviously, on the first subpopulation, the gene with the deleterious effect, not having expressed itself, is effectively neutral and therefore can evolve only by drift. In the second subpopulation, there is evolution by selection. Clearly, the effect of selection on the population as a whole will be less than what it would be if the mentioned gene was expressed at younger ages. This is just another way of saying that the force of selection falls with age.

Now let us consider the effect of genetic drift on the second subpopulation. Being it a fraction of the total population, its N_e will be considerably smaller than the the initial population's N_e. Thus, besides being subject to progressively smaller selection forces as a function of the age of onset, late onset of deleterious genes should be subject to progressively more intense phenomena of genetic drift. This point, not addressed in the current theories about the evolution senescence, might prove to be crucial.

This last aspect makes it fundamental to understand the roles of mutation, selection and drift as a whole in the evolution of senescence, since, at least in part, the force of selection declines with advancing ages precisely due to the decrease on the effective sizes of the subpopulations.

5. Conclusion

For decades, researchers in the field of senescence were divided between the proponents of proximal or mechanistic theories and the proponents of the distal or evolutionary theories of senescence (Masoro & Austad, 2006). Fortunately, recent decades have seen an excellent understanding of the importance of a joint reasoning between "mechanistic" and "evolutionary" thoughts: more and more studies focused on the evolution of senescence seek to understand its physiological mechanisms of onset and progression and researchers focusing on the mechanisms of senescence are increasingly seeking to understand the theoretical evolutionary basis of senescence (Masoro & Austad, 2006, Preface).

The evolutionary study of senescence seeks to explain why this phenomenon exists, providing researchers with mechanistic insights into what could be the proximal causes of senescence and how genetics produces the senescent phenotype (Kirkwood & Austad, 2000). By pointing to non-adaptive origins for senescence, the evolutionary theory may drive mechanistic researchers away from adaptive programs such as apoptosis as a plausible basis of senescence. After all, "nothing in biology makes sense except in the light of evolution" (Dobzhansky, 1973).

6. References

Ameisen, J. C. (2004). Looking for death at the core of life in the light of evolution., *Cell Death Differ* 11(1): 4–10.
URL: *http://www.ncbi.nlm.nih.gov/pubmed/14647240*

Ameisen, J.-C. (2005). [selective "death programs" or pleiotropic "life programs"? looking for programmed cell death in the light of evolution], *J Soc Biol* 199(3): 175–89.
URL: *http://www.ncbi.nlm.nih.gov/pubmed/16471257*

Austad, S. N. (1993). Retarded senescence in an insular population of virginia opossums (*didelphis virginiana*), *Journal of Zoology* 229(4): 695–708.

Baret, P. & Lints, F. A. (1993). Selection for increased longevity in *drosophila melanogaster*: a new interpretation., *Gerontology* 39(5): 252–9.
URL: *http://www.ncbi.nlm.nih.gov/pubmed/8314091*

Bronikowski, A. M., Alberts, S. C., Altmann, J., Packer, C., Carey, K. D. & Tatar, M. (2002). The aging baboon: comparative demography in a non-human primate., *Proc Natl Acad Sci U S A* 99(14): 9591–5.
URL: *http://www.ncbi.nlm.nih.gov/pubmed/12082185*

Broussard, D. R., Risch, T. S., Dobson, F. S. & Murie, J. O. (2003). Senescence and age-related reproduction of female columbian ground squirrels, *Journal of Animal Ecology* 72(2): 212–219.

Bufill, E. & Blesa, R. (2006). [Alzheimer's disease and brain evolution: is alzheimer's disease an example of antagonistic pleiotropy?], *Rev Neurol* 42(1): 25–33.
URL: *http://www.ncbi.nlm.nih.gov/pubmed/16402323*

Carey, J. & Judge, D. (eds) (2000). *Longevity records: life spans of mammals, birds, reptiles, amphibians and fish.*, Odense University Press.

Charlesworth, B. (1980). *Evolution in Age Structured Populations*, Cambridge University Press, Cambridge.

Charlesworth, B. (2001). Patterns of age-specific means and genetic variances of mortality rates predicted by the mutation-accumulation theory of ageing., *J Theor Biol* 210(1): 47–65.
URL: *http://www.ncbi.nlm.nih.gov/pubmed/11343430*

Charmantier, A., Perrins, C., McCleery, R. H. & Sheldon, B. C. (2006). Quantitative genetics of age at reproduction in wild swans: support for antagonistic pleiotropy models of senescence., *Proc Natl Acad Sci U S A* 103(17): 6587–92.
URL: *http://www.ncbi.nlm.nih.gov/pubmed/16618935*

Comfort, A. (1956). *The Biology of Senescence*, Rinehart & Company, Inc., New York.

Cortopassi, G. A. (2002). Fixation of deleterious alleles, evolution and human aging., *Mech Ageing Dev* 123(8): 851–5.
URL: *http://www.ncbi.nlm.nih.gov/pubmed/12044933*

Darwin, C. (1872). The origin of species by means of natural selection, or the preservation of favoured races in the struggle for life.
URL: *http://darwin-online.org.uk*

de Magalhães, J. P., Cabral, J. A. S. & Magalhaes, D. (2005). The influence of genes on the aging process of mice: a statistical assessment of the genetics of aging., *Genetics* 169(1): 265–74.
URL: *http://www.ncbi.nlm.nih.gov/pubmed/15466429*

Dobzhansky, T. (1973). Nothing in biology makes sense except in the light of evolution, *American Biology Teacher* 35: 125–129.

Ericsson, G., Wallin, K., Ball, J. P. & Broberg, M. (2001). Age-related reproductive effort and senescence in free-ranging moose, alces alces, *Ecology* (82): 1613–1620.

Felsenstein, J. (1971). Inbreeding and variance effective numbers in populations with overlapping generations., *Genetics* 68(4): 581–97.
URL: *http://www.ncbi.nlm.nih.gov/pubmed/5166069*

Flatt, T., Tu, M.-P. & Tatar, M. (2005). Hormonal pleiotropy and the juvenile hormone regulation of drosophila development and life history., *Bioessays* 27(10): 999–1010. URL: *http://www.ncbi.nlm.nih.gov/pubmed/16163709*

Fukui, H. H., Pletcher, S. D. & Curtsinger, J. W. (1995). Selection for increased longevity in drosophila melanogaster: a response to baret and lints., *Gerontology* 41(2): 65–8. URL: *http://www.ncbi.nlm.nih.gov/pubmed/7744268*

Futuyma, D. (1998). *Evolution*, Sinauer Associates, Sunderland, MA, U.S.A.

Gavrilov, L. A. & Gavrilova, N. S. (2006). Reliability theory of aging and longevity., *in* E. J. Masoro & S. N. Austad (eds), *Handbook of the Biology of Aging.*, 6th edn, Academic Press, San Diego, pp. 3–40.

Gompertz, B. (1825). On the nature of the function expressive of the law of human mortality and on a new mode of determining life contingencies., *Philos Trans Roy Soc London* (115): 513–85.

Haldane, J. B. S. (1941). *New Paths in Genetics*, Allen and Unwin, London.

Hamilton, W. D. (1966). The moulding of senescence by natural selection., *J Theor Biol* 12(1): 12–45. URL: *http://www.ncbi.nlm.nih.gov/pubmed/6015424*

Hayflick, L. (2000). The future of ageing., *Nature* 408(6809): 267–9. URL: *http://www.ncbi.nlm.nih.gov/pubmed/11089985*

Holliday, R. (1997). Understanding ageing., *Philos Trans R Soc Lond B Biol Sci* 352(1363): 1793–7. URL: *http://www.ncbi.nlm.nih.gov/pubmed/9460062*

Holliday, R. (2005). Ageing and the extinction of large animals., *Biogerontology* 6(2): 151–6. URL: *http://www.ncbi.nlm.nih.gov/pubmed/16034683*

Hughes, K. A., Alipaz, J. A., Drnevich, J. M. & Reynolds, R. M. (2002). A test of evolutionary theories of aging., *Proc Natl Acad Sci U S A* 99(22): 14286–91. URL: *http://www.ncbi.nlm.nih.gov/pubmed/12386342*

Hughes, K. A. & Reynolds, R. M. (2005). Evolutionary and mechanistic theories of aging., *Annu Rev Entomol* 50: 421–45. URL: *http://www.ncbi.nlm.nih.gov/pubmed/15355246*

Kimura, M. & CROW, J. F. (1964). The number of alleles that can be maintained in a finite population, *Genetics* 49: 725–38. URL: *http://www.ncbi.nlm.nih.gov/pubmed/14156929*

Kirkwood, T. B. (1977). Evolution of ageing., *Nature* 270(5635): 301–4. URL: *http://www.ncbi.nlm.nih.gov/pubmed/593350*

Kirkwood, T. B. & Austad, S. N. (2000). Why do we age?, *Nature* 408(6809): 233–8. URL: *http://www.ncbi.nlm.nih.gov/pubmed/11089980*

Kirkwood, T. B. & Holliday, R. (1975a). Commitment to senescence: a model for the finite and infinite growth of diploid and transformed human fibroblasts in culture., *J Theor Biol* 53(2): 481–96. URL: *http://www.ncbi.nlm.nih.gov/pubmed/1195775*

Kirkwood, T. B. & Holliday, R. (1975b). The stability of the translation apparatus., *J Mol Biol* 97(2): 257–65. URL: *http://www.ncbi.nlm.nih.gov/pubmed/1177323*

Kirkwood, T. B. & Holliday, R. (1979). The evolution of ageing and longevity., *Proc R Soc Lond B Biol Sci* 205(1161): 531–46. URL: *http://www.ncbi.nlm.nih.gov/pubmed/42059*

Kirkwood, T. B. & Rose, M. R. (1991). Evolution of senescence: late survival sacrificed for reproduction., *Philos Trans R Soc Lond B Biol Sci* 332(1262): 15–24.
URL: *http://www.ncbi.nlm.nih.gov/pubmed/1677205*

Luckinbill, L. S. & Clare, M. J. (1985). Selection for life span in drosophila melanogaster., *Heredity* 55 (Pt 1): 9–18.
URL: *http://www.ncbi.nlm.nih.gov/pubmed/3930429*

Masoro, E. J. (2006). Are age-associated diseases an integral part of aging?, *in* E. J. Masoro & S. N. Austad (eds), *Handbook of the Biology of Aging.*, 6th edn, Academic Press, London, UK, pp. 43–62.

Masoro, E. J. & Austad, S. N. (2006). *Handbook of the Biology of Aging*, sixth edition edn, Oxford Academic Press, London, UK.

Medawar, P. B. (1946). Old age and natural death, *Modern Quarterly* 1: 30–56.

Medawar, P. B. (1952). *An unsolved problem of biology*, H.K. Lewis and Co., Oxford, UK.

Orell, M. & Belda, E. J. (2002). Delayed cost of reproduction and senescence in the willow tit parus montanus, *Journal of Animal Ecology* 71(1): 55–64.

Orgel, L. E. (1963). The maintenance of the accuracy of protein synthesis and its relevance to ageing., *Proc Natl Acad Sci U S A* 49: 517–21.
URL: *http://www.ncbi.nlm.nih.gov/pubmed/13940312*

Promislow, D. E. L. (1991). Senescence in natural populations of mammals: A comparative study, *Evolution* 45(8): 1869–1887.

Promislow, D. E. L., Fedorka, K. M. & Burger, J. M. S. (2006). Evolutionary biology of aging: Future directions., *in* E. J. Masoro & S. N. Austad (eds), *Handbook of the Biology of Aging.*, 6th edn, Academic Press, London, UK, pp. 217–242.

Reid, J. M., Bignal, E. M., Bignal, S., McCracken, D. I. & Monaghan, P. (2003). Age-specific reproductive performance in red-billed choughs pyrrhocorax pyrrhocorax: patterns and processes in a natural population, *Journal of Animal Ecology* 72(5): 765–776.

Rose, M. & Charlesworth, B. (1980). A test of evolutionary theories of senescence., *Nature* 287(5778): 141–2.
URL: *http://www.ncbi.nlm.nih.gov/pubmed/6776406*

Rose, M. R. (1991). *Evolutionary biology of aging*, 1 edn, Oxford University Press, Oxford, UK.

Rose, M. R. & Charlesworth, B. (1981). Genetics of life history in drosophila melanogaster. i. sib analysis of adult females., *Genetics* 97(1): 173–86.
URL: *http://www.ncbi.nlm.nih.gov/pubmed/6790340*

Saino, N., Ferrari, R. P., Romano, M., Rubolini, D. & MÃÂÿller, A. P. (2003). Humoral immune response in relation to senescence, sex and sexual ornamentation in the barn swallow (hirundo rustica)., *J Evol Biol* 16(6): 1127–34.
URL: *http://www.ncbi.nlm.nih.gov/pubmed/14640404*

Sgro, C. M. & Partridge, L. (1999). A delayed wave of death from reproduction in drosophila., *Science* 286(5449): 2521–4.
URL: *http://www.ncbi.nlm.nih.gov/pubmed/10617470*

Weismann, A. (1889). *Essays upon heredity and kindred biological problems. I*, Vol. 1, 2n ed. edn, Oxford University Press Warehouse, Amen Corner, E. C.

Weismann, A. (1892). *Essays upon heredity and kindred biological problems. II*, Vol. II, 2n ed. edn, Oxford, Clarendon Press, Amen Corner, E. C.

Wessells, R. J., Fitzgerald, E., Cypser, J. R., Tatar, M. & Bodmer, R. (2004). Insulin regulation of heart function in aging fruit flies., *Nat Genet* 36(12): 1275–81.
URL: *http://www.ncbi.nlm.nih.gov/pubmed/15565107*

Westendorp, R. G. & Kirkwood, T. B. (1998). Human longevity at the cost of reproductive success., *Nature* 396(6713): 743–6.
URL: *http://www.ncbi.nlm.nih.gov/pubmed/9874369*

Williams, G. C. (1957). Pleiotropy, natural selection and the evolution of senescence, *Evolution* 11: 398–411.

Wright, S. (1956). Modes of selection, *The American Naturalist* 90(850): 5–24.

2

Programming and Implementation of Age-Related Changes

Jaba Tkemaladze[1], Alexander Tavartkiladze[2] and Konstantin Chichinadze[3,4]
[1]*Georgian Systemic Research Center, Tbilisi,*
[2]*Grigol Robakidze University's Oncology Department, Tbilisi,*
[3]*Life Science Research Center, Tbilisi,*
[4]*I. Javakhishvili Tbilisi State University, Tbilisi,*
Georgia

1. Introduction

From the songs of my childhood I can remember one such song: " The cranes are flying far and they will never come back, the rumour is spread ... . " The author of the song text assumes that the cranes can not come back , but he/she can't bear the thought of losing lovely birds and adds: " it's a rumour !" However ,I am saying once more he lets the thought that maybe the cranes won't be back any more .... I loved this song and while listening sorrow used to enter my soul quietly. Although I knew that birds return from warm countries ... And still there was something ominous in those " never come back." Then , after so long in Dinara Kasradze's monograph ("Quantum satis ", 2005) I read the following: "Each species has it's own quantity of wing waving , after which the bird dies ." ..yhat's where that subconscious sorrow came from ... mine or of that song text author, as we mentioned above... Yes, it's so, the organism tissues , cells "get worn" while acting and get restored too; The only bad thing is that with age , damage of cells exceeds their reparation i. e. the mentioned fact becomes expressed with aging! Every creature turned to have its own strictly defined potential, after which the life expires. Then where is immortality? Only in fairy tales ? "The truth used to be written in fairy tails , the truth , written in a creative and "fairy" way " ,-"In fairy tales , many secrets of nature are explained in "fairy" language ,many things that are unsolved and not clear at all! "... Then where's immortality? The water or spring of immortality ? Maybe it is nearby , here next to us...

2. Hayflick's limit

With the age, damage of molecules exceeds its reparation i. e. the mentioned fact gets more expressed with aging. The question is : What are molecular reasons for wearing cells out. Hayflick's experiments are important in this way: in 1961 Leonard Hayflick demonstrated (at Wister Institute , Philadelphia) that the cells of normal human fetus managed to divide 40-60 times. The scientist in vitro observed human, normal, diploid fibroblast cultures and he saw that these cells have absolutely defined life expectancy - they stop multiplication and get older after 50 times division (doubling). On the contrary, fibroblasts of the patients , ill with Progeria or Verner syndrome (early aging) , used to double 10-12 times - instead of 40-60 . Adopted term , so acceptable nowadays, Hayflick's limit comes from the above . This limit i.e. border was fixed in the culture as for all differential (i.e. mature) cells, also for other multi cell organisms. Maximum number of divisions in one organism is different according to the type of the cell and difference is much more expressed in different organisms. For most of the cells Hayflick's Limit is 52 divisions. The history of Hayflick's experiments is also interesting. He worked together with Paul Moorhead . In one of the tests they mixed equal quantities of normal males' fibroblasts (that had been divided 40 times) and normal females' fibroblasts (that had been divided only 10 times) i.e. men's older cells were mixed with women's younger cells and they received mixed culture."Not mixed" culture (males') was used for control. When division in the controlling culture of the males cells stopped , in that very moment mixed culture was studied and it turned out that only women's cells were left there. This meant: The older cells "remembered" that they were old, even when they were surrounded by young cells.

Hayflick differed the following phases of cell division in the culture: At the beginning of the experiment, he called the "firstly" culture "phase one"; then the period when cells were multiplying was the "phase two". Division (doubling) period after numerous months was called "the third phase phenomenon" - when cell growth was reduced and stopped. The reason for replication aging is uncertain. It is supposed that genes of aging are activated. (They are located in the first and fourth chromosomes), the growth regulating genes are changed or lost (which at last causes growth inhibition in aging cells too. The genes, speeding (accelerating) aging process are revealed. For example, reducing signal transmission by means of Factor 1 , similar to insulin, causes Drosophila, also Nematode c. Elegans and increasing of life expectancy in mice. The reason for such difference in replication intensity is unknown. It can be caused by activating specific genes at an old age. For example, during reduction of replication Kinaza inhibitor genes (p21) are revealed. Replication aging is also induced by increased expression of p 161NK4a - the cell cycle inhibitor and DNA lesion.

3. The telomeres

The essence of genes' chromosome telomere shortening is interesting. The limited replication ability of the cell can be explained by the following: During each division, the chromosomal endings go through the unfinished replication (telomere shortening). After each division telomeres shorten, that ultimately causes stopping the cell division. Telomeres are important in stabilization of terminal parts of the chromosomes and also their fixing at the nucleus matrix. Telomeres gradually get shorter in later passages of the culture and also in older people's cell cultures. Telomeres are the longest in spermatozoids, they are longer in fetus than in an adult. As it seems, DNA loss in the terminal parts of chromosomes and telomere shortening causes deletion of important genes.

De novo synthesis of telomeres is regulated by means of enzym telomeraza. Correlation between the telomere length and telomeraza consistence is found. Hayflick's limit depends on telomere size reduction. Telomeres are DNA short repeated successions (TTAGGG) which are located at the end of chromosomes. If the cell doesn't have active telomeraza (as, for example, most of somatic cells), after each division the telomere size is reduced as DNA polymeraza can't replicate the ends of DNA molecule. In spite of this, taking this event into consideration , telomeres must shorten very slowly-- by 3-6 nucleotides after one cell cycle; According to the Hayflick's limit, after certain number of divisions telomeres shorten by 150-300 nucleotides.

Nowadays, epigenetic theory of aging is delivered by B.A. Galitski (2009),according to which telomere erosion is explained first of all by activation of cell recombinazas, which are activated in respond to the DNA lesion, which is mainly caused by age depression of genome mobile elements. When after certain number of divisions telomeres absolutely disappear, the cell remains at certain stage of the cell cycle or it turns apoptosis mechanism on. It is interesting that A.M. Olivnikov (who was the first to suppose existence of telomeres (1973,1996)) said: "Telomeres action is proved. However, it shouldn't have had direct link with aging".

In spite of theories or certain plurality of opinions, it is clear today that cell aging is caused by a number of factors. Not only endogenous molecular way of cell aging is important, but also harmful exogenous affects during the whole life, which is followed by, so called, cell "wearing". In the course of time lethal changes are gathered in the cell lesions, which lead the cells to death, as ability to respond to lesions is gradually reduced.

Certainly human cells division is not infinite, except embryonic, sex and cancer cells. The cells which have short telomeres, are often "not valid" as their chromosomes are no longer stable. The chromosomes become less protected against various damaging factors, as in norm , their defending telomeres i.e. their endings can't protect them.

During the experiment researchers (group- "Geron Corporation") could change the route of the aging process: They installed genes in the DNA. The genes were responsible for relomeraza synthesis. In these cells Hayflick's limit was doubled (100 divisions) and the life expectancy accordingly increased . According to these scientists' opinion, it is possible to change ordinary human cells that they will be able to divide infinitely. In January 1998 all the media sources informed the world that American scientists were able to force human normal cells to overcome Hayflick's limit: Instead of aging, the cells continued to multiply; and what is remarkable, they didn't transform into cancer cells. Brilliant! Bravo! If it was managed to overcome Hayflick's limit and increase life expectancy so, that cancer transformation didn't take place! And still we must point out that much care and attention is necessary in handling this process!

Telomeraza was discovered in 1984 by Carol W. Greider. For opening of the protecting mechanisms of chromosomes (terminal replication by the help of telomerazas) Elizabeth Blackburn , Carol Grider and Jack Szostak received Nobel Prizes. Human telomeraza's structure was explained by Scott Cohen and his group (Australia, Children's medicine Research Institute, 2007). Telomeraza consists of telomeraza reverse transcriptaza (TERT.), telomeraza RNM (hTR or TERC) and diskerin (two molecules of each substance). TERT, as we pointed out, is the ferment reverse-transcriptaza, which creates one-string DNA on the basis of template one-string RNA (i.e. telomeraza represents a reverse - transcriptaza and a special molecule of RNA is connected with it , which is used as a matrix for the reverse-transcriptaza during lengthening of the telomeres.) Two subunits of the ferment are encoded by two different genes. TERT has a glove shape, which allows it to fix at the chromosome and add to it one-string parts. It is considered that telomeraza is the key to the cell immortality. Thanks to this ferment, cells multiply fast and don't get older.

Telomeraza is the most intensively revealed in the embryonic cells , then-in Germinational cells, very slightly-in adult organism labile cells (cells, that must divide frequently, for example, intestinal epithelial cells), but their detection is difficult in the most of the somatic cells. (More correctly, it doesn't happen at all).

Telomeraza is used in the cosmetic production (The product TA-65 was received from the plant Astragal, which activates telomeraza). Michael Fossel supposed in one of the interviews (2006, int. by D, J Brown), that treatment by telomeraza could be used not only for struggle against cancer , but also for fight against aging i.e. for increasing life expectancy.

When cells come closer to the Hayflick's limit, it is possible to stop aging if deactivation of those genes will happen , which are responsible for creation of the albumin (p53 and pR6) , suppressing cancer . Sooner or later such , changed cells get to the condition , which is called "crisis" i.e. when big part of the cells (in the culture) die (However, sometimes the cell doesn't stop multiplication during the crisis period). As a rule, telomeres are absolutely destroyed at the time and condition of the chromosomes worsens after each division.

4. Cancer

The naked endings of the chromosomes mean splitting of the both DNA strings. Neutralization of such type of lesion happens by connecting split endings. Also endings of different chromosomes can make confluence as they are no longer protected by telomeres. This story as if temporarily relieves telomere absence. However, in the anaphase of the cell division, linked chromosomes come apart quite accidentally, which is followed by a lot of mutations and chromosomal anomaly development and with the process development the cell genome is more and more injured. Ultimately the moment comes when apoptosis is turned on (The gene substance is so much injured) or the mutation is added to the injured genome which activates telomeraza.

After activation of telomeraza some types of cells obtain immortality. Their chromosomes don't become less stable for the number of divisions and the death process doesn't start. Many cancer cells are considered to be immortal as the telomeraza gene is activated in them which allows these cells to divide infinitely and this is the reason for cancer growth.

Hela's cells are good example for cancer cells eternity, which were received in 1951 from Henrietta Lack's cervix cancer tissue. The name of the culture comes from the above. This culture is still used in studies today. Hela's cells are really immortal: They are produced daily in tons and they are descendants of the removed cancer cells from H. Lack's organism.

In spite of the fact, that cancer modeling in the cell culture is effective and it has already been used for years, it is still not exact. At first, it was unknown which was the influence, that causes cell multiplication in this model. Finding the answer became gradually possible: Different mutations were caused in the model cells (Which we meet in different cancer cells in human) which allowed the scientists to reveal several confluent mutations, that was sufficient for creating cancer cells from different types.

Mutations are confluent in different ways in different types of cells. However, in most of these confluences the following is fixed: 1.Activation of telomeraza 2.Cycle damage of the albumin p 53. 3. Activation of Ras , Myc and other protooncogenes . 4.Violation of PP2A phosphataza formation. These changes provide "turning off " the death mechanism by means of damaging chromosome destruction or apoptosis process. With this endless multiplication of the cell begins.

In cell cultures, this model of cancer clears up the telomeraza's role in cancer development. In 90 percent of cancers telomeraza's activation is observed.

In cancers without TERT activation in cells , they mainly used other mechanism of telomeres' protection , which is called ALT (alternative lengthening of telomeres). Details of this mechanism are unknown.

According to Elizabeth Blackburn's works (1985, 2001), telomeraza is also involved in regulation of 70 genes which participate (or probably participate) in cancer origin and development. Furthermore, telomeraza activates glycolysis, which allows cancer cells to use sugar in order to maintain growth and division rate. This speed is equal to the same process rate in embryo.

Fight with the cancer is difficult. If it was possible to receive such a medication, which would cause telomeraza's blockage in cancer cells, then telomere shortening would be renewed. Mutations would appear and cancer cells would die!

Working in this direction has started (Koreans have just created the preparation "Telovak", which in the opinion of creators, activates immune system, that ultimately is directed to suppress telomeraza. After trying it on volunteers, ill with cancer "Telovak" prolonged patients' life by merely one year. It also gifted three months of life to the patients who were on the last stage of cancer. The question is : Maybe not only in cancer cells is telomeraza suppressed? Subsequently, will it ultimately make life shorter?

5. Senescence

Cancer cell is divided infinitely, telomeraza's gene is expressed , " inserted " in it i.e. malignant cell looks like a sex or embryonic cell (Probably, these are the only cells and its product restores normal length of the telomere).

Hereby , we take the liberty of a small comment: Yes ! - Today, immortal cells of cancer are widely discussed , which is said to be the fault of telomerazas . However, it shouldn't be absolutely right. Not all multiplied cancer cells are divided, i.e. among them there are some cells that divide ,so there is the limit too. As regards the culture, it is possible to initiate growth in all the ways . Besides , in vivo , cancer cells gradually cannot divide . Ultimately, they don't have "strength " of even it and die (These are famous facts , pathology -anatomists like to point it out. I remember the talk with Omar Khardzeishvili about this very question) i.e. part of cancer cells die by themselves in the lifetime of the ill organism. (It is an axiom today too); However, the person dies before all the cancer cells die !

In human fibroblasts, Kamozin can increase Hayflicks's Limit by means of reduction of telomeres shortening quality. However, the scientists say , that vertebrates' cells have certain potential of replication. More importantly, A.Melk and his group (2003) saw that in vivo aging is possible without telomere shortening.

The surgeon, Nobeliant Alexsis Carrel said : "All the cells , that grow in the culture , are immortal ; But if they reduced , the number of cell replications fell, it means that the way of cultivation, itself , is to be recovered and improved. " We take again the liberty of adding a comment of our own : "Growing cells in culture are immortal "- The multiplication process itself can be continuous. However, the cells probably change ; In this way a human is also immortal- by means of his/her generation or descendants (since Adam till today). Yes, exactly the same process! And don't let Carrel to deceive us. However, as regards the opinion, that "if the number of cell divisions reduced, then the way of cultivation must be improved. " This is right: It's necessary to improve conditions to increase birth-rate in people!).

Carrel's hypothesis (1921) was strengthened by the fact that the scientist had been growing the chicken's fibroblasts in the culture for 34 years. The part of scientists thought this was possible infinitely in vertebrates. However, there was a certain mistake during the experiment (i.e. in accuracy) and most scientists didn't agree with the results. The mistake was in the fact that Carrel daily added chicken's embryonic, axial cells to the culture; This allowed the new cells to multiply and it wasn't multiplication of the initiate, original cells any more; but in this way cells' multiplication was possible unlimitedly. Some researchers are convinced that Carrel knew about this mistake (in the experimental procedure), but he didn't admit it.

Briefly, Alexsis Carrel's theory didn't succeed and his mistake was evident too. However, Alexei Olivnikov's theory is acceptable and corresponds to Hayflick's experiments: Telomerazas' inhibition is a good remedy in malignant cancer treatment.

Ultimately, in reduction of life expectancy the following is important : 1. Reduction of the cell replication i.e. what Hayflick's limit means ; 2. Telomeres' shortening ; 3. Accumulation of cells and reduction of axial cells number; 4. Changes , revealed in the cell with aging: accumulation of free radicals, influence of radiation, generation of oxygen free radical O2 -, modification of albumins, lipids, nuclein acids , reduction of antioxidant mechanisms (vitamin E, glutation peroxidaza) , accumulation of dot mutations; 5. Lowering of the reparation systems, lowering of the DNA helikaza activity i.e. defect (enzyme is involved in DNA replication, reperation and other functions, that are necessary for DNA perfection. Besides Verner syndrome, defect of this enzym is found in Ataxia-Telangiektazia); 6. Strengthened expression of antioxidant enzyme - SOD and katalaza (it has been seen in Drisophila) ; 7.It is thought that proteasome function can be lowered that is called proteolyses mechanisms. Its responsibility is elimination of abnormal and unnecessary intracellular albumen (Lesion of the organels which is one of the reasons for cell aging).

Consequently, with aging, in all the organ systems, there are physiological and structural changes. The rate of aging process is associated with genetic factors, nutrition specialty, social condition , disease development , associated with age. For example, atherosclerosis, diabetes type 2, osteoarthritis.

The most effective way for increasing life expectancy is limitation of calories. It depends on sirtuins. They have hystondeacetilaza activity and probably they promote expression of different genes , products of which increase life expectancy. This products consist of proteins which increase metabolic activity, reduce apoptosis , stimulates albumens' third structure formation (fights with denaturation), inhibits free radicals of oxygen. Sirtuins also rise sensitivity on insulin and strengthen glucose metabolism. They can also be used in diabetes treatment. It's important that red wine activates sirtuins and this prolongs life.

So, probably nothing is impossible, among them defeating cancer disease and increasing life limit... However...

`In spite of the fact that telomeraza is the key to cell immortality i.e. " infinity" of cancer cell is caused by telomeraza, we know and pointed out above that cancer cells (where expression: of this enzym is high and where exactly telomeraza is "accused " in persistent division of the cells), ultimately the condition is achieved when they don't divide any more (more exactly -can't), so their "infinity" is doubtful! Where is the immortality?

We set an example of studies above where there is indicated that aging is possible without telomere shortening or disappearing i.e. nor immortality or "not aging" reverse-process- the death (which follows the aging) entirely depends on telomeres..... Consequently, it is doubtful again that telomeraza is the key to the cell immortality! Then, where is the immortality!

In our opinion the key to immortality should be that "thing" which causes permanent renovation! Though, it's difficult to imagine it in our dimension , where everything has its beginning and the end , still I remember " immortality water" from fairy tales ! - Where is the immortality?

There are such albumens-stress albumens (which appear during different stresses in the cell and protect it from lesion spreading) - Chaperon , Chaperonins, sirtuin molecules are very important and we pointed out above their protecting and " rescuing " occupation ! Red wine turned to be sirtuins action stimulator! There is Resveratrol in the same red wine -plant estrogen, Melatonin's precursor ! Melatonin is a hormone which acts as an obstacle for lesion, "wearing out "; Melatonin has the most expressed regenerative capacity and consequently the ability to restore/maintain living structures!

In the preamble of my first monograph ("melatonini" 2007) Dinara Kasradze wrote : "And in this system too, the most mysterious turned to be the Epiphysis... Epiphysis with its melatonin! However, it is the main thing where the life takes its origin (metaphor) -i.e. it is the "source " of light and darkness, mist and dawn . It causes sleep and waking up, tiredness and relaxing, wearing out and restoration, spring and autumn, summer and winter too .... It gives the world rhythm, biorhythm which gives us a birth, brings us up, breeds, ages or takes us into other..... It causes health and disease... and probably it is not surprising that in a creative way the "third eye" is called Epiphysis, maybe not entirely creatively. It can be so that overwhelming strength of the Lord is incarnated by means of Melatonin in our dimension.... Biological-cosmic key for immortality and constant transformations may probably lay ...... in the very magic molecule of Melatonin.... Who knows...."

6. Centriole, differentiation, senescence

Death is inevitable for any of living organisms, viruses, plants or animals. However, only multicellular animals, including human beings, die from ageing. A multicellulary organism (animal), containing a multitude of irreversibly differentiated cells, develops from only one cell e the totipotential zygote. The potency to differentiate into a certain tissue or tissues (set by the factor which controls the possibility of repression and activation of genes) determines the individual histological state, or morphogenetic "status'" of a cell. The final morphogenetic status means that such a cell is committed to programmed death (apoptosis). A 'zero' morphogenetic status means that the cell has not been committed to any irreversible pathway, i.e. it remains totipotent. Modification of morphogenetic status of a cell changes the whole spectrum of the tissues into which this cell can differentiate. In ontogenesis, this spectrum consequently narrows (totipotential / pluripotential/ multipotential/ unipotential/ non-potential) until the cell reaches the state of final differentiation (Malaitcev et al., 2002).

6.1 Need for a self-replicating controller

It has been established that morphogenetic status of a cell might be changed only through its division. We consider that cell division, differentiation and apoptosis may be controlled by a single intracellular structure. Naturally, this structure has to be self-replicating. Moreover, it has to have some way of counting or recording cell divisions. In a somatic cell potential candidates for this replicable structure are chromosomes, mitochondria (both contain DNA), and centrioles. To date the structures which might count cell divisions e and, therefore, determine the morphogenetical status of a cell e are generally thought to be telomeres (Bodnar et al., 1998; Greider, 1998; Shay and Wright, 2001). However, some data do not confirm the hypothesis that assigns the role of replication 'clock' to these parts of a chromosome.

Thus, the cells of mice bred by Blasco et al. (1997) did not have telomerase activity. Such mice were viable and could produce up to 6 generations, though the chromosomes of each subsequent generation had increasingly shortened telomeres. Only the sixth generation developed abnormalities caused by the extreme shortening of telomeres. The authors emphasised that 'ageing' of cells took place long before the marginal shortening of telomeres. Rudolph et al. (1999) showed that mice with inactivated genes of telomerase had shorter lifetime and increased frequency of oncological diseases, but physiological and biochemical tests did not reveal any signs of early ageing. Animal cloning disproves telomere hypothesis as well as other hypotheses, which suggest that a morphogenetic factor is located in nuclear DNA. In outline, the main point of cloning procedure is injection of the nucleus of a finally differentiated somatic cell into an oocyte with previously removed or destroyed nucleus. It was undoubtedly proved that the pattern of genes expression is changed to comply with the cytoplasm of a 'host' cell (Hardeman et al., 1986; Dominko et al., 1999). Injection of the nucleus of a frog carcinoma cells into an enucleated egg resulted in, not tumor cells, but normal tadpole development.

The key element of gene network controlling the processes of cell differentiation is an external factor for nuclear genome signal, which activates these groups of interactively expressing genes (Kolchanov et al., 2000). We suggest that the processes of cell ageing, differentiation and division are regulated by cytoplasmic factors. The structure, which regulates processes of irreversible differentiation, determination and modification of morphogenetic status, is most likely to be a centriole (centrosome).

Recent data indicate that centriole and nuclear cycles are not nterdependent (Gorgidze and Vorobjev, 1994). It was shown that nucleus controls the synthesis of 'building material' (basically proteins) used by centrioles, thereby having control over the centriole cycle. When a cell has such 'building material' in excess (i.e. oocytes), cycles of centriole duplication may occur even if mitotic cycle has not been launched (Manandkhar et al., 1990). Phillips and Rattner (1976) also demonstrated that inhibition of RNA or RNA/protein synthesis suppresses the duplication of centrioles in cultured cells.

6.2 The rule of cell division

As a rule, somatic cells contain a diploid set of chromosomes and a pair of centrioles (diplosome). This ratio is maintained by the parallel reproduction of nuclear DNA and centrioles, which takes place while cells are being prepared for the next division. In the process of gametogenesis, the number of chromosomes is reduced and the cell becomes haploid. Studies of spermatogenesis performed in insects (Sciara coprphila, Chrysopa carnea, Bombyx mori) and mammals (Heterohyrax syriacus and Memetes berdmorei) discovered that at the first meiotic division spindle poles contain a pair of centrioles, while at the second division e they contain only one centriole; in mammalians spermatocytes get two centrioles after both first and second divisions (Kriouchkova and nishchenco, 1999). On the contrary, oocytes of mammalians (mouse, rat, and rabbit) and Xenopus laevis do not have visible centriolar structures during both the meiotic divisions (Kriouchkova and Onishchenco, 1999). During the process of fertilization, the chromosomal sets of a spermatozoon and an oocyte are mixed. The chromosomal sets/centrioles ratio, typical for somatic cells, may be restored either in a zygote or early blastomeres. Thus, in sea urchins, the spindle pole at first cell division contains a diplosome, while in mice they are centriole-free at both first and second divisions. Only at third division does a centrosome get a pair of centrioles. The literature reveals several hypotheses, which try to explain the formation of centrioles in embryonic cells. These structures may be obtained exclusively from the paternal cell, as in sea urchins. In most mammalians zygotes, the centrosome is of maternal origin, but the centrioles are formed de novo only in the third cell cycle. It should be noted, however, that human centrioles, as well as centrioles of other mammalian species (sheep, cow, and marsupials), belong to the paternal sex cell (Breed et al., 1994; Nijs et al., 1996; Schatten et al., 1996).

Some studies demonstrate that centrioles act as the regulators that control the course of every phase of a cell cycle influencing the processes that take place at least two phases earlier (Neverova et al., 1996; Maniotis and Schliva, 1991).

6.3 Centrioles and cancer

Centrioles seem to play some role in inheritance of tumorigenic properties. Taking into account aforementioned facts, what molecular mechanism underlying the centriolar activity may control determination of morphogenetic status of a somatic cell? Centrioles may contain differently encoded RNA molecules stacked in a definite order. During mitosis these RNA molecules are probably released one by one into cytoplasm. This process presumably changes the status of repressed and potentially active genes and, subsequently, the morphogenetic status of a cell.

Centrioles may have such molecules controlling morphogenetic status contained in their internal chamber (the area of two electron-dense linear structures). It is also possible that they are packed in definite order. Some molecules may differ from the preceding ones and control different morphogenetic status. Other ones can be exact copies, thus determining the same morphogenetic status. In each mitotic division, one of these molecules is released and 'lost' in the cytoplasm, so the centrioles of daughter cells contain one molecule less than the centrioles of the maternal cell. The number of molecules contained in centrioles decreases after each mitotic division. The last molecule triggers the processes of programmed death (i.e. apoptosis). Thus, the number of hypothetic molecules must correspond to the number of possible mitotic divisions, counting down from the cell having the first morphogenetic status to the last offspring cell having the final morphogenetic status (the "Hayflick limit", Hayflick, 1997).

6.4 Importance of small MW RNA (siRNA)

The best candidate for the role of the carrier of information on morphogenetic status seems to be low molecular weight RNA. Indirect evidence for this hypothesis may be the discovery of new classes of small RNA e so-called interfering RNA (siRNA) and micro-RNA (miRNA) e having regulatory activity. The phenomenon of RNA interference occurs to inhibit selectively gene expression in animal cells (Fire et al., 1998; McManus and Sharp, 2002; Montgomery et al., 1998), including mammalian cells (Wianny and Zornicka-Goets, 2000). MicroRNA appeared to act as a regulator of cell differentiation and development in higher organisms. There direct data confirm the role of these molecules in the process of cell division (Lau et al., 2001; Lee and Ambros, 2001; Lagos-Quintana et al., 2001). In 2002, it was found that the influence of siRNA might not be limited to only temporary 'knockout' of genes on RNA level. Small RNA can modify the chromatin structure and make genes active or silent for quite a long period of time (Zilberman et al., 2003). In addition to post-transcriptional and transcriptional homology-dependent gene silencing links between DNA structure and RNA expose themselves in many other phenomena, like dose-dependent compensation in drosophila and mammalian X-chromosome inactivation (Stuckenholz et al., 1999).

The whole body of existing e though sometimes contradictory e data demonstrating that RNA may be located inside of centrosome or linked to this structure (Heath, 1980; Heidemann et al., 1977; Lambert and Nagy, 2000; Nadejdina et al., 1982; Peterson and Berns, 1980) makes it possible to suggest that small RNA (of 20e300 nucleotides) may be an ideal candidate for the role of the molecule if these are inside the centrioles.

6.5 The centriole as the carrier of controlling siRNA

Filling of centrioles with the molecules of RNA in various species apparently takes place in different cells: those, which develop at the first of second meiotic division of gametogenesis (spermatogenesis or oogenesis, depending on species), or in the first blastomeres at the time of centriole formation de novo. Those molecules are transcribed from nuclear DNA and stacked in a definite order within the centrioles. Following this process subsequent release of RNA molecules during mitosis determines the expression of nuclear DNA. Nuclear genome presumably 'chooses' and 'places' into centrioles the information about the sequence of DNA loci activation in the offspring cells. Absence of centrioles in a stage of preleptotenic chromosome condensation in mice may be good illustration of 'zero' morphogenetic status of cells (Hartung and Stahl, 1977). Intracellular 'morphogenetic clocks' seems to be wound up at that moment returning to the initial 'zero' state.

6.6 Conclusions that can be drawn

Should centrioles indeed control the processes of cell differentiation and programmed death, the following may be concluded:

1. When a cell initially does not have centrioles (this situation denotes the absence of morphogenetical status), it will be totipotential;

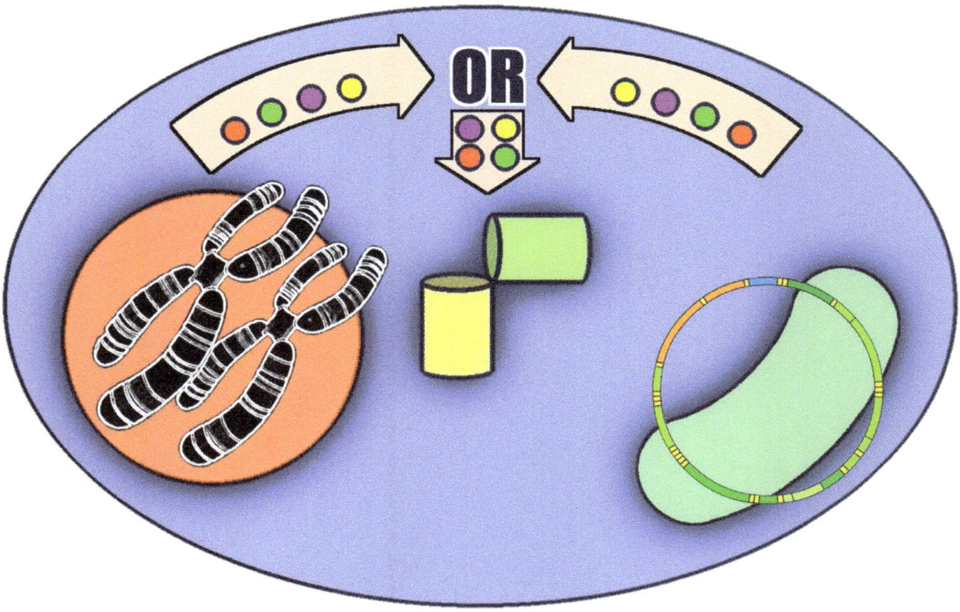

Fig. 1. Causing irrevocable differentiation - Protein or RNA-hypothetical molecular code must be encoded in DNA or mitochondrial or nuclear. It is expected that as many differentiation molecular codes should be encoded in DNA as the quantity of irrevocable differentiation stages, characteristic for each particular species. Molecular structure of the differentiation, complexing at centriol must have the features of matrix - while forming new centriols, it must repeat similar structure.

2. If a cell launches centriole synthesis de novo (the centrioles are considered to be formed de novo as long as the cell is not committed to irreversible differentiation), it will be totipotential and immortal, retaining its initial 'zero' morphogenetic status;

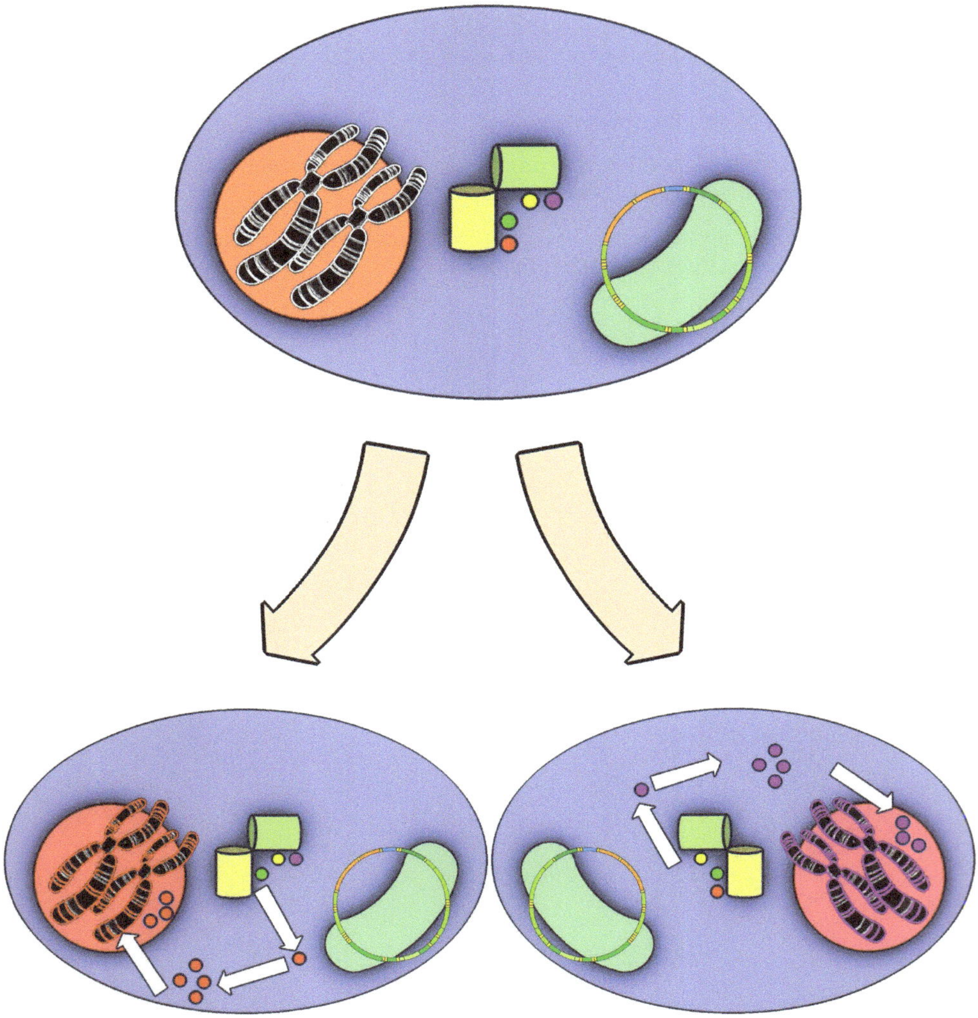

Fig. 2. Cell division, during which release of the molecule, causing differentiation happens, will finish with the irrevocable differentiation of the descendant cells. Probably, the molecule the albumen or RNA differentiation plays the role of the matrix. Ultimately, the small DNA molecule is formed, which enters the nuclear chromosomes, is inserted in them and so changes their structure and function, but not irrevocably.

3. Should a cell not to die due to programmed death when appropriate changes in centrioles do occur, it will be immortal, but not totipotential

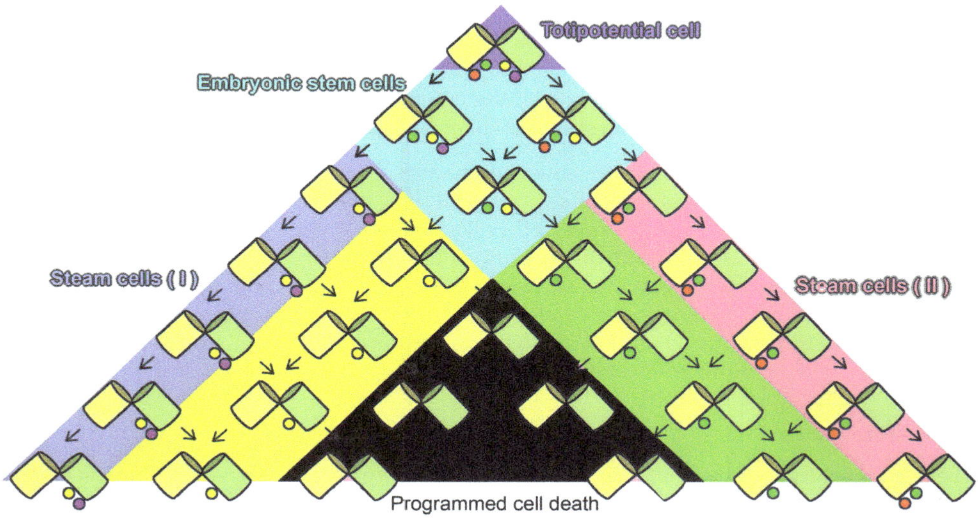

Fig. 3. Maintaining the quantity of the stem cells is explained by the reason, that irrevocable differentiation divisions have asymmetric nature (asymmetric division). Existence of asymmetric division and differentiation structure can also explain the fact that different types of cells give not only different (various), but also similar descendant cells.

Many existing data may be explained on the basis of the proposed hypothesis:

1. Centrioles are initially absent in the cells of higher plants (Sluiman, 1985), zygote and the first blastomeres of some animals (Abumuslimov et al., 1994 Calarco-Gillam et al., 1983). According to our conception they are fully potential and immortal;
2. Centrioles formed de novo in a zygote and the first blastomeres (Maro et al., 1991) belong to the fully potential and immortal cells;
3. Tumor and transformed cells are immortal, but not totipotential. This state may be due to the presence of a certain 'non-zero' morphogenetical status at the moment of transformation.

It is possible that centriolar and nuclear cycles of tumor and transformed cells are irreversibly disengaged. The nucleus looses its ability to perceive some intracellular and extracellular signals, including those which control the morphogenetic status of a cell. The cell becomes immortal, though it has 'non-zero' morphogenetic status.

Based upon the proposed hypothesis, we suggest that centrioles realize their function of 'counting' mitotic divisions. The number of divisions (generations) of a cell is limited; this means that the number of cells of various types is fixed too. Consequently, the possibility of cellular regeneration will be also limited. Sooner or later, regenerating issues experience a lack of cells: then the organism will not be able to further provide its 'homeostatic' support for long-living cells (for instance, neurons). Thus, death from ageing seems to be a phenomenon affecting the whole organism.

Our proposed hypothesis is verifiable, but needs also be tested to destruction. It may be easily checked if the chemical composition of the internal chamber of centrioles is studied more closely. According to the hypothesis, there must be some difference in chemical composition, as well as in ultrastructure (morphology), of the inner chamber of centrioles in different types of somatic cells. Attention should be also paid to monocellular organisms having centrioles.

It should be emphasised that until now the changes of ultrastructure of the centrosome during the processes of maturation of sex cells and their fusion during insemination, as well as in the course of the first embryonic cell cycles, have not yet been studied in any detail. This kind of investigation should throw light upon the problem of centrioles (diplosome) 'inheritance' in embryonic blastomeres. It would be interesting to know how and to what extent the hetero- and homo-gametic states are related to the 'inheritance' of centrioles. It is also important to find out what kind of divisions produce changes in morphogenetic status and how great is the number of such divisions.

6.7 Final remarks

The final resolution of the centriole question, however, can be through their transplantation from differentiated cells of one type into the cells of another type. Today such transplantations are quite possible. To preserve the functional activity of centrioles, they should be extracted immediately after their maturation is finished, i.e. in the middle of metaphase, and injected at the moment the chromosomes are formed. Of course, the centrioles of the recipient cell have to be removed earlier from the spindle poles. Then one could observe whether such 'hybridization' provokes changes in the expression of nuclear genes. In our opinion, the expression of nuclear genes should be changed to resemble that of the donor cell. It would confirm the hypothesis, which states that centrioles (diplosome) are actually the structure controlling the morphogenetical status of a cell.

According to this hypothesis, there must a constant number of stem cells, which is sufficient for regeneration of tissues and organs. However, this doesn't conform with reality. With aging, number of stem cells is reduced. It has to be searched: What is the reason for it? - Influence of the outer factors (e.g.: Gathering mutation in generations, which makes the cell lose the viability) or inner factors, the programmed process (e.g.: Telomeres' shortening in generations, that ends up with transformation in cancer cell or cell death).

7. Centrosomal RNA

In 2006 a group of researchers discovered a specific centrosome-associated RNA, which is called cnRNA. New developments and findings in this direction will clarify the issue whether the differentiation molecules have a protein or RNA nature. It is desirable as well to identify which section of the ovule genome carries the information about these molecules.

8. Conclusion

In spite of the fact, that Telomerase is the key to the cell immortality or that cancer cell "eternity" is caused by Telomerase, we know and have pointed out above that cancer cells (where enzyme expression is high and where every Telomerase is "to be blamed" for constant division of cells), ultimately achieve the condition when they don't divide any more (can't divide is more correct) - i.e. their "eternity" is suspicious! - Where is the immortality?

In our point of view, the "key" to immortality must be "something" which caused constant renovation of the "worn". It's impossible to stop aging, only, it's possible to turn the biological clock back (even for several times). However, it's difficult to imagine in our dimension, where everything has its beginning and end... and still I can remember "water of immortality" from my childhood - Where is the immortality?

The fact that it was possible to induce totipotential qualities in the stem cell, gives us some hope. If we can discover the damaged DNA and replace it with healthy code, it's already possible to return totipotential features to this cell.

If we return this cell to the organism, theoritically, it is expected that the "aged" stem cells "are replaced" by the "rejuvenated" stem cells.

9. References

[1] Abumuslimov SS, Nadezhdina ES, Chentcov YuS. Morphogenesis of centrioles and centrosomes in early mouse development: the Electron microscopic study. Tsitologiia 1994;11:1054e60 [in Russian].

[2] Alliegro M.C., at all. Centrosome-associated RNA in surf clam oocytes. Proc Natl Acad Sci U S A. 2006; 103(24): 9034-9038

[3] Alliegro MC, Satir P., 2009. Origin of the cilium: novel approaches to examine a centriolar evolution hypothesis. Methods Cell Biol. 2009; 94: 53-64

[4] Bartel D., Chen C. Micromanagers of gene expression: the potentially widespread influence of metazoan microRNAs. Nat. Rev. Genet. 2004; 5: 396-400.

[5] Blasco MA, Lee HW, Hande MP, Samper E, Lansdorp PM, DePinho RA, et al. Telomere shortening and tumor formation by mouse cells lacking telomerase RNA. Cell 1997;91:25e34.

[6] Blower MD, Feric E, Weis K, Heald R. Genome-wide analysis demonstrates conserved localization of messenger RNAs to mitotic microtubules. J Cell Biol. 2007; 179(7): 1365-1373.

[7] Blower M.D., Nachury M., Heald R., Weis K. A Rae1-containing ribonucleoprotein complex is required for mitotic spindle assembly. Cell. 2005; 121: 223-234.

[8] Bodnar AG, Ouellette M, Frolkis M, Holt SE, Chiu CP, Morin GB, Harley CB, et al. Extension of life-span by introduction of telomerase into normal human cells. Science 1998 Jan 16;279(5349):334e5.

[9] Breed W, Simerky C, Navara C, Vande Berg J, Schatten G. Distribution of microtubules in eggs and early embtyos of marsupial Monodelphis domestica. Dev Biol 1994;164:230e40.

[10] Calarco-Gillam PD, Siebert MC, Hubble R, Mitchison T, Kirschner M. Centrosome development in early mouse embryos as defined by an autoantibody against pericentriolar material. Cell 1983;35:621e9.

[11] Chang S., Johnston R., Frekjaer-Jensen C. et al. MicroRNAs act sequentially and asymmetrically to control chemosensory laterality in the nematode. Nature. 2004; 430:785-789.

[12] Chapman M.J., Dolan M.F., Margulis L. Centrioles and kinetosomes: form, function, and evolution. Q Rev Biol. 2000; 75(4): 409-429.

[13] Chapman, M.J. and Alliegro, M.A. A symbiotic origin for the centrosome? Symbiosis. 2007; 44: 23-32

[14] Chen C., Li L., Lodish H., Bartel D. MicroRNAs modulate hematopoietic lineange differentiation. Science. 2004; 303: 83-86.
[15] Chu C.Y. and Rana T.M. Small RNAs: regulators and guardians of the genome. J. Cell Physiol. 2007; 213: 412-419.
[16] Dominko T, Mitalipova M, Haley B, Beyhan Z, Memili E, McKusick B, et al. Bovine oocyte cytoplasm supports development of embryos produced by nuclear transfer of somatic cell nuclei from various mammalian species. Biol Reprod 1999;6:1496e502.
[17] Doxsey S., McCollum D., Theurkauf W. Centrosomes in cellular regulation. Ann. Rev. Cell and Dev. Biol. 2005; 21: 411-434.
[18] Etienne-Manneville S. Actin and microtubules in cell motility: which one is in control? Traffic. 2004; 5(7): 470-477.
[19] Filipowicz, W., Bhattacharyya, S.N. and Sonenberg, N. Mechanisms of post-transcriptional regulation by microRNAs: are the answers in sight? Nature Rev. Genet. 2008; 9: 102-114.
[20] Fire A, Xu S, Montgomery MK, Kostas SA, Driver SE, Mello CC. Potent and specific genetic interference by double/stranded RNA in Caenorhabditis elegans. Nature 1998;391:806e11.
[21] Galitskii V.A. Epigenetic nature of ageing. Tsitologiia. 2009; 51(5):388-397. Russian
[22] Gorgidze LA, Vorobjev IA. Centrioles in tissue culture cells are able to replicate in the absence of the nucleus. Tsitologiia 1994;8:837e43 [in Russian].
[23] Gourlay C.W., Ayscough K.R. The actin cytoskeleton: a key regulator of apoptosis and ageing? Nat Rev Mol Cell Biol. 2005; 6(7): 583-589.
[24] Greider CW. Telomerase activity, cell proliferation, and cancer. Proc Natl Acad Sci USA 1998 Jan 6;95(1):90e2.
[25] Hardeman EC, Chiu CP, Minty A, Blau HM. The pattern of actin expression in human fibroblast mouse muscle heterokaryons suggests that human muscle regulatory factors are produces. Cell 1986;47:123e30.
[26] Hiedemann, S.R., Sander, G. and Kirschner, M.W. Evidence for a functional role of RNA in centrioles. Cell. 1977; 10, 337-350.
[27] Hartung M, Stahl A. Preleptotene chromosome condensation in mouse oogenesis. Cytogenet Cell Genet 1977;2:309e19.
[28] Hayflick L. Mortality and Immortality on the cell level. Biochemistry 1997;11:1380e93.
[29] Heath JB. Variant mitosis in lower eucaryotes: indicator of the evolution of mitosis. Int Rev Cytol 1980;64:1e80.
[30] Heidemann SR, Sander G, Kirschner MW. Evidence for a functional role of RNA in centrioles. Cell 1977;10:337e50.
[31] Kim V.N., Han J, Siomi M.C. Biogenesis of small RNAs in animals. Nat Rev Mol Cell Biol. 2009; 10(2): 126-39.
[32] Kingsley E.P., Chan X.Y., Duan Y., Lambert J.D. Widespread RNA segregation in a spiralian embryo. Evol Dev. 2007; 9(6): 527-539.
[33] Kloc M., Zearfoss N.R. and Etkin L.D. Mechanisms of subcellular mRNA localization. Cell. 2002; 108: 533-544.
[34] Kloc M., Etkin L.D. RNA localization mechanisms in oocytes. J Cell Sci. 2005; 118(Pt 2): 269-82.

[35] Klotz, C., Dabauvalle, M.C., Paintrand, M., Weber, T., Bornens, M. and Karsenti, E. Parthenogenesis in Xenopus eggs requires centrosomal integrity. J. Cell Biol. 1990; 110: 405-415.
[36] Kolchanov NA, Anan'ko EA, Kolpakov FA, Podkolodnaia OA, Ignat'eva EV, Goriachkovskaia TN, et al. Gene networks. Mol Biol 2000;4:533e44.
[37] Kriouchkova MM, Onishchenco GE. Structural and functional characteristics of the centrosome in gametogenesis and early embryogenesisi of animals. Intern Rev Cytol 1999;185:107e56.
[38] Lagos-Quintana M, Rauhut R, Lendeckel W, Tuschl T. Identification of novel genes coding for small expressed RNAs. Science 2001; 294:853e8.
[39] Lambert JD, Nagy LM. Asymmetric inheritance of centrosomally localized mRNAs during embryonic cleavages. Nature 2002; 420:682e6.
[40] Lau NC, Lim LP, Weinstein EG, Bartel DP. An abundant class of tiny RNAs with probable regulatory roles in Caenorhabditis elegans. Science 2001;294:858e62.
[41] Lee RC, Ambros V. An extensive class of small RNAs in Caenorhabditis elegans. Science 2001;294:862e4.
[42] Lecuyer E, Yoshida H, Parthasarathy N, Alm C, Babak T, Cerovina T, Hughes TR, Tomancak P, Krause HM. Global analysis of mRNA localization reveals a prominent role in organizing cellular architecture and function. Cell. 2007; 131(1): 174-187.
[43] Lopez de Heredia M, Jansen RP. mRNA localization and the cytoskeleton. Curr Opin Cell Biol. 2004; 16(1): 80-85.
[44] Malaitcev VV, Bogdanov IM, Sukhikh GT. Contemporary introduction of biology of stem cell. Ark Pathol 2002;4:7e11 [in Russian].
[45] Manandkhar G, Khodjakov AL, Onishenko GE. Modified PEGDMSO- Serum method induces high rate of fusion. Cell Biol IntRep 1990;14:82e6.
[46] Maniotis A, Schliva M. Microsurgical removal of centrosomes block cell reproduction and centriole generation in BSC-1 cells. Cell 1991;67:495e504.
[47] Margulis L., Chapman M., Guerrero R., Hall J. The last eukaryotic common ancestor (LECA): acquisition of cytoskeletal motility from aerotolerant spirochetes in the Proterozoic Eon. Proc Natl Acad Sci U S A. 2006; 103(35): 13080-13085
[48] Maro B, Gueth-Hallonet C, Aghion J, Antony C. Cell polarity and microtubule organisation during mouse early embryogenesis. Dev Suppl 1991;1:17e25.
[49] McManus MT, Sharp PA. Gene silencing in mammals by small interfering RNAs. Nat Rev Genet 2002;3:737e47.
[50] Miska E., Alvarez-Saavedra E., Townsend M. et al. Microarray analysis of microRNA expression in the developing mammalian brain. Genome Biol. 2004; 5: R68.
[51] Montgomery MK, Xu S, Fire A. RNA as a target of double/stranded RNA/mediated genetic interference in Caenorhabditis elegans. Proc Natl Acad Sci USA 1998;95:15502e7.
[52] Nadejdina ES, Fais D, Chentsov YuS. Some biomedical aspects of centriole and basal bodies. Macrosomal Funct Cell 2 Sov-Ital. Sympos, Pushchino, 1982, b2, Moscow, p. 163 [in Russian].
[53] Neverova AL, Uzbekov RE, Votchal MS, Vorobjev IA. Effect of microirradiation of the centrosome on cell behaviour. IV. Synthetic activity, spreading and growth of cells with an inactivated centrosome. Tsitologiia 1996;2:145e54 [in Russian].

[54] Nijs M, Vanderswalmen P, Vandamme B, Segal-Bertin G, Lejeune B, Segal L, et al. Fertilizing ability of immotile spermatozoa after intracytoplasmic sperm injection. Hum Reprod 1996;11:2180e5.
[55] Palacios I.M. and St Johnston D. Getting the message across: the intracellular localization of mRNAs in higher eukaryotes. Annu Rev Cell Dev Biol. 2001; 17: 569-614.
[56] Palazzo R.E., Vogel J.M. Isolation of centrosomes from Spisula solidissima oocytes. Methods Cell Biol. 1999; 61: 35-56
[57] Pederson T. The centrosome: built on an mRNA? Nat Cell Biol. 2006; 8(7):652-654
[58] Peterson SP, Berns MW. The centriolar complex. Intern Rev Cytol 1980;64:81e106.
[59] Peterson, S.P., and Berns, M.W. Evidence for centriolar region RNA functioning in spindle formation in dividing PTK2 cells. J. Cell. Sci., 1978; 34; 289-301.
[60] Phillips SG, Rattner JB. Dependence of centriole formation on protein synthesis. J Cell Biol 1976;70:9e19.
[61] Roskelley C.D., Srebrow A., Bissell M.J. A hierarchy of ECM-mediated signalling regulates tissue-specific gene expression. Curr Opin Cell Biol. 1995; 7(5): 736-47.
[62] Rudolph KL, Chang S, Lee H, Blasco M, Gottlieb GJ, Greider C, et al. Longevity, stress response, and cancer in aging telomerasedeficient mice. Cell 1999;96:701e12.
[63] Schatten G. The centrosome and its mode of inheritance: the reduction of the centrosome during gametogenesis and its restoration during fertilization. Dev. Biol. 1994; 165: 299-335.
[64] Schatten H, Schatten G, Mazia D, Balezon R, Simerly C. Behavior of centrosomes during fertilization and cell division in mouse oocytes and sea urchin eggs. Proc Natl Acad Sci U S A 1996;83:105e9.
[65] Shay JW,Wright WE. Ageing and cancdr: the telomere and telomerase connection. Novartis Found Symp 2001;235:116e25.
[66] Sluiman HJ. A cladistic evaluation of the lower and higher green plants (Viridiplantae). Plant Syst Evol 1985;149:217e32.
[67] Stuckenholz C, Kageyama Y, Kuroda MI. Guilt by association: noncoding RNAs, chromosome-specific proteins and dosage compensation in Drosophila. Trends Genet 1999;15:454e8.
[68] Tekotte H., Davis I. Intracellular mRNA localization: motors move messages. Trends Genet. 2002; 18(12): 636-642.
[69] Tkemaladze D., Chichinadze K. Potential role of centrioles in determining the morphogenetic status of animal somatic cells (hypothesis). Cell Biol Int. 2005; 5: 370-374.
[70] Chichinadze K.N., Tkemaladze D.V. Centrosomal hypothesis of cellular ageing and differentiation. Adv Geront. 2008; 21(3): 367-371. Russian.
[71] Tkemaladze J., Chichinadze K. Centriole, Differentiation and Senescence. Rejuven Res. 2010; 13(2-3): 339-342.
[72] Tkemaladze J.V., Chichinadze K.N. Centriolar Mechanisms of Differentiation and Replicative Aging of Higher Animal Cells. Biochemistry (Moscow). 2005; 70(11): 1288-1303.
[73] Wianny F, Zernickab-Goets M. Specific interference with gene function by double-stranded RNA in early mouse development. Nat Cell Biol 2000;2:70e5.
[74] Zilberman D, Cao X, Jacobsen SE. ARGONAUTE4 control of locusspecific siRNA accumulation and DNA and histone methylation. Science 2003;299:716e9.

3

Cellular Degradation Machineries in Age-Related Loss of Muscle Mass (Sarcopenia)

Mikael Altun[1], Max Grönholdt-Klein[1],
Lingzhan Wang[1,2] and Brun Ulfhake[1]
[1]*Department of Neuroscience, Karolinska Institutet, Stockholm,*
[2]*Department of Human Anatomy, School of Medicine,*
Inner Mongolia University for Nationalities, Tongliao City,
[1]*Sweden*
[2]*P. R. China*

1. Introduction

One of the most characteristic features of the aged is a change in body composition and human cross-sectional and longitudinal studies consistently demonstrate gain in fat mass and decline in lean mass (Attaix et al., 2005). Skeletal muscle mass is gradually reduced through both atrophy and loss of myofibers, and along with this connective tissue and intramyocellular lipids increase (Fig. 1) (St-Onge, 2005). Although sarcopenia is widely recognized it remains poorly understood and has not received appropriate attention until quite recently. Sarcopenia is often assumed to have a multi-factorial background (Adamo and Farrar, 2006; Attaix et al., 2005; Paddon-Jones and Rasmussen, 2009; Roth et al., 2006; Solomon and Bouloux, 2006). A sedentary life-style combined with periods of prolonged bed-rest during illnesses has well-established detrimental effects on muscle mass and function (Janssen et al., 2002). Skeletal muscles do not only generate the power to let us move but collectively they are a highly significant component of the systemic metabolic homeostasis machinery. The progressive loss of muscle mass with advancing age is evident in both the 'healthy' aging population and specialist patient populations. The reduced functional muscle mass is associated with increased morbidity, frailty and reduced quality of life (Baumgartner et al., 1998a). The prevalence of sarcopenia increases with about 5% per year and typically begins in the fourth decade of life (Baumgartner et al., 1998b). After the age of 50 years 1-2 % of the muscle mass is lost annually. Individuals with clinically manifest sarcopenia have 4 times greater risk of disability, three times greater risk of balance impairment, and 3 times greater risk of falling (Baumgartner et al., 1998b). Sarcopenia is the single most common etiology to falls and fall-associated fractures in elderly.

1.1 Regulation of muscle mass and tentative mechanisms in sarcopenia

Myofibers are complex structures build-up by merger of aligned myocytes (myotubes as intermediates). Thus, the myofiber has a common plasma membrane confining multiple sub domains each supported by one myonucleus. According to morphometric analysis the ratio myofiber volume to a myonucleus remains fairly constant suggesting an optimal size of the domain supported by the machinery of one myonucleus. The impressive increase in muscle mass during late fetal and postnatal development but also in conditions of repair in adult muscle across life-span is dependent on recruitment of myogenic progenitors from the resident satellite cells (SC) in the stem cell niche (Kadi et al., 2005). SCs remain quiescent until activation through trophic stimuli e.g. IGF-1 and Notch-related signaling (Carlson et al., 2008). The activity of the SC pool is under the influence of several signaling pathways, enhancing or inhibiting SC proliferation, such as Wnt-β–Catenin, DeltaL/Jagged1-Notch or Smad2/3- TGF-β/Activin/Myostatin (Carlson et al., 2008; Zammit, 2008).

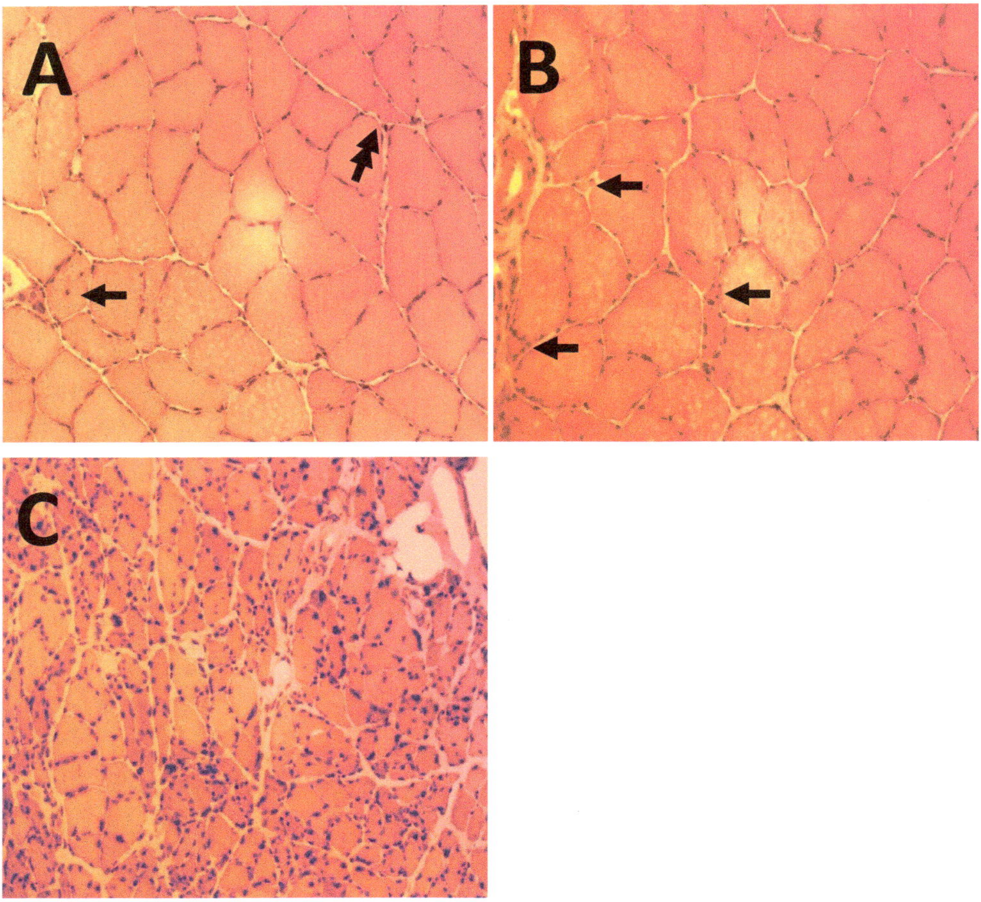

Fig. 1. **Changes in fiber size and occurrence of nuclei with a central location in aged rat soleus muscle.** Eosin-Htx staining of soleus cross sections (8μm) showing (A) young adult muscle where central nuclei (arrow) and very small fibers (double arrow) are infrequent. In early aging (B) fiber size becomes more irregular and the frequency of very small fibers increases (arrows). At advanced age (C) fiber size is highly irregular and centrally located nuclei are frequently occurring. Original micrographs shot with a x20 dry objective.

The signaling events leading to myofiber atrophy converge onto members of the FOXO family of transcription factors, which in an active state induce atrophy by increased proteasome degradation of myofibrillar proteins (Bodine et al., 2001; Sacheck et al., 2007; Sacheck et al., 2004; Sandri et al., 2004; Stitt et al., 2004). Recently, it was discovered that the signaling pathway that activates FOXO also induces an increased degradation through autophagy and lysosomal degradation (Sandri et al., 2006; Solomon and Bouloux, 2006). This "atrophy program" is activated in a range of conditions such as disuse, denervation and systemic diseases (Solomon and Bouloux, 2006). Conversely, myofiber contraction and growth signals (e.g. IGF-1) acting via protein kinase B, mTOR, and pgc-1α induce synthesis of myofibrillar proteins, deactivation of FOXO and, in parallel, adaptation of the energy producing machinery (*idem*). The Activin A/myostatin pathway, acting on Act receptor IIB upstream of Smad, can activate myofibrillar proteolysis (i.e. induce muscle fiber atrophy) and depress SC proliferation (Zhou et al., 2010). This creates a regulatory link between cellular anabolism & catabolism and SC activation & deactivation which should work in concert in conditions with atrophy and hypertrophy, respectively.

A growing body of evidence suggests that loss of muscle mass in elderly occurs through mechanisms more complex than those involved in disuse and disease atrophy in younger individuals (Altun et al., 2010; Edstrom et al., 2007; Edstrom and Ulfhake, 2005). In humans and rodents alike, a number of mechanisms have been suggested to underpin sarcopenia: 1) loss of innervation 2) disuse 3) impaired maintenance and repair including decline in endo-, para- and autocrine signaling (e.g. IGF-1) 4) systemic inflammation 5) imbalance between protein synthesis and degradation and 6) poor nutrition (Adamo and Farrar, 2006; Attaix et al., 2005; Paddon-Jones and Rasmussen, 2009; Roth et al., 2006; Solomon and Bouloux, 2006). However, awaiting more definitive evidence it remains unclear if the mechanism behind loss of muscle mass in elderly is different from that operating in muscle atrophy in young individuals. It should be noted that the progress over the past decade in our understanding changes in muscle mass in health and disease stems mainly from work done in laboratory animals, in particular rodents, and highlights:

1. ***Impaired regeneration of myonuclei from SCs.*** The satellite cell pool has been reported to be reduced in aged individuals and the frequency of myonuclear apoptosis to increase, however, these changes have not been directly associated with the loss of muscle mass in sarcopenia. A reduced SC pool could be consequence of an imbalance of SC recruitment/activation and the replenishment of the SC pool (see above). Furthermore, reports suggest also that age-induced changes in the microenvironment influence utilization of the SC through negative effects on proliferation and differentiation of the SCs and that in the aged the SCs are not activated on demand as in adults (Carlson et al., 2009). Importantly, inhibition of, or failure, to activate SC may augment proliferation of other progenitor cells known to co-exist in the skeletal muscle (Christov et al., 2007). An interplay exists between myogenic cells and fibro/adipogenic cells, and myogenic cells are known to inhibit fibro/adipogenic cells in the SC niche (Rodeheffer, 2010). In animal models this has demonstrated to influence muscle regeneration i.e. the scar tissue formation and fat storage in the skeletal muscle (Fuso et al., 2010), i.e. well-established features of the sarcopenic muscle.
2. ***Loss of innervation.*** Observations in aged human skeletal muscle have shown loss of myofibers, myofiber-atrophy, a selective vulnerability of type IIa fibers and fiber-type grouping suggesting that an underlying mechanism is a progressive age-dependent denervation (Larsson, 1995). Direct observations in animal models have shown both a progressive denervation (Valdez et al., 2010a) and a dramatic increase in fibers re-

expressing the embryonic isoform of myosin (eMyHC; Fig. 2) (Edstrom and Ulfhake, 2005) while the expression levels of adult MyHC isoforms decline (Altun et al., 2007b). In related work using a denervation/re-innervation animal model we have obtained evidence that the re-expression of nicotinic acetylcholine receptor subunit gamma (nAChR-γ) is a reliable marker for muscle denervation (Grönholdt-Klein et al., in preparation) and in animal models on sarcopenia we consistently find a dramatic increase in nAChR-γ expression. Earlier work also showed that spinal motoneurons innervating aged atrophic skeletal muscles show a regenerative phenotype suggesting an impaired contact with the target myofibers (Johnson et al., 1995). Combined, these observations suggest denervation as a significant component of sarcopenia.

3. ***Protein degradation and proteotoxicity.*** Accumulation of damaged proteins is a hallmark of aging that is believed to reflect increased imbalance between generation and scavenging of radicals, and/or decreased ability to degrade damaged proteins (Harriman et al., 1970). A number of studies have reported on an accumulation of oxidatively damaged proteins in aged rodent muscle (Cai et al., 2004; Clavel et al., 2006) as well as increased levels of chaperones that selectively bind unfolded proteins (Clavel et al., 2006; Ferrington et al., 2005). Several groups have published observations suggesting a decreased proteasomal proteolysis as a mechanism for the buildup of worn-out and otherwise damaged proteins in aged rodent skeletal muscle (Ferrington et al., 2005), however, it is hard to reconcile such a reduced rate of proteolysis with the general loss of skeletal muscle mass characteristic of aging (sarcopenia).

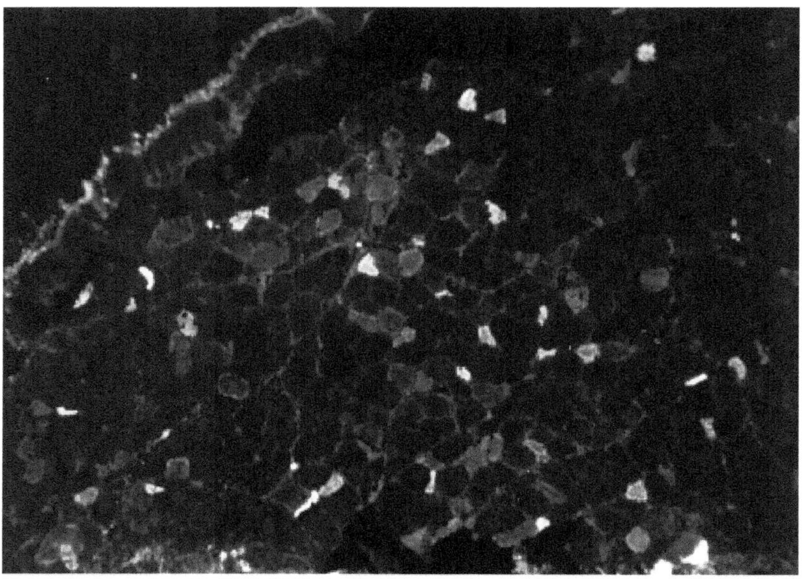

Fig. 2. **Re-expression of the embryonic myosin (eMyHC) in the soleus muscle of an aged mouse.** Immunofluorescence micrograph of an 8 μm thick cross-section through the mid portion of a soleus muscle from a 28 month-old C57BL/6J mouse. Light profiles often with an irregular out-line and small size indicate fibers expressing embryonic MyHC probably caused by an age-related loss of innervation. Original magnification x20.

This chapter will focus on the main cellular degradation machineries and our current understanding of how age-related changes in these systems impact skeletal muscle integrity.

2. Regulated proteolysis

Regulated proteolysis is instrumental for such diverse cellular processes as signaling, cell cycle progression, and apoptosis. Protein and organelle degradation enables recycling of the building blocks and is also an important survival response to starvation, whereby proteins are degraded to supply the organism with fuel for energy production. The accumulation of aggregates of misfolded proteins is a hallmark of cells and tissues of aged organisms, and protein aggregation occurs when damaged or partially unfolded proteins are not efficiently degraded. Disposal of proteins is in most conditions a selective and coordinated process, handled mainly by two cellular proteolytic systems, the ubiquitin-proteasomal system (UPS) and autophagy-lysosomal system (ALS).

2.1 The ubiquitin proteasomal system (UPS)

Work over the last 25 years has established the importance of regulating protein ubiquitination in a wide range of cellular functions including cell cycle control, transcriptional regulation, and diverse aspects of cell signaling. Lysine[48] polyubiquitination targets proteins for degradation by the proteasome, a highly selective mechanism by which multiple cellular processes are being modulated. Collectively, the proteasome and enzymes involved in ubiquitination are referred to as the ubiquitin proteasomal system (UPS) (Fig. 3).

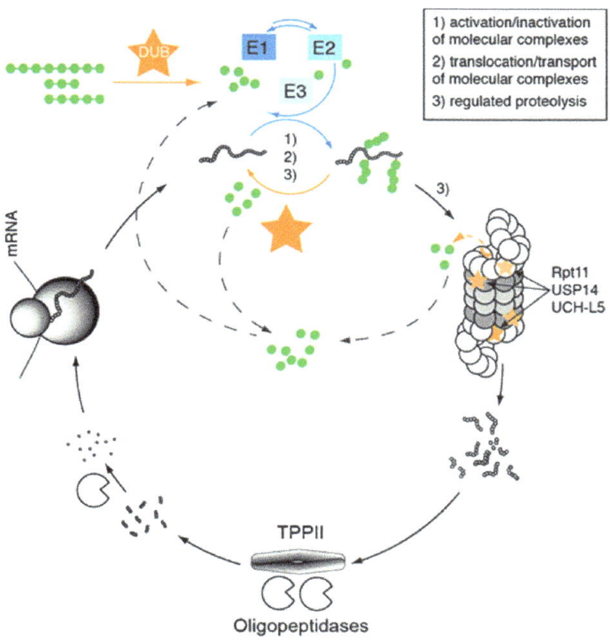

Fig. 3. **The ubiquitin proteasomal system.** Cartoon of the different components of the ubiquitin-proteasomal system described in the text.

However, ubiquitination does not only associate with proteasomal degradation, other forms of ubiquitination play roles in such diverse processes as transcriptional regulation, endocytosis, DNA modifications and stabilization of protein complexes (Harper and Schulman, 2006). Ubiquitin (Ub) is transcribed as precursor proteins from 4 genes and after cleavage to monomeric Ub, Ub-conjugation occurs through an enzymatic cascade where the E3-ligases (>600 E3 ligases have been identified so far) provide for target specificity. Prior to cleavage in the proteasome of a lys^{48} poly-Ub conjugated protein, the Ub-recognition signal is removed (Fig. 3). This serves to enhance passage of the targeted protein into the catalytic chamber and will also free ubiquitin for reuse and hinder the proteasomes to be preoccupied with destroying Ub [itself] (Hanna et al., 2007; Koulich et al., 2008). Hydrolysis of bound Ub and processing of Ub precursor proteins is accomplished by members of the large group of deubiquitinating enzymes (DUBs) (>90 members identified so far); DUBs are cysteine proteases that can be divided into distinct subfamilies based on sequence similarities and likely mechanism of action (reviewed in Nijman et al., 2005). Since DUBs can reshuffle ubiquitin from poly- to mono- or multi-ubiquitin chains, they are not only important for protein half-life but probably all cellular processes that involve ubiquitination of proteins (Clague and Urbe, 2006).

2.2 Activation of the UPS in sarcopenia

Because available data on the integrity of the UPS in the elderly was controversial we examined in 30-month old Sprague-Dawley rats the effects of aging on the content and activity of 26S proteasomes, proteasome-associated regulatory proteins and various other components of the UPS, including multiple DUBs (Altun et al., 2010). Muscles of these aged animals undergo marked atrophy compared to muscles of young adult animals and this age-related atrophy could be impeded if the animals were maintained on a restricted diet (Altun et al., 2007a). Analysis of proteasome protein content and proteasome degradation capacity in these animals revealed a 2-3 fold increase in proteasomes in the aged muscle. The increase in 26S proteasomes was suppressed completely in aged animals maintained on a restricted diet (30% of the consumption recorded for rats having free access to food). Since muscle wasting was reduced in these animals, these findings suggest that the buildup of proteasomes contributes to the loss of muscle mass in aged animals. In support of this notion we found that the levels of proteasome subunits in aged skeletal muscle were inversely correlated with muscle weight (subunits β1: $r=-0.71$, $p<0.05$; β5: $r=-0.69$; $p<0.05$), and no such inverse relationship was found in muscles of adult rats or aged animals maintained on dietary restriction (Altun et al., 2010). However, the underlying mechanism of this accumulation remains unclear. In contrast to muscles atrophying due to disuse, fasting or various systemic diseases, the age-related accumulation of proteasomes occurred without any increase in corresponding mRNAs. Thus, the accumulation of proteasomes in aged muscle must be due to enhanced subunit translation, more efficient assembly of the 26S particles, or slower degradation of the 26S particles (see below under 2.4).

Previous reports suggested that age-related decrease in the proteasome's peptidase activities were due to oxidative modifications (Bulteau et al., 2000; Conconi et al., 1996; Ferrington et al., 2005; Grune et al., 2001; Hayashi and Goto, 1998; Keller et al., 2000). However, we could not observe such defect in degrading capacity towards a range of substrates including ubiquitinated native proteins (Fig. 4). Instead, our data indicates that proteasome content increases during sarcopenia, and that these particles retain their full ability to function in protein breakdown. Together these findings suggest that the enhanced capacity of the UPS may be a response to the increased generation of damaged polypeptides. Strong support for this conclusion is our finding of elevated levels of the ubiquitin ligases, CHIP and E6AP, in muscle of aged animals. CHIP ubiquitinates misfolded or mutated proteins bound to Hsp70 or Hsp90 (Connell et al., 2001), and recently a similar function in cellular quality control has been reported for E6AP (Mishra et al., 2009). Moreover, most of these conclusions about a negative impact of aging on proteasomal degradation in muscle were based exclusively on measurements of the 20S core particle. Although some degradation of unfolded or denatured proteins may occur by the free 20S (Jariel-Encontre et al., 2008), the bulk of proteasome-mediated degradation, even of oxidatively damaged proteins (Medicherla and Goldberg, 2008), seems to require the 26S proteasome and ubiquitylation of the substrate.

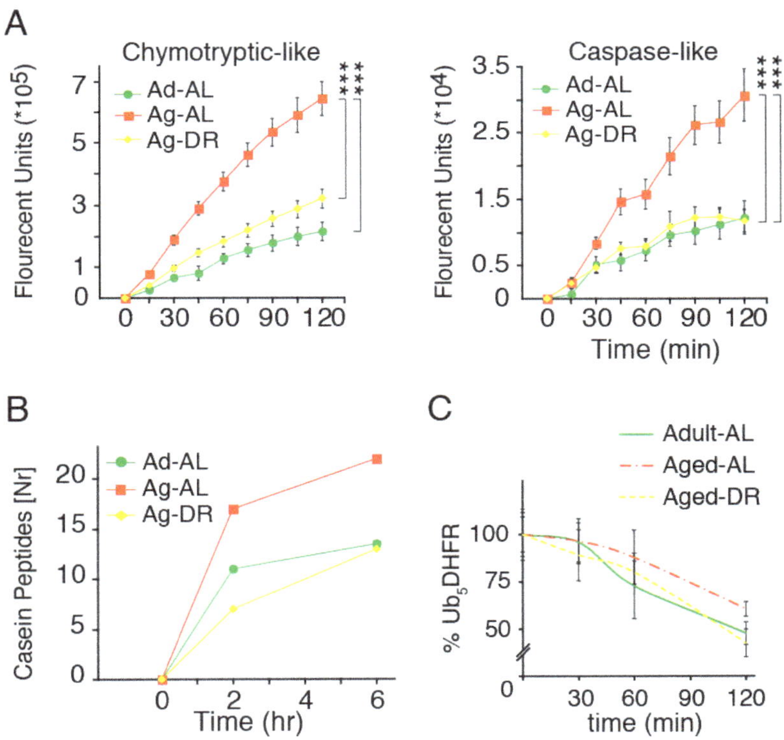

Fig. 4. **Degradation capacity towards different substrates of proteasomes isolated from triceps surae muscle of adult rats (Ad-AL), aged (Ag-AL) and aged rats maintained on dietary restriction (Ag-DR)** (original data reproduced from Altun et al., 2010). The higher content per unit muscle mass of proteasomes in aged (Ag-AL) muscles is paralleled by a corresponding increase in degradation of synthetic peptides designed for the chymotryptic and caspase sites of the proteasome (A) as well as casein (B). In (C) degradation of a native ubiquitinylated protein is shown illustrating that we could not detect any impairment of the degradation capacity of 26S proteasomes isolated from aged rats muscle. Further details on experimental design in Altun at al., 2010.

A decline in the cell's pool of free ubiquitin can limit the rate of proteolysis by the proteasome (Hanna et al., 2003; Kimura et al., 2009) and free ubiquitin is in turn released during degradation of ubiquitylated proteins by the DUBs associated with the 26S proteasome. Thus, in order to achieve high degradation rates, the capacity for deubiquitylation probably needs to be increased too. Using mechanism-based probes to assess the state of DUB activity and expression (Altun et al., 2010), eleven DUB-enzymes were found to be strongly up-regulated in the muscles of aged rats; including USP14 and Uch37 which are known to be associated with the 26S proteasome. USP14 and UCH37 trim off ubiquitin from the polyubiquitin chain and release ubiquitin monomers for re-use. In addition, USP14 controls gate opening into the 20S proteasome and facilitates substrate degradation (Peth et al., 2009). On the other hand, the yeast homolog of USP14, Ubp6, regulates the overall rate of proteolysis and appears critical in replenishing the free ubiquitin pool in yeast (Hanna et al., 2003; Hanna et al., 2007). Similarly, USP5 hydrolyses anchorless ubiquitin chains (Reyes-Turcu et al., 2006; Reyes-Turcu et al., 2008) and appears to work downstream of Rpn11, an intrinsic DUB on the proteasome, to prevent the binding of free ubiquitin chains to the 19S, which would inhibit proteolysis. Together these data illustrate important roles for deubiquitylation in the aged muscle, probably ensuring a supply of ubiquitin for enhanced proteolysis but perhaps also serving additional regulatory functions.

Loss of muscle mass can occur through increased protein degradation but also decreased protein synthesis or through some combination of these responses. With aging, sarcopenia develops over months to years depending on the species, unlike the rapid loss of muscle weight induced by fasting, disuse, and in various catabolic diseases, where marked atrophy (20-50% loss) can occur in rodents in several days. In these latter types of rapid atrophy, as described above, there is a common program of changes in the transcription of a set of atrophy-related genes (Lecker et al., 2006; Lecker et al., 2004; Sacheck et al., 2007). Several of the biochemical changes observed in aged atrophic muscle clearly distinguish them from those undergoing rapid atrophy in adult animals. Upon denervation or fasting, the atrophy-specific ubiquitin ligases, Atrogin-1/MAFbx and MuRF1, are induced by members of the FOXO family of transcription factors, and this induction is essential for the rapid weight loss (Bodine et al., 2001; Lecker et al., 2006; Sandri et al., 2004). Inhibition of FOXOs prevents their induction and the loss of muscle mass upon denervation, fasting, or glucocorticoid treatment (Sandri et al., 2004). In contrast, in the aged muscles, mRNAs for Atrogin-1/MAFbx, MuRF1 and the Ub-conjugating enzyme, E2-14K, were unchanged or lower than adult levels (Edstrom et al., 2006). Also, treatment with the glucocorticoid dexamethasone failed to induce Atrogin-1/MAFbx or MuRF1 or to cause muscle wasting, as it does in adult animals (Bodine et al., 2001; Gomes et al., 2001). However, MuRF1 protein increased in aged muscle, while Atrogin-1/MAFbx protein decreased (in accord with the mRNA data). Another distinction between sarcopenia and rapid atrophy is that the ubiquitin ligases, CHIP and E6AP, increased markedly in the aged muscle, though they do not rise in rapidly atrophying muscles. Their induction may reflect adaptations in the aged muscle to eliminate more efficiently misfolded proteins.

In summary, the finding of increased content of proteasomes and other UPS components (e.g. MuRF1) argues that proteolysis also increases in these muscles and may contribute to muscle wasting in the aged rats. Dietary restriction decreased levels of proteasomes and several other UPS components toward the levels in adult animal and partially inhibited the development of sarcopenia.

2.3 Accumulation of ubiquitinated proteins in sarcopenic muscles

During muscle atrophy in adult animals induced by fasting, disuse or disease, there is a marked increase in protein ubiquitination and proteolysis, especially of myofibrillar components (Lecker et al., 1999; Mitch and Goldberg, 1996). If the rate of ubiquitination exceeds the rate of Ub-dependent proteolysis, Ub-protein will accumulate; this will also occur if proteins are damaged or rearranged (multimeric aggregates) in a way that hinders normal proteasomal degradation. Immunoblotting of muscle extracts with an anti-ubiquitin antibody reveal that the muscles from aged rats contained greater amounts of ubiquitylated proteins than those from adult rats (Fig.5B; see also (Altun et al., 2010)). An increase in free ubiquitin and ubiquitin conjugates with aging was also reported by others in rat muscles (Cai et al., 2004; Clavel et al., 2006). Since the 26S proteasomes isolated from aged muscle are fully active in degrading ubiquitylated proteins, the elevated levels of ubiquitylated proteins strongly suggest higher overall rates of protein ubiquitylation that exceed the rates of conjugate degradation or the formation of nondegradable Ub-protein aggregates. Immunohistochemical analysis (using the same antibody towards ubiquitin) show that some of the fibers in aged skeletal muscles contain Ub+ deposits (Fig. 5A).

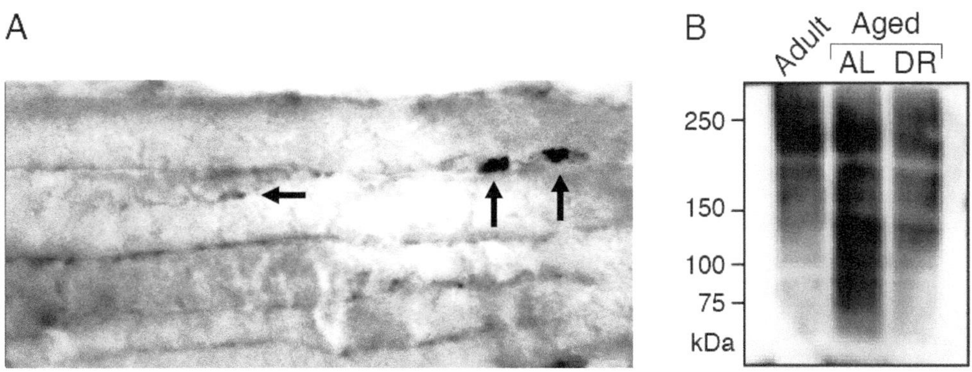

Fig. 5. **Amount of ubiquitylated protein is increased in aged skeletal muscle.** (A) shows micrograph of an aged soleus muscle cut along fiber length and processed with an antibody that recognizes ubiquitylated proteins using the ABC peroxidase technique. Ub+ protein deposits appear in black (arrows) and vary in size but are usually localized to the peripheral domain of the fiber (original micrograph was taken with an x20 dry objective). (B) shows a immunoblot of muscle extract from adult, aged and aged rats maintained on dietary restriction (DR) using the same antibody as in (A). The amount of Ub-proteins is clearly higher in aged fed ad libitum, while aged rats on DR are more similar to adults.

2.4 The autophagy lysosomal system (ALS)

Ubiquitination is also involved in processes leading to abrogation of signal transduction and targeting of the receptors for destruction by the autophagosome lysosomal system (ALS; Fig. 6). One example is the monoubiquitin-dependent endocytosis and degradation of epidermal growth factor receptor (Haglund et al., 2003; Mizuno et al., 2005). Degradation through the LS involves not only proteins but also damaged cell organelles and large complexes (Cuervo et al., 2005; Terman and Brunk, 2006; Terman et al., 2006). Multiple vesicles that constantly fuse and fission constitute the lysosomal system. Lysosomal degradation is initiated through several pathways based on the substrate delivering mechanism (Fig. 6): Endocytosis (clathrin mediated pinching-off of membrane patches with/without inclusion of extracellular material) occurs and the resulting particle is referred to as an endosomes (carrying rab5). Endosomes can then develop into late endosomes, which are equipped with enzymes delivered by secretory vesicles from the trans-Golgi-network and sorted by mannose-6-phosphate receptors, which are markers for late endosomes. Late endosomes may also fuse with lysosomes (then they lose the mannose-6-phosphate receptors).

A second route is chaperone mediated autophagy (CMA), which targets proteins carrying a pentapeptide (KFERQ) motif that is recognized by the chaperon Hsc73 and through binding to the LAMP-2A receptor (lysosome associated membrane protein 2A; a marker of the lysosome), the targeted protein is taken up and degraded in the lysosomal lumen (Fig. 6). The major route for organelles and proteins targeted for lysosomal degradation is via autophagy, a process controlled by the autophagy-conjugase complexes Atg12-Atg5 and Atg8-Atg3 (Atg, autophagy related genes) whereby in-bulk cytoplasm becomes entrapped in a double membrane forming an autophagosome (autophagic vacuole, AV) (Terman and Brunk, 2004).

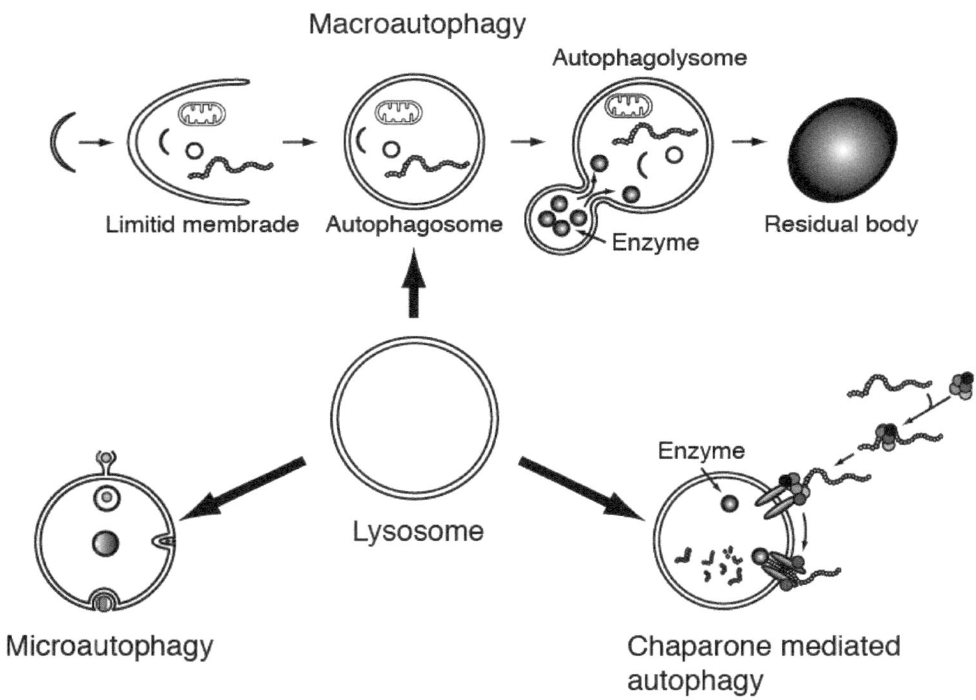

Fig. 6. **Schematic drawing of the autophagy-lysosomal system.** The different cargo routes to lysosomal degradation are illustrated and described in the text.

Subsequently, the autophagosome fuses with a lysosome for degradation under the formation of an autophagolysosome (Fig. 6). With the formation of the AV, the Atg conjugase complexes disassociate, but the Atg 8 (the mammalian homologue is MAP-LC3) remains in the ALS until degradation. While, Atg8 may be a marker for AVs and [by its elimination also] the completion of this degradation route, the autophagy conjugase complexes indicate ongoing ALS formation/autophagy. The role of the autophagy-lysosomal system was not widely recognized until it was demonstrated that cathepsins (D and B+L, respectively; (Felbor et al., 2002)) and autophagy (Hara et al., 2006) were non-redundant for a normal development and that cells rely on a basal level of autophagy to keep them free of worn-out organelles and aberrant proteins.

2.5 Signs of distress of the ALS during aging

There is solid evidence for the engagement of autophagy in major human degenerative diseases and in normal aging. A decreased capacity of the lysosomes to degrade waste will cause a build-up of damaged organelles and proteins disturbing cell homeostasis; and e.g. accumulation of dysfunctional mitochondria in aging cardiomyocytes, myofibers and neurons has been reported. Recently, several studies have shown that the induction of autophagy, by intervention of the above discussed signaling pathways, increase cellular clearance of aberrant proteins also when the normal degradation route operates via the proteasome or through CMA. The load on the ALS in normal aging is evident from the dramatic accumulation of lipofuscin by time (Fig. 7 G-H) and the concomitant upregulation of markers of lysosomal proteolysis. Lipofuscin forms due to iron-catalyzed intralysosomal peroxidation and its accumulation in the lysosomal compartment seems to set down the capacity of autophagy (Brunk and Terman, 2002). Oxidative stress, even mild stress, accelerates the formation/build-up of lipofuscin (Kurz et al., 2008) and stress can cause lysosomes to become leaky due to intralysosomal reactive iron mediated Fenton- reactions with production of hydroxyl radicals and resulting labilization of the lysosomal membrane. Even though the pH is suboptimal, several of the lysosomal enzymes are apparently proteolytically active if released to the cytosol and may then trigger cell degeneration/death (Chu, 2006).

Analysis of atrophic skeletal muscle in aged rats (Altun et al., 2007b) revealed increased levels of molecules involved in iron transport (transferrin), binding (ferritin), as well as iron response element binding protein-1 activity; combined these observations suggest increased load on the iron-handling machinery. Measurements of iron levels revealed a significant accumulation in the aged skeletal muscle, providing further support for iron loading in senescence. Iron-loading increases the risk for common hallmarks of aging such as DNA damage, protein oxidation and misfolding, and lipid peroxidation and may accelerate the accumulation of lipofuscin (see above). Iron-load is not exclusive to aging-related muscle wasting, but is also evident in experimental disuse atrophy (Kondo et al., 1992) and iron-restriction has been shown to be beneficiary in certain inherited myopathies (Bornman et al., 1998).

3. Cross talk between the UPS and the ALS

Until recently the UPS and the ALS were considered as independent pathways for protein degradation. The UPS offers a fast and highly specific mechanism to remove selected proteins. However, the targeted protein must be a monomer and unfolded to be able to enter the proteolytic lumen of the proteasome. The ALS represents a partly overlapping, partly complementary degradation route, taking care of folded and aggregated proteins as well as large complexes and organelles with heterogenic building blocks (carbohydrates, lipids and proteins). However, several lines of evidence indicate intersections between these two pathways. In neurons treated with proteasome inhibitors, aggregate-prone proteins normally degraded by the UPS are degraded by autophagy (Cuervo et al., 2004). Two recent papers also provide evidence that induction of the UPS and LS may occur via a common pathway in skeletal muscle (Mammucari et al., 2007; Zhou et al., 2010).

Several studies report that with increasing age the capacity for lysosomal proteolysis is reduced in postmitotic cells including myofibers, while the capacity to degrade proteins through the UPS is enhanced (see above). Given that transcription of proteasome subunits and the expression of proteins involved in the assembly of the 20S were unaltered in in aged rodents, we hypothesized that proteasome particles accumulate in muscles during aging because they are not degraded at the same rate as in young adults. It is currently unknown which mechanism is responsible for the degradation of proteasomes, although the lysosomal pathway was suggested to accomplish this task in hepatocytes (Cuervo et al., 1995). To test whether this mechanism may be present in skeletal muscle cells, we treated adult rats (Fig.7A-C) and rat muscle derived cell line L6, respectively, with the lysosomal inhibitor chloroquine followed by quantitation of proteasome levels and activity (Fig.7A-F). In these experiments, we observed an accumulation of proteasomes accompanied by increased active site labeling in the muscle extracts from both chloroquine treated animals and L6 cell extracts.

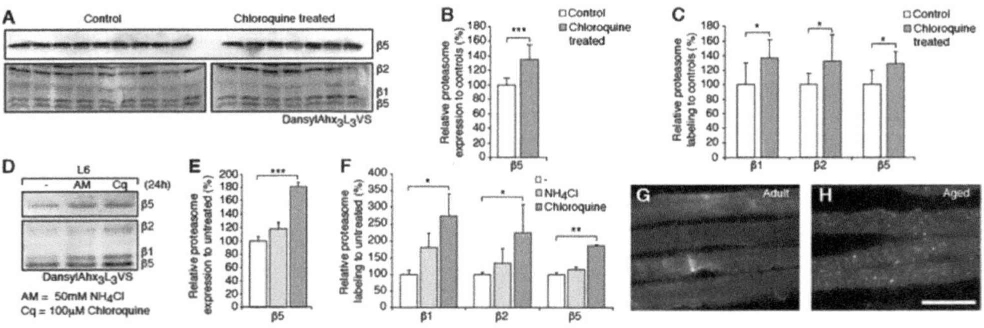

Fig. 7. **Compromised lysosomal proteolysis increases proteasome content and activity.** (**A-C**) Adult animals treated with chloroquine (50mg/kg) for 16 days (1 injection/day). (**A**) Immunoblot against proteasome subunit b5 to assay proteasome content (top) and proteasome labeling (bottom) using the active site directed probe DansylAhx$_3$L$_3$VS. (**B**) Individual bands were quantified for b5 content and (**C**) active site directed probe DansylAhx$_3$L$_3$VS labeling of the b1, b2 and b5 catalytic subunits. (**D-F**) Rat muscle derived cell line L6 was treated when confluence reached about 70% with 50mM NH_4Cl or 100mM chloroquine for 24 hour before lysis. Immunoblot against proteasome subunit b5 to assay proteasome content (**D** top panel, **E**) and active site labeling of the b1, b2 and b5 catalytic subunits (**D** lower panel, **F**). b-actin was used as loading control for immunoblotting (data not shown). All error bars are standard deviations. Statistical significance: * $p<0.05$, ** $p<0.01$ and *** $p<0.001$. (**G, H**) Lipofuscin accumulation in aged skeletal muscle. Unstained longitudinal sections of soleus muscle form adult (A) and aged (B) rats show increased content of lipofuscin autofluorescence in aged sarcopenic muscle. The bar represents 100μm

As in the aged rat muscles, the accumulation of proteasomes following lysosomal inhibition was not induced by a transcriptional up-regulation (data not shown) and therefore results likely from a reduced clearance of proteasomes due to an impairment of lysosomal function. Consistent with this, we observed lipofuscin accumulation in the aged rat muscle tissue (Fig.7 G and H). In addition, Cuervo and colleagues reported that autophagy declines in liver with aging but this effect is reduced by dietary restriction (Hanna et al., 2003; Watts et al., 2004). Activation of autophagy in muscle by dietary restriction could explain our findings that this regime prevented the increase in proteasomes with age without any change in proteasome mRNA levels. Combined these observations suggest the ALS as a candidate pathway for proteasomal degradation awaiting more definitive evidence.

4. Concluding remarks

Human cross-sectional and longitudinal studies have consistently demonstrated a gain in fat mass and decline in lean mass during aging. Although aging-related muscle wasting, or sarcopenia, is widely recognized it still remains poorly understood. The reduced functional muscle mass is associated with increased morbidity and reduced quality of life. To maintain integrity muscle myofibers rely on degradation pathways to keep clean from worn-out organelles and damaged proteins. Our current understanding is that the autophagy-lysosomal system is in distress possibly driven by an age-dependent accumulation of reactive iron, while contrary to the widespread view, the complementary pathway for degradation of proteins, the UPS, is enhanced. The age-dependent increase in the muscle specific ubiquitin-ligase MuRF and ubiquitin-dependent proteasomal proteolysis is expected to occur in the progression of sarcopenia since this route is nonredundant for degradation of myofibrillar proteins. In addition, the UPS in aged skeletal muscle shows adaptations to an increased demand on degradation of aberrant proteins and an accumulation of ubiquitinated proteins. Combined these stigmata suggest that aged myofibers may be at risk to enter a state of proteotoxicity.

Normal skeletal muscles have a good capacity to regenerate following wasting conditions such as disuse. The muscle regenerative response relies on signaling that evokes satellite cell replication and asymmetric division generating offsprings that will differentiate to myocytes via the myoblast stage. Such cells are then incorporated into the myofiber allowing it to grow (for example in response to an exercise stimulus). Poor capacity to regenerate muscle tissue at advanced age may depend on impaired signaling, exhaustion of the SC pool or changes in the extracellular matrix or stem-cell niche impeding the regeneration and incorporation of myoblasts into existing/regenerating myofibers (for references see Introduction). Assessments of the regenerative drive in aged sarcopenic muscle have, however, shown that myogenic differentiation factors are upregulated and that there are overt signs of incorporation of new nuclei into existing fibers (Fig. 1B,C; (Edstrom and Ulfhake, 2005)). Still, regeneration fails and tissue atrophy progresses.

The triggering mechanism for the age-dependent fiber atrophy and fiber loss remains enigmatic. However, several lines of evidence converge towards support of the "neurogenic" theory (Gutman and Hanzlikova, 1972), which stipulates that sarcopenia is driven by a successive drop-out of motoneurons. Early evidence in favor of this theory was the observation of fiber-type grouping and also histological examination revealing regressive changes at the neuro-muscular junctions. The strongest argument against this theory is the absence of unbiased evidence of a significant age-dependent loss of motoneurons. However, as discussed at length elsewhere (Johnson et al., 1995) the denervation may be a peripheral process primarily involving subdomains of the motor axon's terminal arborization within a motor unit (see also (Valdez et al., 2010b)). Failure to maintain the distant axon arbor could cause branches and terminals to degenerate leaving myofibers vacated from innervation. As this process progresses more and more fibers will be denervated and read-outs validating this process are the increase in expression of the nAChR-γ subunit (Gronholdt-Klein et al., in preparation) and embryonic myosin (Edstrom and Ulfhake, 2005); and it should be noted that these proteins are re-expressed by young adult myofibers upon denervation. As denervation becomes significant, it will trigger the atrophy program by which MuRF1 and other enzymes will accelerate proteasomal myofibrillar protein degradation (see above). Recently, further direct histological evidence for this process was obtained by Valdez and coworkers (Valdez et al., 2010b). These authors also showed that exercise and dietary restriction slow-down the progression of age-dependent denervation, at least the latter observation is consistent with our biochemical data on the UPS (see above). It will be important to seek proof of principle for the neurogenic theory in experimental animal models followed by validation in humans since it will impact the development of strategies to impede sarcopenia in humans.

5. Acknowledgements

This work was supported by a grant from the Swedish research council (VR 10820, to B. Ulfhake), Swedish research council post doc grant to M. Altun and a guest researcher stipendship from the PR China government to L. Wang.

6. References

Adamo, M.L., and Farrar, R.P. (2006). Resistance training, and IGF involvement in the maintenance of muscle mass during the aging process. Ageing Res Rev *5*, 310-331.

Altun, M., Bergman, E., Edstrom, E., Johnson, H., and Ulfhake, B. (2007a). Behavioral impairments of the aging rat. Physiol Behav *92*, 911-923.

Altun, M., Besche, H.C., Overkleeft, H.S., Piccirillo, R., Edelmann, M.J., Kessler, B.M., Goldberg, A.L., and Ulfhake, B. (2010). Muscle wasting in aged, sarcopenic rats is associated with enhanced activity of the ubiquitin proteasome pathway. J Biol Chem *285*, 39597-39608.

Altun, M., Edstrom, E., Spooner, E., Flores-Moralez, A., Bergman, E., Tollet-Egnell, P., Norstedt, G., Kessler, B.M., and Ulfhake, B. (2007b). Iron load and redox stress in skeletal muscle of aged rats. Muscle Nerve *36*, 223-233.

Attaix, D., Mosoni, L., Dardevet, D., Combaret, L., Mirand, P.P., and Grizard, J. (2005). Altered responses in skeletal muscle protein turnover during aging in anabolic and catabolic periods. Int J Biochem Cell Biol *37*, 1962-1973.

Baumgartner, R.N., Koehler, K.M., Gallagher, D., Romero, L., Heymsfield, S.B., Ross, R.R., Garry, P.J., and Lindeman, R.D. (1998a). Epidemiology of sarcopenia among the elderly in New Mexico. American Journal of Epidemiology *147*, 755-763.

Baumgartner, R.N., Koehler, K.M., Gallagher, D., Romero, L., Heymsfield, S.B., Ross, R.R., Garry, P.J., and Lindeman, R.D. (1998b). Epidemiology of sarcopenia among the elderly in New Mexico. Am J Epidemiol *147*, 755-763.

Bodine, S.C., Latres, E., Baumhueter, S., Lai, V.K., Nunez, L., Clarke, B.A., Poueymirou, W.T., Panaro, F.J., Na, E., Dharmarajan, K., *et al.* (2001). Identification of ubiquitin ligases required for skeletal muscle atrophy. Science *294*, 1704-1708.

Bornman, L., Rossouw, H., Gericke, G.S., and Polla, B.S. (1998). Effects of iron deprivation on the pathology and stress protein expression in murine X-linked muscular dystrophy. Biochem Pharmacol *56*, 751-757.

Brunk, U.T., and Terman, A. (2002). Lipofuscin: mechanisms of age-related accumulation and influence on cell function. Free Radic Biol Med *33*, 611-619.

Bulteau, A.L., Petropoulos, I., and Friguet, B. (2000). Age-related alterations of proteasome structure and function in aging epidermis. Exp Gerontol *35*, 767-777.

Cai, D., Lee, K.K., Li, M., Tang, M.K., and Chan, K.M. (2004). Ubiquitin expression is up-regulated in human and rat skeletal muscles during aging. Arch Biochem Biophys *425*, 42-50.

Carlson, M.E., Silva, H.S., and Conboy, I.M. (2008). Aging of signal transduction pathways, and pathology. Exp Cell Res *314*, 1951-1961.

Carlson, M.E., Suetta, C., Conboy, M.J., Aagaard, P., Mackey, A., Kjaer, M., and Conboy, I. (2009). Molecular aging and rejuvenation of human muscle stem cells. EMBO Mol Med *1*, 381-391.

Christov, C., Chretien, F., Abou-Khalil, R., Bassez, G., Vallet, G., Authier, F.J., Bassaglia, Y., Shinin, V., Tajbakhsh, S., Chazaud, B., *et al.* (2007). Muscle satellite cells and endothelial cells: close neighbors and privileged partners. Mol Biol Cell *18*, 1397-1409.

Chu, C.T. (2006). Autophagic stress in neuronal injury and disease. J Neuropathol Exp Neurol *65*, 423-432.

Clague, M.J., and Urbe, S. (2006). Endocytosis: the DUB version. Trends Cell Biol *16*, 551-559.

Clavel, S., Coldefy, A.S., Kurkdjian, E., Salles, J., Margaritis, I., and Derijard, B. (2006). Atrophy-related ubiquitin ligases, atrogin-1 and MuRF1 are up-regulated in aged rat Tibialis Anterior muscle. Mech Ageing Dev *127*, 794-801.

Conconi, M., Szweda, L.I., Levine, R.L., Stadtman, E.R., and Friguet, B. (1996). Age-related decline of rat liver multicatalytic proteinase activity and protection from oxidative inactivation by heat-shock protein 90. Arch Biochem Biophys *331*, 232-240.

Connell, P., Ballinger, C.A., Jiang, J., Wu, Y., Thompson, L.J., Hohfeld, J., and Patterson, C. (2001). The co-chaperone CHIP regulates protein triage decisions mediated by heat-shock proteins. Nat Cell Biol *3*, 93-96.

Cuervo, A.M., Bergamini, E., Brunk, U.T., Droge, W., Ffrench, M., and Terman, A. (2005). Autophagy and aging: the importance of maintaining "clean" cells. Autophagy *1*, 131-140.

Cuervo, A.M., Palmer, A., Rivett, A.J., and Knecht, E. (1995). Degradation of proteasomes by lysosomes in rat liver. Eur J Biochem *227*, 792-800.

Cuervo, A.M., Stefanis, L., Fredenburg, R., Lansbury, P.T., and Sulzer, D. (2004). Impaired degradation of mutant alpha-synuclein by chaperone-mediated autophagy. Science *305*, 1292-1295.

Edstrom, E., Altun, M., Bergman, E., Johnson, H., Kullberg, S., Ramirez-Leon, V., and Ulfhake, B. (2007). Factors contributing to neuromuscular impairment and sarcopenia during aging. Physiol Behav *92*, 129-135.

Edstrom, E., Altun, M., Hagglund, M., and Ulfhake, B. (2006). Atrogin-1/MAFbx and MuRF1 are downregulated in aging-related loss of skeletal muscle. J Gerontol A Biol Sci Med Sci *61*, 663-674.

Edstrom, E., and Ulfhake, B. (2005). Sarcopenia is not due to lack of regenerative drive in senescent skeletal muscle. Aging Cell *4*, 65-77.

Felbor, U., Kessler, B., Mothes, W., Goebel, H.H., Ploegh, H.L., Bronson, R.T., and Olsen, B.R. (2002). Neuronal loss and brain atrophy in mice lacking cathepsins B and L. Proc Natl Acad Sci U S A *99*, 7883-7888.

Ferrington, D.A., Husom, A.D., and Thompson, L.V. (2005). Altered proteasome structure, function, and oxidation in aged muscle. Faseb J *19*, 644-646.

Fuso, A., Ferraguti, G., Grandoni, F., Ruggeri, R., Scarpa, S., Strom, R., and Lucarelli, M. (2010). Early demethylation of non-CpG, CpC-rich, elements in the myogenin 5'-flanking region: a priming effect on the spreading of active demethylation. Cell Cycle *9*, 3965-3976.

Gomes, M.D., Lecker, S.H., Jagoe, R.T., Navon, A., and Goldberg, A.L. (2001). Atrogin-1, a muscle-specific F-box protein highly expressed during muscle atrophy. Proc Natl Acad Sci U S A *98*, 14440-14445.

Grune, T., Shringarpure, R., Sitte, N., and Davies, K. (2001). Age-related changes in protein oxidation and proteolysis in mammalian cells. J Gerontol A Biol Sci Med Sci *56*, B459-467.

Gutman, B., and Hanzlikova, V. (1972). Age changes in the neuromuscular system. Scientechnica Ltd, Bristol, 1-20.

Haglund, K., Sigismund, S., Polo, S., Szymkiewicz, I., Di Fiore, P.P., and Dikic, I. (2003). Multiple monoubiquitination of RTKs is sufficient for their endocytosis and degradation. Nat Cell Biol *5*, 461-466.

Hanna, J., Leggett, D.S., and Finley, D. (2003). Ubiquitin depletion as a key mediator of toxicity by translational inhibitors. Mol Cell Biol *23*, 9251-9261.

Hanna, J., Meides, A., Zhang, D.P., and Finley, D. (2007). A ubiquitin stress response induces altered proteasome composition. Cell *129*, 747-759.

Hara, T., Nakamura, K., Matsui, M., Yamamoto, A., Nakahara, Y., Suzuki-Migishima, R., Yokoyama, M., Mishima, K., Saito, I., Okano, H., *et al.* (2006). Suppression of basal autophagy in neural cells causes neurodegenerative disease in mice. Nature *441*, 885-889.

Harper, J.W., and Schulman, B.A. (2006). Structural complexity in ubiquitin recognition. Cell *124*, 1133-1136.

Harriman, D.G., Taverner, D., and Woolf, A.L. (1970). Ekbom´s syndrome and burning paraesthesiae. A biopsy study by vital staining and electron microscopy of the intramuscular innervation with a note on age changes in motor nerve endings in distal muscles. Brain *93*, 393-406.

Hayashi, T., and Goto, S. (1998). Age-related changes in the 20S and 26S proteasome activities in the liver of male F344 rats. Mech Ageing Dev *102*, 55-66.

Janssen, I., Heymsfield, S.B., and Ross, R. (2002). Low relative skeletal muscle mass (sarcopenia) in older persons is associated with functional impairment and physical disability. J Am Geriatr Soc *50*, 889-896.

Jariel-Encontre, I., Bossis, G., and Piechaczyk, M. (2008). Ubiquitin-independent degradation of proteins by the proteasome. Biochim Biophys Acta *1786*, 153-177.

Johnson, H., Mossberg, K., Arvidsson, U., Piehl, F., Hökfelt, T., and Ulfhake, B. (1995). Increase in alpha-CGRP and GAP-43 in aged motoneurons: A study of peptides, growth factors, and ChAT mRNA in the lumbar spinal cord of senescent rats with symptoms of hindlimb incapacities. J Comp Neurol *359*, 69-89.

Kadi, F., Charifi, N., Denis, C., Lexell, J., Andersen, J.L., Schjerling, P., Olsen, S., and Kjaer, M. (2005). The behaviour of satellite cells in response to exercise: what have we learned from human studies? Pflugers Arch *451*, 319-327.

Keller, J.N., Huang, F.F., and Markesbery, W.R. (2000). Decreased levels of proteasome activity and proteasome expression in aging spinal cord. Neuroscience *98*, 149-156.

Kimura, Y., Yashiroda, H., Kudo, T., Koitabashi, S., Murata, S., Kakizuka, A., and Tanaka, K. (2009). An inhibitor of a deubiquitinating enzyme regulates ubiquitin homeostasis. Cell *137*, 549-559.

Kondo, H., Miura, M., Kodama, J., Ahmed, S.M., and Itokawa, Y. (1992). Role of iron in oxidative stress in skeletal muscle atrophied by immobilization. Pflugers Arch *421*, 295-297.

Koulich, E., Li, X., and DeMartino, G.N. (2008). Relative structural and functional roles of multiple deubiquitylating proteins associated with mammalian 26S proteasome. Mol Biol Cell *19*, 1072-1082.

Kurz, T., Terman, A., Gustafsson, B., and Brunk, U.T. (2008). Lysosomes in iron metabolism, ageing and apoptosis. Histochem Cell Biol *129*, 389-406.

Larsson, L. (1995). Motor units: remodeling in aged animals. J Gerontol A Biol Sci Med Sci *50 Spec No*, 91-95.

Lecker, S.H., Goldberg, A.L., and Mitch, W.E. (2006). Protein degradation by the ubiquitin-proteasome pathway in normal and disease states. J Am Soc Nephrol *17*, 1807-1819.

Lecker, S.H., Jagoe, R.T., Gilbert, A., Gomes, M., Baracos, V., Bailey, J., Price, S.R., Mitch, W.E., and Goldberg, A.L. (2004). Multiple types of skeletal muscle atrophy involve a common program of changes in gene expression. Faseb J *18*, 39-51.

Lecker, S.H., Solomon, V., Mitch, W.E., and Goldberg, A.L. (1999). Muscle protein breakdown and the critical role of the ubiquitin-proteasome pathway in normal and disease states. J Nutr *129*, 227S-237S.

Mammucari, C., Milan, G., Romanello, V., Masiero, E., Rudolf, R., Del Piccolo, P., Burden, S.J., Di Lisi, R., Sandri, C., Zhao, J., *et al.* (2007). FoxO3 controls autophagy in skeletal muscle in vivo. Cell Metab *6*, 458-471.

Medicherla, B., and Goldberg, A.L. (2008). Heat shock and oxygen radicals stimulate ubiquitin-dependent degradation mainly of newly synthesized proteins. J Cell Biol *182*, 663-673.

Mishra, A., Godavarthi, S.K., Maheshwari, M., Goswami, A., and Jana, N.R. (2009). The ubiquitin ligase E6-AP is induced and recruited to aggresomes in response to proteasome inhibition and may be involved in the ubiquitination of Hsp70-bound misfolded proteins. J Biol Chem *284*, 10537-10545.

Mitch, W.E., and Goldberg, A.L. (1996). Mechanisms of muscle wasting. The role of the ubiquitin-proteasome pathway. N Engl J Med *335*, 1897-1905.

Mizuno, E., Iura, T., Mukai, A., Yoshimori, T., Kitamura, N., and Komada, M. (2005). Regulation of epidermal growth factor receptor down-regulation by UBPY-mediated deubiquitination at endosomes. Mol Biol Cell *16*, 5163-5174.

Nijman, S.M., Luna-Vargas, M.P., Velds, A., Brummelkamp, T.R., Dirac, A.M., Sixma, T.K., and Bernards, R. (2005). A genomic and functional inventory of deubiquitinating enzymes. Cell *123*, 773-786.

Paddon-Jones, D., and Rasmussen, B.B. (2009). Dietary protein recommendations and the prevention of sarcopenia. Curr Opin Clin Nutr Metab Care *12*, 86-90.

Peth, A., Besche, H.C., and Goldberg, A.L. (2009). Ubiquitinated proteins activate the proteasome by binding to Usp14/Ubp6, which causes 20S gate opening. Molecular cell *36*, 794-804.

Reyes-Turcu, F.E., Horton, J.R., Mullally, J.E., Heroux, A., Cheng, X., and Wilkinson, K.D. (2006). The ubiquitin binding domain ZnF UBP recognizes the C-terminal diglycine motif of unanchored ubiquitin. Cell *124*, 1197-1208.

Reyes-Turcu, F.E., Shanks, J.R., Komander, D., and Wilkinson, K.D. (2008). Recognition of polyubiquitin isoforms by the multiple ubiquitin binding modules of isopeptidase T. J Biol Chem *283*, 19581-19592.

Rodeheffer, M.S. (2010). Tipping the scale: muscle versus fat. Nat Cell Biol *12*, 102-104.

Roth, S.M., Metter, E.J., Ling, S., and Ferrucci, L. (2006). Inflammatory factors in age-related muscle wasting. Curr Opin Rheumatol *18*, 625-630.

Sacheck, J.M., Hyatt, J.P., Raffaello, A., Jagoe, R.T., Roy, R.R., Edgerton, V.R., Lecker, S.H., and Goldberg, A.L. (2007). Rapid disuse and denervation atrophy involve transcriptional changes similar to those of muscle wasting during systemic diseases. Faseb J *21*, 140-155.

Sacheck, J.M., Ohtsuka, A., McLary, S.C., and Goldberg, A.L. (2004). IGF-I stimulates muscle growth by suppressing protein breakdown and expression of atrophy-related ubiquitin ligases, atrogin-1 and MuRF1. Am J Physiol Endocrinol Metab *287*, E591-601.

Sandri, M., Lin, J., Handschin, C., Yang, W., Arany, Z.P., Lecker, S.H., Goldberg, A.L., and Spiegelman, B.M. (2006). PGC-1alpha protects skeletal muscle from atrophy by suppressing FoxO3 action and atrophy-specific gene transcription. Proc Natl Acad Sci U S A *103*, 16260-16265.

Sandri, M., Sandri, C., Gilbert, A., Skurk, C., Calabria, E., Picard, A., Walsh, K., Schiaffino, S., Lecker, S.H., and Goldberg, A.L. (2004). Foxo transcription factors induce the atrophy-related ubiquitin ligase atrogin-1 and cause skeletal muscle atrophy. Cell *117*, 399-412.

Solomon, A.M., and Bouloux, P.M. (2006). Modifying muscle mass - the endocrine perspective. J Endocrinol *191*, 349-360.

St-Onge, M.P. (2005). Relationship between body composition changes and changes in physical function and metabolic risk factors in aging. Curr Opin Clin Nutr Metab Care *8*, 523-528.

Stitt, T.N., Drujan, D., Clarke, B.A., Panaro, F., Timofeyva, Y., Kline, W.O., Gonzalez, M., Yancopoulos, G.D., and Glass, D.J. (2004). The IGF-1/PI3K/Akt pathway prevents expression of muscle atrophy-induced ubiquitin ligases by inhibiting FOXO transcription factors. Mol Cell *14*, 395-403.

Terman, A., and Brunk, U.T. (2004). Myocyte aging and mitochondrial turnover. Exp Gerontol *39*, 701-705.

Terman, A., and Brunk, U.T. (2006). Oxidative stress, accumulation of biological 'garbage', and aging. Antioxid Redox Signal *8*, 197-204.

Terman, A., Gustafsson, B., and Brunk, U.T. (2006). The lysosomal-mitochondrial axis theory of postmitotic aging and cell death. Chem Biol Interact *163*, 29-37.

Valdez, G., Tapia, J.C., Kang, H., Clemenson, G.D., Jr., Gage, F.H., Lichtman, J.W., and Sanes, J.R. (2010a). Attenuation of age-related changes in mouse neuromuscular synapses by caloric restriction and exercise. Proc Natl Acad Sci U S A *107*, 14863-14868.

Valdez, G., Tapia, J.C., Kang, H., Clemenson, G.D., Jr., Gage, F.H., Lichtman, J.W., and Sanes, J.R. (2010b). Attenuation of age-related changes in mouse neuromuscular synapses by caloric restriction and exercise. Proc Natl Acad Sci U S A *107*, 14863-14868.

Watts, G.D., Wymer, J., Kovach, M.J., Mehta, S.G., Mumm, S., Darvish, D., Pestronk, A., Whyte, M.P., and Kimonis, V.E. (2004). Inclusion body myopathy associated with Paget disease of bone and frontotemporal dementia is caused by mutant valosin-containing protein. Nat Genet *36*, 377-381.

Zammit, P.S. (2008). All muscle satellite cells are equal, but are some more equal than others? J Cell Sci *121*, 2975-2982.

Zhou, X., Wang, J.L., Lu, J., Song, Y., Kwak, K.S., Jiao, Q., Rosenfeld, R., Chen, Q., Boone, T., Simonet, W.S., *et al.* (2010). Reversal of cancer cachexia and muscle wasting by ActRIIB antagonism leads to prolonged survival. Cell *142*, 531-543.

4

The Nucleolus and Ribosomal Genes in Aging and Senescence

Nadine Hein[1,*], Elaine Sanij[1,2,*],
Jaclyn Quin[1,3], Katherine M. Hannan[1],
Austen Ganley[4,#] and Ross D. Hannan[1,3,5,6,#]
[1]*Division of Cancer Research, Peter MacCallum Cancer Centre, Melbourne,*
[2]*Department of Pathology, University of Melbourne, Melbourne,*
[3]*Department of Biochemistry and Molecular Biology, University of Melbourne, Melbourne,*
[4]*Institute of Natural Sciences, Massey University, Auckland,*
[5]*Department of Biochemistry and Molecular Biology, Monash University, Melbourne,*
[6]*School of Biomedical Sciences, The University of Queensland, Melbourne,*
[1,2,3,5,6]*Australia*
[4]*New Zealand*

1. Introduction

The nucleolus forms around the tandem repeats of the ribosomal RNA (rRNA) genes (rDNA) that are transcribed by RNA polymerase I (Pol I), giving rise to the production of rRNAs. These represent the nucleic acid backbone of the functional ribosomes in the cytoplasm, and as such rDNA transcription dictates the cells' protein translational capacity. More recently it has become apparent that the epigenetic status of these rDNA repeats and the integrity of the nucleolus can modulate cellular homeostasis beyond ribosome biogenesis. Such roles include mediating the titration of tumor suppressors and oncogenes, modulating the heterochromatic state of many RNA Polymerase II (Pol II) transcribed genes, and importantly, regulating the process of aging and senescence. This chapter will focus on the molecular and cellular evidence that the nucleolus and the rDNA repeats play critical roles in the control of aging and cellular senescence in yeast and mammals.

2. Introduction to rDNA transcription and the nucleolus

This section will provide a brief overview of the regulation of rDNA transcription, however, for more details refer to (Tschochner & Hurt, 2003; McStay & Grummt, 2008).

---

* These authors contributed equally to this work.
# Corresponding Authors

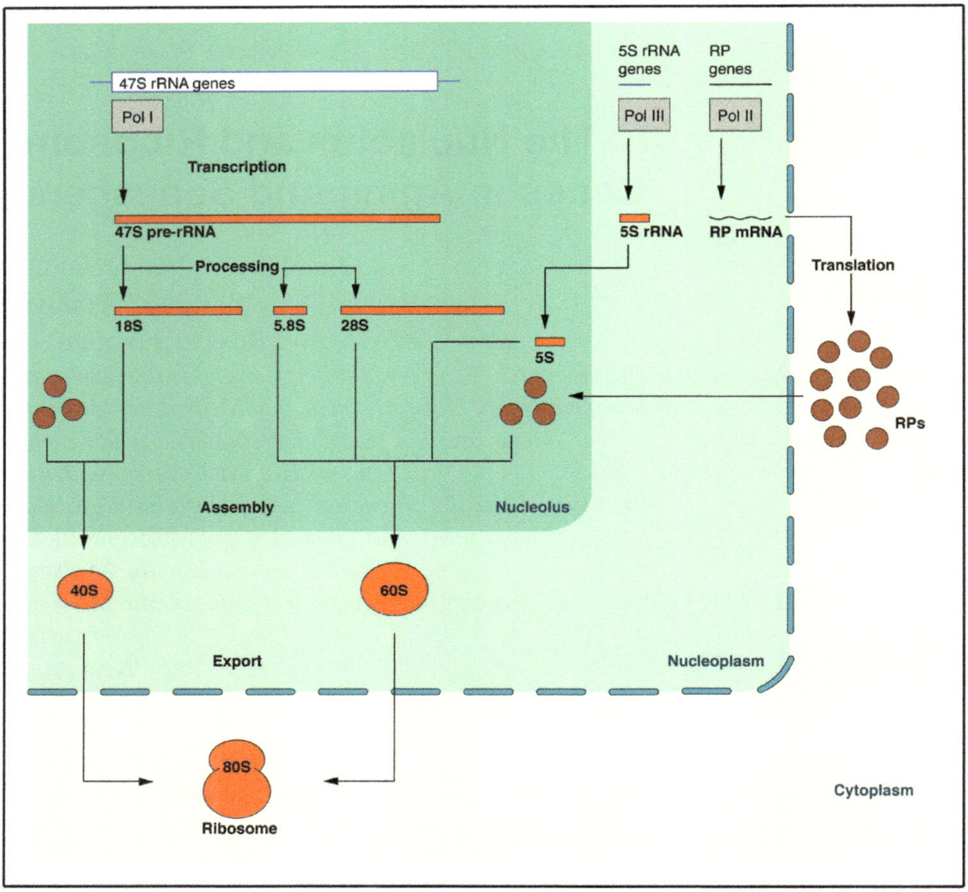

Fig. 1. Overview of ribosome biogenesis in mammalian cells.

2.1 Ribosome biogenesis

Ribosome biogenesis dictates the capacity of a cell to grow and proliferate, and is one of the most energy consuming processes in eukaryotic cells (Grummt & Pikaard, 2003). The synthesis of a ribosome is a highly complex, yet exquisitely coordinated process, which utilizes all three DNA-dependent RNA polymerases (Pol I, Pol II and Pol III) to produce approximately equimolar amounts of numerous ribosomal proteins (RP) and four rRNA (Fig. 1). Transcription of the Pol I-transcribed rRNA genes (the rDNA) has traditionally been considered the major rate-limiting step in ribosome biogenesis. Consistent with this dogma, any perturbations in the cellular environment, such as nutrient withdrawal, altered growth factor signaling, cell cycle cues and stress, are directly accompanied by modulation of the rate of rDNA transcription. However, in addition to the two key components of the ribosome (rRNAs and RPs) a multitude of non-ribosomal proteins and non-coding RNAs have been identified as essential for various steps in the generation of new ribosomes. These steps not only include Pol I transcription of the pre-RNA precursor but its subsequent processing and modifications, the nuclear import of RP's, and the final assembly of the large and small ribosomal subunits followed by their export to the cytoplasm (Grummt & Pikaard, 2003; Moss et al., 2007). The process of ribosome biogenesis is fundamental to cellular life and consequently is highly conserved, and this is illustrated by the numerous similarities between yeast and mammals.

In budding yeast (*Saccharomyces cerevisiae)* and mammals the rRNA gene is transcribed exclusively by Pol I in the nucleolus. While in yeast this generates a 35S rRNA precursor which is processed into the mature 18S, 5.8S and 26S rRNAs (Fig. 2a), in mammals a 47S rRNA precursor is generated and processed to give 18S, 5.8S and 28S rRNAs (Fig. 2b). There are also numerous similarities in the other component of the ribosome, the RPs, which in both cases are transcribed by Pol II in the nucleoplasm. In growing yeast it has been established that ~40 nascent ribosomes leave the nucleolus every second, 80% of the total RNA is rRNA, and ~50% of total protein consists of RPs (Tschochner & Hurt, 2003). Overall the process of ribosome biogenesis consumes between 60-80% of the cells total energy both in yeast and mammals (Moss & Stefanovsky, 2002). Thus for both systems, even minor perturbations to ribosome biogenesis are likely to have major repercussions for the cell.

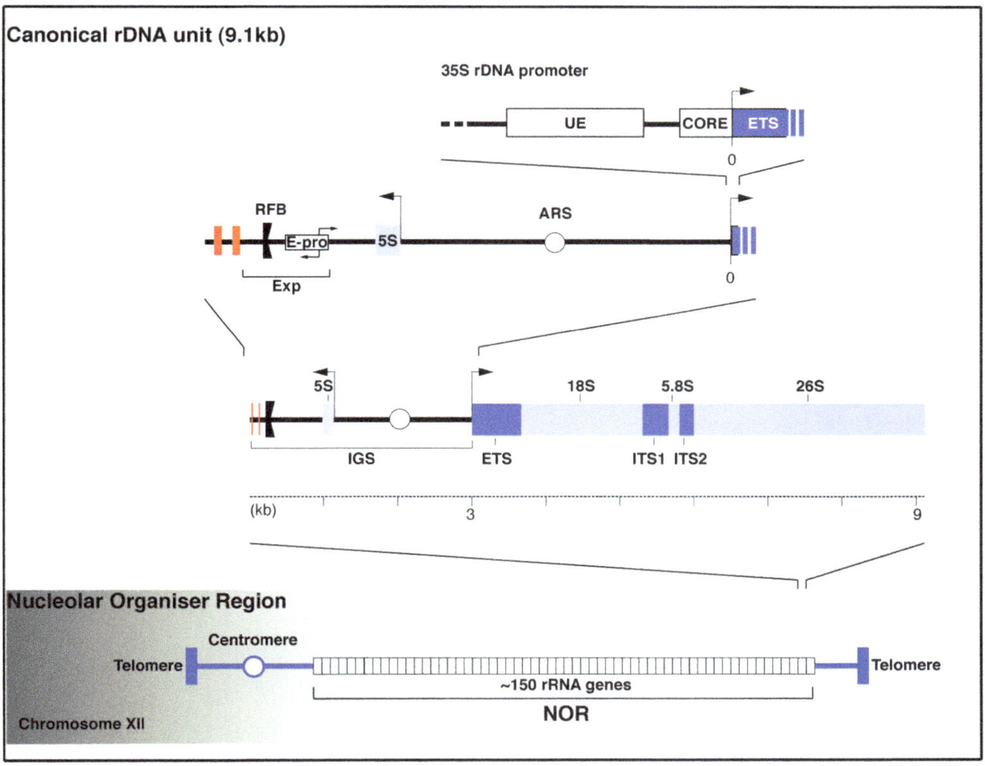

Fig. 2a. Organization of ribosomal RNA genes in yeast.

2.2 Organization of eukaryotic ribosomal RNA genes

The number of rDNA units per cell varies among eukaryotes, from 40 to ~19,000 in animals and from 150 to 26,000 in plants and correlates positively with genome size (Richard et al., 2008). In human cells there are up to 200 hundred copies of the ribosomal genes per haploid genome which are arranged in a head to tail orientation in clusters of tandem repeats. In yeast a single cluster consisting of ~150 copies of rRNA genes termed the nucleolar organizer region (NOR) is located on chromosome XII (Fig. 2a) and comprises over 10% of the whole genome (Kobayashi, 2011). During interphase a single nucleolus forms around this cluster. Two transcriptional regulatory DNA elements have been identified (Kulkens et al., 1991; Musters et al., 1989): the upstream element (UE), which is the binding site for the upstream activating factor (UAF); and the core promoter (CORE), which recruits the core factor (CF) (Elion & Warner, 1984). In addition to these regulatory DNA elements several non-coding regions have been identified within the gene and the IGS, including an origin of replication, called the ribosomal autonomous replicating sequence (ARS), and an expansion sequence containing a replication fork barrier (RFB) plus a bidirectional promoter (E-pro) that is required for rDNA amplification (Fig. 2a) (Brewer et al., 1992; Kobayashi et al., 1992).

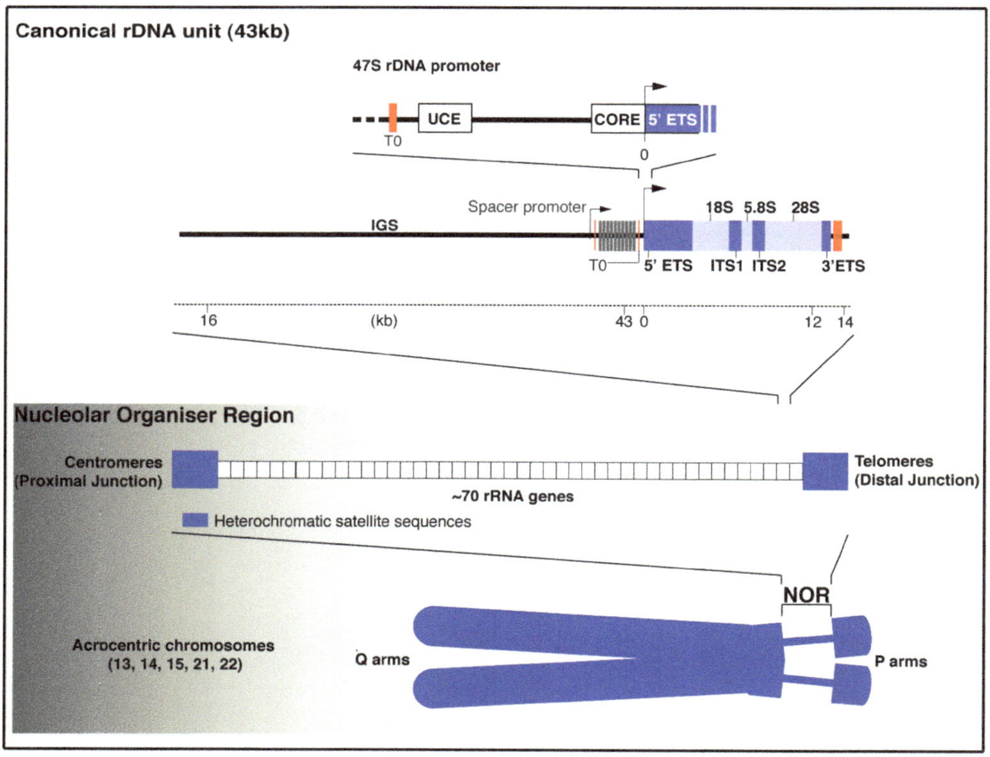

Fig. 2b. Organization of ribosomal RNA genes in mammalian cells.

In contrast to yeast, the rDNA repeats of higher eukaryotes are located in multiple NORs (Fig. 2b). For example, humans have five NORs located on the short arms of the acrocentric chromosomes (chromosomes 13, 14, 15, 21 and 22) that contain ~70 copies of the rRNA genes (Sakai et al., 1995). During interphase two or more NORs coalesce to form multiple nucleoli in exponentially growing cells (McStay & Grummt, 2008).

A single rDNA unit, containing a transcribed region followed by intergenic spacer (IGS) is referred to as a canonical rDNA unit. However, non-canonical units, which form a palindromic structure are present and arranged as a mosaic with the canonical units in human NORs (Caburet et al., 2005). These non-canonical units are likely to be non-functional and it is believed that they are silenced since their transcription would result in antisense transcripts that could use the RNAi machinery to degrade the 47S precursor rRNA. However, this has not been formally tested.

The rDNA repeats were not sequenced during the human genome project due to their repetitive nature and high copy number, thus the full sequence of only two repeats have been reported (Kuo et al., 1996). A canonical human rDNA repeat (43 kb) consists of a 13 kb coding region that encodes the 47S pre-rRNA, which is rapidly processed at the external transcribed spacer (ETS) and internal transcribed spacer (ITS) regions and ~30 kb of IGS (Fig. 2b). The IGS harbors several DNA regulatory elements, including the 47S rDNA promoter. The rDNA promoter has a bipartite architecture composed of a CORE and an upstream control element (UCE) located 100 bp upstream (McStay & Grummt, 2008). One or more terminator elements are located at the 5` and 3` ends of each mammalian rDNA repeat. Intriguingly, the IGS also contains one or more regions that are almost identical to the 47S rDNA promoter, and these are termed spacer promoters. Recent studies demonstrated that non-coding IGS transcripts play a role in the epigenetic control of rRNA gene silencing through modulation of the activity of the nucleolar remodeling complex (NoRC), the rDNA silencing complex (Mayer et al., 2006) (see below for more detail).

2.3 Epigenetic regulation of Pol I transcription

It is well established that even in exponentially growing cells only a subset of rRNA genes are active. In yeast, active and inactive rRNA genes are randomly distributed in the single NOR (French et al., 2003). However in higher eukaryotes it is thought that active and silent rRNA genes are clustered, thus generating active and silent NORs. Silent NORs are condensed and do not contribute to the formation of nucleoli during interphase. It is likely that active NORs may contain a mosaic of active and inactive units however this remains a matter for discussion (McStay & Grummt, 2008).

2.3.1 Epigenetic silencing of rDNA

While the precise mechanism controlling the silencing of rRNA genes is yet to be fully elucidated, there appears to be a marked difference between yeast and higher organisms in this process. For example, *S. cerevisiae* lacks the two major repressive methylation marks (CpG dinucleotide and H3K9 methylation) that are associated with the silencing of higher eukaryotic genes. Indeed, little is known about what regulates the number of active/inactive repeat units in yeast, although the histone deacetylase, Rpd3, is thought to play a role (Sandmeier et al, 2003), and TOR signaling regulates Pol I transcription and alters nucleolar compaction (Tsang et al, 2007). Instead, the majority of studies looking at rDNA silencing in yeast have looked at the silencing of Pol II genes in the rDNA, and it is this form of silencing that is most strongly linked to senescence and aging, particularly through the Pol II-dependent E-pro promoter in the rDNA (Kobayashi & Ganley, 2005).

Silencing of E-pro is regulated by Sir2p, a member of the Sirtuin family of NAD+ dependent protein deacetylases. Sir2p also regulates the silencing of the telomeres and the mating loci (Guarente, 1999). The Sir2p analog in mammals SIRT1 is a key component of the energy-dependent nucleolar silencing complex (eNoSC), which has been reported to repress rRNA gene transcription in response to altered intracellular energy status (Murayama et al., 2008). Sir2p together with Net1 and Cdc14 are part of a well-known epigenetic regulator of the yeast rDNA locus, the regulator of nucleolar silencing and telophase exit (RENT) complex (Huang & Moazed, 2003). In addition, Sir2p in a complex with accessory proteins such as condensin or cohesin can repress recombination events within the rDNA repeats (Huang et al., 2006; Machin et al., 2004). More recently it has become clear that the stability of the rDNA repeats and their accurate replication depends on the proportion of the epigenetically silenced rRNA genes (Kobayashi, 2011).

In higher eukaryotes methylation of CpG dinucleotides is a common modification associated with establishing stable transcriptional repression. This covalent modification is catalyzed by the DNA methyltransferases DNMT1, DNMT3a and DNMT3b, and is maintained during cell division (Klose & Bird, 2006). Stanchev et al (1979) demonstrated that CpG dinucleotide methylation marks are predominantly present in the promoter and enhancer regions of inactive rRNA genes (Fig. 3). In murine cells, methylation of a single CpG dinucleotide within the UCE (position -133 relative to the start of transcription) impairs the association of the Pol I transcription factor, upstream binding factor (UBF), to the rDNA and thus inhibits the assembly of the preinitiation complex (PIC) at the promoter (McStay & Grummt, 2008). In contrast, the human rDNA promoter contains ~25 CpG islands none of which are completely methylated or non-methylated. This suggests that the overall level of methylation rather than a binary on/off switch, dictates the transcriptional status of the rDNA. NoRC is the major complex involved in CpG methylation silencing of rDNA repeats (Santoro et al., 2002) (Fig. 3). The evolutionary logic underlying the additional complexity of rDNA silencing in higher eukaryotes compared to yeast is not clear but it potentially relates to regulation of cell differentiation and multicellular development.

2.3.2 Active rDNA repeats and activation of Pol I transcription

Regulation of rDNA transcription can occur at multiple levels, through regulatory elements defined by the primary DNA sequence as described above and also via the structure of the chromatin, which determines the accessibility of the DNA. Similar to regulation of Pol II and Pol III transcription, post-translational modification of the histones, such as acetylation, methylation, phosphorylation and ubiquitination, represent a key mechanism for the regulation of transcription by Pol I of active rDNA (Fig. 3).

In yeast it has been shown that the chromatin of actively transcribed rRNA genes is largely devoid of histone molecules, and instead is associated with the high-mobility group protein Hmo1, which interacts with the Pol I subunit Rpa49, binds across the entire 35S rDNA sequence and stabilize open rRNA gene chromatin (Hall et al., 2006; Merz et al., 2008).

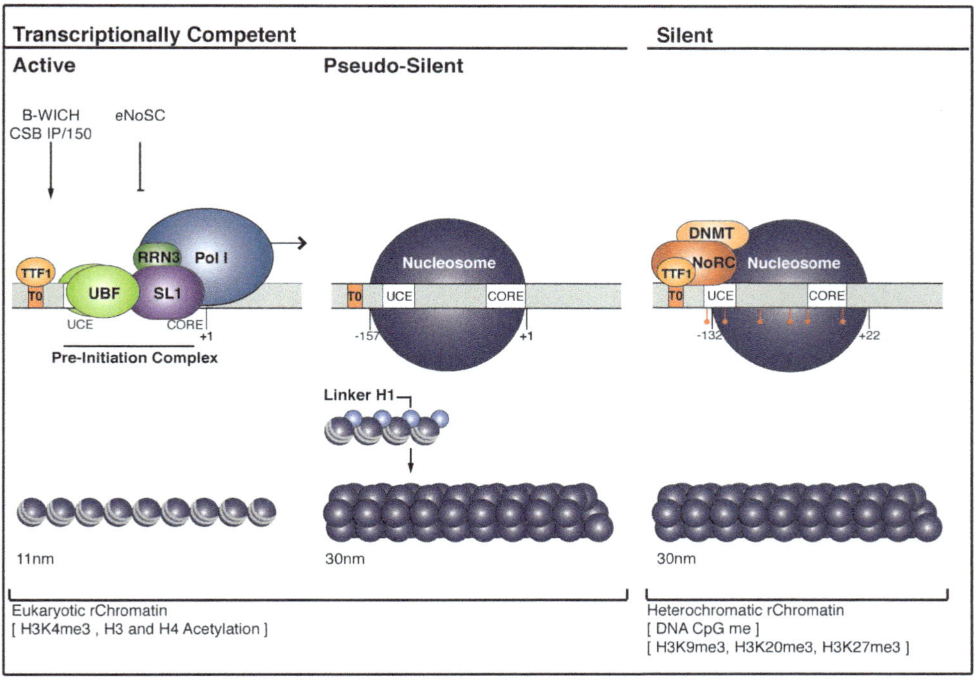

Fig. 3. Regulation of Pol I transcription in mammalian cells.

In mammals transcriptionally active rRNA genes lack repressive histone modifications such as H3K9, H3K20 and H3K27 methylation and CpG DNA methylation (Conconi et al., 1989). Furthermore, they are associated with markers for active genes including H3K4 methylation and acetylation of histone H3 and H4 (Fig. 3). Importantly, transcriptionally active mammalian rDNA are characterized by the presence of UBF, which is enriched at the promoter and the transcribed regions of the repeat, and to a lesser extent at the IGS (Fig. 3) (Sanij et al., 2008; Wright et al., 2006). UBF seems to play multiple roles at the rDNA including transcriptional initiation, promoter escape and elongation control (Stefanovsky et al., 2006). Most likely these functions relate to the essential role UBF plays in maintaining active genes in an open, uncondensed configuration, which is achieved, in part, through the ability of UBF to outcompete histone linker H1, thus preventing the formation of higher order chromatin (Sanij & Hannan, 2009; Sanij et al., 2008). Of note, active rRNA genes are around tenfold less condensed then adjacent DNA and remain uncondensed during mitosis (Heliot et al., 1997). This is undoubtedly due to the continual association of UBF and a subset of the Pol I transcription machinery with the rDNA repeats, which maintains them in an under-condensed configuration to allow the rapid resumption of rDNA transcription as cells re-enter the cell cycle (Prieto & McStay, 2007; Roussel et al., 1996).

One complex that has been described to promote the formation of an active chromatin environment for Pol I transcription is the chromatin remodeling complex B-WICH, which is composed of the William syndrome transcription factor (WSTF), SNF2h and nuclear myosin (NM1) has been described to promote the formation of an active chromatin environment for Pol I transcription (Vintermist et al., 2011) (Fig. 3). The remodeling activity of the B-WICH complex is restricted to a specific 200 bp region around the promoter, which includes the UCE, CORE and transcriptional start site (Vintermist et al., 2011). An ATP-dependent chromatin remodeling complex (CSB IP/150) also promotes transcription of active rRNA genes. CSB IP/150 consists of the Crockayne syndrome protein B (CSB), TFIIH and TIF1B (Bradsher et al., 2002) (Fig. 3).

A key similarity between yeast and mammals is that the rate of rRNA transcription is regulated in response to stress signals and the availability of nutrients as sensed by the TOR pathway. In mammals the Pol I transcription factors, UBF and Pol I-specific transcription initiation factor 1A (TIF-1A)/RRN3 (Hannan et al., 2003; Mayer et al., 2004) have been reported to be activated by TOR kinase. Similar finding have been made in yeast for Hmo1 and Rrn3p. Specifically binding of Hmo1 to the rDNA is TORC1 dependent, and nutrient starvation or rapamycin (inhibitor of mTOR) treatment prevents this association (Berger et al., 2007). In the absence of Hmo1 the histone H4 deacetylase, Rpd3, can associate with the rDNA, resulting in rDNA condensation and a reduction of nucleolar size (Tsang et al., 2003).

The basal Pol I transcription machinery in yeast involves two multiprotein complexes, the UAF consisting of Rrn5, Rrn9, and Rrn10 and UAF30 (Keys et al., 1996; Siddiqi et al., 2001) and the CF, consisting of Rrn7p, Rrn11p, and Rrn6p (Steffan et al., 1996). Both complexes interact with the TATA-box-binding protein TBP. Binding of UAF to the promoter is essential for the recruitment of CF, once the UAF-CF is established active Pol I is recruited to initiate transcription. Initiation in yeast and mammals requires the essential Pol I-associated factor Rrn3p. Rrn3p is only found associated with a small fraction of Pol I, and in yeast this association requires Pol I phosphorylation. Upon initiation of transcription Pol I enters into the elongation phase of transcription and Rrn3p is released.

As with yeast, in mammals the initiation of transcription of active rRNA genes requires the assembly of a PIC at the promoter, although some of the components are species specific (Grummt, 2003; Moss et al., 2007). In the mammalian system the PIC (Fig. 3) contains the selectivity factor I (SL1), a complex itself of 4 or more TATA association factors (TAFs) unique to Pol I transcription plus the TBP that is utilized by all three Pol's (Learned et al., 1985; Zomerdijk et al., 1994). Our current understanding is that SL1 is recruited to the promoter by UBF, and consequently stabilizes UBF interaction with the rDNA promoter. Upon formation of a stable UBF/SL1 complex, active Pol I (defined by its association with RRN3 (Hempel et al., 1996; Yuan et al., 2002) is then recruited to complete the PIC. Following Pol I transcription mediated initiation and promoter clearance, RRN3 is thought to be released to be recycled for another round of transcription. Various steps in transcription including initiation or elongation, are also regulated in response to extracellular signals such as nutrients, amino acids, ATP, stress, which is mediated by signaling pathways, including the PI3K/AKT/mTOR, RAS/RAF/ERK and JNK pathways (Chan et al., 2011; Hannan et al., 2003; Mayer et al., 2005; Stefanovsky et al., 2001). More recently the transcription termination factor (TTF1), which was originally identified for its role in the termination of Pol I transcription, has been implicated in modulating DNA looping at the rDNA repeat thus facilitating a specific interaction between the promoter and terminator elements of actively transcribed rDNA repeats (Sander & Grummt, 1997). Thus by creating an rDNA loop, TTF1 is thought to promote efficient re-initiation of the Pol I complex at the rDNA promoter (Grummt et al., 1985; Henderson & Sollner-Webb, 1986).

2.3.3 Pseudo-silenced rDNA repeats in higher eukaryotes

Interestingly, in mammals not all the transcriptionally active rRNA genes of interphasic nucleoli are transcribed at any one time (Sanij & Hannan, 2009). Transcriptionally competent genes can be subdivided into two categories; active genes and pseudo-silent genes (Fig. 3). Active genes are undermethylated, bound by the cytoarchitectural chromatin remodeling factor UBF and are highly transcribed, whereas pseudo-silent rRNA genes are undermethylated and bound by linker-histone H1, but not by UBF, and thus are not transcriptionally active. This pseudo-silenced conformation of rDNA repeats, when induced by RNA inference mediated knock down of UBF, is stably propagated throughout the cell cycle of many generations in the absence of changes in CpG methylation and can be reversed by restoration of UBF to wild-type level (Sanij et al., 2008). Importantly, pseudo-silencing seems to be a physiologically relevant phenomenon. For example, terminal differentiation of various cell types is associated with decreased UBF expression and a concomitant increase in the number of pseudo-silent rRNA genes (Poortinga et al., 2004; Poortinga et al., 2011; Sanij & Hannan, 2009). Moreover, the transition from a pre-malignant to malignant state is also associated with a decrease in the proportion of pseudo-silenced rRNA genes (Hannan RD and Bywater M, unpublished observation).

3. rDNA stability and aging in yeast

S. cerevisiae and its unique genetic and biochemical attributes have proven to be an outstanding model organism to analyze many aspects of eukaryotic ribosome biogenesis. This is evident as much of our current understanding of the link between the nucleolus/rDNA transcription with aging and senescence comes from studies utilizing the experimental advantages of budding yeast.

3.1 Maintenance of rDNA copy number in yeast

The highly repetitive nature of eukaryotic rDNA makes it one of the most fragile and dynamic regions of the genome, as recombination events within these repeats can cause either loss or gain of rDNA copies. Typically, such recombination events are highly regulated and are essential for maintenance of the rDNA copy number and the evolutionary stability of the rDNA repeats (Hawley & Marcus, 1989). Two mechanisms have been shown to be utilized for the repair of DNA double strand breaks (DSB) in the rDNA: homologous recombination (HR) and the single strand annealing (SSA) (Fishman-Lobell et al., 1992). Both these repair pathways can cause loss of rDNA copies, however the HR pathway can also result in a gain of copies. Despite the fluctuation resulting from the loss or gain of rDNA copies, an rDNA maintenance system provides a mechanism for the cell to keep copy number at a uniform level and ensure genomic stability (Fig. 4) (Kobayashi, 2006). Part of this maintenance system utilizes amplification of the rDNA, which rectifies the loss of rDNA copies (Kobayashi et al., 2004). Interestingly two of the key players in the regulation of rDNA copy number are also known as "aging associated genes", Fob1p and Sir2p (Fig. 4).

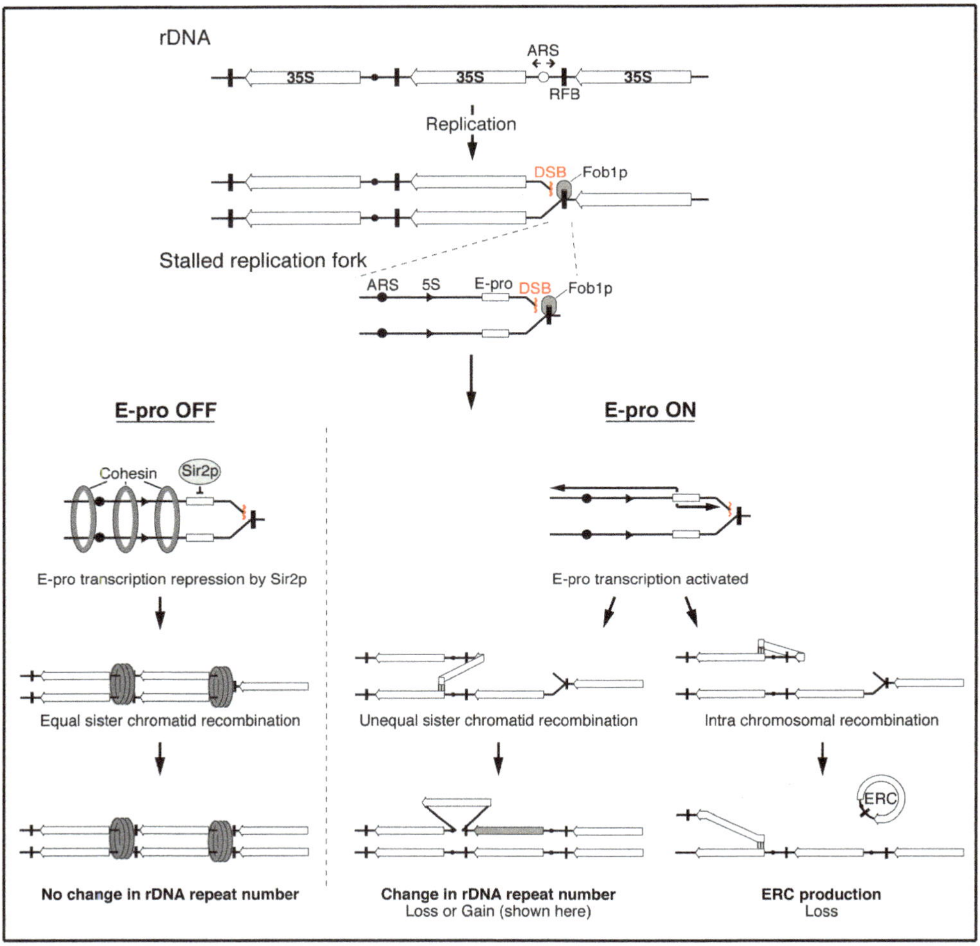

Fig. 4. Maintenance of rDNA copy number in yeast. Following initiation of replication from an ARS, Fob1p binding at the RFB inhibits replication fork progression. The outcome of the repair of these DSBs is dependent on the copy number of the rDNA. "E-pro OFF" illustrates cells containing wild-type copy number with no change in rDNA copy number. "E-pro ON" illustrates cells where rDNA copy number is altered, resulting in the production of an ERC.

During S-phase, initiation of DNA replication occurs at the origins of a subset of rDNA repeats (Pasero et al., 2002). The protein Fob1p binds in a sequence-specific manner to the rDNA at the RFB site, stalling the replication fork in one direction (Kobayashi, 2003). DNA

DSB can occur at these paused replication forks, and they are repaired using HR. The outcome of this repair is dependent on the copy number of the rDNA (Kobayashi & Ganley, 2005). In cells containing wild-type copy number, Sir2p represses noncoding Pol II-dependent transcription at the bidirectional promoter located in the IGS (Fig. 4: E-pro OFF), enabling the association of cohesin with the IGS. Cohesin is a chromosome-associated multisubunit complex that connects sister chromatids and plays an essential role in the correct segregation of chromosomes during cell division and post-replicative DNA repair (Merkenschlager, 2010). Cohesin complex association with the IGS prevents the broken end from using a non-cognate repeat as the template for HR, thereby ensuring repair through equal sister chromatid recombination, with no rDNA copy number change. If the number of rDNA copies is reduced, however, a transcription-dependent rDNA amplification mechanism is activated whereby Sir2p repression is lifted, thus activating bidirectional E-pro transcription (Fig. 4: E-pro ON). This non-coding transcription promotes the dissociation of cohesin from the IGS, allowing the broken end to use an unequal repeat as the repair template, resulting in a change of copy number. Copy number can either increase or decrease depending on whether the repeat used as the template for repair is upstream or downstream of the broken repeat (Ganley et al., 2005; Kobayashi & Ganley, 2005; Santangelo et al., 1988). If the template for repair is the same sister chromatid, a circular pop-out molecule, called an extra-chromosomal ribosomal circle (ERC) is formed. When rDNA copy number reaches wild-type levels E-pro transcription is silenced by Sir2p again, and rDNA amplification is inhibited. Sir2p mutant yeast cells can accumulate up to 300 copies due to non-restricted rDNA amplification (Kobayashi et al., 2004). Fob1p also meditates recombination events that are important for sequence homogenization of rDNA repeats and thus maintenance of the rRNA genes with identical or similar sequences (Ganley & Kobayashi, 2007). Whilst the mechanism by which cells monitor their rDNA copy number remains to be determined, it is clear that the maintenance of numerous copies of the rDNA is very important for genomic stability.

3.2 Silenced rDNA copies and DNA damage

A recent landmark study (Ide et al., 2010) demonstrated that the number of silenced rDNA copies determines the cells sensitivity to DNA damage inducing agents such as ultraviolet (UV) radiation and methyl methanesulfonate (MMS) (Fig. 5). By using low-rDNA copy number strains (20 copies), the ratio of actively transcribed rRNA genes increased, and these strains were deficient in their ability to repair DNA damage during S-phase. Low copy number strains mutated in the Pol I subunit Rpa135p or Rrn3p were not impaired in their DNA repair capacity and consequently do not exhibit a higher sensitivity to DNA damaging agents. These findings suggest that rDNA transcription determines the sensitivity to DNA damage by inhibiting DNA repair (Ide et al., 2010). The authors also reported that this transcription-dependent sensitivity resulted from the inability of the multi subunit complex condensin, that is important for establishing and maintaining chromosome condensation, to associate with actively-transcribing rDNA units. This results in premature sister-chromatid separation, which impairs accurate sister-chromatid recombination required for DNA repair. Strikingly the major site of condensin complex occupation in the genome is the NOR. Consistent with this, the binding of mitotic condensin to the rDNA was shown to be reduced when Pol I transcription was elevated, and this impaired proper DNA repair and chromatid cohesion, thus resulting in increased rDNA instability (Wang et al., 2006).

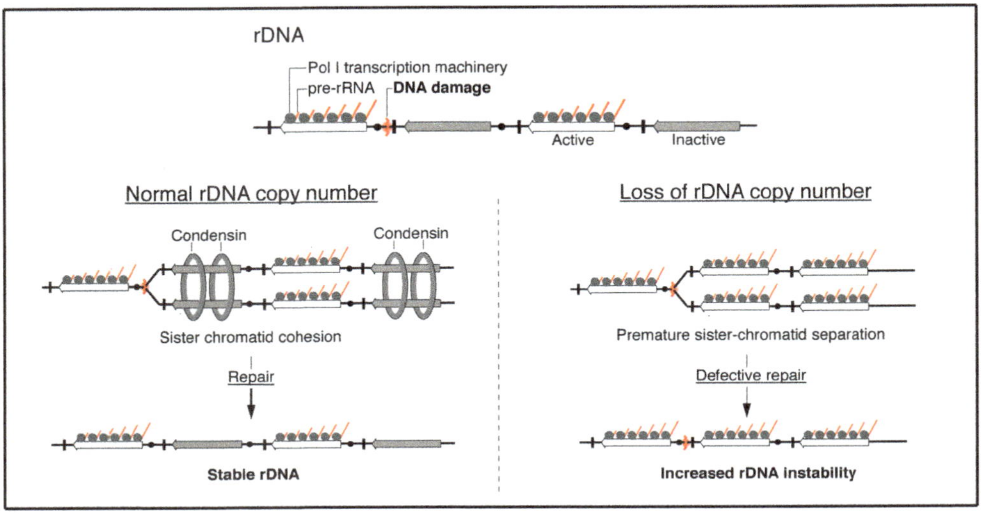

Fig. 5. Silenced rDNA copies and DNA damage in yeast. In cells with normal rDNA copy number, a proportion of genes are transcriptionally silent. Following DNA damage, condensin binds to untranscribed rDNA copies and facilitates sister chromatid cohesion, allowing repair of the rDNA by HR between sister chromatids at the stalled replication fork. Loss of rDNA copy number results in increased rates of transcription from the remaining rDNA repeats. Pol I transcription hinders the association of condensin with the rDNA, resulting in premature sister chromatid separation and preventing repair of the rDNA.

3.3 Aging in budding yeast

Aging can generically be defined as a progressive functional decline, or a gradual deterioration of physiological function and loss of viability (Partridge & Mangel, 1999). In *S. cerevisiae* an aged phenotype becomes apparent when the mother cell gets beyond ~10 asymmetric cell divisions, and this includes enlargement of the cell and vacuole, extension of the cell division cycle, and sterility. Furthermore aged wild-type cells display an extremely enlarged and often fragmented nucleolus (Shore, 1998). The average lifespan of wild-type yeast is ~ 20 buddings, after which time the mother cell dies (Jazwinski, 2001). In contrast, daughter cells are born with a full budding capacity independent of the age of the mother cell. Studies in the late 90`s implicated a role for the nucleolus in yeast aging (reviewed in Guarente, 1997). As mentioned above Sirtuins are a protein family of NAD+ dependent protein deacetylases, and they are linked to the process of yeast aging by acting as silencing factors at a site termed the *AGE* locus, prolonging life span. It has been shown that Sir3p relocalizes to the nucleolus with age and that deletion of Sir2, 3 and 4 can abbreviate yeast life span (Kennedy et al., 1997). Later studies revealed that Sir2p could repress recombination in the rDNA locus, maintain rDNA stability and promote yeast longevity (Kaeberlein et al., 1999), suggesting a role for rDNA stability in the regulation of life span. Intriguingly, extra copies of Sir2 orthologs are capable of extending the lifespan of both worms and flies (Bauer et al., 2009; Tissenbaum & Guarente, 2001), suggesting an evolutionary role for Sir2 in regulation of longevity. As discussed earlier in section 2, the mammalian Sir2p orthologue, SIRT1 is part of the eNoSC complex, which mediates epigenetic silencing of rDNA in response to varying intracellular energy status (Murayama et al., 2008). It has been proposed that the SIRT1-eNoSC complex and epigenetic regulation of rDNA may provide a novel regulatory pathway for mammalian aging, which is associated with lower metabolic rates (Salminen & Kaarniranta, 2009).

There is a clear prediction for aging factors in *S. cerevisiae,* as cell division is asymmetrical and the daughter cell receives a full lifespan, thus any aging factor must be preferentially sequestered in the mother cell and not passed on to the daughter. Indeed, Sinclair & Guarente (1997) demonstrated that aging wild-type yeast accumulated ERC, and these accumulated exclusively in mother cells. ERC accumulated even more rapidly in mutants (*sgs1*) that exhibit premature aging. In addition, accumulation of other extra-chromosomal genetic elements (i.e. plasmids) in the mother were shown to induce senescence. It was proposed that the accumulation of extra-chromosomal elements, including ERC and episomes, in the mother titrates genomic factors important for the maintenance of a young phenotype. A recent study suggested that it is not the ERC themselves that are the aging factor, but instead the rDNA recombination process that produces the ERC (Ganley et al., 2009). This study used strains with altered rDNA replication efficiencies. ERC exist in the cell effectively as plasmids because they harbour a replication origin. In the absence of selection, plasmid stability correlates with replication origin strength. Thus by altering rDNA replication strength, ERC production could be separated from their maintenance, and a strain with very little ERC accumulation was shown to age quickly when rDNA recombination was high. This study also reported that other episomes can induce genomic instability (Ganley et al., 2009), reconciling their results with those of Guarente and colleagues. ERCs have also been identified in *Drosophila* and humans (Gagnon-Kugler et al., 2009; Peng & Karpen, 2007) however its origin and role in aging is yet to be determined.

Other aging theories propose that senescence is caused by an accumulation of DNA damage or cytoplasmic senescence factors that remain within the mother cell due to asymmetrical segregation. These theories are supported by the observation showing that oxidized (damaged) proteins predominantly accumulate in mother rather than daughter cells (Erjavec et al., 2007). Nucleolar rDNA is proposed to be particularly sensitive to the presence of elevated levels of oxidized proteins, as this leads to an impaired protein turnover and defects in DNA repair. Furthermore, the asymmetrical segregation of oxidized proteins is Sir2p dependent (Erjavec et al., 2007), leads to rDNA instability and the accumulation of ERC in the mother cell, which then promotes cellular senescence (Kobayashi, 2008).

3.4 A specific role for rDNA in aging

The role of rDNA in aging is most clearly demonstrated in yeast with rDNA instability and cellular aging strongly correlating with rDNA copy number (Burkhalter & Sogo, 2004; Kobayashi et al., 2004). A recent review proposed the "rDNA theory" for aging. Specifically that dysfunction of DNA repair and the replication proteins, predominantly within the nucleolus of the mother cell, is a cause of increased rDNA instability. Because the nucleolus is the most sensitive cellular component to age-related DNA damage, the stability of the rDNA will in turn dictate the stability of the whole genome (Kobayashi,

2008). In both yeast and humans mutations within DNA repair genes result in a reduced lifespan (Park et al., 1999). In humans mutations associated with a premature aging phenotype (Werner and Bloom syndrome) are prominently found in RecQ homolog helicases, which are involved in rDNA repair (Ellis et al., 1995; Yu et al., 1996). In yeast, mutations in genes involved in rDNA transcription and elongation have been identified as modulators of rDNA stability and longevity (Heo et al., 1999; Hoopes et al., 2002; Merker & Klein, 2002). In conclusion whilst the findings in yeast clearly link the stability of rDNA, to aging and cellular senescence, relatively few studies in mammals investigating this link have been reported, predominantly due to the difficulties in studying this complicated part of the genome.

The idea that the rDNA has other, extra-coding functions has received increasing attention over the last few years (Kobayashi, 2011). In yeast, due to the fact that the rDNA cluster comprises ~10% of the genome, rDNA copy number and stability can influence the effective concentration of proteins and protein complexes located within the nucleus through titration. For example, studies investigating Sirtuins revealed that Sir2p is released from the nucleolus upon loss of rDNA copies. Sir2p, together with Sir3p, Sir4p and RAP1 can mediate silencing of telomeres and the mating-type loci. Intriguingly, depletion of ~50% of the rDNA repeats caused an increase in telomeric and mating-type gene silencing, suggesting that the effective concentration of Sir2p at different genomic loci plays a critical role in epigenetic regulation. The fact that mammalian rDNA comprises only ~0.3% of the genome raises the question of whether the association of genomic factors with the rDNA is sufficient to titrate silencing complexes to a similar extent to that reported in yeast. However, the mammalian nucleolus has been reported to influence various cellular functions via sequestering or releasing factors important for various cellular processes that regulate senescence and aging and this is described in more detail in section 4.

4. The nucleolus and senescence in mammals

Senescence is considered a manifestation of organismal aging at a cellular level, although this remains mechanistically unproven (Guarente, 1997; McCormick & Campisi, 1991). However, a number of studies have shown that senescent cells accumulate within mammalian tissues with increasing chronological age (Dimri et al., 1995; Y. Li et al., 1997; Pawelec et al., 1999). As discussed above, in yeast the rDNA locus, and hence the nucleolus, has been implicated in the regulation of longevity and senescence (Kobayashi, 2008). Consistent with this, a nontraditional role for the mammalian nucleolus is also now emerging that involves sequestration and release of tumor suppressors or oncogenes, cell cycle regulators and factors involved in modulating telomerase function (Olson et al., 2002; Olson, 2004; Olson & Dundr, 2005). Thus, similar to yeast, the mammalian nucleolus has been proposed to play an active role in senescence (Mehta et al., 2007; Olson et al., 2000).

Changes in nucleolar morphology are detected in aging cells (Mehta et al., 2007). While pre-senescent cells show a higher number of smaller nucleoli (Bemiller & Lee, 1978), senescent cells have a single prominent nucleolus. Cellular senescence however, does not always correlate with a concomitant decrease in rRNA gene transcription (Halle et al., 1997; Machwe et al., 2000). Even so, studies from many laboratories indicate that inhibition of rRNA synthesis and ribosome biogenesis and subsequent changes in nucleolar structure and function induce cell cycle arrest, implicating the nucleolus in regulation of cell survival and proliferation (Boisvert et al., 2007; Boulon et al., 2010). Indeed, a number of proteins regulated by their localization to the nucleolus, including the tumor suppressor protein ARF (alternative reading frame; p19ARF in mouse, p14ARF in human) and nucleophosmin (NPM), are involved in cellular senescence by mediating p53 stability (Colombo et al., 2002; Daniely et al., 2002). Therefore, it is likely that the nucleolus plays an active role in establishing and maintaining the senescent phenotype.

Senescent cells are growth arrested in the transition from G1 to S-phase of the cell cycle (Sherwood et al., 1988). The role of the G1-S and G2-M cell cycle checkpoints is to ensure that the cell accurately duplicates its genome and successfully divide into the two daughter cells. The tumor suppressor factors p53 and retinoblastoma protein (RB) regulate these cell cycle checkpoints, which upon activation induce cell cycle arrest, senescence or apoptosis. Correct cell division requires increased protein synthesis that, in turn, is achieved by upregulation of ribosome biogenesis, thus these processes are tightly linked and potentially regulated by common mechanisms (Montanaro, 2008). For example, rDNA transcription and subsequent assembly of the nucleoli during G1 have been shown to be prerequisites for G1-S progression (Pardee, 1989; Sirri et al., 2002). Conversely, decreases in rates of rDNA transcription and disassembly of the nucleoli are observed during mitosis (Grummt, 1999; Pyronnet et al., 2001; Sirri et al., 2002). In fact, the nucleolus can sense and respond to cellular stress by modifying its size and content throughout interphase (Nalabothula et al., 2010). Consequently, it is not surprising that the nucleolus has been recognized as a central regulatory link between ribosome biogenesis and cell cycle progression (Carmo-Fonseca et al., 2000). Certainly, an ever increasing number of nucleolar proteins have been reported to play multiple roles in regulating ribosome biogenesis and cell cycle progression (Boisvert et al., 2007). Although, the exact molecular mechanisms responsible for mediating this crosstalk remain largely unknown, the p53 pathway is prevailing as an important link between ribosome biogenesis and the cell cycle (Pestov et al., 2001).

Early evidence for p53 as a key mediator of the crosstalk between ribosome biogenesis and cell cycle progression came from studies showing that inhibition of Bop1 (block of proliferation), a factor involved in rRNA synthesis and assembly, led to a p53-dependent G1 checkpoint arrest (Pestov et al., 2001). This is consistent with the notion that ongoing ribosome synthesis acts as a checkpoint at the G1-S boundary. In support of this inhibition of rRNA synthesis by microinjecting antibodies to UBF, disruption of the *TIF-IA* gene by Cre-dependent HR, or low doses of actinomycin D (Act D), leads to perturbations in nucleolar structure and function, p53-dependent G1-S cell cycle arrest and apoptosis (Montanaro et al., 2007; Rubbi & Milner, 2003; Yuan et al., 2005).

We have previously shown that inhibition of Pol I transcription by CX-5461, a small molecule inhibitor of initiation of rRNA synthesis, induces senescence in solid tumor cell lines (Drygin et al., 2011). Moreover, CX-5461 treatment or low doses of Act D induces premature senescence in TERT immortalized primary human fibroblasts (BJ-TERT) (Fig. 6) (Hahn et al., 1999). Within 24 hours of treatment with Act D or CX-5461, the protein levels of p53 and its transcriptional target, p21, are upregulated and sustained for a further 24 hours (Fig. 6a). Inhibition of rRNA synthesis correlates with the appearance of single nucleoli, as visualized by fluorescence *in situ* hybridization (FISH) of rDNA (Fig. 6b), while in interphasic control cells the rDNA repeats are present as multiple nucleoli. The induction of p53 and nucleolar disorganization correlates with the subsequent appearance of γH2A.X foci indicative of DNA damage, which is associated with senescence (Gire et al., 2004). After 96 hours of Pol I transcription inhibition, BJ-TERT cells appear bigger in size with a flat cell morphology and display acidic β-galactosidase activity, characteristic phenotypic markers of senescence (Fig. 6c) (Dimri et al., 1995). UBF depletion in BJ-TERT cells is also associated with nucleolar disorganization and premature senescence, consistent with its role in establishing and maintaining nucleolar structure. Intriguingly, UBF depletion does not lead to decreased rates of rRNA synthesis, suggesting that the premature senescence may be a consequence of nucleolar disruption (Sanij E and Hannan RD, manuscript in preparation).

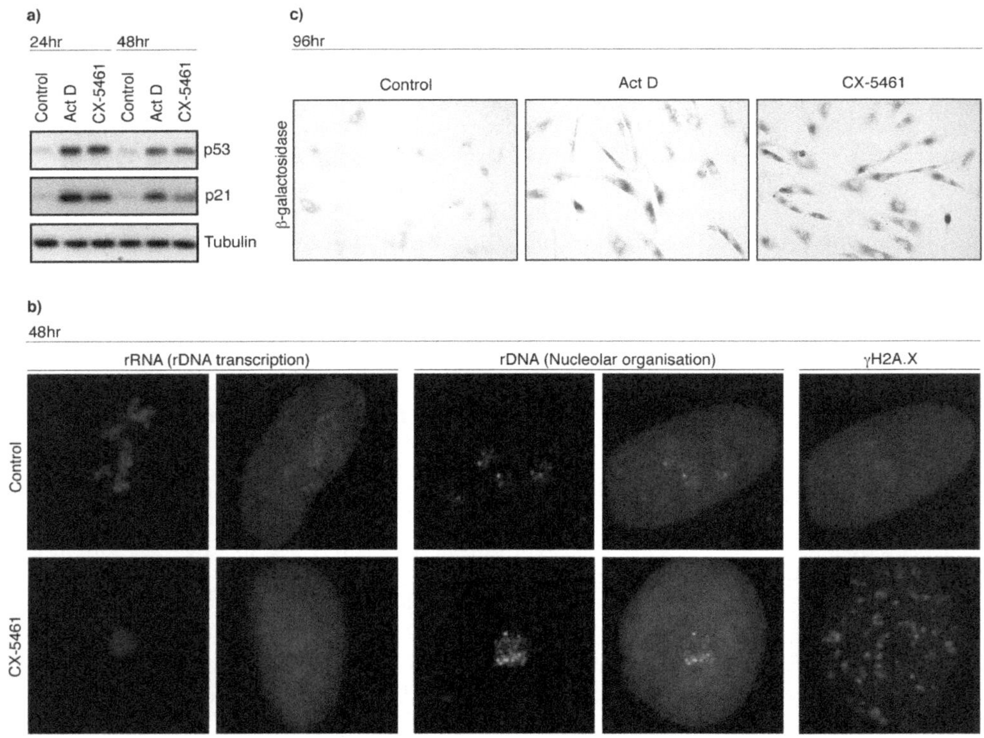

Fig. 6. Inhibition of Pol I transcription induces senescence. Following inhibition of Pol I transcription by treatment with either 5nM Act D or 1μM CX-5461, BJ-TERT fibroblasts display hallmarks of senescence. a) Western blot analysis of p53 and p21 after treatment with either Act D or CX-5461. b) Following 48hr CX-5461 treatment, from left to right: RNA FISH to the 5′ETS region of the 47S pre-rRNA (red); DNA FISH to the rDNA (green) combined with immunofluorescent analysis for γH2A.X (red). gDNA is visualized by DAPI (blue)
c) β-galactosidase staining of BJ-TERT cells after 96hr treatment with Act D or CX-5461.

Recent studies exploring the nucleolar proteome have revealed the involvement of the nucleolus in multiple biological processes including senescence, regulation of telomerase function, cell cycle regulation, and stress signaling (Boisvert et al., 2010; Boisvert et al., 2007). Furthermore, nucleolar structure and function are intimately linked with the regulation of p53 stabilization and activation of the p53 pathway, which firmly places the integrity of rDNA transcription and ribosome biogenesis at the centre of control of the cell cycle progression. In addition, over the last few years, significant advances have been made in understanding the higher order organization of nuclear structures, including the nucleolus, and their importance for the regulation of nuclear functions (Nemeth & Langst, 2011). Here, we address the impact of these recent results and discuss the molecular mechanisms underlying nucleolar function in the regulation of cell cycle progression and senescence.

4.1 The nucleolus in control of cell cycle regulation and senescence

An increasing number of nucleolar proteins have been reported to play multiple roles in regulating ribosome biogenesis and cell cycle progression. For instance, components of the DNA replication initiation machinery, origin of replication complex (ORC) and minichromosome maintenance (MCM) proteins, have been purified from human nucleoli (Boisvert et al., 2007; Couté et al., 2006) and were shown, in yeast, to associate with several 60S ribosomal synthesis factors that are required for pre-rRNA processing and are also essential for initiation of DNA replication by mediating the association of the ORC and MCM proteins at replication origins (Zhang et al., 2002). However, the most direct role for nucleoli in regulation of the cell cycle is via the sequestration or release of proteins directly involved in cell cycle progression.

The tumor suppressor protein RB is an important regulator of senescence (Campisi & di Fagagna, 2007). RB is generally active during senescence, indeed its enforced expression has been shown to induce senescence (Narita et al., 2003). RB was initially reported to accumulate in the nucleolus and to have a repressive role in Pol I transcription (Cavanaugh et al., 1995; Hannan et al., 2000). Subsequently, Nucleolin (NCL), a multifunctional nucleolar protein essential for rRNA processing (Mongelard & Bouvet, 2007) was reported to associate with hypophosphorylated (active) RB (pRB) during the G1 phase of the cell cycle (Grinstein et al., 2006). pRB mediates a cell cycle checkpoint between G_1 and S phase (Bartek et al., 1996) by targeting members of the E2F family of transcriptional activators (Chellappan et al., 1991) that are essential for cellular proliferation. While pRB has been reported to reside in the nucleoli in a cell type dependent manner (Angus et al., 2003) hyperphosphorylated RB (ppRB) is eliminated from the nucleolus until late S or G2 phase. Import of ppRB into the nucleolus in late S or G2 phase is mediated by its interaction with nucleolar NPM (Takemura et al., 2002). Although the functional significance of nucleolar retention and release of RB is currently unresolved, it has been proposed that its retention could represent a negative regulatory mechanism to sequester RB to prevent checkpoint activation during the cell cycle (Angus et al., 2003).

By far the most convincing studies linking the dynamic release of nucleolar proteins and cell cycle progression were performed in yeast. Cdc14p is a protein phosphatase that is crucial for promoting exit from mitosis into the G1 phase (Jin et al., 2008; Visintin et al., 1998; Zachariae et al., 1998). Cdc14p activation is also a prerequisite for successful chromosome segregation as it is necessary for condensin enrichment at the rDNA, which triggers rDNA segregation and ensures the completion of chromosome segregation (D'Amours et al., 2004). In G1 or during S phase, Cdc14p is sequestered to the nucleolus by its inhibitor Net1/Cfi1p, a component of the multifunctional protein complex RENT, where it remains inactive until the onset of anaphase, thereby preventing the premature onset of mitotic exit (Shou et al., 1999). The release of Cdc14p from the nucleolus is mediated through the sequential action of two regulatory networks: FEAR (CDC Fourteen Early Anaphase Release) and MEN (Mitotic Exit Network) that lead to Net1/Cfi1p phosphorylation reducing its affinity for Cdc14p and thus disassociation of the complex (Shou et al., 2002; Yoshida & Toh-e, 2002). The roles of the two human orthologs of Cdc14p (CDC14A and CDC14B) are not yet established. CDC14B has been shown to translocate from the nucleolus to the nucleoplasm following genotoxic stress in G2, leading to the activation of the APC/CDH1 complex and the establishment of a DNA damage induced G2 checkpoint (Bassermann et al., 2008). Nevertheless, a wider role in promoting mammalian cell cycle progression has been proposed for CDC14B including the governing of cell cycle re-entry after G2 block (De Wulf & Visintin, 2008).

Another example of nucleolar sequestration-mediated regulation of specific cellular activity during the cell cycle is the regulation of telomerase, the enzyme that adds telomeric repeats sequences to the ends of chromosomes (Fig. 7). Telomeres are composed of TTAGGG repeats that form a 3′ overhang of 100-400 nucleotides forming a T-loop structure that is stabilized by telomeric proteins (Griffith et al., 1999). Telomeres maintain chromosome integrity by protecting against end shortening and end-to-end fusions (de Lange, 2005; Sahin & Depinho, 2010). If telomere length is not maintained, the telomeres will reach a critically short length, triggering the cell to undergo replicative senescence (Harley et al., 1990; Stewart et al., 2003). Although the telomeres and the nucleolus are separate subnuclear domains, multiple telomeric components have been detected in the nucleolus suggesting an underling regulatory connection between the nucleolus and telomeres (Tsai, 2009).

The ribonucleoprotein (RNP) telomerase is composed of the telomerase RNA component (TERC) and the telomerase reverse transcriptase (TERT), which catalyzes *de novo* repeat addition by utilizing TERC as a template (Greider & Blackburn, 1989). Outside of S phase, TERC and TERT are localised within distinct nucleoplasmic foci separate from telomeres. In early S-phase, TERT moves to nucleoli while Cajal bodies containing TERC accumulate at the periphery of nucleoli (Fig. 7). Nucleolar transportation of TERC and TERT has been proposed as a prerequisite step in the process of telomerase RNP biogenesis (Etheridge et al., 2002; Narayanan et al., 1999; Yang et al., 2002) and/or the transport of active telomerase, which occurs during mid-S phase (Tomlinson et al., 2006). Nucleolar localization of telomerase has also been reported to be mediated by NCL, which interacts with the active telomerase complex and is involved either in the assembly or maturation of telomerase. Nucleolar NCL-telomerase complexes are exported and maintained in the nucleoplasm and delivered to the telomeres (Khurts et al., 2004). Cell-cycle dependent nucleolar localization of telomerase is lost in transformed cells or following DNA damage (Wong et al., 2002).

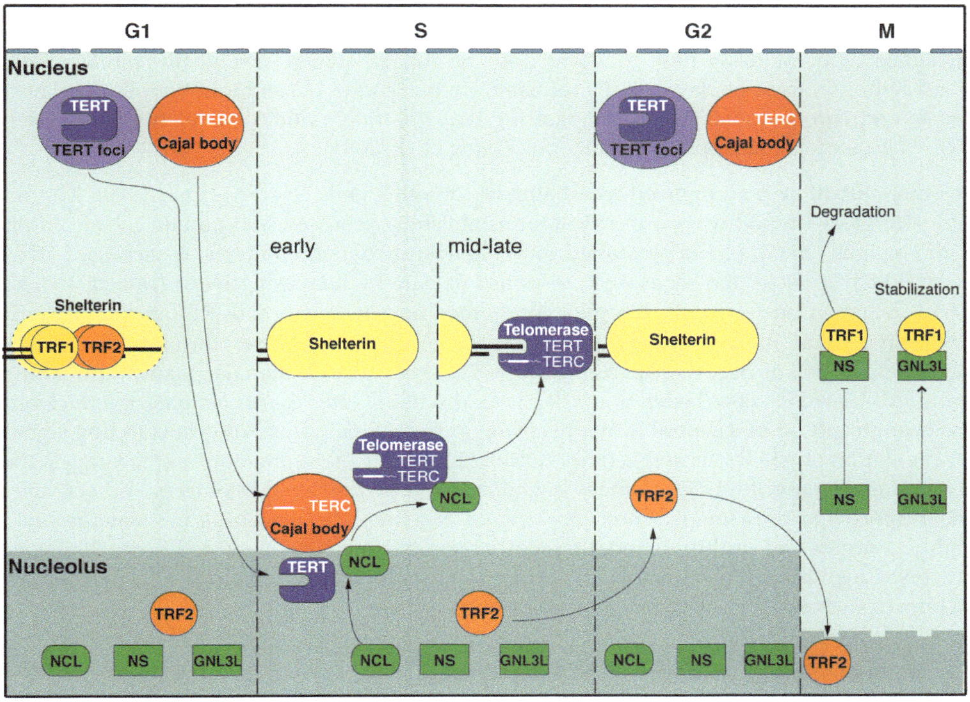

Fig. 7. Nucleolar regulation of telomere stability. Telomerase and the shelterin complex are regulated by the nucleolus through sequestration and release. In early S-phase, TERT moves to the nucleoli and Cajal bodies containing TERC accumulate at the periphery of the nucleoli. NCL-telomerase complexes are then exported to the telomeres during mid-S phase. TRF1 is regulated by NS and GNL3L in opposing manners. TRF2 is sequestered in the nucleolus during G1 and S phase, released to the nucleoplasm in G2, and returned to the nucleoli at cytokinesis.

In addition to telomerase, components of the shelterin telomere binding complex, including telomeric repeat binding factors 1 (TRF1) and TRF2 have been reported as regulated in part by localization to the nucleolus (Fig. 7) (Lin et al., 2008; Tsai, 2009; Wong et al., 2002; Zhang et al., 2004). TRF1, which is required for establishing 'closed' structures at the telomeres that are inaccessible to telomerase, has been shown to be regulated by a number of nucleolar proteins, including nucleostemin (NS) and guanine nucleotide binding protein-like 3-like (GNL3L). NS binding to TRF1 enhances its degradation, while GNL3L binding stabilizes TRF1 (Zhu et al., 2009). Since the majority of TRF1 resides in the nucleoplasm, nucleolar retention of NS and GNL3L renders them inactive in modulating TRF1 activity. However, their nucleoplasmic localization during mitosis or in response to nucleolar stress may allow modulation of TRF1 in regulating telomere capping (Tsai, 2009).

In addition, TRF2, the component of shelterin considered responsible for the formation of the protective telomeric T-loop structure required for protecting the telomeres, localizes to the nucleolus at G0 and S but diffuses into the nucleoplasm in G2 and returns to the nucleolus at cytokinesis (Fig. 7). Low dose of Act D, which specifically inhibits Pol I transcription, causes a delay in TRF2 release from nucleoli in G2 and mitotic cells displaying end-to-end chromosomal fusions, suggesting that the timely nucleolar retention/release of TRF2 regulates its nucleoplasmic function (Zhang et al., 2004).

Telomere attrition is recognized as a hallmark of aging cells (Harley et al., 1990). The p53 and p16INK4a-RB pathways are critical for establishing senescence in human cells (Campisi & di Fagagna, 2007). p53 is presumed to sense dysfunctional telomeres as damaged DNA, upon which it elicits the senescence response in part by increasing expression of the p21 CDKI, which in turn prevents the phosphorylation and inactivation of RB (Sherr & Roberts, 1999). In several mouse models, inappropriate p53 activity, either through deregulated expression of p53 or in response to constitutive stress like DNA damage, leads to premature aging (Maier et al., 2004; Tyner et al., 2002). As discussed later in this chapter, the nucleolus has been proposed as a central hub for sensing major cellular stress and transmitting signals for regulation of p53 levels and activity (Olson, 2004). It is therefore tempting to suggest that nucleoli may sense the DNA damage signal induced by damaged telomeres and activate a p53 response to implement senescence. Taken together, the nucleolus has emerged as a highly complex and multifunctional regulatory compartment involved in diverse biological processes including the regulation of proliferation and the execution of anti-proliferative responses such as cell cycle arrest and senescence.

4.2 The nucleolus as a sensor of cellular stress

One of the most intriguing roles proposed for the nucleolus is as a sensor of cellular stress and a means to couple cellular stress to the p53 pathway (Rubbi & Milner, 2003), a key regulator of senescence and longevity (Fig. 8) (Vigneron & Vousden, 2010) In this paradigm, under normal conditions, the nucleolus contributes to the maintenance of low p53 levels, while in response to cellular stress p53 levels and activity are dramatically elevated through the actions of select nucleolar proteins. Key to the nucleolar control of p53 is the oncogene MDM2 (mouse double minute 2; or HDM2 in humans). In proliferating cells, p53 activity is kept under surveillance by MDM2, via two complimentary mechanisms: (i) MDM2 acts as an E3 ubiquitin ligase directly transfering ubiquitin onto p53 thereby targeting it for 26S proteosomal degradation (Haupt et al., 1997; Kubbutat et al., 1997); and (ii) the direct binding of MDM2 to the N-terminal domain of p53 inhibits its transcriptional activity by abrogating its interaction with the basal Pol II transcription machinery (Momand et al., 1992; Oliner et al., 1993). The two best characterized mechanisms by which cellular stress modulates MDM2/p53 pathway in a nucleolar specific manner are in response to oncogenes (oncogenic stress) and to perturbations that alter ribosome biogenesis (nucleolar stress).

4.2.1 ARF and oncogenic / replicative stress

The *Ink4/Arf* locus encodes two tumor-suppressor proteins, p16INK4a and p19ARF, that govern the antiproliferative functions of RB and p53 proteins, respectively (Fig. 8). ARF binds to MDM2 and sequesters it into the nucleolus, thereby preventing negative-feedback regulation of p53 by MDM2, leading to the activation of p53 in the nucleoplasm (Honda & Yasuda, 1999; Palmero et al., 1998; Zindy et al., 1998). Under normal conditions, ARF is expressed at very low levels and is sequestered into to the nucleolus, due to its association with the nucleolar protein NPM (Gjerset & Bandyopadhyay, 2006; Korgaonkar et al., 2005). This prevents its interaction with MDM2. In contrast during replicative senescence of MEFs or stress induced by activation of oncogenes such as c-MYC and H-RAS, ARF rapidly accumulates to sufficient quantities and is able to bind and sequester MDM2 leading to p53 activation (Palmero et al., 1998; Sharpless et al., 2001; Weber et al., 1999). In addition ARF directly suppresses rRNA synthesis and processing to modulate ribosome biogenesis (Ayrault et al., 2006; Lessard et al., 2010; Sugimoto et al., 2003). Although, the importance of the later for modulation of p53 is unclear, a ribosome biogenesis-dependent-ARF pathway may complement ARF's function in modulating the p53 pathway.

4.2.2 Nucleolar stress

The protein content of the nucleolus has been shown to change dramatically under various stress conditions (Boulon et al., 2010). It is now clear that, the nucleolar proteome undergoes distinct spatial and temporal alterations in response to different stress insults, suggesting that the nucleolus responds to different stress stimuli in a unique and specific manner (Moore et al., 2011). The landmark study by Rubbi and Milner (2003) proposed that disruption of nucleolar structure and function and subsequent release of nucleolar components into the nucleoplasm as a common denominator in most or possibly all p53-inducing stresses (Reviewed in Olson, 2004). Consistent with this, it is now recognized that inactivation of rDNA transcription, RP synthesis, rRNA processing, and the assembly and nucleolar export of the 40S and 60S ribosomal subunits (Zhang & Lu, 2009) are established mechanisms for causing nucleolar disruption and activation of the p53 pathway. From these observations a model of nucleolar surveillance of ribosome biogenesis (also termed nucleolar stress) has been proposed to integrate a diverse array of metabolic irregularities and oncogenic stimuli whereby the rate or efficiency of ribosome production serves as a signal by which cells could regulate cell-cycle progression via controlling p53 levels (Fig. 8) (Boulon et al., 2010; Deisenroth & Zhang, 2010; Ruggero & Pandolfi, 2003; Shcherbik & Pestov, 2010). In addition, due to the repetitive nature of the rRNA genes as well as the high rate of transcription by Pol I complexes, the rDNA is considered unstable and has been proposed to act as a potential sensor for DNA damage (Boisvert & Lamond, 2010). Signals associated with stalled polymerases and/or reduced rRNA transcription could activate p53 and possibly other DNA damage response pathways (Boisvert & Lamond, 2010; Kobayashi, 2008) leading to cell cycle arrest or programmed cell death (Drygin et al., 2009). Under conditions of "nucleolar stress", p53 stabilization can be achieved via different mechanisms including posttranslational modifications, protein-protein interactions and increases in the translation rate of p53 mRNA. Of these mechanisms perhaps the best documented is the role of RPs which are able to interact directly with MDM2 leading to p53 stabilization in response to ribosomal stress. Interestingly, while the RPs are required for p53 response to ribosomal stress, ARF in this context is not required, suggesting that different cellular conditions, oncogenic stress or ribosomal stress modulate the binding of either ARF or RPs to MDM2 and subsequent activation of p53 (Pan et al., 2011).

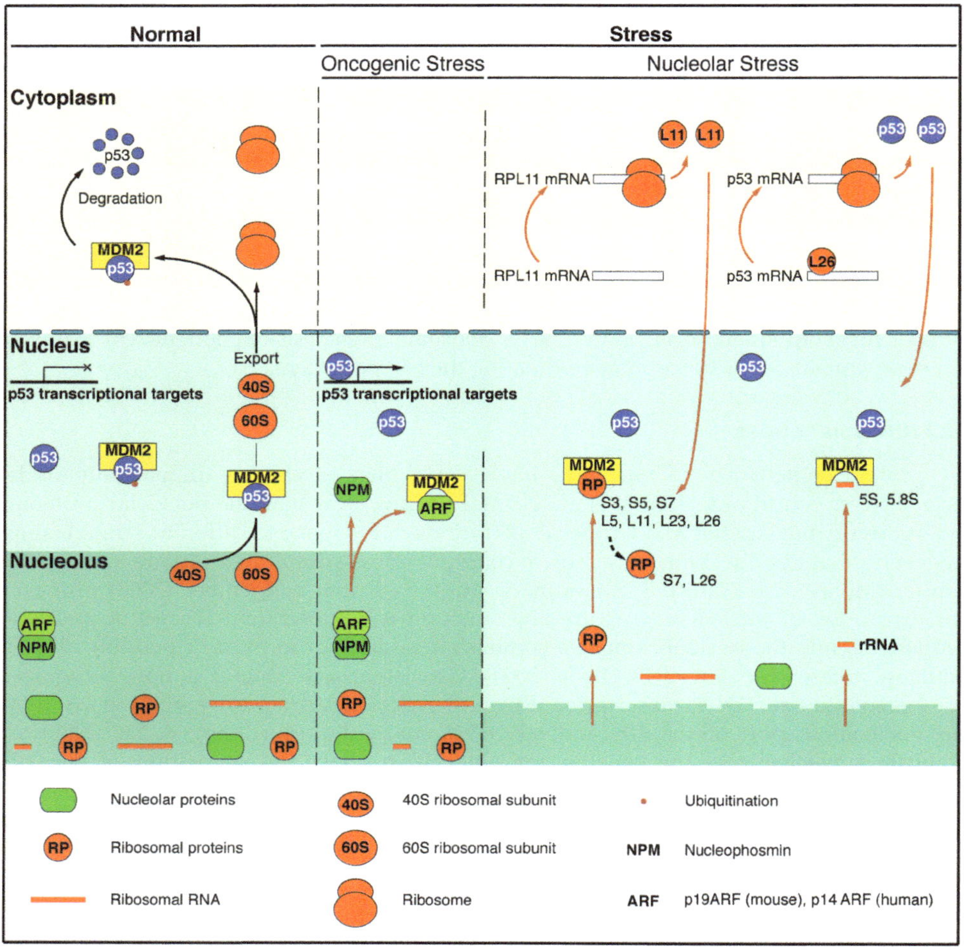

Fig. 8. The nucleolus as a sensor of cellular stress. Under normal conditions, p53 activity is maintained at low levels by MDM2, via two mechanisms: First, MDM2 ubiquitinates p53 thereby promoting its degradation; second, the binding of MDM2 to p53 abrogates its interaction with Pol II transcription machinery. Following oncogenic stress, ARF binds MDM2 and sequesters it in the nucleolus. Under nucleolar stress, p53 can be activated by the following mechanisms: (i). The co-transport of p53 and/or MDM2 with the ribosomal subunits to the cytoplasm is impaired; (ii). RPs interact directly with MDM2; (iii). 5.8S and 5S rRNA interact directly with MDM2; (iv). RPL26 binds p53mRNA and enhances its translation; (v). Increased RPL11 mRNA translation results in enhanced interaction between RPL11 and MDM2.

An increasing number of RPs (RPS3, RPS5, RPS7, RPL5, RPL11, RPL23, RPL26) as well as the 5.8S and 5S rRNAs are capable of interacting with MDM2 leading to p53 stabilization (Deisenroth & Zhang, 2010; Fontoura et al., 1992; Fumagalli et al., 2009; Ofir-Rosenfeld et al., 2008; Riley & Maher, 2007; Zhang & Lu, 2009). A number of different models have been proposed for how these various interactions might regulate p53 (Fig. 8). Under normal growth conditions, RPs are synthesized in equimolar amounts with the rRNAs and assembled with large and small ribosomal subunits in the nucleolus and transported to the cytoplasm to form functional ribosomes. In one model, so called "riding the ribosome", the interaction of p53 and/or MDM2 with the ribosomal subunits may facilitate p53/MDM2 transport from the nucleolus to the cytoplasm thus preventing p53 from interacting with its target genes in the nucleoplasm and/or promoting its ubiquitin-mediated degradation in the cytoplasm (Boulon et al., 2010). Conversely, stress signaling that impairs production and export of ribosome subunits, would be predicted to decrease p53/MDM2 transport to the cytoplasm, thus allowing p53 to activate transcription of its target genes in the nucleoplasm (Boulon et al., 2010). In a second and perhaps better described model, conditions that inhibit rRNA transcription or stall ribosome synthesis and assembly in the nucleolus are postulated to create a pool of free RPs (such as RPL5, RPL11 and RPL23) that are directly interact and sequester MDM2 resulting in suppression of p53 ubiquitination (Daniely et al., 2002; Deisenroth & Zhang, 2010; Lindstrom & Nister, 2010; Pestov et al., 2001; Warner & McIntosh, 2009; Zhang & Lu, 2009). However, Horn and Vousden (2008) observed a synergistic suppression of MDM2 activity through cooperation of RPL11 and RPL5, suggesting they have distinct roles in inhibiting MDM2 function. In addition, binding sites for ARF, RPL5, and RPL11 on MDM2 do not appear to overlap (Lindstrom et al., 2007; Zhang et al., 2003) suggesting that ARF and RPL5/L11 may respond to different stimuli and converge at the point of MDM2 inactivation (Shcherbik & Pestov, 2010).

RPL26 is so far unique in its ability to bind the 5`untranslated region of the p53 mRNA and enhance its translation. Its interaction with MDM2 triggers its own ubiquitination and degradation, which in turn causes downregulation of p53 mRNA translation (Ofir-Rosenfeld et al., 2008). The diverse roles of RPs in the regulation of the MDM2-p53 pathway is further supported by the finding that knockdown of RPS6 not only affects 40S ribosomal biogenesis but also enforces RPL11 mRNA translation. This leads to an enhanced interaction between RPL11 and MDM2 leading to the accumulation and activation of p53 (Fumagalli et al., 2009). Since multiple RPs have separate mechanisms for activating p53, it is plausible that they may have distinct roles in sensing different types of signals leading to activation of nucleolar stress response. In summary there is now a robust set of data demonstrating that the nucleolus and rDNA transcription indirectly play an important role in the regulation of tumor suppressors and oncogenes such as ARF, MDM2 and p53 and thus perturbation in the nucleolus are predicted to have profound effects on cellular functions that are controlled by these factors. In this manner the nucleolus can be considered as a sensitive cellular stress detector integrating various perturbations in homeostasis and converting them to appropriate responses such as cell cycle arrest and senescence.

4.3 Alterations in genome organization in and around the nucleolus are associated with senescence

The nucleus is compartmentalized into substructures that perform distinct nuclear activities (Lanctot et al., 2007). These nuclear structures include the nucleoli, nuclear envelope, nuclear bodies, nuclear matrix and chromosome territories (Cremer & Cremer, 2001). Organization and spatial location of chromosomes and their interactions with other nuclear substructures ensures that transcription is correctly regulated (Misteli, 2004). The periphery of the nucleolus consists of satellite DNA repeats, which are proposed to play a role in the formation of perinucleolar heterochromatin (Manuelidis, 1984), and its been suggested they serve as a distinct nuclear space with a primary function in maintaining repressive chromatin states (Nemeth & Langst, 2011; van Koningsbruggen et al., 2010). For example the inactive X chromosome (Xi) must continuously visit the perinucleolar compartment during S phase to maintain its epigenetic status (Zhang et al., 2007). Other conserved chromosomal regions have also been shown to interact with the nucleolus including a fraction of the human centromeres (Bridger & Bickmore, 1998; Leger et al., 1994; Park & De Boni, 1992). Furthermore, chromosome mobility studies demonstrated that nucleolar-associated chromatin is significantly less mobile then other genomic regions. Specifically disruption of the nucleolar structure enhanced chromatin mobility, thus implying the nucleoli plays an active role in constraining chromatin movement and maintaining the three-dimensional organization of the genome within the nucleus (Chubb et al., 2002). This is further supported by the identification of specific interactions between repetitive and non-repetitive loci within the yeast genome including specific repeated elements that interact with rRNA genes. Therefore, it has been proposed that genomic architecture is organized by restricting the mobility of these repeat elements relative to the nucleolar interaction point (O'Sullivan et al., 2009). Similar results have been obtained using live cell imaging, with frequent interactions observed between the nucleolar and non-nucleolar chromatin (Berger et al., 2008). Intriguingly, extensive disorganisation of nuclear architecture, at the level of whole chromosomes, is associated with the transition from proliferative to senescent states. The human non-rDNA bearing chromosome 18, for instance, exhibits altered spatial positioning, changing from the apical edge of the nucleus in proliferating cells to nucleoli in senescent cells (Bridger et al., 2000). It is plausible that the reorganisation of the genome as the cells enter senescence is responsible for extensive changes in the transcriptional status of the genome (Foster & Bridger, 2005). Consitent with this, we have recently found that reductions in UBF levels lead to disruptions in nucleolar structure and acrocentric chromosome organization and induces premature senescence in primary human fibroblasts (Huang S, Hannan RD, manuscript in preparation). Together, the data suggest a functional role for nucleoli in the organization of the genome and in the regulation of cellular senescence.

5. Conclusion

The nucleolus is a highly evolutionary conserved subnuclear compartment traditionally associated with rDNA transcription and ribosome-subunit production. However it is now apparent that the nucleolus is dynamic in nature and its organization, size and protein composition changes dramatically during the cell cycle and under different cellular conditions including stress. Consistent with this dynamic nature the nucleolus has now been implicated in regulating additional important cellular processes beyond ribosome-subunit synthesis, including cell-cycle control, stress responses, senescence and aging. The fundamental role the rDNA repeats play in aging of fission yeast is now overwhelming. Similarly, in higher eukaryotes, nucleolar function in coupling ribosome subunit biogenesis and cell-cycle progression, through the activity of the tumor suppressor protein p53, places the nucleolus at the centre of coordinating cellular stress response and determining cell fate such as survival and senescence. It is likely we have only begun to scratch the surface of the detail by which eukaryotes have evolved to utilise the unique subnuclear domain, the nucleolus, to control fundamental cellular process such as aging and senescence.

6. References

Angus, S.P., Solomon, D.A., Kuschel, L., Hennigan, R.F. & Knudsen, E.S. (2003). Retinoblastoma tumor suppressor: Analyses of dynamic behavior in living cells reveal multiple modes of regulation. Molecular and Cellular Biology *23*, 8172-8188.

Ayrault, O., Andrique, L., Fauvin, D., Eymin, B., Gazzeri, S. & Seite, P. (2006). Human tumor suppressor p14ARF negatively regulates rRNA transcription and inhibits UBF1 transcription factor phosphorylation. Oncogene *25*, 7577-7586.

Bartek, J., Bartkova, J. & Lukas, J. (1996). The retinoblastoma protein pathway and the restriction point. Current Opinion in Cell Biology *8*, 805-814.

Bassermann, F., Frescas, D., Guardavaccaro, D., Busino, L., Peschiaroli, A. & Pagano, M. (2008). The Cdc14B-Cdh1-Plk1 axis controls the G2 DNA-damage-response checkpoint. Cell *134*, 256-267.

Bauer, J.H., Morris, S.N., Chang, C., Flatt, T., Wood, J.G. & Helfand, S.L. (2009). dSir2 and Dmp53 interact to mediate aspects of CR-dependent lifespan extension in D. melanogaster. Aging (Albany NY) *1*, 38-48.

Bemiller, P.M. & Lee, L.H. (1978). Nucleolar changes in senescing WI-38 cells. Mech Ageing Dev *8*, 417-427.

Berger, A.B., Cabal, G.G., Fabre, E., Duong, T., Buc, H., Nehrbass, U., Olivo-Marin, J.C., Gadal, O. & Zimmer, C. (2008). High-resolution statistical mapping reveals gene territories in live yeast. Nat Methods *5*, 1031-1037.

Berger, A.B., Decourty, L., Badis, G., Nehrbass, U., Jacquier, A. & Gadal, O. (2007). Hmo1 is required for TOR-dependent regulation of ribosomal protein gene transcription. Mol Cell Biol *27*, 8015-8026.

Boisvert, F.M., Lam, Y.W., Lamont, D. & Lamond, A.I. (2010). A quantitative proteomics analysis of subcellular proteome localization and changes induced by DNA damage. Mol Cell Proteomics *9*, 457-470.

Boisvert, F.M. & Lamond, A.I. (2010). p53-Dependent subcellular proteome localization following DNA damage. Proteomics.

Boisvert, F.M., van Koningsbruggen, S., Navascues, J. & Lamond, A.I. (2007). The multifunctional nucleolus. Nat Rev Mol Cell Biol *8*, 574-585.

Boulon, S., Westman, B.J., Hutten, S., Boisvert, F.M. & Lamond, A.I. (2010). The nucleolus under stress. Mol Cell *40*, 216-227.

Bradsher, J., Auriol, J., Proietti de Santis, L., Iben, S., Vonesch, J.L., Grummt, I. & Egly, J.M. (2002). CSB is a component of RNA pol I transcription. Mol Cell *10*, 819-829.

Brewer, B.J., Lockshon, D. & Fangman, W.L. (1992). The arrest of replication forks in the rDNA of yeast occurs independently of transcription. Cell *71*, 267-276.

Bridger, J.M. & Bickmore, W.A. (1998). Putting the genome on the map. Trends Genet *14*, 403-409.

Bridger, J.M., Boyle, S., Kill, I.R. & Bickmore, W.A. (2000). Re-modelling of nuclear architecture in quiescent and senescent human fibroblasts. Curr Biol *10*, 149-152.

Burkhalter, M.D. & Sogo, J.M. (2004). rDNA enhancer affects replication initiation and mitotic recombination: Fob1 mediates nucleolytic processing independently of replication. Mol Cell *15*, 409-421.

Bystricky, K., Laroche, T., van Houwe, G., Blaszczyk, M. & Gasser, S.M. (2005). Chromosome looping in yeast: telomere pairing and coordinated movement reflect anchoring efficiency and territorial organization. J Cell Biol *168*, 375-387.

Caburet, S., Conti, C., Schurra, C., Lebofsky, R., Edelstein, S.J. & Bensimon, A. (2005). Human ribosomal RNA gene arrays display a broad range of palindromic structures. Genome Res *15*, 1079-1085.

Campisi, J. & di Fagagna, F.D. (2007). Cellular senescence: when bad things happen to good cells. Nat Rev Mol Cell Biol *8*, 729-740.

Carmo-Fonseca, M., Mendes-Soares, L. & Campos, I. (2000). To be or not to be in the nucleolus. Nat Cell Biol *2*, E107-112.

Cavanaugh, A.H., Hempel, W.M., Taylor, L.J., Rogalsky, V., Todorov, G. & Rothblum, L.I. (1995). Activity of RNA polymerase I transcription factor UBF blocked by Rb gene product. Nature *374*, 177-180.

Chan, J.C., Hannan, K.M., Riddell, K., Ng, P.Y., Peck, A., Lee, R.S., Hung, S., Astle, M.V., Bywater, M., Wall, M., Poortinga, G., Jastrzebski, K., Sheppard, K.E., Hemmings, B.A., Hall, M.N., Johnstone, R.W., McArthur, G.A., Hannan, R.D. & Pearson, R.B. (2011). AKT Promotes rRNA Synthesis and Cooperates with c-MYC to Stimulate Ribosome Biogenesis in Cancer. Sci Signal *4*, ra56.

Chellappan, S.P., Hiebert, S., Mudryj, M., Horowitz, J.M. & Nevins, J.R. (1991). The E2F transcription factor is a cellular target for the RB protein. Cell *65*, 1053-1061.

Chubb, J.R., Boyle, S., Perry, P. & Bickmore, W.A. (2002). Chromatin motion is constrained by association with nuclear compartments in human cells. Curr Biol *12*, 439-445.

Colombo, E., Marine, J.C., Danovi, D., Falini, B. & Pelicci, P.G. (2002). Nucleophosmin regulates the stability and transcriptional activity of p53. Nat Cell Biol *4*, 529-533.

Conconi, A., Widmer, R.M., Koller, T. & Sogo, J.M. (1989). Two different chromatin structures coexist in ribosomal RNA genes throughout the cell cycle. Cell *57*, 753-761.

Couté, Y., Burgess, J.A., Diaz, J.-J., Chichester, C., Lisacek, F., Greco, A. & Sanchez, J.-C. (2006). Deciphering the human nucleolar proteome. Mass Spectrometry Reviews *25*, 215-234.

Cremer, T. & Cremer, C. (2001). Chromosome territories, nuclear architecture and gene regulation in mammalian cells. Nat Rev Genet *2*, 292-301.

D'Amours, D., Stegmeier, F. & Amon, A. (2004). Cdc14 and condensin control the dissolution of cohesin-independent chromosome linkages at repeated DNA. Cell *117*, 455-469.

Daniely, Y., Dimitrova, D.D. & Borowiec, J.A. (2002). Stress-dependent nucleolin mobilization mediated by p53-nucleolin complex formation. Mol Cell Biol *22*, 6014-6022.

de Lange, T. (2005). Shelterin: the protein complex that shapes and safeguards human telomeres. Genes Dev *19*, 2100-2110.

De Wulf, P. & Visintin, R. (2008). Cdc14B and APC/C tackle DNA damage. Cell *134*, 210-212.

Deisenroth, C. & Zhang, Y. (2010). Ribosome biogenesis surveillance: probing the ribosomal protein-Mdm2-p53 pathway. Oncogene *29*, 4253-4260.

Dimri, G.P., Lee, X., Basile, G., Acosta, M., Scott, G., Roskelley, C., Medrano, E.E., Linskens, M., Rubelj, I., Pereira-Smith, O. & et al. (1995). A biomarker that identifies senescent human cells in culture and in aging skin in vivo. Proc Natl Acad Sci U S A *92*, 9363-9367.

Drygin, D., Lin, A., Bliesath, J., Ho, C.B., O'Brien, S.E., Proffitt, C., Omori, M., Haddach, M., Schwaebe, M.K., Siddiqui-Jain, A., Streiner, N., Quin, J.E., Sanij, E., Bywater, M.J., Hannan, R.D., Ryckman, D., Anderes, K. & Rice, W.G. (2011). Targeting RNA polymerase I with an oral small molecule CX-5461 inhibits ribosomal RNA synthesis and solid tumor growth. Cancer Res *71*, 1418-1430.

Drygin, D., Siddiqui-Jain, A., O'Brien, S., Schwaebe, M., Lin, A., Bliesath, J., Ho, C.B., Proffitt, C., Trent, K., Whitten, J.P., Lim, J.K., Von Hoff, D., Anderes, K. & Rice, W.G. (2009). Anticancer activity of CX-3543: a direct inhibitor of rRNA biogenesis. Cancer Res *69*, 7653-7661.

Elion, E.A. & Warner, J.R. (1984). The major promoter element of rRNA transcription in yeast lies 2 kb upstream. Cell *39*, 663-673.

Ellis, N.A., Groden, J., Ye, T.Z., Straughen, J., Lennon, D.J., Ciocci, S., Proytcheva, M. & German, J. (1995). The Bloom's syndrome gene product is homologous to RecQ helicases. Cell *83*, 655-666.

Erjavec, N., Larsson, L., Grantham, J. & Nystrom, T. (2007). Accelerated aging and failure to segregate damaged proteins in Sir2 mutants can be suppressed by overproducing the protein aggregation-remodeling factor Hsp104p. Genes Dev *21*, 2410-2421.

Etheridge, K.T., Banik, S.S., Armbruster, B.N., Zhu, Y., Terns, R.M., Terns, M.P. & Counter, C.M. (2002). The nucleolar localization domain of the catalytic subunit of human telomerase. J Biol Chem *277*, 24764-24770.

Fishman-Lobell, J., Rudin, N. & Haber, J.E. (1992). Two alternative pathways of double-strand break repair that are kinetically separable and independently modulated. Mol Cell Biol *12*, 1292-1303.

Fontoura, B.M., Sorokina, E.A., David, E. & Carroll, R.B. (1992). p53 is covalently linked to 5.8S rRNA. Mol Cell Biol *12*, 5145-5151.

Foster, H.A. & Bridger, J.M. (2005). The genome and the nucleus: a marriage made by evolution. Genome organisation and nuclear architecture. Chromosoma *114*, 212-9.

French, S.L., Osheim, Y.N., Cioci, F., Nomura, M. & Beyer, A.L. (2003). In exponentially growing Saccharomyces cerevisiae cells, rRNA synthesis is determined by the summed RNA polymerase I loading rate rather than by the number of active genes. Mol Cell Biol *23*, 1558-1568.

Fumagalli, S., Di Cara, A., Neb-Gulati, A., Natt, F., Schwemberger, S., Hall, J., Babcock, G.F., Bernardi, R., Pandolfi, P.P. & Thomas, G. (2009). Absence of nucleolar disruption after impairment of 40S ribosome biogenesis reveals an rpL11-translation-dependent mechanism of p53 induction. Nat Cell Biol *11*, 501-508.

Gagnon-Kugler, T., Langlois, F., Stefanovsky, V., Lessard, F. & Moss, T. (2009). Loss of human ribosomal gene CpG methylation enhances cryptic RNA polymerase II transcription and disrupts ribosomal RNA processing. Mol Cell *35*, 414-425.

Ganley, A.R., Hayashi, K., Horiuchi, T. & Kobayashi, T. (2005). Identifying gene-independent noncoding functional elements in the yeast ribosomal DNA by phylogenetic footprinting. Proc Natl Acad Sci U S A *102*, 11787-11792.

Ganley, A.R., Ide, S., Saka, K. & Kobayashi, T. (2009). The effect of replication initiation on gene amplification in the rDNA and its relationship to aging. Mol Cell *35*, 683-693.

Ganley, A.R. & Kobayashi, T. (2007). Highly efficient concerted evolution in the ribosomal DNA repeats: total rDNA repeat variation revealed by whole-genome shotgun sequence data. Genome Res *17*, 184-191.

Gire, V., Roux, P., Wynford-Thomas, D., Brondello, J.M. & Dulic, V. (2004). DNA damage checkpoint kinase Chk2 triggers replicative senescence. EMBO J *23*, 2554-2563.

Gjerset, R.A. & Bandyopadhyay, K. (2006). Regulation of p14ARF through subnuclear compartmentalization. Cell Cycle *5*, 686-690.

Greider, C.W. & Blackburn, E.H. (1989). A telomeric sequence in the RNA of Tetrahymena telomerase required for telomere repeat synthesis. Nature *337*, 331-337.

Griffith, J.D., Comeau, L., Rosenfield, S., Stansel, R.M., Bianchi, A., Moss, H. & de Lange, T. (1999). Mammalian telomeres end in a large duplex loop. Cell *97*, 503-514.

Grinstein, E., Shan, Y., Karawajew, L., Snijders, P.J., Meijer, C.J., Royer, H.D. & Wernet, P. (2006). Cell cycle-controlled interaction of nucleolin with the retinoblastoma protein and cancerous cell transformation. J Biol Chem *281*, 22223-22235.

Grummt, I. (1999). Regulation of mammalian ribosomal gene transcription by RNA polymerase I. Prog Nucleic Acid Res Mol Biol *62*, 109-154.

Grummt, I. (2003). Life on a planet of its own: regulation of RNA polymerase I transcription in the nucleolus. Genes Dev *17*, 1691-1702.

Grummt, I., Maier, U., Ohrlein, A., Hassouna, N. & Bachellerie, J.P. (1985). Transcription of mouse rDNA terminates downstream of the 3' end of 28S RNA and involves interaction of factors with repeated sequences in the 3' spacer. Cell *43*, 801-810.

Grummt, I. & Pikaard, C.S. (2003). Epigenetic silencing of RNA polymerase I transcription. Nat Rev Mol Cell Biol *4*, 641-649.

Guarente, L. (1997). Link between aging and the nucleolus. Genes Dev *11*, 2449-2455.

Hahn, W.C., Counter, C.M., Lundberg, A.S., Beijersbergen, R.L., Brooks, M.W. & Weinberg, R.A. (1999). Creation of human tumour cells with defined genetic elements. Nature *400*, 464-468.

Hall, D.B., Wade, J.T. & Struhl, K. (2006). An HMG protein, Hmo1, associates with promoters of many ribosomal protein genes and throughout the rRNA gene locus in Saccharomyces cerevisiae. Mol Cell Biol *26*, 3672-3679.

Halle, J.P., Muller, S., Simm, A. & Adam, G. (1997). Copy number, epigenetic state and expression of the rRNA genes in young and senescent rat embryo fibroblasts. Eur J Cell Biol *74*, 281-288.

Hannan, K.M., Brandenburger, Y., Jenkins, A., Sharkey, K., Cavanaugh, A., Rothblum, L., Moss, T., Poortinga, G., McArthur, G.A., Pearson, R.B. & Hannan, R.D. (2003). mTOR-dependent regulation of ribosomal gene transcription requires S6K1 and is mediated by phosphorylation of the carboxy-terminal activation domain of the nucleolar transcription factor UBF. Mol Cell Biol *23*, 8862-8877.

Hannan, K.M., Hannan, R.D., Smith, S.D., Jefferson, L.S., Lun, M. & Rothblum, L.I. (2000). Rb and p130 regulate RNA polymerase I transcription: Rb disrupts the interaction between UBF and SL-1. Oncogene *19*, 4988-4999.

Harley, C.B., Futcher, A.B. & Greider, C.W. (1990). Telomeres shorten during ageing of human fibroblasts. Nature *345*, 458-460.

Haupt, Y., Maya, R., Kazaz, A. & Oren, M. (1997). Mdm2 promotes the rapid degradation of p53. Nature *387*, 296-299.

Hawley, R.S. & Marcus, C.H. (1989). Recombinational controls of rDNA redundancy in Drosophila. Annu Rev Genet *23*, 87-120.

Heliot, L., Kaplan, H., Lucas, L., Klein, C., Beorchia, A., Doco-Fenzy, M., Menager, M., Thiry, M., O'Donohue, M.F. & Ploton, D. (1997). Electron tomography of metaphase nucleolar organizer regions: evidence for a twisted-loop organization. Mol Biol Cell *8*, 2199-2216.

Hempel, W.M., Cavanaugh, A.H., Hannan, R.D., Taylor, L. & Rothblum, L.I. (1996). The species-specific RNA polymerase I transcription factor SL-1 binds to upstream binding factor. Mol Cell Biol *16*, 557-563.

Henderson, S. & Sollner-Webb, B. (1986). A transcriptional terminator is a novel element of the promoter of the mouse ribosomal RNA gene. Cell *47*, 891-900.

Henras, A.K., Capeyrou, R., Henry, Y. & Caizergues-Ferrer, M. (2004). Cbf5p, the putative pseudouridine synthase of H/ACA-type snoRNPs, can form a complex with Gar1p and Nop10p in absence of Nhp2p and box H/ACA snoRNAs. RNA *10*, 1704-1712.

Heo, S.J., Tatebayashi, K., Ohsugi, I., Shimamoto, A., Furuichi, Y. & Ikeda, H. (1999). Bloom's syndrome gene suppresses premature ageing caused by Sgs1 deficiency in yeast. Genes Cells *4*, 619-625.

Honda, R. & Yasuda, H. (1999). Association of p19(ARF) with Mdm2 inhibits ubiquitin ligase activity of Mdm2 for tumor suppressor p53. EMBO J *18*, 22-27.

Hoopes, L.L., Budd, M., Choe, W., Weitao, T. & Campbell, J.L. (2002). Mutations in DNA replication genes reduce yeast life span. Mol Cell Biol *22*, 4136-4146.

Horn, H.F. & Vousden, K.H. (2008). Cooperation between the ribosomal proteins L5 and L11 in the p53 pathway. Oncogene *27*, 5774-5784.

Huang, J., Brito, I.L., Villen, J., Gygi, S.P., Amon, A. & Moazed, D. (2006). Inhibition of homologous recombination by a cohesin-associated clamp complex recruited to the rDNA recombination enhancer. Genes Dev *20*, 2887-2901.

Huang, J. & Moazed, D. (2003). Association of the RENT complex with nontranscribed and coding regions of rDNA and a regional requirement for the replication fork block protein Fob1 in rDNA silencing. Genes Dev *17*, 2162-2176.

Ide, S., Miyazaki, T., Maki, H. & Kobayashi, T. (2010). Abundance of ribosomal RNA gene copies maintains genome integrity. Science *327*, 693-696.

Jazwinski, S.M. (2001). New clues to old yeast. Mech Ageing Dev *122*, 865-882.

Jin, F., Liu, H., Liang, F., Rizkallah, R., Hurt, M.M. & Wang, Y. (2008). Temporal control of the dephosphorylation of Cdk substrates by mitotic exit pathways in budding yeast. Proc Natl Acad Sci U S A *105*, 16177-16182.

Kaeberlein, M., McVey, M. & Guarente, L. (1999). The SIR2/3/4 complex and SIR2 alone promote longevity in Saccharomyces cerevisiae by two different mechanisms. Genes Dev *13*, 2570-2580.

Kennedy, B.K., Gotta, M., Sinclair, D.A., Mills, K., McNabb, D.S., Murthy, M., Pak, S.M., Laroche, T., Gasser, S.M. & Guarente, L. (1997). Redistribution of silencing proteins from telomeres to the nucleolus is associated with extension of life span in S. cerevisiae. Cell *89*, 381-391.

Keys, D.A., Lee, B.S., Dodd, J.A., Nguyen, T.T., Vu, L., Fantino, E., Burson, L.M., Nogi, Y. & Nomura, M. (1996). Multiprotein transcription factor UAF interacts with the upstream element of the yeast RNA polymerase I promoter and forms a stable preinitiation complex. Genes Dev *10*, 887-903.

Khurts, S., Masutomi, K., Delgermaa, L., Arai, K., Oishi, N., Mizuno, H., Hayashi, N., Hahn, W.C. & Murakami, S. (2004). Nucleolin interacts with telomerase. Journal of Biological Chemistry *279*, 51508-51515.

Klose, R.J. & Bird, A.P. (2006). Genomic DNA methylation: the mark and its mediators. Trends Biochem Sci *31*, 89-97.

Kobayashi, T. (2003). The replication fork barrier site forms a unique structure with Fob1p and inhibits the replication fork. Mol Cell Biol *23*, 9178-9188.

Kobayashi, T. (2006). Strategies to maintain the stability of the ribosomal RNA gene repeats--collaboration of recombination, cohesion, and condensation. Genes Genet Syst *81*, 155-161.

Kobayashi, T. (2008). A new role of the rDNA and nucleolus in the nucleus--rDNA instability maintains genome integrity. Bioessays *30*, 267-272.

Kobayashi, T. (2011). Regulation of ribosomal RNA gene copy number and its role in modulating genome integrity and evolutionary adaptability in yeast. Cell Mol Life Sci *68*, 1395-1403.

Kobayashi, T. & Ganley, A.R. (2005). Recombination regulation by transcription-induced cohesin dissociation in rDNA repeats. Science *309*, 1581-1584.

Kobayashi, T., Hidaka, M., Nishizawa, M. & Horiuchi, T. (1992). Identification of a site required for DNA replication fork blocking activity in the rRNA gene cluster in Saccharomyces cerevisiae. Mol Gen Genet *233*, 355-362.

Kobayashi, T., Horiuchi, T., Tongaonkar, P., Vu, L. & Nomura, M. (2004). SIR2 regulates recombination between different rDNA repeats, but not recombination within individual rRNA genes in yeast. Cell *117*, 441-453.

Korgaonkar, C., Hagen, J., Tompkins, V., Frazier, A.A., Allamargot, C., Quelle, F.W. & Quelle, D.E. (2005). Nucleophosmin (B23) targets ARF to nucleoli and inhibits its function. Molecular and Cellular Biology *25*, 1258-1271.

Kubbutat, M.H., Jones, S.N. & Vousden, K.H. (1997). Regulation of p53 stability by Mdm2. Nature *387*, 299-303.

Kulkens, T., Riggs, D.L., Heck, J.D., Planta, R.J. & Nomura, M. (1991). The yeast RNA polymerase I promoter: ribosomal DNA sequences involved in transcription initiation and complex formation in vitro. Nucleic Acids Res *19*, 5363-5370.

Kuo, B.A., Gonzalez, I.L., Gillespie, D.A. & Sylvester, J.E. (1996). Human ribosomal RNA variants from a single individual and their expression in different tissues. Nucleic Acids Res *24*, 4817-4824.

Lam, Y.W., Lamond, A.I., Mann, M. & Andersen, J.S. (2007). Analysis of nucleolar protein dynamics reveals the nuclear degradation of ribosomal proteins. Curr Biol *17*, 749-760.

Lanctot, C., Cheutin, T., Cremer, M., Cavalli, G. & Cremer, T. (2007). Dynamic genome architecture in the nuclear space: regulation of gene expression in three dimensions. Nat Rev Genet *8*, 104-115.

Learned, R.M., Cordes, S. & Tjian, R. (1985). Purification and characterization of a transcription factor that confers promoter specificity to human RNA polymerase I. Mol Cell Biol *5*, 1358-1369.

Leger, I., Guillaud, M., Krief, B. & Brugal, G. (1994). Interactive computer-assisted analysis of chromosome 1 colocalization with nucleoli. Cytometry *16*, 313-323.

Lessard, F., Morin, F., Ivanchuk, S., Langlois, F., Stefanovsky, V., Rutka, J. & Moss, T. (2010). The ARF tumor suppressor controls ribosome biogenesis by regulating the RNA polymerase I transcription factor TTF-I. Mol Cell *38*, 539-550.

Li, Y., Yan, Q. & Wolf, N.S. (1997). Long-term caloric restriction delays age-related decline in proliferation capacity of murine lens epithelial cells in vitro and in vivo. Invest Ophthalmol Vis Sci *38*, 100-107.

Lin, J., Jin, R., Zhang, B., Chen, H., Bai, Y.X., Yang, P.X., Han, S.W., Xie, Y.H., Huang, P.T., Huang, C. & Huang, J.J. (2008). Nucleolar localization of TERT is unrelated to telomerase function in human cells. J Cell Sci *121*, 2169-2176.

Lindstrom, M.S., Jin, A., Deisenroth, C., White Wolf, G. & Zhang, Y. (2007). Cancer-associated mutations in the MDM2 zinc finger domain disrupt ribosomal protein interaction and attenuate MDM2-induced p53 degradation. Mol Cell Biol *27*, 1056-1068.

Lindstrom, M.S. & Nister, M. (2010). Silencing of ribosomal protein S9 elicits a multitude of cellular responses inhibiting the growth of cancer cells subsequent to p53 activation. PLoS One *5*, e9578.

Machin, F., Paschos, K., Jarmuz, A., Torres-Rosell, J., Pade, C. & Aragon, L. (2004). Condensin regulates rDNA silencing by modulating nucleolar Sir2p. Curr Biol *14*, 125-130.

Machwe, A., Orren, D.K. & Bohr, V.A. (2000). Accelerated methylation of ribosomal RNA genes during the cellular senescence of Werner syndrome fibroblasts. FASEB J *14*, 1715-1724.

Maier, B., Gluba, W., Bernier, B., Turner, T., Mohammad, K., Guise, T., Sutherland, A., Thorner, M. & Scrable, H. (2004). Modulation of mammalian life span by the short isoform of p53. Genes Dev *18*, 306-319.

Manuelidis, L. (1984). Different central nervous system cell types display distinct and nonrandom arrangements of satellite DNA sequences. Proc Natl Acad Sci U S A *81*, 3123-3127.

Mayer, C., Bierhoff, H. & Grummt, I. (2005). The nucleolus as a stress sensor: JNK2 inactivates the transcription factor TIF-IA and down-regulates rRNA synthesis. Genes Dev *19*, 933-941.

Mayer, C., Schmitz, K.M., Li, J., Grummt, I. & Santoro, R. (2006). Intergenic transcripts regulate the epigenetic state of rRNA genes. Mol Cell *22*, 351-361.

Mayer, C., Zhao, J., Yuan, X. & Grummt, I. (2004). mTOR-dependent activation of the transcription factor TIF-IA links rRNA synthesis to nutrient availability. Genes Dev *18*, 423-434.

McCormick, A. & Campisi, J. (1991). Cellular aging and senescence. Current Opinion in Cell Biology *3*, 230-234.

McStay, B. & Grummt, I. (2008). The epigenetics of rRNA genes: from molecular to chromosome biology. Annu Rev Cell Dev Biol *24*, 131-157.

Mehta, I.S., Figgitt, M., Clements, C.S., Kill, I.R. & Bridger, J.M. (2007). Alterations to nuclear architecture and genome behavior in senescent cells. In Biogerontology: Mechanisms and Interventions, pp. 250-263.

Merkenschlager, M. (2010). Cohesin: a global player in chromosome biology with local ties to gene regulation. Curr Opin Genet Dev *20*, 555-561.

Merker, R.J. & Klein, H.L. (2002). hpr1Delta affects ribosomal DNA recombination and cell life span in Saccharomyces cerevisiae. Mol Cell Biol *22*, 421-429.

Merz, K., Hondele, M., Goetze, H., Gmelch, K., Stoeckl, U. & Griesenbeck, J. (2008). Actively transcribed rRNA genes in S. cerevisiae are organized in a specialized chromatin associated with the high-mobility group protein Hmo1 and are largely devoid of histone molecules. Genes Dev *22*, 1190-1204.

Misteli, T. (2004). Spatial positioning; a new dimension in genome function. Cell *119*, 153-156.

Momand, J., Zambetti, G.P., Olson, D.C., George, D. & Levine, A.J. (1992). The mdm-2 oncogene product forms a complex with the p53 protein and inhibits p53-mediated transactivation. Cell *69*, 1237-1245.

Mongelard, F. & Bouvet, P. (2007). Nucleolin: a multiFACeTed protein. Trends Cell Biol *17*, 80-86.

Montanaro, L., Mazzini, G., Barbieri, S., Vici, M., Nardi-Pantoli, A., Govoni, M., Donati, G., Trere, D. & Derenzini, M. (2007). Different effects of ribosome biogenesis inhibition on cell proliferation in retinoblastoma protein- and p53-deficient and proficient human osteosarcoma cell lines. Cell Prolif *40*, 532-549.

Moore, H.M., Bai, B., Boisvert, F.M., Latonen, L., Rantanen, V., Simpson, J.C., Pepperkok, R., Lamond, A.I. & Laiho, M. (2011). Quantitative proteomics and dynamic imaging of the nucleolus reveals distinct responses to UV and ionizing radiation. Mol Cell Proteomics.

Moss, T., Langlois, F., Gagnon-Kugler, T. & Stefanovsky, V. (2007). A housekeeper with power of attorney: the rRNA genes in ribosome biogenesis. Cell Mol Life Sci *64*, 29-49.

Moss, T. & Stefanovsky, V.Y. (2002). At the center of eukaryotic life. Cell *109*, 545-548.

Murayama, A., Ohmori, K., Fujimura, A., Minami, H., Yasuzawa-Tanaka, K., Kuroda, T., Oie, S., Daitoku, H., Okuwaki, M., Nagata, K., Fukamizu, A., Kimura, K., Shimizu, T. & Yanagisawa, J. (2008). Epigenetic control of rDNA loci in response to intracellular energy status. Cell *133*, 627-639.

Musters, W., Knol, J., Maas, P., Dekker, A.F., van Heerikhuizen, H. & Planta, R.J. (1989). Linker scanning of the yeast RNA polymerase I promoter. Nucleic Acids Res *17*, 9661-9678.

Nalabothula, N., Indig, F.E. & Carrier, F. (2010). The Nucleolus Takes Control of Protein Trafficking Under Cellular Stress. Mol Cell Pharmacol *2*, 203-212.

Narayanan, A., Lukowiak, A., Jady, B.E., Dragon, F., Kiss, T., Terns, R.M. & Terns, M.P. (1999). Nucleolar localization signals of box H/ACA small nucleolar RNAs. EMBO J *18*, 5120-5130.

Narita, M., Nunez, S., Heard, E., Lin, A.W., Hearn, S.A., Spector, D.L., Hannon, G.J. & Lowe, S.W. (2003). Rb-mediated heterochromatin formation and silencing of E2F target genes during cellular senescence. Cell *113*, 703-716.

Nemeth, A. & Langst, G. (2011). Genome organization in and around the nucleolus. Trends Genet *27*, 149-156.

O'Sullivan, A.C., Sullivan, G.J. & McStay, B. (2002). UBF binding in vivo is not restricted to regulatory sequences within the vertebrate ribosomal DNA repeat. Mol Cell Biol *22*, 657-668.

O'Sullivan, J.M., Sontam, D.M., Grierson, R. & Jones, B. (2009). Repeated elements coordinate the spatial organization of the yeast genome. Yeast *26*, 125-138.

Ofir-Rosenfeld, Y., Boggs, K., Michael, D., Kastan, M.B. & Oren, M. (2008). Mdm2 regulates p53 mRNA translation through inhibitory interactions with ribosomal protein L26. Mol Cell *32*, 180-189.

Oliner, J.D., Pietenpol, J.A., Thiagalingam, S., Gyuris, J., Kinzler, K.W. & Vogelstein, B. (1993). Oncoprotein MDM2 conceals the activation domain of tumour suppressor p53. Nature *362*, 857-860.

Olson, M.O., Dundr, M. & Szebeni, A. (2000). The nucleolus: an old factory with unexpected capabilities. Trends Cell Biol *10*, 189-196.

Olson, M.O., Hingorani, K. & Szebeni, A. (2002). Conventional and nonconventional roles of the nucleolus. Int Rev Cytol *219*, 199-266.

Olson, M.O.J. (2004). Sensing Cellular Stress: Another New Function for the Nucleolus? Sci STKE *2004*, pe10-.

Olson, M.O.J. & Dundr, M. (2005). The moving parts of the nucleolus. Histochem Cell Biol *123*, 203-216.

Palmero, I., Pantoja, C. & Serrano, M. (1998). p19ARF links the tumour suppressor p53 to Ras. Nature *395*, 125-126.

Pan, W., Issaq, S. & Zhang, Y. (2011). The in vivo role of the RP-Mdm2-p53 pathway in signaling oncogenic stress induced by pRb inactivation and Ras overexpression. PLoS One *6*, e21625.

Pardee, A.B. (1989). G1 events and regulation of cell proliferation. Science *246*, 603-608.

Park, P.C. & De Boni, U. (1992). Spatial rearrangement and enhanced clustering of kinetochores in interphase nuclei of dorsal root ganglion neurons in vitro: association with nucleolar fusion. Exp Cell Res *203*, 222-229.

Park, P.U., Defossez, P.A. & Guarente, L. (1999). Effects of mutations in DNA repair genes on formation of ribosomal DNA circles and life span in Saccharomyces cerevisiae. Mol Cell Biol *19*, 3848-3856.

Partridge, L. & Mangel, M. (1999). Messages from mortality: the evolution of death rates in the old. Trends Ecol Evol *14*, 438-442.

Pasero, P., Bensimon, A. & Schwob, E. (2002). Single-molecule analysis reveals clustering and epigenetic regulation of replication origins at the yeast rDNA locus. Genes Dev *16*, 2479-2484.

Pawelec, G., Wagner, W., Adibzadeh, M. & Engel, A. (1999). T cell immunosenescence in vitro and in vivo. Exp Gerontol *34*, 419-429.

Peng, J.C. & Karpen, G.H. (2007). H3K9 methylation and RNA interference regulate nucleolar organization and repeated DNA stability. Nat Cell Biol *9*, 25-35.

Pestov, D.G., Strezoska, Z. & Lau, L.F. (2001). Evidence of p53-dependent cross-talk between ribosome biogenesis and the cell cycle: effects of nucleolar protein Bop1 on G(1)/S transition. Mol Cell Biol *21*, 4246-4255.

Poortinga, G., Hannan, K.M., Snelling, H., Walkley, C.R., Jenkins, A., Sharkey, K., Wall, M., Brandenburger, Y., Palatsides, M., Pearson, R.B., McArthur, G.A. & Hannan, R.D. (2004). MAD1 and c-MYC regulate UBF and rDNA transcription during granulocyte differentiation. EMBO J *23*, 3325-3335.

Poortinga, G., Wall, M., Sanij, E., Siwicki, K., Ellul, J., Brown, D., Holloway, T.P., Hannan, R.D. & McArthur, G.A. (2011). c-MYC coordinately regulates ribosomal gene chromatin remodeling and Pol I availability during granulocyte differentiation. Nucleic Acids Res *39*, 3267-3281.

Prieto, J.L. & McStay, B. (2007). Recruitment of factors linking transcription and processing of pre-rRNA to NOR chromatin is UBF-dependent and occurs independent of transcription in human cells. Genes Dev *21*, 2041-2054.

Prieto, J.L. & McStay, B. (2008). Pseudo-NORs: a novel model for studying nucleoli. Biochim Biophys Acta *1783*, 2116-2123.

Pyronnet, S., Dostie, J. & Sonenberg, N. (2001). Suppression of cap-dependent translation in mitosis. Genes Dev *15*, 2083-2093.

Richard, G.F., Kerrest, A. & Dujon, B. (2008). Comparative genomics and molecular dynamics of DNA repeats in eukaryotes. Microbiol Mol Biol Rev *72*, 686-727.

Riley, K.J. & Maher, L.J., 3rd (2007). p53 RNA interactions: new clues in an old mystery. RNA *13*, 1825-1833.

Roussel, P., Andre, C., Comai, L. & Hernandez-Verdun, D. (1996). The rDNA transcription machinery is assembled during mitosis in active NORs and absent in inactive NORs. J Cell Biol *133*, 235-246.

Rubbi, C.P. & Milner, J. (2003). Disruption of the nucleolus mediates stabilization of p53 in response to DNA damage and other stresses. EMBO J *22*, 6068-6077.

Ruggero, D. & Pandolfi, P.P. (2003). Does the ribosome translate cancer? Nat Rev Cancer *3*, 179-192.

Sahin, E. & Depinho, R.A. (2010). Linking functional decline of telomeres, mitochondria and stem cells during ageing. Nature *464*, 520-528.

Sakai, K., Ohta, T., Minoshima, S., Kudoh, J., Wang, Y., de Jong, P.J. & Shimizu, N. (1995). Human ribosomal RNA gene cluster: identification of the proximal end containing a novel tandem repeat sequence. Genomics *26*, 521-526.

Salminen, A. & Kaarniranta, K. (2009). SIRT1 regulates the ribosomal DNA locus: Epigenetic candles twinkle longevity in the Christmas tree. Biochemical and Biophysical Research Communications *378*, 6-9.

Sander, E.E. & Grummt, I. (1997). Oligomerization of the transcription termination factor TTF-I: implications for the structural organization of ribosomal transcription units. Nucleic Acids Res *25*, 1142-1147.

Sanij, E. & Hannan, R.D. (2009). The role of UBF in regulating the structure and dynamics of transcriptionally active rDNA chromatin. Epigenetics *4*, 374-382.

Sanij, E., Poortinga, G., Sharkey, K., Hung, S., Holloway, T.P., Quin, J., Robb, E., Wong, L.H., Thomas, W.G., Stefanovsky, V., Moss, T., Rothblum, L., Hannan, K.M., McArthur, G.A., Pearson, R.B. & Hannan, R.D. (2008). UBF levels determine the number of active ribosomal RNA genes in mammals. J Cell Biol *183*, 1259-1274.

Santangelo, G.M., Tornow, J., McLaughlin, C.S. & Moldave, K. (1988). Properties of promoters cloned randomly from the Saccharomyces cerevisiae genome. Mol Cell Biol *8*, 4217-4224.

Santoro, R., Li, J. & Grummt, I. (2002). The nucleolar remodeling complex NoRC mediates heterochromatin formation and silencing of ribosomal gene transcription. Nat Genet *32*, 393-396.

Sharpless, N.E., Bardeesy, N., Lee, K.H., Carrasco, D., Castrillon, D.H., Aguirre, A.J., Wu, E.A., Horner, J.W. & DePinho, R.A. (2001). Loss of p16Ink4a with retention of p19Arf predisposes mice to tumorigenesis. Nature *413*, 86-91.

Shcherbik, N. & Pestov, D.G. (2010). Ubiquitin and ubiquitin-like proteins in the nucleolus: multitasking tools for a ribosome factory. Genes Cancer *1*, 681-689.

Sherr, C.J. & Roberts, J.M. (1999). CDK inhibitors: positive and negative regulators of G1-phase progression. Genes Dev *13*, 1501-1512.

Sherwood, S.W., Rush, D., Ellsworth, J.L. & Schimke, R.T. (1988). Defining cellular senescence in IMR-90 cells: a flow cytometric analysis. Proc Natl Acad Sci U S A *85*, 9086-9090.

Shore, D. (1998). Cellular senescence: lessons from yeast for human aging? Curr Biol *8*, R192-195.

Shou, W., Azzam, R., Chen, S.L., Huddleston, M.J., Baskerville, C., Charbonneau, H., Annan, R.S., Carr, S.A. & Deshaies, R.J. (2002). Cdc5 influences phosphorylation of Net1 and disassembly of the RENT complex. BMC Mol Biol *3*, 3.

Shou, W., Seol, J.H., Shevchenko, A., Baskerville, C., Moazed, D., Chen, Z.W., Jang, J., Charbonneau, H. & Deshaies, R.J. (1999). Exit from mitosis is triggered by Tem1-dependent release of the protein phosphatase Cdc14 from nucleolar RENT complex. Cell *97*, 233-244.

Siddiqi, I.N., Dodd, J.A., Vu, L., Eliason, K., Oakes, M.L., Keener, J., Moore, R., Young, M.K. & Nomura, M. (2001). Transcription of chromosomal rRNA genes by both RNA polymerase I and II in yeast uaf30 mutants lacking the 30 kDa subunit of transcription factor UAF. EMBO J *20*, 4512-4521.

Sinclair, D.A. & Guarente, L. (1997). Extrachromosomal rDNA circles--a cause of aging in yeast. Cell *91*, 1033-1042.

Sirri, V., Hernandez-Verdun, D. & Roussel, P. (2002). Cyclin-dependent kinases govern formation and maintenance of the nucleolus. Journal of Cell Biology *156*, 969-981.

Sirri, V., Urcuqui-Inchima, S., Roussel, P. & Hernandez-Verdun, D. (2008). Nucleolus: the fascinating nuclear body. Histochem Cell Biol *129*, 13-31.

Stancheva, I., Lucchini, R., Koller, T. & Sogo, J.M. (1997). Chromatin structure and methylation of rat rRNA genes studied by formaldehyde fixation and psoralen cross-linking. Nucleic Acids Res *25*, 1727-1735.

Stefanovsky, V., Langlois, F., Gagnon-Kugler, T., Rothblum, L.I. & Moss, T. (2006). Growth factor signaling regulates elongation of RNA polymerase I transcription in mammals via UBF phosphorylation and r-chromatin remodeling. Mol Cell *21*, 629-639.

Stefanovsky, V.Y., Pelletier, G., Hannan, R., Gagnon-Kugler, T., Rothblum, L.I. & Moss, T. (2001). An immediate response of ribosomal transcription to growth factor stimulation in mammals is mediated by ERK phosphorylation of UBF. Mol Cell *8*, 1063-1073.

Steffan, J.S., Keys, D.A., Dodd, J.A. & Nomura, M. (1996). The role of TBP in rDNA transcription by RNA polymerase I in Saccharomyces cerevisiae: TBP is required for upstream activation factor-dependent recruitment of core factor. Genes Dev *10*, 2551-2563.

Stewart, S.A., Ben-Porath, I., Carey, V.J., O'Connor, B.F., Hahn, W.C. & Weinberg, R.A. (2003). Erosion of the telomeric single-strand overhang at replicative senescence. Nat Genet *33*, 492-496.

Sugimoto, M., Kuo, M.L., Roussel, M.F. & Sherr, C.J. (2003). Nucleolar Arf tumor suppressor inhibits ribosomal RNA processing. Mol Cell *11*, 415-424.

Takemura, M., Ohoka, F., Perpelescu, M., Ogawa, M., Matsushita, H., Takaba, T., Akiyama, T., Umekawa, H., Furuichi, Y., Cook, P.R. & Yoshida, S. (2002). Phosphorylation-dependent migration of retinoblastoma protein into the nucleolus triggered by binding to nucleophosmin/B23. Exp Cell Res *276*, 233-241.

Tissenbaum, H.A. & Guarente, L. (2001). Increased dosage of a sir-2 gene extends lifespan in Caenorhabditis elegans. Nature *410*, 227-230.

Tomlinson, R.L., Ziegler, T.D., Supakorndej, T., Terns, R.M. & Terns, M.P. (2006). Cell cycle-regulated trafficking of human telomerase to telomeres. Molecular Biology of the Cell *17*, 955-965.

Trumtel, S., Leger-Silvestre, I., Gleizes, P.E., Teulieres, F. & Gas, N. (2000). Assembly and functional organization of the nucleolus: ultrastructural analysis of Saccharomyces cerevisiae mutants. Mol Biol Cell *11*, 2175-2189.

Tsai, R.Y.L. (2009). Nucleolar modulation of TRF1 A dynamic way to regulate telomere and cell cycle by nucleostemin and GNL3L. Cell Cycle *8*, 2912-2916.

Tsang, C.K., Bertram, P.G., Ai, W., Drenan, R. & Zheng, X.F. (2003). Chromatin-mediated regulation of nucleolar structure and RNA Pol I localization by TOR. EMBO J *22*, 6045-6056.

Tschochner, H. & Hurt, E. (2003). Pre-ribosomes on the road from the nucleolus to the cytoplasm. Trends Cell Biol *13*, 255-263.

Tyner, S.D., Venkatachalam, S., Choi, J., Jones, S., Ghebranious, N., Igelmann, H., Lu, X., Soron, G., Cooper, B., Brayton, C., Hee Park, S., Thompson, T., Karsenty, G., Bradley, A. & Donehower, L.A. (2002). p53 mutant mice that display early ageing-associated phenotypes. Nature *415*, 45-53.

van Koningsbruggen, S., Gierlinski, M., Schofield, P., Martin, D., Barton, G.J., Ariyurek, Y., den Dunnen, J.T. & Lamond, A.I. (2010). High-resolution whole-genome sequencing reveals that specific chromatin domains from most human chromosomes associate with nucleoli. Mol Biol Cell *21*, 3735-3748.

Vigneron, A. & Vousden, K.H. (2010). p53, ROS and senescence in the control of aging. Aging (Albany NY) *2*, 471-474.

Vintermist, A., Bohm, S., Sadeghifar, F., Louvet, E., Mansen, A., Percipalle, P. & Ostlund Farrants, A.K. (2011). The chromatin remodelling complex B-WICH changes the chromatin structure and recruits histone acetyl-transferases to active rRNA genes. PLoS One *6*, e19184.

Visintin, R., Craig, K., Hwang, E.S., Prinz, S., Tyers, M. & Amon, A. (1998). The phosphatase Cdc14 triggers mitotic exit by reversal of Cdk-dependent phosphorylation. Mol Cell *2*, 709-718.

Wang, B.D., Butylin, P. & Strunnikov, A. (2006). Condensin function in mitotic nucleolar segregation is regulated by rDNA transcription. Cell Cycle *5*, 2260-2267.

Warner, J.R. & McIntosh, K.B. (2009). How Common Are Extraribosomal Functions of Ribosomal Proteins? Molecular Cell *34*, 3-11.

Weber, J.D., Taylor, L.J., Roussel, M.F., Sherr, C.J. & Bar-Sagi, D. (1999). Nucleolar Arf sequesters Mdm2 and activates p53. Nat Cell Biol *1*, 20-26.

Wong, J.M.Y., Kusdra, L. & Collins, K. (2002). Subnuclear shuttling of human telomerase induced by transformation and DNA damage. Nat Cell Biol *4*, 731-U735.

Wright, J.E., Mais, C., Prieto, J.L. & McStay, B. (2006). A role for upstream binding factor in organizing ribosomal gene chromatin. Biochem Soc Symp, 77-84.

Yang, Y., Chen, Y., Zhang, C., Huang, H. & Weissman, S.M. (2002). Nucleolar localization of hTERT protein is associated with telomerase function. Exp Cell Res *277*, 201-209.

Yoshida, S. & Toh-e, A. (2002). Budding yeast Cdc5 phosphorylates Net1 and assists Cdc14 release from the nucleolus. Biochem Biophys Res Commun *294*, 687-691.

Yu, C.E., Oshima, J., Fu, Y.H., Wijsman, E.M., Hisama, F., Alisch, R., Matthews, S., Nakura, J., Miki, T., Ouais, S., Martin, G.M., Mulligan, J. & Schellenberg, G.D. (1996). Positional cloning of the Werner's syndrome gene. Science *272*, 258-262.

Yuan, X., Zhao, J., Zentgraf, H., Hoffmann-Rohrer, U. & Grummt, I. (2002). Multiple interactions between RNA polymerase I, TIF-IA and TAF(I) subunits regulate preinitiation complex assembly at the ribosomal gene promoter. EMBO Rep *3*, 1082-1087.

Yuan, X., Zhou, Y., Casanova, E., Chai, M., Kiss, E., Grone, H.J., Schutz, G. & Grummt, I. (2005). Genetic inactivation of the transcription factor TIF-IA leads to nucleolar disruption, cell cycle arrest, and p53-mediated apoptosis. Mol Cell *19*, 77-87.

Zachariae, W., Schwab, M., Nasmyth, K. & Seufert, W. (1998). Control of cyclin ubiquitination by CDK-regulated binding of Hct1 to the anaphase promoting complex. Science *282*, 1721-1724.

Zhang, L.F., Huynh, K.D. & Lee, J.T. (2007). Perinucleolar targeting of the inactive X during S phase: evidence for a role in the maintenance of silencing. Cell *129*, 693-706.

Zhang, S., Hemmerich, P. & Grosse, F. (2004). Nucleolar localization of the human telomeric repeat binding factor 2 (TRF2). J Cell Sci *117*, 3935-3945.

Zhang, Y. & Lu, H. (2009). Signaling to p53: ribosomal proteins find their way. Cancer Cell *16*, 369-377.

Zhang, Y., Yu, Z., Fu, X. & Liang, C. (2002). Noc3p, a bHLH protein, plays an integral role in the initiation of DNA replication in budding yeast. Cell *109*, 849-860.

Zhang, Y.P., Wolf, G.W., Bhat, K., Jin, A., Allio, T., Burkhart, W.A. & Xiong, Y. (2003). Ribosomal protein L11 negatively regulates oncoprotein MDM2 and mediates a p53-dependent ribosomal-stress checkpoint pathway. Molecular and Cellular Biology *23*, 8902-8912.

Zhu, Q., Meng, L., Hsu, J.K., Lin, T., Teishima, J. & Tsai, R.Y. (2009). GNL3L stabilizes the TRF1 complex and promotes mitotic transition. J Cell Biol *185*, 827-839.

Zindy, F., Eischen, C.M., Randle, D.H., Kamijo, T., Cleveland, J.L., Sherr, C.J. & Roussel, M.F. (1998). Myc signaling via the ARF tumor suppressor regulates p53-dependent apoptosis and immortalization. Genes Dev *12*, 2424-2433.

Zomerdijk, J.C., Beckmann, H., Comai, L. & Tjian, R. (1994). Assembly of transcriptionally active RNA polymerase I initiation factor SL1 from recombinant subunits. Science *266*, 2015-2018.

The Quest for Immortality in Triatomines: A Meta-Analysis of the Senescence Process in Hemimetabolous Hematophagous Insects

Paula Medone[1], Jorge Rabinovich[1], Eliana Nieves[1], Soledad Ceccarelli[1], Delmi Canale[2], Raúl L. Stariolo[2] and Frédéric Menu[3]
[1]*Parasitological and Vector Studies Center (CEPAVE), National University of La Plata, La Plata,*
[2]*Vectors Reference Center, National Coordination of Vectors Control, Direction of Transmitted Vectors Diseases, National Ministry of Health, Córdoba,*
[3]*Laboratory of Biometry and Biology Evolutionary, CNRS, UMR 5558, Lyon University F-69000 Lyon, Claude Bernard University, Lyon 1 F-69622, Villeurbanne,*
[1,2]*Argentina*
[3]*France*

1. Introduction

There are different views on senescence as a process. In its most general conception it represents the change in the biology of an organism as it ages. However this process may be viewed either at the physiological or at the demographic level. In the former sense senescence deals with changes affecting cells and tissues of the organism and their function and its effect on the organism as a whole (somatic senescence). In the demographic sense (actuarial senescence) the emphasis is in the population's survival decrease as a function of age (Promislow 1991, Tatar et al., 1993); this very general definition does not necessary imply somatic senescence (a physiological deterioration) because the organism may suffer an increased age-specific mortality rate because of an increased reproductive effort (Roff, 2002); the decrease in the reproductive performance with age may be termed reproductive senescence. Williams (1957), based on evolutionary arguments, claims that natural selection will frequently maximize vigor in youth at the expense of vigor later on and thereby he identifies senescence as a declination in vigor during adult life, using the term vigor as associated with a reproductive probability distribution. Here we are interested in this second approach to senescence, and we adhere to the definition of Rose (1991, cited in Roff 2002): "a persistent decline in the age-specific fitness components of an organism due to internal physiological deterioration".

According to Charlesworth & Partridge (1997) there are two main theories trying to explain the senescence process: (1) natural selection is less effective at reducing the frequency of later-acting mutations in populations, and so ageing is expected to evolve, and this is known as the "mutation accumulation" theory of ageing; (2) mutations that increase fitness at younger ages (perhaps because they increase fertility) but at the expense of decreasing fitness at later ages (perhaps because they increase the death rate) can be incorporated into a population because natural selection will act more strongly on the earlier, beneficial effect. This is the reasoning behind the "antagonistic pleiotropy" or "trade-off" theory of ageing (Williams, 1957).

Abrams & Ludwig (1995), based on an extension of the "disposable soma" model (Kirkwood & Holliday, 1979), provide an explicit realization of the trade-off idea which postulates a conflict between the allocation of resources to reproduction and to the repair of somatic damage. A reduction in damage repair at a given age is assumed to cause an elevated death rate at all subsequent ages. Given a functional relationship between repair allocation and reproductive rate at a given age, the age-specific pattern of allocation to repair versus reproduction that maximizes life-time reproductive success can be determined, yielding a prediction of the age-specific pattern of mortality for the optimal life history (Charlesworth & Partridge, 1997).

Longevity and senescence patterns in mammals and birds are very variable according to the life-history of these organisms (Gaillard et al., 2004), and variation within and between phyla can be expected. Despite phylogenetic similarity is reciprocal to taxonomic level of relatedness (Cheverud et al., 1985) it is also possible that species phylogenetically related show different senescence and/or longevity patterns. So a comparative approach focused on the frequency of senescence in closely related species may contribute to our knowledge of the senescence process. Comparative studies in insects are scarce as compared to mammals and birds; within the insects, demographic analyses of senescence have been carried out mainly for the Diptera (Styer et al., 2007; Curtsinger et al., 1992; Carey et al., 1992, 2005; Fukui et al., 1993), and the Coleoptera (Tatar et al., 1993), and very few studies have considered hemimetabolous species (Dingle, 1966, Chaves et al., 2004a,b, Rabinovich et al., 2010).

In this chapter we investigate the frequency of senescence in a closely related species group of insects: the Triatominae (Hemiptera: Reduviidae). We compiled from the bibliography and resorted to personal data to have phylogeny and life history traits of 27 species reared under laboratory controlled and comparable conditions, and analyzed mortality and fecundity through several death and reproductive parameters. In particular we investigated: (a) species patterns of mortality with respect to age (from cohort studies that followed all individuals from the egg stage until the death of the last individual), (b) the relationship between mortality and different life-history traits (size, reproductive allocation), and (c) the relationship between mortality and environmental factors. Being the 27 selected triatomine species close relatives (they belonged to only five different genera), for our comparative study we included in the analysis a correction for the possible effect of the degree of phylogenetic relatedness.

The advantage of working on laboratory data is that we can estimate intrinsic mortality and fecundity rates without confounding effects resulting from extrinsic factors acting on mortality and fecundity (predation, accidental deaths, starvation, etc.). Even if work with triatomines has the advantage that under natural conditions all life stages occur in a single type of environment and have similar biological requirements, there is always the disadvantage that laboratory data does not reflect natural condition: insects are fed *ad libitum*, and predators, parasites and pathogens are kept out, so that it does not constitute the best of conditions to detect trade-offs between reproductive effort and mortality. However, we think (in agreement with Mueller et al., 2005) that the identification of which aspects of the environment matter in the evolution of trade-offs can only be obtained by performing experiments in which these environmental variables are carefully manipulated; additionally, if even under such stable and near optimal conditions we are able to detect trade-offs, then our conclusions become much more robust.

Although life history traits such as fecundity, juvenile and adult survival, fasting capacity, developmental time, mortality patterns, and life span have been estimated under controlled conditions in the laboratory for a variety of triatomine species (about 500 scientific articles have been written on these aspects since 1910), very few studies have considered recent evolutionary ecology concepts (although see Menu et al., 2010) to shed some light on the trade-off aspects of life history traits. Understanding the mortality pattern in this group of insects is important both for academic and human health reasons. In the former sense we will provide elements to contribute to the theory of senescence and we discuss our results within of this theoretical background; in particular we will analyze the senescence pattern looking into the relationship between reproductive effort and mortality. In the latter sense, our analyses will provide information about a group of insects that are the vectors of Chagas disease, and represent a health threat estimated in 28 million people, living mostly in Latin America (see WHO, 2007).

2. Materials and methods

2.1 Demographic parameters

The basic information for the demographic parameters was obtained from a database compiled by one of us (JER). The original dataset comprised information on 534 case studies of triatomines representing 71 species; however, many of those cases had only partial information on demographic parameters and were not adequate for the present study. From the ones with complete information (55 cases) we selected 29 species that had relatively homogenous rearing conditions in the laboratory, to render them adequate for meta-analysis. The 29 species were later reduced to 27 species because two of them were not in the phylogenetic tree used for the application of the Phylogenetic Independent Contrast method (see section 2.4). The final list of species is shown in Table 1.

For each species we calculated basic life table parameters following Carey (2001): (i) age-specific survival or fraction alive at age x ($l_x = N_x / N_0$) (where N_x= number of individuals alive at age x), (ii) age-specific period survival or fraction alive at age x surviving to $x + 1$ ($p_x = l_{x+1} / l_x$), (iii) age-specific period mortality (probability of dying over the one-week interval used ($q_x = 1 - p_x$), and (iv) the force of mortality or instantaneous mortality rate ($\mu_x = -\ln(p_x)$). For more details on the definitions and formulae of the life table parameters see Rabinovich & Nieves (2011). The original values obtained from the laboratory (N_x, or number of individuals alive at age x, and M_x or total number of eggs laid by all females of the cohort aged x) were processed with a special program (called TriTV) developed by one of us (JER) that calculates the life table statistics and population growth rate parameters. As it is usual with laboratory cohorts, data of the last time-units of the cohorts are based on a very small number of individuals; thus the mortality rate estimated over the last living individual were excluded from the analysis, due to the unreliability of the mortality rate estimates (Carey et al., 1992). The TriTV software was programmed in Delphi language, and is available under request to the second author (the interested user should have an adequate command of the Spanish language).

We considered of interest the division of the female's adult life in two, three or four periods of equal length, to analyze possible lag effects between reproductive effort and mortality. For that purpose, the reproductive effort (m_x= number of female eggs per female per unit time) and the instantaneous mortality rates (μ_x) were averaged for each of the periods in which the female's adult life was divided. We also accumulated mortality from the egg to the last adult female ($\sum\mu_x$) and the female's *per capita* fecundity ($\sum m_x$) to look for a relationship of one with respect to the other along the female´s adult life (after scaling both between 0 and 1).

2.2 Mortality pattern models

For several reasons (see Rabinovich et al., 2010) we preferred the use of mortality analysis over survival analysis; in particular because we agree with Carey (2001) in that, despite mortality and survival are intimately related, death can be considered as an event whereas survival is a "non-event", that is, the absence of the mortality event. For the analysis of the age-specific mortality pattern of the 27 triatomine species selected for this study the following two mortality models were used with the formulation proposed by Carey (2001): the Gompertz model

$$\mu_x = a\, e^{\,bx}$$

and the Logistic model

$$\mu_x = (n\, x^{n-1})/(g^n + x^n)$$

We also used a third model based on reliability theory as proposed by Gavrilov & Gavrilova (2001), but simplified to two parameters, and which, for simplicity, will be called hereafter "Gavrilovs". Its formulation is given as:

$$\mu_x = n\, k^n\, x^{n-1}$$

In all models x is the age (in our case in weekly time-units), and for the interpretation of the model's parameters, one of them usually represents the "base" mortality rate, and the other the shape of the function (more directly related to the rate of increase of mortality with age, also called the ageing parameter in the case of the Gompertz model). The reason we selected these three models among about a dozen available models for the analysis of mortality patterns, is that Rabinovich et al. (2010), fitting seven models to the instantaneous mortality rate to another triatomine species (*R. neglectus*), found that these three models offered the best fit to the data.

2.3 Model fitting to the data

We used several tools and procedures to fit the three models to the age-specific instantaneous mortality data (μ_x): (i) the R language (R Development Core Team, 2007), (ii) the Kolmogorov-Zurbenko Adaptive smoothing package (kza) in R language (used with parameter q= 2) (Zurbenko et al., 1996), (iii) the Statistica software (StatSoft, 2009),

(iv) the "Online Curve Fitting and Surface Fitting Web Site" (accessible at http://ZunZun.com/) (Christopoulos & Lew, 2000), and (v) the Solver utility of the Excel 2007 program. The reason for the variety of procedures resorted to for fitting the data, is that -even after smoothing- the age-specific instantaneous mortality data was quite irregular, and frequently a given software was not able to succeed in fitting the data to the models while another did. This would happen even if the parameter starting values for fitting the data would have been successfully estimated with the Solver utility of the Excel 2007 program. All software products used, if successful in fitting the data, provided the standard deviation of each parameter and the probability for deciding on the significance of the parameters' estimates. When more than one model fitted the data of the same species, the sum of squares ($SSQ = \Sigma(\mu_{obs} - \mu_{mod})^2$) was calculated and the model with the smallest SSQ was selected. We did not use the Akaike model selection criterion (Akaike, 1974) because the three models tested had the same number of parameters, so the SSQ goodness of fit values would not be affected by any penalty due to the number of degrees of freedom of each model.

2.4 Phylogenetic Independent Contrasts

Comparative studies, either among life-history traits or between those traits and environmental variables, frequently imply resorting to statistical methods such as regressions, correlations and contingency tables, which assume that the data is drawn from a common and independent distribution. However, the comparison among species that are related among each other with different degrees of phylogenetic relatedness does not necessary comply with this assumption. Methods are available to correct for this violation of the assumption of independence. For such purpose we used the Phylogenetic Independent Contrast method (Paradis, 2006) that corrects this violation when derived from phylogeny. The application of this method requires the topology of the phylogenetic tree, and an estimate of the degree of relatedness, generally expressed by the length of the branches of the phylogenetic tree. Calculations were carried out in language R using package "ape".

The phylogenetic tree used for this purpose was the one provided by Silva de Paula et al. (2005), who analyzed the Reduviidae phylogeny by aligning groups of sequences using Clustal-X under gap opening/gap extension penalties, and treating the gaps as missing. For Cladistic analysis Silva de Paula et al. (2005) used the programs PAUP and MacClade to derive trees based both on maximum parsimony (MP) and on maximum likelihood (ML); parsimony bootstrap values were conducted with PAUP employing heuristic search with 100 bootstrap replicates. Decay index for the strict consensus trees (Bremer support) were retained using the decay commands performed by the MacClade software with the heuristic search command activated, and executed with PAUP. This phylogenetic tree covered 57 species of Triatominae, and included 27 species of the 29 species we had selected from the demographic dataset (see section 2.1); thus *Triatoma breyeri* and *Panstrongylus lignarius* were deleted from the analysis, which remained composed by 27 species (Table 1).

3. Results

The following are the main results obtained from our analyses.

3.1 Demographic parameters

Table 1, provides a list of the species analyzed and some of their main environmental information (average annual temperature and precipitation) and provides information about rearing conditions of each species analyzed. In the last column the initial number of eggs (both sexes) with which each cohort was initiated is given; as the sex ratio at the egg stage cannot be established, it was assumed to be 50% for each sex (see Rabinovich et al., 2010 for a justification of this procedure). Initial cohort size was between 35 and 500 eggs (mean: 167 eggs/cohort). Because the senescence analysis was based on female mortality, the initial number for each species was assumed to be one half of original total number of eggs. Feeding frequency of all cohorts was once a week.

Species	Country of origin	Area of distribution (km^2)	Average annual temp. (°C)	Average precipitation (mm/year)	Laboratory rearing conditions			
					Temp. (°C)	R.H (%)	Feeding source	Initial Number of eggs
Dipetalogaster maximus	Mexico	36,074	22.2	198.42	28	70	pigeon	35
Eratyrus mucronatus	Venezuela	6,257,724	25.69	2340.64	26	60	chicken	200*
Panstrongylus geniculatus	Venezuela	12,040,606	24.45	1843.01	26	60	chicken	100
Panstrongylus herreri	Peru	354,387	21.46	1764.42	28	70	pigeon	105
Panstrongylus megistus	Brazil	3,739,358	22.55	1281.21	28	70	pigeon	96*
Rhodnius nasutus	Brazil	403,797	25.49	989.55	26	60	chicken	500*
Rhodnius neglectus [a]	Brazil	2,381,373	23.3	1318.39	26	60	chicken	500*
Rhodnius neivai	Venezuela	155,568	24.73	1347.3	26	60	chicken	100
Rhodnius prolixus	Colombia	5,281,236	25.46	2286.3	28	70	pigeon	102
Rhodnius robustus	Venezuela	3,133,393	25.68	2132.36	26	60	chicken	500*
Triatoma delpontei	Argentina	705,366	20.58	733.59	28	70	pigeon	108*
Triatoma dimidiata	Ecuador	2,176,134	24.26	1807.86	28	70	pigeon	45
Triatoma eratyrusiformis	Argentina	540,347	16.72	407.53	28	70	pigeon	112*
Triatoma garciabesi	Argentina	930,966	19.11	668.67	28	70	pigeon	132*
Triatoma guasayana	Argentina	1,412,079	19.38	747.88	28	70	pigeon	108*
Triatoma infestans [b]	Chile	5,198,083	19.26	957.3	26	60	chicken	500*
Triatoma maculata [c]	Venezuela	152,506	24.26	1421.51	26	60	chicken	50*
Triatoma matogrossensis	Argentina	205,258	22.3	815.29	28	70	pigeon	108
Triatoma mazzotti	Mexico	267,872	20.82	876.65	28	70	pigeon	56
Triatoma pallidipennis	Mexico	1,631,377	16.29	621.44	28	70	pigeon	138*
Triatoma patagonica	Argentina	53,047	20.5	915.06	28	65	pigeon	90
Triatoma platensis	Argentina	1,653,623	18.22	734	28	70	pigeon	126*
Triatoma protracta	USA	2,082,288	16.27	404.55	28	70	pigeon	68*
Triatoma pseudomaculata	Brazil	2,650,721	23.7	1198.89	28	70	pigeon	225*
Triatoma rubrovaria	Argentina	470,659	18.57	1472.36	28	70	pigeon	201*
Triatoma sordida	Brazil	4,408,479	22.25	1170.12	28	70	pigeon	120*
Triatoma vitticeps	Brazil	524,660	21.06	1352.76	28	70	pigeon	135*

Table 1. Natural environmental and laboratory rearing conditions of triatomine species selected for the analysis. R. H.=relative humidity. * Pool of several cohorts. Published sources: [a] Rabinovich & Nieves (2010), [b] Rabinovich (1972), [c] Feliciangeli & Rabinovich (1985); for *T. patagonica* original data from Dr. Elena Visciarelli; the rest of species: original data from this study.

Fig. 1 shows the instantaneous mortality and the fecundity rates of the 27 triatomine species analyzed. The senescence pattern is clear: instantaneous mortality rates stay at extremely low levels during most of the juvenile stages (except in the egg stage for some species) and during the first part of the adult stage, increasing greatly in the old ages (i.e., for old females). Most triatomine species seem to invest more in reproduction during the earlier and intermediate periods (but in general with lower effort) of the female's life. Only three species (*R. nasutus, R. neivai* and *D. maximus*) seem to invest more in the intermediate ages. However, six species (*T. eratyrusiformis, T. mazzotti, T. platensis, T. sordida, E. mucronatus,* and *P. geniculatus*) invest in reproduction during all periods of the reproductive life of the female. The difference in the reproductive effort pattern seem to be related to particular genera.

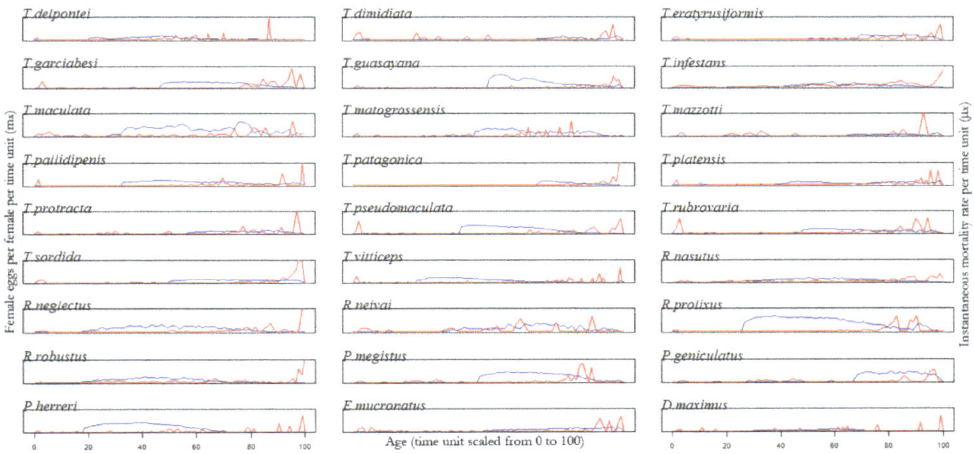

Fig. 1. Age-specific fecundity and instantaneous mortality rates for 27 triatomine species.

Age-specific fecundity (blue lines) and instantaneous mortality rates (red lines) for the 27 species analyzed. All species have a common scale for fecundity and mortality. Fecundity is represented in a scale from 0 to 25 ♀ eggs/♀/unit-time (not shown for better clarity); mortality is represented in a scale from 0 to 1 per unit-time; the x-axis is the age, originally in weeks, but scaled from 0 to 100 to have a common scale for all species.

3.2 Reproductive effort and mortality relationship

Fig. 2 shows the female adult accumulated mortality ($\sum \mu_x$) with respect the female adult accumulated fecundity ($\sum m_x$), both scaled from 0 to 1, in the form of a scatterplot ($\sum \mu_x$ vs $\sum m_x$); to facilitate interpretation of the different patterns, the scatterplots for all species were drawn in the same scale and with a 45° line to be used as a frame of reference. From those graphs we defined six groups with respect of the $\sum \mu_x$ vs $\sum m_x$ observed patterns. Despite the relationships differ among the species of the six groups, the species classed in the first five groups invest strongly in reproduction before a high mortality has been accumulated to a high degree (most of the curves stay below the 45° line) although with increased initial female adult mortality from Group 1 to Group 5; while Group 6 (R. neivai) has a different pattern with a curve that stays above this line during a long period, indicating no significant reproductive accumulation effort in relation to the accumulated mortality rate

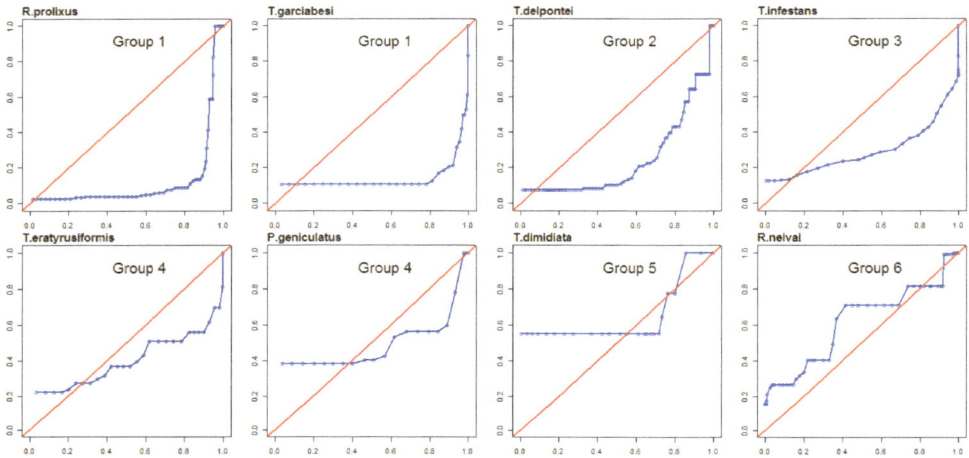

Fig. 2. Grouping of the accumulated mortality and accumulated fecundity relationship

Female adult accumulated instantaneous mortality rate ($\sum\mu_x$) (y- axis) with respect to the female adult accumulated fecundity ($\sum m_x$) (x- axis) (blue line), both scaled from 0 to 1. The 45° line was drawn as a reference (red line). Only selected species representative of each group are shown. For the criteria for grouping and description of the groups see text.

The species classed in first group (n=11) are: *T. garciabesi, T. guasayana, T. pallidipennis, T. vitticep, T. pseudomaculata, T. sordida, R. neglectus, R. prolixus, E. mucronatus, D. maximus* and *P. herreri;* in the second group (n=8): *T. delpontei, T. platensis, T. rubrovaria, T. pallidipennis, T. protracta, T. matogrossensis, T. mazzotti* and *P. megistus;* in the third group (n=4): *T. infestans, T. maculata, R. nasutus* and *R. robustus;* in fourth group (n=2): *T. eratyrusiformis* and *P. geniculatus;* in fifth group (n=1): *T. dimidiata,* and in last group (n=1): *R. neivai.*

3.3 Mortality pattern models

Table 2 shows the coefficients of the fit of the kza- smoothed instantaneous mortality rate (μ_x) to the three mortality models tested (Gavrilovs, Gompertz, and Logistic). Asterisks indicate a statistically significant coefficient with $p< 0.05$. The squared residuals (SSQ) are also shown. Certain mortality models could not be fitted to the instantaneous mortality rate data of some species; additionally, no single model fitted the mortality pattern of all species (Fig. 3, Table 2).

The fit is statistically significant for 14 out of 15 successful fits, for 10 out of 18 successful fits, and for 20 out of 21 successful fits, for the Gavrilovs, Gompertz, and Logistic models, respectively. From the results of the model fitting, Fig. 3 shows three main types of mortality patterns. A first group (Group I) shows a mortality pattern that we called "Late-spiky senescence" and is defined by a very long period in which the mortality rates are low and stable, with a high increase in the mortality rate at the end of the adult female life. In a second group, called "Gradual-medium senescence" (Group II), the increase in the mortality rate with age is more progressive than in the Group I and the mortality rate values stay at an intermediate level. In these two groups, no stabilization of mortality rate occurs at old age.

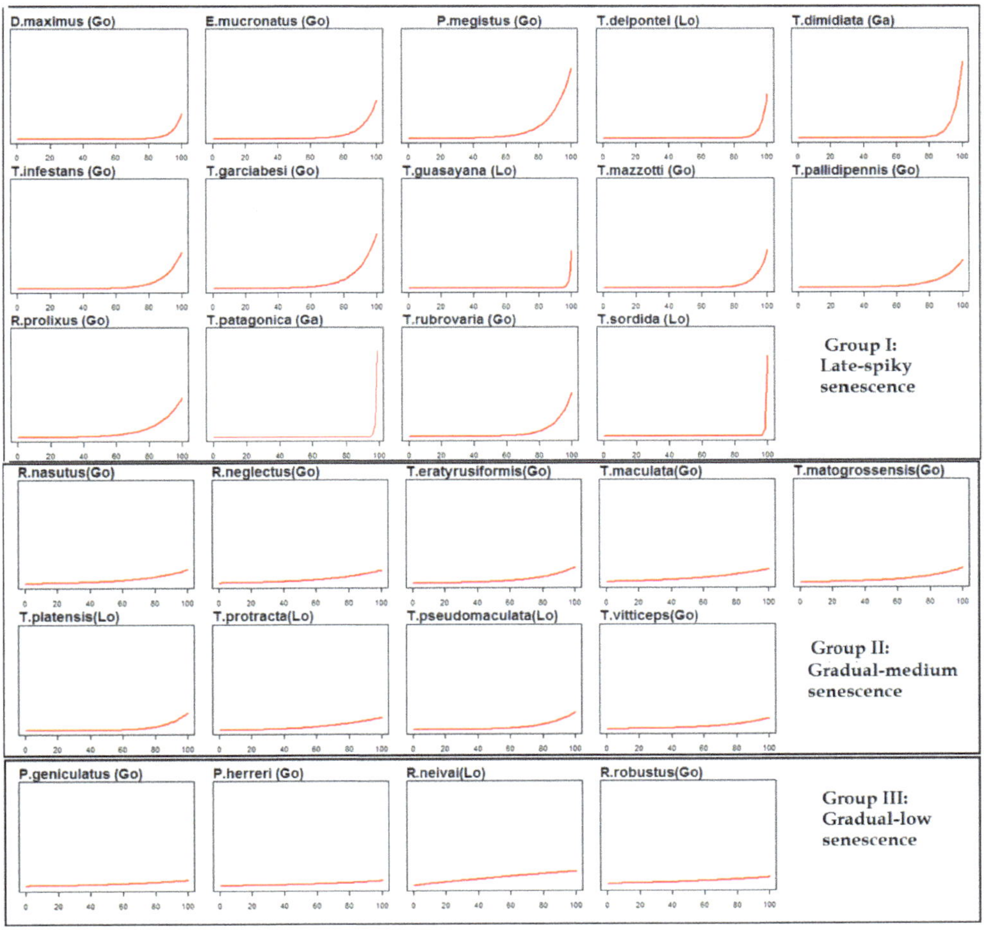

Fig. 3. Age specific mortality patterns of 27 species of triatomines.

The abbreviations in parenthesis indicate the best model that fitted the age-specific instantaneous mortality rate (Go= Gompertz, Lo= Logistic, and Ga= Gavrilovs). Axes have a common scale: from 0 to 1 in the ordinates (not shown) and from 0 to 100 in the abscissa; the latter covers from the egg stage to the last female adult alive.

	Gavrilovs model			Gompertz model			Logistic model		
Species	K	n	SSQ	a	b	SSQ	n	g	SSQ
D. maximus	6.7E-03*	18.78*	1.7E-01	1.5E-09*	1.6E-01*	1.7E-01	18.94*	149.8*	1.7E-01
E. mucronatus	-	-	-	1.6E-06*	1.4E-01*	1.8E-01	11.76*	115.3*	1.7E-01
P. geniculatus	6.7E-03*	18.78*	5.1E-02	1.1E-02*	9.3E+00*	4.8E-02	0.57	3310.3	5.0E-02
P. herreri	-	-	-	7.4E-03*	1.6E-02*	7.2E-02	-	-	-
P. megistus	1.1E-02*	8.18*	3.1E-01	7.9E-05*	1.4E-01*	3.3E-01	8.52*	85.7*	3.0E-01
R. nasutus	4.9E-03*	3.46*	9.4E-02	3.6E-03*	4.5E-02*	8.7E-02	3.56*	200.1*	9.4E-02
R. neglectus	-	-	-	5.0E-03*	3.9E-02*	7.6E-02	2.64*	260.3*	7.1E-02
R. neivai	2.0E-03	0.82	4.6E-01	-	-	-	0.93*	374.1*	4.6E-01
R. prolixus	8.5E-03*	6.32*	7.2E-02	2.6E-04	9.7E-02*	1.7E-01	6.49*	115.8*	1.6E-01
R. robustus	-	-	-	2.1E-02*	1.6E-02*	3.2E-01	-	-	-
T. delpontei	6.1E-03*	29.39*	3.2E-01	-	-	-	29.59*	162.7*	3.2E-01
T. dimidiata	1.2E-02*	28.09*	6.1E-01	-	-	-	-	-	-
T. eratyrusiformis	-	-	-	1.3E-03	5.1E-02*	1.3E-01	3.65*	225.2*	1.4E-01
T. garciabesi	1.2E-02*	8.49*	7.8E-02	4.1E-05	1.5E-01*	8.7E-02	8.72*	85.6*	7.6E-02
T. guasayana	8.9E-03*	88.89*	7.4E-02	-	-	-	89.28*	111.7*	7.4E-02
T. infestans	2.3E-02*	11.85*	5.8E-02	6.0E-06	3.3E-01*	5.4E-02	-	-	-
T. maculata	4.2E-03*	2.10*	1.3E-01	1.0E-02*	4.0E-02*	1.3E-01	2.18*	222.8*	1.3E-01
T. matogrossensis	-	-	-	2.6E-03*	4.1E-02*	7.7E-02	3.07*	270.4*	7.9E-02
T. mazzotti	-	-	-	2.8E-08	3.3E-01*	7.9E-02	15.69*	63.6*	7.7E-02
T. pallidipennis	-	-	-	5.8E-05	7.0E-02*	4.3E-01	-	-	-
T. patagonica	1.3E-02*	137.98*	2.0E-02	-	-	-	-	-	-
T. platensis	5.8E-03*	7.71*	7.9E-02	-	-	-	7.78*	172.5*	7.9E-02
T. protracta	-	-	-	-	-	-	2.96*	192.7*	1.5E-01
T. pseudomaculata	-	-	-	-	-	-	5.64*	163.4*	1.2E-01
T. rubrovaria	1.1E-02*	10.33*	1.8E-01	8.2E-06	1.7E-01*	1.8E-01	10.59*	86.6*	1.8E-01
T. sordida	-	-	-	-	-	-	146.65*	60.1*	4.3E-02
T. vitticeps	-	-	-	1.9E-03	2.8E-02*	2.0E-01	3.29*	406.0*	1.9E-01

Table 2. Fit to the Gavrilovs, Gompertz, and Logistic mortality models.

Parameter values of the mortality models fitted to the laboratory instantaneous mortality rates (μ_x). A "*" indicates a statistically significant coefficient of $p< 0.05$. $SSQ= \Sigma(\mu_{obs} - \mu_{mod})^2$. A "-" indicates that the model could not be fitted to the data.

The third group (Group III), which we called "Gradual-low senescence", is defined by a low and progressively increasing instantaneous mortality rate with age. In this group, particularly in *R. neivai*, the mortality rate seems to stabilize at old ages. The senescence pattern variation does not seem to be homogeneous for a given genus: all senescence pattern groups seem to occur in all genera, particularly in the dominant ones (*Triatoma, Rhodnius*).

3.4 Phylogenetic Independent Contrasts

The results of a series of phylogenetic independent contrasts among several dependent variables related to mortality and other variables related to the longevity and reproduction of the triatomines are shown in Tables 3-10. For a better illustration about the phylogenetic relationships we present (Fig. 4) the phylogenetic tree provided by Silva de Paula et al. (2005) but restricted to the 27 species used in our analysis. In Fig. 4 numbers close to the bifurcations identify each node of the phylogenetic tree. Numbers above the branches indicate the length of each branch, estimated by a branch length/decay index (Bremer support); it is proportional to the distance to ancestors. To the right of each species there is a label with a roman numeral corresponding to the Group into which each species has been classified as following a certain age-specific mortality pattern (Group I= "Late-spiky senescence"; Group II= "Gradual-medium senescence", and Group III= "Gradual-low senescence"; see Section 3.3 for definition of each group).

No significant relationship exists between the female's post reproductive period (FPRP) and the total length of adults or the age of first reproduction (non scaled, weeks) (Table 3). The relationship is statistically significant and positive between FPRP and the age of first reproduction scaled by total longevity. A statistically significant and negative correlation exists between FPRP and the female's reproductive period (scaled by longevity), the female reproductive life period (weeks, i.e., not scaled), and fecundity (expressed as ♀ eggs/♀/life).

Independent variables	**Coefficient**	**Std Dev**	**t value**	**p**
Female's reproductive period (longevity scaled)	-0.12288	0.05419	-2.26754	0.03226
Total length of adults (mm)	-0.00362	0.00192	-1.88994	0.07042
Female reproductive life period (weeks)	-0.00113	0.00043	-2.61769	0.01481
Fecundity (♀eggs/♀/life)	-0.00007	0.00002	-2.84123	0.00881
Age of first reproduction (weeks) (α)	0.00068	0.00107	0.64255	0.52636
α scaled by total longevity	0.11125	0.04857	2.29050	0.03070

Table 3. Female's post-reproductive period regressed on several independent variables

Simple lineal regression results, using the Phylogenetic Independent Contrast method, of the effect of several independent variables on the female's post-reproductive period (scaled in terms of lifespan) using 27 species of triatomines.

Therefore, the female's post reproductive period decreases with reproductive effort, which means that females live less after their last reproduction in those triatomine species in which the reproductive effort is high.

We show in Table 4 a statistically significant and negative correlation between the mean instantaneous mortality rate (called hereafter $\bar{\mu}_x$) and the average total female longevity since the egg stage, but no statistically significant relationship with the female's total length.

Independent variable	**Coefficient**	**Std. Error**	**t value**	**p**
Total average ♀ longevity (from the egg stage)	-0.00070	0.00014	-4.86	5.0E-05
Total length (mm)	-0.00076	0.00105	-0.73	0.4740

Table 4. Mean instantaneous mortality rate ($\bar{\mu}_x$) regressed on longevity and total length
Simple lineal regression results, using the Phylogenetic Independent Contrast method, of the effect of average female longevity and total body length as independent variables on the mean instantaneous mortality rate ($\bar{\mu}_x$).

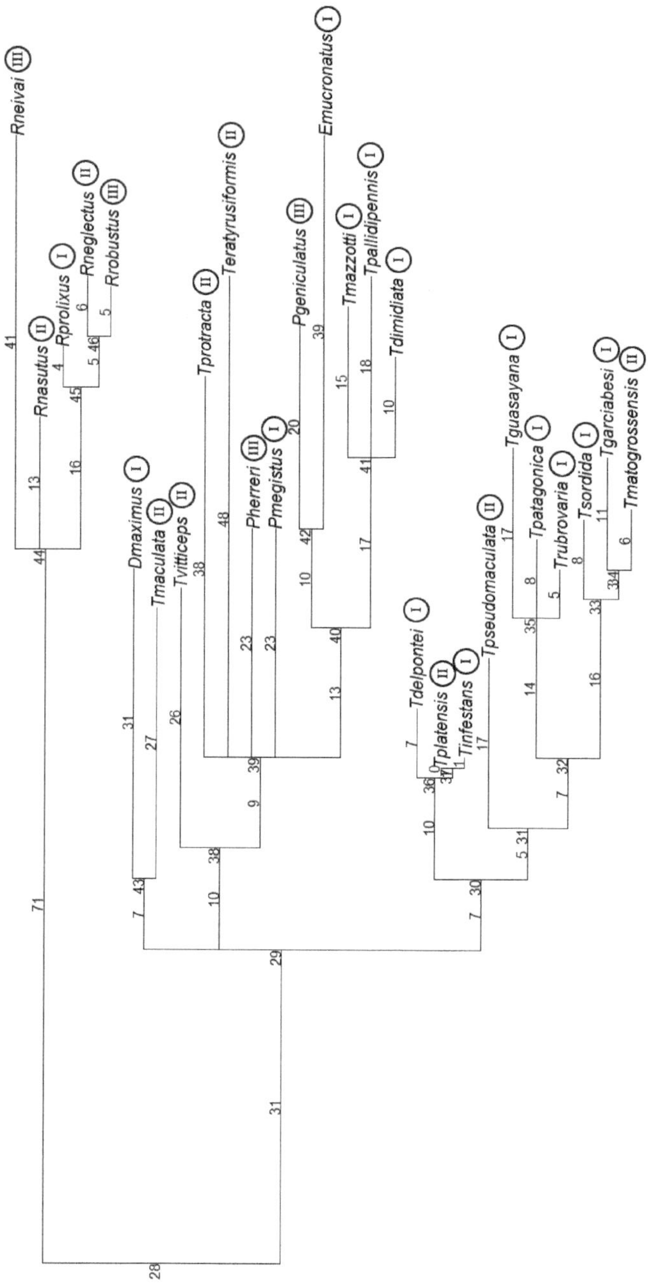

Fig. 4. Phylogenetic tree of the 27 triatomine species analyzed left from the original phylogenetic tree based on 72 species of Reduviidae proposed by Silva de Paula (2005)

Two indicators of mortality, total mean female mortality (1/e0) and the mean female mortality after the first reproduction ($1/e_{\alpha}$), are significantly correlated to two reproductive effort indicators (total eggs per female per life and reproductive weeks of females), decreasing with higher reproductive effort (Table 5). The mortality indicator $1/e_0$ is statistically significant and negatively correlated with both reproductive effort indicators used. Therefore, the total mean female mortality seems to decrease with the reproductive effort.

	Dependent variables: total mean ♀ mortality (1/e0)			
Independent variable	**Coeff.**	**Std. Err.**	**T value**	**p**
Total eggs/♀/life	-2.7E-05	2.5E-06	-11.05	4.1E-11
Reproductive weeks of ♀	-0.00092	0.00012	-7.96	2.6E-08
	Dependent variable: mean ♀ mortality after first reproduction ($1/e_{\alpha}$)			
Independent variable	**Coeff.**	**Std. Err.**	**T value**	**p**
Total eggs/♀/life	-3.0E-05	4.5E-06	-6.72	4.8E-07
Reproductive weeks of ♀	-0.00127	9.4E-05	-13.54	5.2E-13

Table 5. Two indicators of mortality rate regressed on two reproductive effort variables

Simple lineal regression results, using the Phylogenetic Independent Contrast method, of the effect of reproductive effort indicators as independent variables on two measures of mortality using 27 species of triatomines. The parameter e_0 is the expectation of life at birth (weeks).

Independent variable	**Coefficient**	**Std**	**t value**	**p**
Average ♀eggs/♀/week	-0.01099	0.07154	-0.15368	0.87979
Average ♀eggs/♀/life	0.00076	0.00133	0.56869	0.57746
α scaled by total longevity	-0.67365	0.89178	-0.75539	0.46099
α scaled by lifespan	-0.98590	0.87855	-1.12219	0.27834
Age of first reproduction (weeks) (α)	-0.01470	0.01375	-1.06885	0.30100
Total female longevity (weeks)	0.00478	0.00692	0.69122	0.49933
Total length (mm)	-0.03625	0.02683	-1.35086	0.19554

Table 6. The parameter b of the Gompertz model regressed on several independent variables

Simple lineal regression results, using the Phylogenetic Independent Contrast method, of the effect of several independent variables on the parameter b of the Gompertz mortality model. The phylogenetic tree used was reduced to the 18 species with $\bar{\mu}_x$ data that could be fitted to the Gompertz model. α represents the female's age (weeks) of first reproduction.

No statistically significant relationship was observed between the parameter *b* of the Gompertz mortality model (which is related to the shape of the mortality pattern) and the life history traits indicated in Table 6. Additionally, no statistically significant correlation was observed between the parameter *b* of the Gompertz mortality model and the geographic and/or climatic variables indicated in the table (Table 7).

Independent variable	Coefficient	Std Dev	t value	p
Surface area	-0.02572	0.02877	-0.89415	0.38449
Average annual precipitation (mm)	0.00027	0.00023	1.19358	0.25004
95% Lower precipitation (mm)	0.00029	0.00043	0.66386	0.51623
95% Upper precipitation (mm)	0.00021	0.00014	1.52282	0.14732
Colwell's precipitation index of predictability	2.40489	2.66950	0.90088	0.38101
Precipitation coefficient of variation (%)	0.03058	0.02010	1.52139	0.14768
NDVI coefficient of variation (%)	-0.11072	0.12554	-0.88195	0.39086
Maximum rain of the rainiest month (mm)	0.00405	0.00520	0.77925	0.44721
Minimum rain of the driest month (mm)	0.00204	0.00170	1.20070	0.24734

Table 7. The *b* Gompertz model parameter regressed on geographic/climatic variables

Simple lineal regression results, using the Phylogenetic Independent Contrast method, of the effect of several climatic variables used as independent variables on the parameter *b* of the Gompertz model. The phylogenetic tree was reduced to the 18 species with $\overline{\mu}_x$ that could be fitted to the Gompertz model. The NDVI is the Normalized Difference Vegetation Index.

Table 8 shows that from several geographic and/or climatic variables analyzed only the surface area and various precipitation indicators are correlated (negatively) with the mean mortality rate (defined here as $1/e_0$), while with the coefficient of variation (in %) of the Normalized Difference Vegetation Index (NDVI, a common indicator of live green vegetation obtained from satellite data) it was found to be positively correlated. In other words, species' geographical range size and some climatic factors, mainly the ones related to precipitation, seem to be related to mean mortality.

However, mortality ($1/e_0$) is probably not a good estimator of mean mortality because in our data the mortality rate varies greatly with age and this violates the hypothesis underlying this parameter as estimator of mean mortality rate. In consequence, we also used a more reliable estimator of the mean mortality rate from our data: the average for different time periods of the female's adult life ($\overline{\mu}_x$) (see Tables 9 and 10).

Independent variables	Coefficient	Std Err	t value	p
Surface area (km^2)	-0.00140	0.00022	-6.43226	9.8E-07
Modal latitude (degrees)	0.00034	0.00022	1.54316	0.13536
Modal longitude (degrees)	0.00025	0.00021	1.20700	0.23900
Average altitude (m)	1.1E-05	1.1E-05	1.05245	0.30266
Average annual temperature (°C)	-0.00060	0.00180	-0.33135	0.74315
Minimum annual temperature (°C)	0.00162	0.00137	1.18693	0.24641
Maximum annual temperature (°C)	0.00225	0.00210	1.07133	0.29425
Average annual precipitation (mm)	-1.5E-05	4.2E-06	-3.56725	0.00149
95% Lower precipitation (mm)	-2.6E-05	7.7E-06	-3.33849	0.00264
95% Upper precipitation (mm)	-9.3E-06	2.7E-06	-3.47074	0.00190
Colwell's rain index of predictability	-0.15913	0.04713	-3.37684	0.00240
Rain coefficient of variation (%)	0.00114	0.00044	2.61404	0.01494
Temperature amplitude (°C)	0.00047	0.00121	0.38703	0.70201
Temperature coefficient of variation (%)	0.00072	0.00055	1.30837	0.20265
Average annual NDVI	-0.05010	0.03350	-1.49561	0.14727
NDVI coefficient of variation (%)	0.00804	0.00263	3.06192	0.00520
Average AET (mm)	-0.00023	0.00015	-1.47877	0.15169
Minim temperature of coldest month (°C)	0.00088	0.00093	0.94976	0.35133
Maximum temperature of the warmest month (°C)	-9.8E-05	0.00272	-0.03617	0.97144
Maximum rain of the rainiest month (mm)	-0.00033	0.00008	-3.99956	0.00050
Minimum rain of the driest month (mm)	-0.00011	3.2E-05	-3.26415	0.00317
Number of dry months/year	0.00197	0.00108	1.83082	0.07907
Number of humid months/year	0.00214	0.00181	1.18209	0.24829
Number of super-humid months/year	-0.00200	0.00089	-2.25957	0.03282

Table 8. Average mortality ($1/e_0$) regressed on various geographic/climatic variables

Simple lineal regression results, using the Phylogenetic Independent Contrast method, of the effect of several climatic variables used as independent variables on the average mortality ($1/e_0$) using 27 species of triatomines.

The results obtained from the regressed of mean instantaneous mortality rate $\bar{\mu}_x$ were statistically non-significant with respect to the mean reproductive effort (first line in Table 9). However, if we divide the individual female adult life in two, three or four periods of equal length, some correlations between $\bar{\mu}_x$ for a given period and the mean reproductive effort for the same period are statistically significant and positive (Table 9 except the first line). Furthermore, we observe several significant positive correlations between $\bar{\mu}_x$ for a given period and the reproductive effort in the previous period(s) (Table 10). For instance, when divided in two periods, the $\bar{\mu}_x$ value in the second period is significantly and positively correlated to the reproductive effort during the first period (first line in Table

10) or during the first and second ones (second line in Table 10). Similarly, when divided in three periods, the $\bar{\mu}_x$ value of the third period is significantly and positively correlated to the reproductive effort during the all three periods (lines 4 and 6 in Table 10). Similarly for four periods.

Independent variable	Dependent variable	Coefficient	Std Err	t value	p
$\bar{m}_x$ 1/1	$\bar{\mu}_x$1/1	0.00021	0.001126	0.187	0.85354
$\bar{m}_x$ 1/2	$\bar{\mu}_x$1/2	-0.00556	0.00047	-11.96	7.7E-12
$\bar{m}_x$ 2/2	$\bar{\mu}_x$2/2	0.01840	0.00501	3.67	0.00115
$\bar{m}_x$ 1/3	$\bar{\mu}_x$1/3	-0.00266	0.00027	-9.75	5.3E-10
$\bar{m}_x$ 2/3	$\bar{\mu}_x$2/3	-0.00696	0.00212	-3.29	0.00300
$\bar{m}_x$ 3/3	$\bar{\mu}_x$3/3	0.02206	0.00813	2.71	0.01190
$\bar{m}_x$ 1/3	$\bar{\mu}_x$2/3	-0.00437	0.00062	-7.00	2.0E-07
$\bar{m}_x$ 2/3	$\bar{\mu}_x$3/3	0.01603	0.00327	4.91	4.7E-05
$\bar{m}_x$ 1/4	$\bar{\mu}_x$1/4	-0.00216	0.00022	-9.60	7.3E-10
$\bar{m}_x$ 2/4	$\bar{\mu}_x$2/4	-0.01076	0.00103	-10.41	1.4E-10
$\bar{m}_x$ 3/4	$\bar{\mu}_x$3/4	0.00910	0.00678	1.34	0.37520
$\bar{m}_x$ 4/4	$\bar{\mu}_x$4/4	0.02308	0.00877	2.63	0.01438
$\bar{m}_x$ 1/4	$\bar{\mu}_x$2/4	-0.00769	0.00066	-11.73	1.2E-11
$\bar{m}_x$ 2/4	$\bar{\mu}_x$3/4	0.01196	0.00148	8.07	2.0E-08
$\bar{m}_x$ 3/4	$\bar{\mu}_x$4/4	0.00822	0.00607	1.35	0.23480

Table 9. Average mortality and fecundity variables lagged by periods

Simple lineal regression results, using the Phylogenetic Independent Contrast method, of the effect of average fecundity ($\bar{m}_x$) on the average instantaneous mortality rates ($\bar{\mu}_x$) by periods of equal length (see text) using 27 species of triatomines.

Independent variable	Dependent variable	Coefficient	Std Err	t value	p
$\bar{m}_x$ 1/2	$\bar{\mu}_x$2/2	0.00528	0.00121	4.38	0.00019
$\bar{m}_x$ 1+2/2	$\bar{\mu}_x$2/2	0.01071	0.00197	5.45	1.2E-05
$\bar{m}_x$ 1+2/3	$\bar{\mu}_x$2/3	-0.00583	0.00102	-5.73	5.7E-06
$\bar{m}_x$ 1+2/3	$\bar{\mu}_x$3/3	0.00960	0.00203	4.73	7.5E-05
$\bar{m}_x$ 2+3/3	$\bar{\mu}_x$3/3	0.02390	0.00489	4.89	4.9E-05
$\bar{m}_x$ 1+2+3/3	$\bar{\mu}_x$3/3	0.01430	0.00276	5.18	2.3E-05
$\bar{m}_x$ 1+2/4	$\bar{\mu}_x$2/4	-0.00903	0.00079	-11.37	2.3E-11
$\bar{m}_x$ 1+2/4	$\bar{\mu}_x$3/4	0.00970	0.00128	7.56	6.5E-08
$\bar{m}_x$ 2+3/4	$\bar{\mu}_x$3/4	0.02099	0.00267	7.85	3.3E-08
$\bar{m}_x$ 2+3/4	$\bar{\mu}_x$4/4	0.00450	0.00436	1.03	0.21760
$\bar{m}_x$ 1+2+3/4	$\bar{\mu}_x$3/4	0.01455	0.00176	8.25	1.3E-08
$\bar{m}_x$ 1+2+3/4	$\bar{\mu}_x$4/4	0.00231	0.00301	0.77	0.47810
$\bar{m}_x$ 1+2+3+4/4	$\bar{\mu}_x$4/4	0.00374	0.00383	0.98	0.31270

Table 10. Accumulated average mortality and fecundity variables lagged by periods

Simple lineal regression results, using the Phylogenetic Independent Contrast method, of the effect of accumulated average fecundity on the accumulated average instantaneous mortality rates by periods of equal length (see text) using 27 species of triatomines.

4. Discussion

The following are our concepts on triatomine senescence in the light of evolutionary ecology.

4.1 Senescense in the triatomines

Since the first attempt to explain evolution of ageing was made by Weismann (1891, cited in Kirkwood & Holliday 1979) senescence has been a major topic of research in evolutionary ecology, both from an experimental and a theoretical approach. Being such a general area of investigation it is not surprising that it has relied strongly on the use of a comparative approach across a wide range of taxa. The phylogenetic variation in rates of senescence has been considered a consequence of a combination of factors that decrease the rate of decline in reproductive probability (and intensifying selection against senescence) and factors that increase this rate (relaxing selection against senescence). We applied a comparative meta-analyses approach within a single subfamily (Triatominae) correcting, for the first time in this group, the regression analyses by the phylogeny.

Williams (1957) has argued that the rate of senescence shown by any species will reflect the balance between a direct adverse selection of senescence as an unfavorable character, and an indirect, favorable selection through the age-related bias in the selection of pleiotropy genes. Thus positive variations in fecundity increase adult mortality rate, and affect other life-history traits (e.g., the shape of the distribution of reproductive effort with age) and thereby influence the evolution of senescence and phylogenetic variation. This theory predicts that (a) rapid morphogenesis should be associated with rapid senescence, (b) that senescence should always be a generalized deterioration of many organs and systems, and (c) that post-reproductive periods should be short and infrequent in any wild population. The latter prediction seems to have been confirmed in the case of triatomines (at least in the laboratory). This is another reason why, despite having selected for our study a relatively low taxonomic level (the species of only five genera, so we can expect a strong degree of phylogenetic relatedness), it is important to have the phylogeny included in these comparisons.

Our study shows that the 27 triatomine species analyzed present a senescence pattern that does not decrease at older ages. Mortality rate stays very low during most of the juvenile stages (except in the egg stage for some species) and during the first part of adult stage, and then increases greatly in the old ages in most species. This senescence pattern observed in triatomine species are in contrast with patterns reported for various Diptera: *Drosophila melanogaster* (Curtsinger et al., 1992; Pletcher & Curtsinger, 1998), *Ceratitis capitata* (Carey et al., 1992, 1998), *Anastrepha ludens* (Carey et al., 2005), and *Aedes aegypti* (Styer et al., 2007).

The mortality pattern of triatomines (Chaves et al., 2004a, Rabinovich et al 2010, and this study) seems to be more similar to that reported in *Oncopeltus fasciatus* (Dingle, 1966), *Dysdercus fasciatus* (Dingle, 1966), and *Callosobruchus maculatus* (Tatar et al., 1993) and is consistent with models of senescence based on the decline in physiological functions with age (known as the "disposable soma" theory for the evolution of senescence; Kirkwood & Holliday, 1979; Kirkwood & Rose, 1991; Kirkwood & Austad, 2000; Kirkwood, 2002), which can be considered a consequence of the equilibrium predictions of the antagonistic-pleiotropy and mutation accumulation hypotheses (Abrams & Ludwig, 1995).

In our comparative analysis we have used several measures of mortality. Although the use of the Gompertz model to describe the acceleration of mortality with age has been a matter of debate (Nusbaum et al., 1996), for a comparative study this model seems more sound than, e.g., maximum lifespan, particularly because of the sensitivity of maximum lifespan to the initial numbers of a cohort. Nusbaum et al. (1996) have analyzed the evolutionary relationships among several measures of mortality (Gompertz parameters, and average and maximum longevity) in 50 related populations of *D. melanogaster*; they included populations that had been selected for postponed aging and in their conclusions they give credit to a redundancy among these measures of aging, and consider that both the maximum lifespan and the Gompertz equation as adequate indices of aging in evolutionary research. This gives support to their use in our study.

4.2 High variation in senescence pattern

Our results show that the mortality pattern with age varies greatly among triatomine species (showing late-spiky, gradual-medium, and gradual-low senescence) despite our study was based on cohorts reared under very homogeneous environmental conditions; the present challenge is to understand which are the ultimate factors underlying such diversity of mortality patterns in triatomines (Fig. 3). These different patterns are observed in all the species we studied, even in the more phylogenetic related species within any given genus (Fig. 4). We do not dismiss the possibility that the small initial number of eggs of some cohorts may have played a role in the estimation of demographic parameters (e.g. threshold mortality) (Carey, 2001), and thus in the high degree of mortality patterns variability; however, we still think that the variability found is genuine and could reflects multiple underlying causes.

Despite most of the studied species show "late-spiky" (n= 14) or "gradual-medium" senescence (n= 9), the species in the "gradual-low" group (n= 4), particularly *R. neivai*, seem to show a weak late-life mortality plateau. Some demographic findings point to the existence of a late-life mortality plateau in a few dipteran insects (Mueller & Rose, 1996), with both antagonistic pleiotropy and mutation accumulation as driving population genetic mechanisms; this late-life attribute is a switch from accelerating mortality to a relatively stable mortality (Rauser et al., 2006); such plateaus seem to depend on the collection of high numbers of late-life data (Carey et al., 1992). One plausible explanation of the "plateau" behavior was introduced by Vaupel et al. (1979) assuming a life-long heterogeneity in the mortality rates: more robust subgroups survive to later ages, slowing the rate of decline in average survival probabilities at late ages among large cohorts. The older remaining individuals from the cohort are expected to be much more robust so that the mortality rate becomes a very shallow function of age, resembling a plateau. However this demographic heterogeneity model does not seem to be a reasonable explanation of demographic patterns and it has only a weak biological basis (Mueller et al., 2003).

With rare exceptions the Triatominae are exclusively hematophagous, and they also show strict hematophagy across all developmental stages. This may be a potential key to the explanation of the observed mortality pattern: hematophagy from vertebrate blood leads to the digestion of vertebrate hemoglobin resulting in the production of large amounts of heme, a potentially cytotoxic molecule that can exert biological damage (Graca-Souza et al., 2006). The evolution of hematophagy has resulted in many adaptations developed by blood-feeding insects and ticks to counteract those deleterious effects. Antioxidant enzymes and urates are known to play a major role in the protection of cells against free radical damage, and massive amounts of urates have been found in the haemolymph of *R. prolixus* (Souza et al., 1997). Given the strict hematophagy of triatomines these mechanisms that reduce the accumulation of oxidative substances associated with aging (Graca-Souzaet al., 2006) may be one of the basic mechanisms that may help explain the senescence patterns observed in the triatomines.

The relationships between average instantaneous mortality rates $\bar{\mu}_x$ and average fecundity by periods show a delay in the effect of the latter on the former, indicating an important reproductive investment before a high accumulated instantaneous mortality rate occurs. The negative relationship between the average instantaneous mortality rate $\bar{\mu}_x$ and average female longevity suggests that species with high longevity can invest in reproduction with a relatively low mean mortality and then a smaller senescence rate than the species with low longevity. This type of relationship between reproductive effort and the effects on age-specific mortality rate is similar to the one found in the beetle *C. maculatus* (Tatar et al., 1993) and confirms the observation of Sulbaran & Chaves (2006) in the kissing bug *R. prolixus*.

Additionally, the female's post reproductive period decreases with reproductive effort indicating that longevity after their last reproduction is shorter in species in which the reproductive effort is high. The above results suggest a trade-off between mortality rate and reproductive effort: a large investment in the reproduction during the first and/or intermediate part of the female adult life that seems to result in an increase in high mortality, and then a high senescence rate at the end of the adult life. Such trade-off could contribute to explain the variation between species in the mean mortality and in the post-reproductive period duration but probably not the difference in the senescence pattern since no significant relationship exists between the parameter *b* of the Gompertz model and other variables used as indicators of reproductive effort.

4.3 Triatominae senescence in the light of evolutionary ecology

The balance between extrinsic and intrinsic mortality rates is an important factor underlying the evolution and the diversity of senescence patterns. Williams (1957) proposed that organisms living in environments with high extrinsic mortality rate may evolve towards high senescence rates. Physiological senescence results from an optimal equilibrium between energy allocation in somatic maintenance and other competitive traits as reproduction (Kirkwood, 2005). In habitats in which life expectancy is short resulting from extrinsic factors (as high predation, starvation, etc.), the maintenance of costly mechanisms to guarantee the reparation of metabolic deterioration is not evolutionary stable.

Conversely, species living in habitats with low extrinsic mortality may increase energy allocation in somatic maintenance until old age and increase longevity by natural selection.

In triatomines no quantitative data exists concerning extrinsic mortality factors in the field. The levels of extrinsic mortality in relation to the habitat (domestic, peri-domestic and sylvatic), is still little known in triatomines. However, our results show a negative relationship between mortality rate and female longevity supporting Williams' (1957) prediction, contrary to other empirical observations that do not support this prediction (Promislow, 1991; Ricklefs, 1998; Reznicket et al., 2004), which has also been criticized on conceptual grounds due to the difficulty in separating extrinsic and intrinsic mortality rates (Williams & Day, 2003).

In a recent comparative analysis of mammal and bird survival senescence Jones et al. (2008) arrived to generalizations such as that mammals senesce faster than similarly sized birds. Furthermore, McCoy & Gillooly (2008) developed a model of natural mortality (relating body size and temperature to biological rate processes) and tested it with extensive field data from plants, invertebrates, fish, birds and mammals; their results indicate that much of the heterogeneity in natural mortality rates can be predicted, explicitly and quantitatively, despite the high diversity of extrinsic sources of mortality in natural systems, something that suggests that mortality rates may be governed by common rules.

We show that the geographical range size of species and some climatic factors, mainly the ones related to precipitation, seem to be related to total mean mortality ($1/e_0$). These results suggest that the influence of geographic and climatic factors on the senescence pattern deserves further investigation. However, can the potential senescence that is observed and measured in the laboratory be expressed in natural habitats where extrinsic mortality occurs? In the field, triatomines could die before intrinsic mortality and decreasing fecundity occurs, due to extrinsic factors such as predation, parasitism, extreme climatic conditions, etc. Recent studies in vertebrates (Adams, 1985; Gaillard et al., 1993; Reznick et al., 2002; Rebke et al., 2010) show that senescence is observed under natural conditions, but that kind of information is lacking for triatomines and needs to be investigated. Williams (1957) claimed that greater rates of extrinsic mortality (age- and condition-independent) favored more rapid senescence, but Abrams (1991) showed that the effects of the "extrinsic" mortality affect differentially the rate of senescence as a function of the degree of density-dependence. Abrams (1991) also showed that mortality patterns, contrary to Williams' (1957) predictions, are possible when density-dependence is present, and acts primarily on the survival or fertility of later ages, or when most of the variation in mortality rates is due to variation in non-extrinsic mortality.

There are few laboratory evaluations of the density-dependent processes in triatomines. Rodríguez & Rabinovich (1980) showed that in *R. prolixus* density had a significant effect on the development rate of second, third, and fourth instars, but not on the survivorship of either the first or fifth instars, or even of the adults, nor on the instantaneous population parameters, or the age-specific parameters. Influence of density-dependence on the senescence pattern in triatomines still needs more research.

How organisms distribute their "investment" in reproduction with respect to age is a major question in the senescence theory. It will be optimal for an organism to "invest" more in its reproduction when it becomes old-aged (Williams, 1966). An extension of William's (1966) theory is that old individuals may increase their reproductive effort during their last reproductive occasion(s) because it is their last chance to reproduce. This extension was called the "hypothesis of terminal investment" (Ricklef, 2000; Coulson & Fairweather, 2001; Ricklefs, 2008). Thus senescent individuals may invest more in reproduction than the non-senescent individuals. Our observations in the triatomines do not support these predictions. Indeed most of the reproductive effort (about 50 to 90%) occurs in the first and intermediate part of the adult life and before the high mortality that takes place at older ages. Furthermore, the duration of the post-reproductive period decreases with reproductive effort suggesting a trade-off between reproduction allocation and future survival. Recently William's (1966) prediction and the concept of terminal investment have been criticized because of scarce empirical support and for resorting to two hypotheses based on quite unrealistic assumptions: (i) organisms must have a fixed longevity, which implies that individual trajectories cannot influence longevity, and (ii) in order to increase their reproductive effort at the end of their life during their last reproductive occasion(s), individuals must be able to anticipate their future, i.e., they have to use cues indicating when their death may occur (Gadgil & Bossert, 1970).

Environmental heterogeneity, and particularly high environmental uncertainty (stochastic effects), affect circumstances that are very important in insects, such as encounter rates associated to suitable oviposition sites, food and refuge availability, physiological state (e.g., reserves for producing oocytes, egg maturation rate, and somatic maintenance costs), and expected reproductive success that can lead to different patterns of behavior and rates of mortality and reproduction (Partridge & Mangel, 1999). Environmental stochasticity and/or density-dependence processes can select bet-hedging dormancy including a development delay in insects (e.g., Menu et al., 2000; Gourbière & Menu, 2009; Rajon et al., 2009). Such a risk spreading strategy has been postulated to exist in triatomines (Menu et al., 2010) and we recommend that this approach be incorporated in future studies for understanding senescence in insects.

Mortality rates of triatomine vectors have epidemiological importance through the demography of its populations (Chaves et al., 2004a). Particularly, it has been shown in these insects that important changes in life-history traits take place when reared under laboratory conditions; for example, in *T. infestans*, *T. pseudomaculata*, *T. brasiliensis*, and *P. megistus* after only four years in the laboratory, the reduction in the number of eggs within the first month of oviposition and the average female life span was in average 37.5 and 45.9%, respectively (Perlowagora-Szumlewicz, 1976). Similar changes, including a reduction in life history traits and population growth parameters, during four successive and separate generations of *T. infestans* of Argentina (reared in the laboratory from a first generation from a sylvatic individual) have also been found (G. Martí, personal communication).

The adaptation to a domestic environment causes similar effects on triatomines as the adaptation to laboratory conditions (Forattini, 1980): indoor conditions of rural houses provide better chances of feeding, and are characterized by smaller predation risks, and smaller fluctuations of temperature and relative humidity that exclude potentially deleterious extreme values. In our study we did not analyze possible adaptations to the laboratory because the precise number of laboratory generations was known for only a few cohorts; however, it is recommended that, when estimating trade-offs, and in particular the senescence pattern, the influence of such changes should be taken into account.

Furthermore, as the data for our comparative analyses did not indicate the mean body length of the individuals used in the experiments we used the estimates provided by Galíndez Girón et al. (1998) based on the types and paratypes of triatomine measurements in museum collections. As the individual's body size distribution can vary with respect to each population, it is recommended that future cohort studies in the laboratory measure the body length of the individuals used. In consequence, we must take with some reservation the lack of a statistically significant correlation between life history traits and total length as obtained in this study and based on 27 triatomine species.

5. Conclusions

Most studies in triatomines have investigated physiology, population genetics, phylogeny (see Gourbière et al., 2011 for a review) and ecology of a given species (e.g., Dumonteil et al., 2002; Gourbière et al., 2008; Barbu et al., 2011) but very few studies have investigated the evolution of life history traits in the light of evolutionary ecology concepts (Menu et al., 2010). Our study is the first comparative analysis to the senescence pattern and its relationship with life history traits in triatomines.

Our results indicate that triatomines show both an actuarial and a reproductive senescence, with a high diversity of mortality patterns, even within a given genus. We believe that the relationship between life history traits, and particularly the trade-off between reproductive effort and future survival, is central to understand this diversity in mortality patterns. In order to identify ultimate and proximate factors underlying this diversity, we need longitudinal studies conducted in the field in order to estimate if the potential senescence observed in the laboratory can be expressed under natural (sylvatic and domestic) conditions. The analysis of senescence in relation to other life history traits in triatomines has not only academic value but also impinges in the areas of vector population management and epidemiology.

6. Acknowledgments

We are extremely grateful to the Vectors Reference Center of the National Ministry of Health, Córdoba, Argentina and to Dr. Elena Visciarelli for allowing the use of original data from 21 species of triatomines and from *T. patagonica*, respectively. We thank to Mr. Waldo Hasperué for his collaboration in the data processing with the computer program TriTV. This work has been supported by the French National Research Agency (grant reference «ANR-08-MIE-007»), the CNRS and the *Agencia Nacional de Promoción Científica y Tecnológica* of Argentina (grant PICT2008-0035).

7. References

Abrams P. A. & Ludwig, D. (1995). Optimality theory, Gompertz' law, & the disposable soma theory of senescence. *Evolution*, Vol.49, pp. 1055-1066

Abrams, P. A. (1991). Life History and the Relationship between Food Availability and Foraging Effort. *Ecology*, Vol.72, No.4, pp. 1242-1252

Adams, C. E. (1985). Reproductive senescence, In: *Reproduction in Mammals*, C. R. Austin & R. V. Short (Eds.), 210-233. Cambridge, Univ. Press

Akaike, H. (1974). A new look at statistical model identification. *IEEE Transactions on Automatic Control,* Vol.19, pp. 716–723

Barbu, C.; Dumonteil, E. & Gourbière, S. (2011). Evaluation of Spatially Targeted Strategies to Control Non-Domiciliated *Triatoma dimidiata* Vector of Chagas Disease. *PLoS Negl Trop Dis,* Vol.5, No.5, pp. e1045

Carey, J. R. (2001). Insect Biodemography. *Annual Review of Entomology,* Vol.46, pp. 79-110

Carey, J. R.; Liedo, P.; Müller, H.-G.; Wang, J.-L. & Vaupel, J. W. (1998). Dual Modes of Aging in Mediterranean Fruit Fly Females. *Science, New Series,* Vol.281, No.5379, pp. 996-998

Carey, J. R.; Liedo, P.; Muller, H-G.; Wang, J-L.; Senturke, D. & Harshman, L. (2005). Biodemography of a long-lived tephritid: Reproduction & longevity in a large cohort of female Mexican fruit flies, *Anastrepha ludens. Experimental Gerontology,* Vol.40, pp. 793–800

Carey, J. R.; Liedo, P.; Orozco, D. & Vaupel, J. W. (1992). Slowing of mortality rates at older ages in large medfly cohorts. *Science,* Vol. 258, pp. 457–461

Charlesworth, B. & Partridge, L. (1997). Ageing: Levelling of the Grim Reaper. *Current Biology,* Vol.7, pp. 440-442

Chaves, L. F.; Hernández, M. J.; Revilla, T. A.; Rodríguez, D. J. & Rabinovich, J. E. (2004a). Mortality profiles of *Rhodnius prolixus* (Heteroptera: Reduviidae), vector of Chagas disease. *Acta Tropica,* Vol.92, pp. 119–125

Chaves, L. F.; Zamora, E. & Aldana, E. (2004b). Mortality profile of female *Rhodnius robustus* (Heteroptera: Reduviidae). *Revista de Saúde Pública,* Vol.38, pp. 466–468

Cheverud, J. M.; Dow, M. M. & Leutenegger, W. (1985). The Quantitative Assessment of Phylogenetic Constraints in Comparative Analyses: Sexual Dimorphism in Body Weight Among Primates. *Evolution,* Vol.39, No.6, pp. 1335-1351

Christopoulos, A. & Lew, M. J. (2000). Beyond Eyeballing: Fitting Models to Experimental Data. Critical Reviews in *Biochemistry and Molecular Biology,* Vol.35, No.5, pp. 359–391

Coulson, J. C. & Fairweather, J. A. (2001). Reduced reproductive performance prior to death in the Black-legged Kittiwake: senescence or terminal illness? *Journal of Avian Biology,* Vol.32, pp. 146-152

Curtsinger, J. W.; Fukui, H. H.; Townsend, D. R. & Vaupel, J. W. (1992). Demography of genotypes: failure of the limited life-span paradigm in *Drosophila melanogaster. Science,* Vol.258, pp. 461–463

Dingle, H. (1966). The effect of population density on mortality & sex ratio in the milkweed bug, *Oncopeltus,* & the cotton stainer, *Dysdercus* (Heteroptera). *American Naturalist,* Vol.100, pp. 465–470

Dumonteil, E.; Gourbière, S.; Barrera-Perez, M.; Rodriguez-Félix, E.; Ruiz-Piña, H.; Banos-Lopez, O.; Ramirez-Siera, M. J.; Menu, F. & Rabinovich, J. E. (2002). Geographic distribution of *Triatoma dimidiata* and transmission dynamics of *Trypanosoma cruzi* in the Yucatan peninsula Mexico. *American Journal of Tropical Medicine and Hygiene,* Vol.67, pp. 176-183

Feliciangeli, M. D. & Rabinovich, J. E. (1985). Vital Statistics of Triatominae (Hemiptera: Reduviidae) under Laboratory Conditions. II. *Triatoma maculata. Journal of Medical Entomology,* Vol.22, pp. 43-48

Forattini, O. P. (1980). Biogeography, origin, and distribution of triatominae domiciliarity in Brazil. *Revista de Saúde Pública,* Vol.14, pp. 265-299.

Fukui, H. H.; Xiu, L. & Curtsinger, J. W. (1993). Slowing of age-specific mortality rates in *Drosophila melanogaster. Experimental Gerontology,* Vol.28, pp. 585–599

Gadgil, M. & Bossert, W. H. (1970). Life Historical Consequences of Natural Selection. *Am. Nat.,* Vol.104, No.935, pp. 1-24.

Gaillard, J.-M. ; Viallefont, A. ; Loison, A. & Festa-Bianchet, M. (2004). Assessing Senescence Patterns in Populations of Large Mammals. *Animal Biodiversity & Conservation,* Vol.27, No.1, pp. 47-58

Galíndez Girón, I.; Carcavallo, R. U.; Jurberg, J.; Galvao, C.; Lent, H.; Barata, J. M. S.; Pinto Serra, O. & Valderrama, A. (1998). Morfology and external anatomy. A: General, In: *Atlas of Chagas' Disease Vectors in the Americas,* R. U. Carcavallo, I. Galíndez Girón, J. Jurberg & H. Lent (Comps.), 53-73, FioCruz, Rio de Janeiro, Brazil

Gavrilov, L. A & Gavrilova, N. S. (2001) The reliability theory of aging and longevity. *Journal of Theoretical Biology,* Vol.213, No.4, pp. 527-545

Ghalambor, C. K.; Reznick, D. N. & Walker, J. A. (2004). Constraints on Adaptive Evolution: the Functional Trade-Off between Reproduction and Fast-Start Swimming Performance in the Trinidadian Guppy (*Poecilia reticulata*). *Am. Nat.,* Vol.164, No.1, pp. 38-50

Gourbière, S. & Menu, F. (2009). Adaptive dynamics of dormancy duration variability: evolutionary trade-off and priority effect lead to suboptimal adaptation. *Evolution,* Vol.63, No.7, pp. 1879-1892

Gourbière, S. ; Dumonteil, E. ; Rabinovich, J. E. ; Minkoue, R. & Menu, F. (2008). Demographic and Dispersal Constraints for Domestic Infestation by Non-Domicilated Chagas Disease Vectors in the Yucatan Peninsula Mexico. *American Journal of Tropical Medicine and Hygiene,* Vol78, pp. 133-139

Gourbière, S.; Dorn, P.; Tripet, F. & Dumonteil, E. (2011). Genetics and evolution of triatomines: from phylogeny to vector control. *Heredity,* pp. 1-13

Graca-Souza, A. V.; Maya-Monteiro, C.; Paiva-Silva, G. O.; Braz, G. R. C.; Paes, M. C.; Sorgine, M. H. F.; Oliveira, M. F. & Oliveira, P. L. (2006). Adaptations against heme toxicity in blood-feeding arthropods. *Insect Biochemistry and Molecular Biology,* Vol.36, pp. 322-335

Jones, O. R.; Gaillard, J.-M.; Tuljapurkar, S.; Alho, J. S.; Armitage, K. B.; Becker, P. H.; Bize, P.; Brommer, J.; Charmantier, A.; Charpentier, M.; Clutton-Brock, T.; Dobson, F. S.; Festa- Bianchet, M.; Gustafsson, L.; Jensen, H.; Jones, C. G. & Lillandt, B.-G. (2008). Senescence rates are determined by ranking on the fast-slow life-history continuum. *Ecology Letters,* Vol.11, pp. 664-673

Kirkwood, T. B. L. & Austad, S. N. (2000) Why do we age? *Nature* Vol.408, pp. 233-238

Kirkwood, T. B. L. & Holliday, R. (1979). The Evolution of Ageing & Longevity, *Proceedings of the Royal Society of London, Series B, Biological Sciences* Vol.205, No.1161, pp. 531-546

Kirkwood, T. B. L. & Rose, M. R. (1991). Evolution of senescence: late survival sacrificed for reproduction. *Transactions of the Royal Society, London B,* Vol.332, pp. 15-24

Kirkwood, T. B. L. (2002). Evolution of Ageing. *Mechanisms of Ageing & Development* Vol.123, pp. 737-745.

Kirkwood, T. B. L. (2005). Understanding the odd science of aging. *Cell* Vol.120, pp. 437-447

McCoy, M. W. & Gillooly, J. F. (2008). Predicting natural mortality rates of plants and animals. *Ecology Letters,* Vol.11, pp. 710-716

Menu F.; Roebuck, J. P. & Viala, M. (2000). Bet-hedging diapause strategies in stochastic environment. *The American Naturalist,* Vol.155, pp. 724-734

Menu, F.; Ginoux, M.; Rajon, E.; Lazzari, C. R. & Rabinovich, J. E. (2010). Adaptive Developmental Delay in Chagas Disease Vectors: An Evolutionary Ecology Approach. *PLoS Neglected Tropical Diseases,* Vol.4, No.5, pp. 1-10

Mueller, L. D. & Rose, M. R. (1996). Evolutionary theory predicts late-life mortality plateaus. *Proceedings of the National Academy of Sciences* Vol.93, pp. 15249-15253

Mueller, L. D.; Drapeau, M. D.; Adams, C. S.; Hammerle, C. W.; Doyal, K. M.; Jazayeri, A. J.; Ly, T.; Beguwala, S. A.; Mamidi, A. R. & Rose, M. R. (2003). Statistical Tests of Demographic Heterogeneity Theories. *Experimental Gerontology* Vol.38, pp. 373-386

Mueller, L. D.; Rauser, C. L. & Rose, M. R. (2005). Population Dynamics, Life History, & Demography: Lessons From *Drosophila*. *Advances in Ecological Research,* Vol.37, pp. 77-99

Nusbaum, T. J.; Mueller, L. D. & Rose, M. R. (1996). Evolutionary Patterns Among Measures of Aging. *Experimental Gerontology,* Vol.31, No.4, pp. 507-516

Paradis, E. (2006). *Analysis of Phylogenetics and Evolution with R.* Springer-Verlag, Heidelberg. ISBN 978-0-387-32914-7

Partridge, L. & Mangel, M. (1999). Messages from Mortality: the Evolution of Death Rates in the Old . *Trends in Ecology & Evolution,* Vol.14, No.11, pp. 438-442

Perlowagora-Szumlewicz, A. (1976). Laboratory Colonies of Triatominae, Biology, and Population Dynamics. *Proceedings of an International Symposium, New Approaches in American Trypanosomiasis Research,* pp. 63-82. Belo Horizonte, Minas Gerais, Brazil, March 18-21, 1975

Pletcher, S. D. & Curtsinger, J. W. (1998). Mortality Plateaus and the Evolution of Senescence: Why are Old-Age Mortality Rates so Low? *Evolution,* Vol.52, No.2, pp. 454-464

Promislow, D. E. L. (1991). Senescence in Natural Populations of Mammals: A Comparative Study. *Evolution,* Vol.45, No.8, pp. 1869-1887

R Development Core Team (2007) *R: A Language and Environment for Statistical Computing.* R Foundation for Statistical Computing, Vienna, Austria. Available from <http//www.R-project.org>

Rabinovich J. E. & Nieves, E. L. (2011). Vital Statistics of Triatominae (Hemiptera: Reduviidae) Under Laboratory Conditions III. *Rhodnius neglectus. Journal of Medical Entomology,* Vol.48, No.4, pp. 775-787

Rabinovich, J. E. (1972). Vital statistics of Triatominae (Hemiptera: Reduviidae) under laboratory conditions. I. *Triatoma infestans* Klug. *Journal of Medical Entomology,* Vol.9, No.4, pp. 351-370

Rabinovich, J. E.; Nieves, E. L. & Chaves, L. F. (2010). Age-specific mortality analysis of the dry forest kissing bug, *Rhodnius neglectus. Entomologia Experimentalis et Applicata,* Vol.135, No.3, pp. 252-262

Rajon, E.; Venner, S. & Menu, F. (2009). Spatially heterogeneous stochasticity and the adaptive diversification of dormancy. *Journal of Evolutionary Biology,* Vol.22, pp. 2094-2103

Rauser, C. L.; Mueller, L. D. & Rose, M. R. (2006). The Evolution of Late Life. *Ageing Research Reviews,* Vol.5, pp. 14-32

Rebke, M.; Coulson, T.; Becker, P. H. & Vaupel, J. W. (2010). Reproductive improvement and senescence in a longlived bird. *Proceedings of the National Academy of Sciences,* Vol.107, pp. 7841-7846

Reznick, D.; Bryant, M. J. & Bashey, F. (2002). r- and K- selection revisited: the role of population regulation in life history evolution. *Ecology,* Vol.83, No.6, pp. 1509-1520

Ricklefs, R. E. (1998). Evolutionary Theories of Aging: Confirmation of a Fundamental Prediction, with Implications for the Genetic Basis and Evolution of Life Span. *Am. Nat.,* Vol.152, No.1, pp. 24-44

Ricklefs, R. E. (2000). Intrinsic aging-related mortality in birds. *Journal of Avian Biology,* Vol.31, pp. 103-111

Ricklefs, R. E. (2008) The evolution of senescence from a comparative perspective. *Functional Ecology,* Vol.22, pp. 379-392

Rodríguez, D. & Rabinovich, J. E. (1980). The Effect of Density on some Population Parameters of *Rhodnius prolixus* (Hemiptera: Reduviidae) under Laboratory Conditions. *Journal of Medical Entomology,* Vol.17, No.2, pp. 165-171

Roff, D. A. (2002). *Life History Evolution. Sinauer Associates,* Inc. Sunderland, Massachusetts, USA.

Silva de Paula, A.; Diotaiuti, L. & Schofield, C. J. (2005). Testing the sister-group relationship of the Rhodniini and Triatomini (Insecta: Hemiptera: Reduviidae: Triatominae). *Molecular Phylogenetics and Evolution,* Vol.35, pp. 712-718

Souza, A. V. G.; Petretski, J. H.; Demasi, M.; Bechara, E. J. H.; Oliveira, P. L. (1997). Urate protects a blood-sucking insect against hemin induced oxidative stress. *Free Rad. Biol. Med.,* Vol.22, pp. 209–214

StatSoft (2009) STATISTICA (Data Analysis Software System), version 9.0. Statsoft, Tulsa, OK, USA.

Styer, L. M.; Carey, J. R.; Wang, J.-L. & Scott, T. W. (2007). Mosquitoes do senesce: departure from the paradigm of constant mortality. *American Journal of Tropical Medicine & Hygiene,* Vol.76, pp. 111–117

Sulbaran, J. E. & Chaves, L. F. (2006). Spatial complexity and the fitness of the kissing bug, *Rhodnius prolixus*. *Journal of Applied Entomology,* Vol.130, No.1, pp. 51-55

Tatar M.; Carey, J. R. & Vaupel, J. W. (1993). Long-term cost of reproduction with & without accelerated senescence in *Callosobruchus maculatus*: analysis of age-specific mortality. *Evolution,* Vol.47, pp. 1302-1312

Vaupel, J. W. Manton, K. G. & Stallard, E. (1979). The impact of heterogeneity in individual frailty on the dynamics of mortality. *Demography* Vol.16, No.3, pp. 439-454.

WHO (World Health Organization). 2007. Reporte sobre la enfermedad de Chagas. Grupo de trabajo científico 17–20 de abril de 2005, Buenos Aires, Argentina..

Williams, G. C. (1957). Pleiotropy, Natural Selection, & the Evolution of Senescence. *Evolution,* Vol.11, No.4, pp. 398-411

Williams, G. C. (1966). Natural Selection, the Costs of Reproduction, and a Refinement of Lack's Principle. *Am. Nat.,* Vol.100, No.916, pp. 687-690

Williams, P. D. & Day, T. (2003). Antagonistic Pleiotropy, Mortality Source Interactions, and the Evolutionary Theory of Senescence. *Evolution,* Vol. 57, No. 7 pp. 1478-1488

Zurbenko, I.; Porter, P. S.; Rao, S. T.; Ku, J. Y.; Gui, R.; Eskridge, R. E. (1996). Detecting Discontinuities in Time Series of Upper-air Data: Development and Demonstration of an Adaptive Filter Technique. *Journal of Climate,* Vol.9, No.12, pp. 3548-3560

6

Macromolecular Synthesis in the Urinary and Reproductive Systems

Tetsuji Nagata[1,2]
[1]Department of Anatomy and Cell Biology, Shinshu University School of Medicine, Matsumoto
[2]Shinshu Institute of Alternative Medicine and Welfare, Nagano
Japan

1. Introduction

This chapter deals with the third parts of the application of microscopic radioautography to the organ systems, including the urinary organs, the reproductive organs and the endocrine organs.

2. Macromolecular synthesis in the urinary system

The urinary system consists of the kidney and the urinary tract. We studied only the macromolecular synthesis in the kidneys of several groups of aging mice by LM and EM RAG, while the localization of an anti-allergic agent was observed in the urinary bladders of adult rats (Nagata 2005).

2.1 Macromolecular synthesis in the kidney

We studied only the DNA, RNA and glucides syntheses in the kidneys of several groups of aging mice by LM and EM RAG.

2.1.1 The DNA synthesis in the kidney

The kidneys of mammals microscopically consist of the nephrons, which can be divided into two portions, the renal corpuscles and the uriniferous tubules. The renal corpuscles are composed of the glomeruli which are covered with the Bowman's capsules. They are localized in the outer zone of the kidney, the renal cortex, while the uriniferous tubules are composed of two portions, the proximal portions and the distal portions which can further be divided into several portions which run from the outer zone of the kidney, the renal cortex, to the inner zone, the medulla. We studied the DNA synthesis by ^{3}H-thymidine radioautography in 3 groups of ddY mouse embryos from prenatal day 13 (Fig. 16A), day 15 (Fig. 16B) to day 19 in vitro, as well as perinatal mice from embryonic day 19 to postnatal day 1, 8, 30, 60 and 365 (1 year) in vivo (Hanai 1993, Hani et al. 1993, Hanai and Nagata 1994a,b).

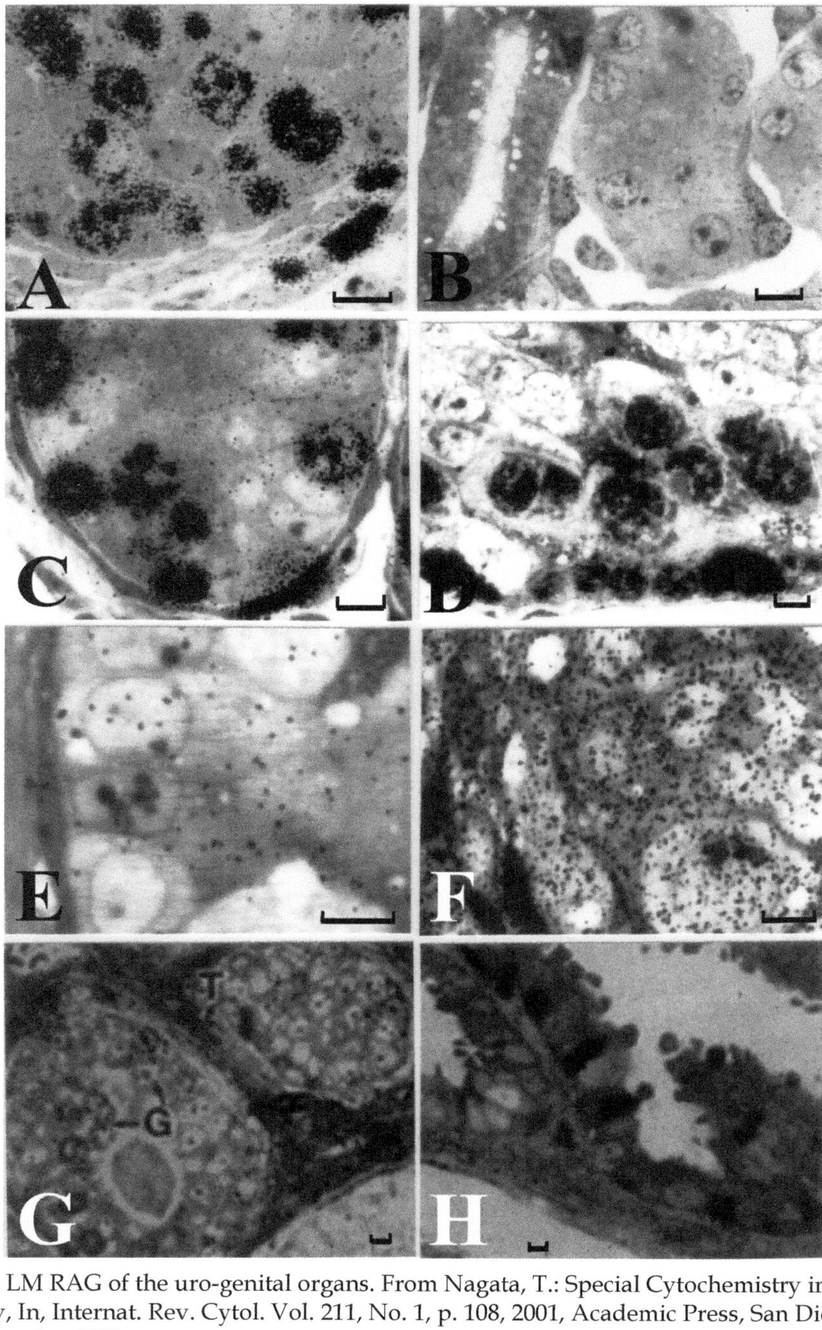

Fig. 16. LM RAG of the uro-genital organs. From Nagata, T.: Special Cytochemistry in Cell Biology, In, Internat. Rev. Cytol. Vol. 211, No. 1, p. 108, 2001, Academic Press, San Diego, USA, London, UK.

Fig. 16A. LM RAG of the metanephros of a prenatal day 13.5 mouse embryo labeled with ^{3}H-thymidine, showing DNA synthesis. x1,200.
Fig. 16B. LM RAG of the metanephros cortex of a prenatal day 15.5 mouse embryo labeled with ^{3}H-thymidine, showing DNA synthesis. x1,200.
Fig. 16C. LM RAG of the testis of a postnatal day 7 male mouse labeled with ^{3}H-thymidine, showing DNA synthesis. x800.
Fig. 16D. LM RAG of the testis of a postnatal year 1 male mouse labeled with ^{3}H-thymidine, showing DNA synthesis. x800.
Fig. 16E. LM RAG of the testis of a postnatal day 3 male mouse labeled with ^{3}H-uridine in vitro, showing RNA synthesis. x1,500.
Fig. 16F. LM RAG of the testis of a male mouse at postnatal day 1 labeled with ^{3}H-leucine in vitro, showing protein synthesis. x1,125.
Fig. 16G. LM RAG of the ovary of a postnatal day 3 female mouse labeled with ^{3}H-thymidine in vitro, showing DNA synthesis in granulosa cells (G) and theca cells (T). x400.
Fig. 16H. LM RAG of the oviduct of a postnatal day 30 female mouse labeled with ^{3}H-thymidine in vitro, showing DNA synthesis in epithelial cells. x400.

The labeling indices by LM RAG in glomeruli (28 to 32%) and uriniferous tubules (31 to 33%) in the superficial layer were higher than those of labeling indices (10 to 12%) and (8 to 16%) in the deeper layer from the late fetal to the suckling period, then decreased with aging from weaning to senescence (Fig. 17). EM RAG revealed the same results (Hanai and Nagata 1994a,b,c). At the same time, immunocytochemical staining of PCNA/cyclin was carried out in the same animals in several aging groups as ^{3}H-thymidine RAG (Hanai 1993, Hanai et al. 1993). The results from the PCNA/cyclin positive indices in respective aging groups were almost the same as the labeling indices with ^{3}H-thymidine RAG. The incorporation of ^{3}H-thymidine was formerly observed by EM RAG in mitochondrial matrix of cultured kidney cells from chickens and mice in vitro demonstrating mitochondrial DNA synthesis (Nagata et al. 1967b).

2.1.2 The RNA synthesis in the kidney

The RNA synthesis by incorporation of ^{3}H-uridine into the kidneys of aging mice was studied by LM and EM RAG (Hanai and Nagata 1994a,b, Nagata 2002). When the kidneys of several groups of aging mice from embryo to postnatal 1 year were radioautographed with ^{3}H-uridine either in vitro (embryonic day 15, 19 and postnatal day 1) and in vivo (embryonic day 19, postnatal day 1, 7, month 1, 2, 12), RNA synthesis was observed in all the cells of the kidney at various ages. The numbers of silver grains demonstrating the incorporation of ^{3}H-uridine in glomeruli (34.6 per cell) and uriniferous tubules (56.4 per cell) were higher in the superficial layer than those (15.6 and 18.6 per cell) in the deeper layer at embryonic day 15 and decreased gradually with aging. These results demonstrated the aging changes of RNA synthesis in the kidney.

2.1.3 Glucide synthesis in the kidney

The incorporations of ^{3}H-glucosamine in the kidneys of aging mice were studies by LM RAG (Joukura 1996, Joukura and Nagata 1995) and EM RAG (Joukura et al. 1996). Silver grains were observed over all the cell type nephrons at embryonic day 19, i.e., glomerular epithelial cells, endothelial cells, mesangial cells, Bowman's capsular cells and tubule cells.

In newborn and suckling stages, from postnatal day 1, 3, 7 to 14, both the renal corpuscles and urinary tubules were well differentiated and the number of silver grains increased (Figs. 36 C, D, E F, G, H in Nagata 2002). The results from grain counting revealed that the numbers of silver grains in both the renal corpuscles and the uriniferous tubules were less in the embryonic stage, but increased postnatally and reached peaks at day 1 and 3, then decreased to senescence at 1 year. These results showed that glucide synthesis in the kidney cells also changed with aging of animals.

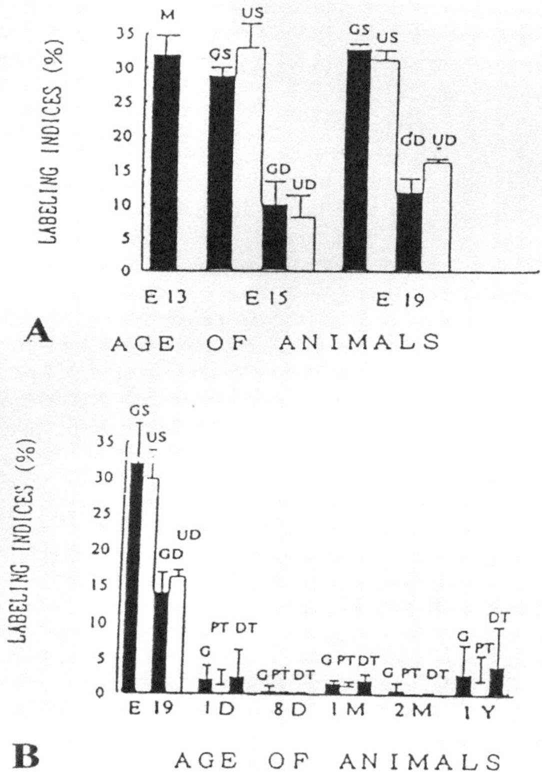

Fig. 17. Histogram showing aging changes of average labeling indices in respective cell types of the kidneys of aging mice labeled with ^{3}H-thyidine. Mean ± Standard Deviation. From Nagata, T.: Radioautographology, General and Special. In, Prog. Histochem. Cytochem. Vol. 37, No. 2, p. 156, 2002, Urban & Fischer, Jena, Germany

Fig. 17A. Labeling indices of the glomeruli and the uriniferous tubules of of mouse embryo from prenatal day 13 and 19.

Fig. 17B. Labeling indices of the glomeruli and the proximal and distal tubules of mouse embryo from prenatal day 19 to postnatal year 1. Abbreviations: GS=glomeruli of the superficial layer. US=uriniferous tubules of the superficial layer. GD=glomeruli of the deeper layer. UD=uriniferous tubules of the deeper layer. G=glomeruli. PT=proximal tubules. DT=distal tubules.

2.2 Localization of drugs in the urinary tract

The urinary tract is composed of the ureter, the urinary bladder and the urethra. We studied the urinary bladder of adult rats by LM RAG after oral administration of ^{3}H-tranilast, an anti-allergic agent produced by Kissei Pharmaceutical Co. (Momose et al. 1989, Nishigaki et al. 1987, 1990a,b). It was found that this agent specifically localized over the transitional epithelium and the endothelium of the veins in the mucosa of normal adult rats. However, any study on the DNA synthesis in the ureter, the urinary bladder and the urethra was not carried out.

3. Macromolecular synthesis in the reproductive system

The reproductive system or genital organs can be divided into two parts, the male genital organs and female genital organs. We studied the DNA and RNA syntheses and protein synthesis in several groups of aging mice, both male and female, by LM and EM RAG (Nagata 2002).

3.1 Macromolecular synthesis in the male genital organs

The male genital organs consist of the testis and its excretory ducts such as ductuli efferentes, ductus epididymidis, ductus deferens, ejaculatory ducts, auxiliary glands and penis. We studied both DNA and RNA syntheses in these organs of several groups of ddY aging mice by LM and EMRAG using macromolecular precursors.

3.1.1 The DNA synthesis in the male genital organs

Among the male genital organs, the testis was the main target of the scientific interests. Formerly, Clermont (1958, 1963) demonstrated using ^{3}H-thymidine radioautography that several stages of development of the spermatogonia were found at different levels in the germinal epithelium of mature men and rodents, with the most primitive germ cells found at the base and the more differentiated cells located at higher levels. We studied the DNA synthesis in the testis of several groups of aging mice.

3.1.1.1 The DNA synthesis in the testis

The structure of the testis of mammals is a compound tubular gland enclosed in tunica albuginea, a thick fibrous capsule. The parenchyma of the testis is composed of around 250 pyramidal compartments in men and animals, named lobules. Each lobule is made of convoluted seminiferous tubules, consisting of many spermatogenic cells differentiatiang to sperms among the supporting cells of Sertoli in the seminiferous epithelium, surrounded by the interstitial cells of Leydig. We first studied the macromolecular synthesis in the testis of aging male ddY mice at various ages (Gao 1993, Gao et al. 1994,1995a,b). When testicular tissues were labeled with ^{3}H-thymidine and observed by LM and EM RAG, many spermatogonia and myoid cells as well as Leydig cells were labeled with ^{3}H-thymidine at various ages from embryonic day 19 to postnatal day 1, 3, 7 (Fig. 16C), 14, month 1, 2, 6, 12 (Fig. 16D) and 24 (2 years). Silver grains were localized over the nuclei and several mitochondria of the spermatogonia showing DNA synthesis. Among of the aging groups, we counted the numbers of mitochondria per cell profile area, the numbers of labeled mitochondria per cell of the spermatogonia from 4 aging groups, prenatal embryonic day 19, postnatal day 3, and adults at month 1 and 6, and the labeling indices were calculated.

The results showed that the LI of the spermatogonia increased from embryonic day 19 (17%) to postnatal day 7 (25%) and month 1 (30%), reaching the maximum, then decreased to month 6 (20%) to year 2.

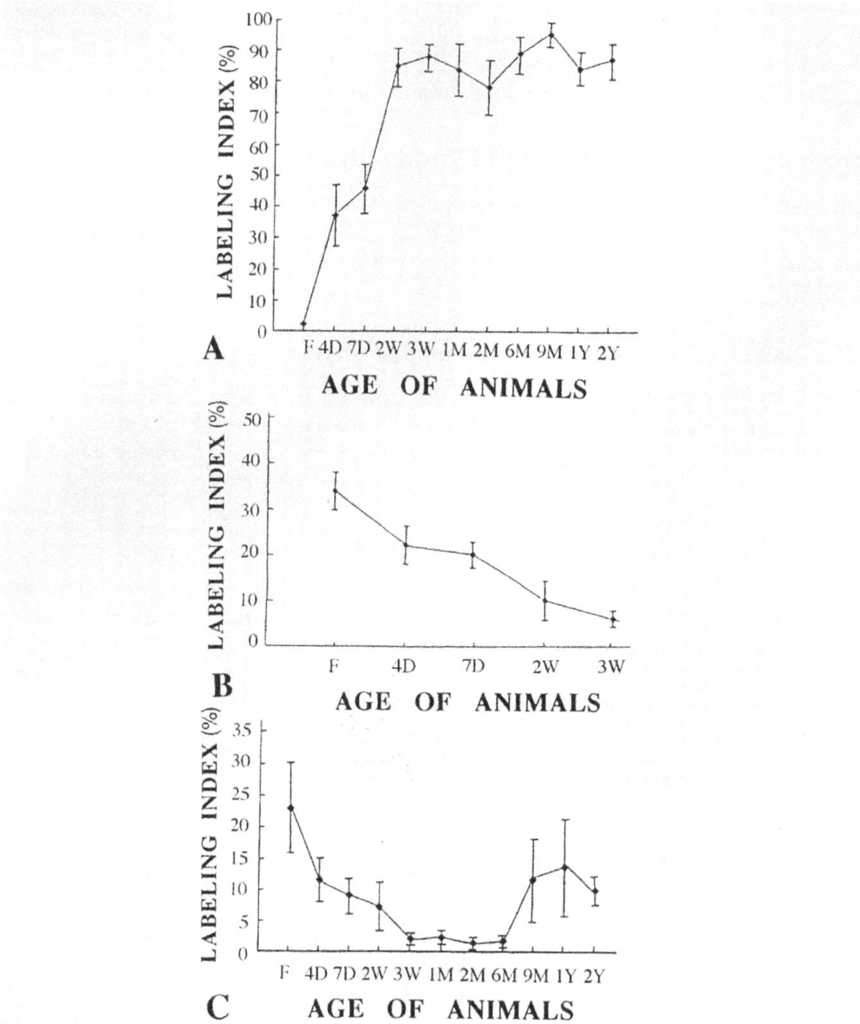

Fig. 18. Transitional curves of the labeling indices of respective cell types in the testis of aging mice labeled with ^3H-thymidine, showing DNA synthesis. Mean ± Standard Deviation. From Nagata, T.: Radioautographology, General and Special. In, Prog. Histochem. Cytochem. Vol. 37, No. 2, p. 160, 2002, Urban & Fischer, Jena, Germany

Fig. 18A. The spermatogonia.

Fig. 18B. The Sertoli cells.

Fig. 18C. The myoid cells.

At embryonic and neonatal stages, DNA synthesis of spermatocytes was weak and only a few labeled spermatogonia could be observed during the perinatal stages. The labeled spermatocytes were recognized at postnatal day 4 and 7 (Fig. 16C) and the number of labeled spermatogonia and spermatocytes increased from 2 and 3 weeks, keeping high level from month 1 to year 1 and 2 until senescence (Fig. 18A). However, the Sertoli's cells (Fig. 18B) and myoid cells (Fig. 18C) labeled with ^{3}H-thymidine were frequently observed at perinatal stages from embryo to postnatal day 7, while the labeling indices of both cells decreased from young adulthood (postnatal 2 weeks) to senescence (Gao 1993, Gao et al. 1994,1995a). The interstitial cells of Leydig in the testis surrounding the seminiferous tubules shall be described in the following section of the endocrine system in detail.

3.1.2 The RNA synthesis in the male genital organs

Among of the male genital organs, we studied the RNA synthesis in the testis of several groups of aging mice.

3.1.2.1 The RNA synthesis in the testis

We studied the RNA syntheses in aging mouse testis by LM and EM RAG, demonstrating the incorporations of ^{3}H-uridine into various cells of the seminiferous tubules (Gao 1993, Gao et al. 1994, Nagata 2002). The RNA synthesis of various cells in the seminiferous tubules was studied using ^{3}H-uridine. Silver grains due to ^{3}H-uridine demonstrating RNA synthesis were observed over the nuclei and cytoplasm of all spermatogonia, spermatocytes, Sertoli's cells, myoid cells of immature mice at perinatal stages at day 1 and 3 (Fig. 16E), as well as in mature and senescent mice from month 1, 6 to year 1 and 2. The synthetic activities of spermatogonia, Sertoli's cells and myoid cells as shown by grain counting with ^{3}H-uridine, as expressed by grain counting, were low (2-8 grain counts per 10 mm^2) at the embryonic and neonatal stages but increased at adult stages and maintained high levels (10-20 grain counts per 10 mm^2) until senescence. These results showed that DNA synthesis in myoid cells and Sertoli's cells increased at the perinatal stages and decreased from postnatal 2 weeks as described previously (Fig. 16A), while the RNA synthesis (Fig. 16E) in spermatogonia increased from postnatal 2 weeks together with DNA and protein syntheses (Fig. 16F) to senescence.

3.1.3 The protein synthesis in the male genital organs

We studied the protein synthesis of the reproductive system in both the male and female reproductive organs.

3.1.3.1 The protein synthesis in the testis

We studied the protein syntheses in aging mouse testis by LM and EM RAG, demonstrating the incorporations of ^{3}H-leucine into various cells of the seminiferous tubules (Gao 1993, Gao et al. 1994, Nagata 2002). The protein synthesis of various cells in the seminiferous tubules was first studied after administration of ^{3}H-leucine into aging male mice at various ages from perinatal to senescence at postnatal 2 years. Silver grains due to ^{3}H-leucine incorporation demonstrating protein synthesis were observed over the nuclei and cytoplasm of all the cells, spermatogonia, spermatocytes, Sertoli's cells, myoid cells of all male mice at respective stages from perinatal to senescence. The synthetic activities of spermatogonia, Sertoli's cells and myoid cells as shown by the number of silver grains due to ^{3}H-leucine, as expressed by grain counting, were low at the embryonic and neonatal stages but increased at adult stages and maintained high levels until senescence. These results showed that DNA synthesis in Sertoli's cells (Fig. 18B) and myoid cells (Fig. 18C) that increased at the perinatal stages and decreased from postnatal 2 weeks, while the DNA synthesis in spermatogonia increased from postnatal 2 weeks (Fig. 18A) together with RNA and protein syntheses to senescence.

3.2 Macromolecular synthesis in the female genital organs

The female genital organs consist of the ovary, the oviduct, the uterus, the vagina and the external genitals. We studied the macromolecular synthesis in the ovary, oviduct and uterus of several litters of ddY mice in aging.

3.2.1 The DNA synthesis in the female genital organ

Among the female genital organs, we studied the DNA synthesis in the ovary, oviduct and uterus of several litters of ddY mice in aging.

3.2.1.1 The DNA synthesis in the ovary

The ovary consists of the germinal epithelium covering the surface and the stroma containing many developing ovarian follicles depending upon the age of animals.

The nucleic acids, DNA and RNA, syntheses in the developing virgin mice ovaries of 6 litters, each 3 individuals, consisting of 36 female mice at various ages in respective precursors were studied by ^{3}H-thymidine and ^{3}H-uridine radioautography (Li 1994, Li and Nagata 1995, Li et al. 1992). The ^{3}H-thymidine incorporations were active in all surface epithelial cells, stromal and follicular cells of the ovaries between postnatal days 1 to 7 and decreased from day 14 (Fig. 16G) and maintained a lower level to day 60, while ^{3}H-uridine incorporations were active in all surface epithelial cells, stromal and follicular cells of the ovaries between postnatal days 1 to 7 and maintained medium levels from day 14 on.

The labeling indices with ^{3}H-thymidine showing DNA synthetic activity were high in all the surface epithelial cells, follicular cells and stromal cells of mice at neonatal stage from postnatal day 1 to 7, but decreased from day 40 to day 60 at mature stage (Fig. 19A). The grain counts showing RNA synthetic activity were high at neonatal stage from day 1 to day 7, and maintained medium levels from day 14 to day 60 at mature stage.

3.2.1.2 The DNA synthesis in the oviduct

The nucleic acids, DNA and RNA, syntheses in the oviducts of developing virgin mice at various ages were studied by ^{3}H-thymidine and ^{3}H-uridine radioautography (Li 1994, Li and Nagata 1995). The silver grains with ^{3}H-thymidine showing the DNA synthesis were observed over many nuclei in all surface epithelial cells, stromal and smooth muscle cells at neonatal stage between postnatal day 1 to 3 and decreased from day 7 to 30 (Fig. 16H) and 60, while the silver grains showing the RNA synthesis with ^{3}H-uridine were observed over the nuclei and cytoplasm of all the epithelial and stromal cells from postnatal day 1 to day 60. The labeling indices with ^{3}H-thymidine were high at neonatal stage from postnatal day 1 to 3 but decreased from day 7 to day 60 (Fig. 19C). The grain counts with ^{3}H-uridine were high at neonatal stage from postnatal day 1 to 3 and increased from day 7 to day 14 and decreased from day 30 to day 60. These results demonstrated an unparalleled alternation of DNA and RNA syntheses in the oviduct (Li and Nagata 1995).

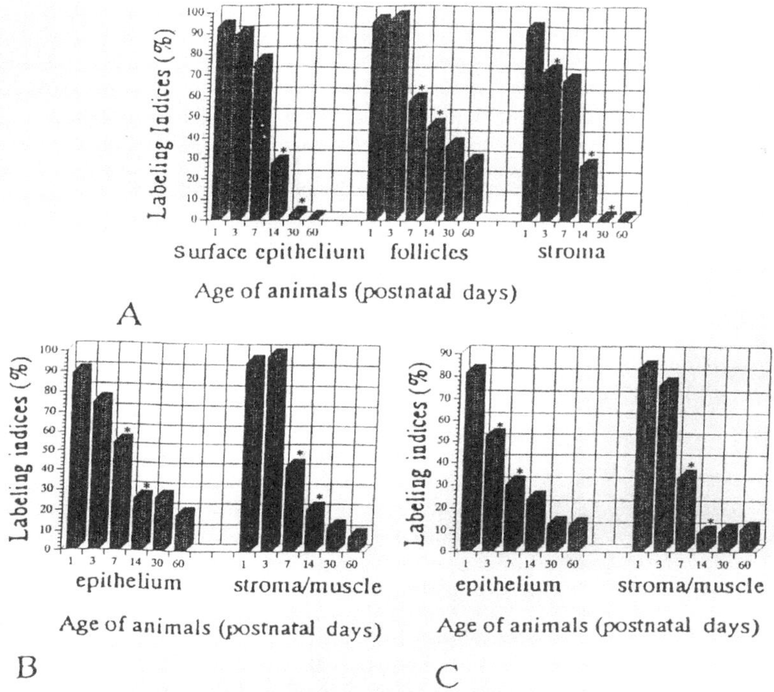

Fig. 19. Histogram showing aging changes of average labeling indices in respective cell types of female genital organs of aging mice labeled with ^{3}H-thymidine. Mean ± Standard Deviation. From Nagata, T.: Radioautographology, General and Special. In, Prog. Histochem. Cytochem. Vol. 37, No. 2, p. 166, 2002, Urban & Fischer, Jena, Germany
Fig. 19A. The ovary.
Fig. 19B. The uterus.
Fig. 19C. The oviduct.

3.2.1.3 The DNA synthesis in the uterus

The silver grains with ^{3}H-thymidine showing DNA synthesis of the uterus was observed over some of the nuclei of all the cells in the epithelia, stroma and smooth muscles from postnatal day 1 to 60 (Li 1994, Li and Nagata 1995). The labeling indices with ^{3}H-thymidine (Fig. 19B) were high (80-95%) at postnatal day 1 and decreased from day 3 to 60 (>10%). The silver grains showing RNA synthesis of the uterus were observed over all the nuclei and cytoplasm of all the cells in the uterine epithelia, stroma and smooth muscles from day 1 to 60. The number of silver grains in the uterine epithelium increased from postnatal day 1 to 7 and decreased from day 14 to 60, while they increased in the stroma from day 1 to 3 and decreased from day 7 to 60.

These results from the female genital organs showed that both DNA and RNA syntheses, as expressed by labeling indices and grain counting, were active in all kinds of cells, such as surface epithelial cells, stromal cells and follicular cells of the ovaries between postnatal days 1 to 7, then they decreased from day 14 to 60. However, the DNA synthesis in the epithelial cells and the stromal cells of both the uteri and the oviducts was active at postnatal day 1 and 3 and decreased from day 7 to 60. The RNA synthesis in the uteri and oviducts was active at postnatal day 1, increased from day 1 to day 14, and decreased from day 30 to 60. The unparalleled alteration of the DNA and RNA syntheses was shown between the ovary and the uterus or oviduct (Li and Nagata 1995).

We also studied PCNA/cyclin immunostaining in the ovary, oviduct and uterus (Li 1994). It was demonstrated that PCNA/cyclin positive cells were observed in the ovarian follicular epithelium, ovarian interstitial cells, tubal epithelial cells, tubal interstitial cells, uterine epithelial cells and uterine interstitial cells. The positive cells increased from postnatal 1 day to 3 and 7 days, then decreased from 14 days to senescence. These results accorded well with the results obtained from the ^{3}H-thymidine radioautography (Li 1994, Li and Nagata 1995). Moreover, the mucosubstance synthesis incorporating sulfuric acid was also carried out (Oliveira et al. 1991, Li et al. 1995).

3.2.1.4 The DNA synthesis in gametogenesis

The gametogenesis consists of both spermatogenesis in male germ cells and the oogenesis in female germ cells, leading to the implantation and further development of blastcysts. The macromolecular synthesis, DNA, RNA and protein synthesis, in both the testis and the ovary were already described in the sections of male and female reproductive systems (3.7.1. and 3.7.2.) previously.

3.2.1.5 The DNA synthesis in implantaion

In order to detect the changes of DNA, RNA and protein synthesis of the developing blastcysts in mouse endometrium during activation of the implantation, ovulations of female BALB/C strain mice were controlled by pregnant mare serum gonadotropin and human chorionic gonadotropin, then pregnant female mice were ovariectomized on the 4th day of pregnancy (Yamada 1993, Yamada and Nagata 1992a,b, 1993). The delay implantation state was maintained for 48 hrs and after 0 to 18 hrs of estrogen supply ^{3}H-thymidine was injected. The three regions of the endometrium, i. e. the interinplantation site, the antimesometrial and mesometrial sides of implantation site, were taken out and processed for LM and EM RAG. It was well known that the uterus of the rodent becomes receptive to blastcyst implantation only for a restricted period. This is called the implantation window which is intercalated between refractory states of the endometrium whose cycling is regulated by ovarian hormones (Yoshinaga 1988). We studied the changes of DNA synthesis by ^{3}H-thymidine (Yamada and Nagata 1992a,b) incorporations in the endometrial cells of pregnant-ovariectomized mice after time-lapse effect of nidatory estradiol. As the results, the endometrial cells showed topographical and chronological differences in the nucleic acid synthesis. The cells labeled with ^{3}H-thymidine increased after nidatory estradiol effects in the stromal cells around the blastocyst, but not in the epithelial cells. The results suggested that the presence of the blastocysts in the uterine lumen induced selective changes in the behavior of endometrial cells after nidatory estradiol effect showing the changes of DNA synthesis.

As for a lower vertebrate, cell proliferation and migration of scleroblasts and their precursor cells during ethisterone-induced anal-fin process formation of the medaka, orizias latipes, was studied by LM RAG labeled with ^{3}H-thymidine (Uwa and Nagata 1976). The results showed that the labeling index in the posterior margin of the joint plate rapidly increased and the scleroblast population in the central portion increased simultaneously from the 3rd to 5th day of ethisterone treatment. These results indicated that the scleroblasts and their precursor cells migrated from the peripheral portion to the central portion along the proxi-distal axis of the joint plate.

3.2.2 The RNA synthesis in the female genital organs

We studied the RNA synthesis of female reproductive organs of aging mice after the administration of ^{3}H-uridine at various ages.

3.2.2.1 The RNA synthesis in the ovary

The ovary consists of the germinal epithelium covering the surface and the stroma containing many developing ovarian follicles depending upon the age of animals.

The nucleic acids, DNA and RNA, syntheses in the developing virgin mice ovaries of 6 litters, each 3 individuals, consisting of 36 female mice at various ages in 2 groups were studied by ^{3}H-thymidine and ^{3}H-uridine radioautography (Li 1994, Li and Nagata 1995, Li et al. 1992). The ^{3}H-thymidine incorporations were active in all surface epithelial cells, stromal and follicular cells of the ovary between postnatal days 1 to 7 and decreased from day 14 (Fig. 16G) and maintained a lower level to day 60, while ^{3}H-uridine incorporations were active in all surface epithelial cells, stromal and follicular cells of the ovary between postnatal days 1 to 7 and maintained medium levels from day 14 on.

The labeling indices with ^{3}H-thymidine showing DNA synthetic activity were high in all the surface epithelial cells, follicular cells and stromal cells of mice at neonatal stage from postnatal day 1 to 7, but decreased from day 40 to day 60 at mature stage. The grain counts showing RNA synthetic activity were high at neonatal stage from day 1 to day 7, and maintained medium levels from day 14 to day 60 at mature stage.

3.2.2.2 The RNA synthesis in the oviduct

The nucleic acids, DNA and RNA, syntheses in the oviducts of developing virgin mice at various ages were studied by ^{3}H-thymidine and ^{3}H-uridine radioautography (Li 1994, Li and Nagata 1995). The silver grains with ^{3}H-thymidine showing the DNA synthesis were observed over many nuclei in all surface epithelial cells, stromal and smooth muscle cells at neonatal stage between postnatal day 1 to 3 and decreased from day 7 to 30 (Fig. 16H) and 60, while the silver grains showing the RNA synthesis with ^{3}H-uridine were observed over the nuclei and cytoplasm of all the epithelial and stromal cells from postnatal day 1 to day 60. The labeling indices with ^{3}H-thymidine were high at neonatal stage from postnatal day 1 to 3 but decreased from day 7 to day 60. The grain counts with ^{3}H-uridine were high at neonatal stage from postnatal day 1 to 3 and increased from day 7 to day 14 and decreased from day 30 to day 60. These results demonstrated an unparalleled alternation of DNA and RNA syntheses in the oviduct (Li and Nagata 1995).

3.2.2.3 The RNA synthesis in the uterus

The silver grains with ^{3}H-uridine showing RNA synthesis of the uterus was observed over almost all the nuclei and cytoplasm of all the cells in the epithelia, stroma and smooth muscles from postnatal day 1 to 60 (Li 1994, Li and Nagata 1995). The labeling indices with

^{3}H-thymidine were high (80-95%) at postnatal day 1 and decreased from day 3 to 60 (>10%). The silver grains showing RNA synthesis of the uterus were observed over all the nuclei and cytoplasm of all the cells in the uterine epithelia, stroma and smooth muscles from day 1 to 60. The number of silver grains in the uterine epithelium increased from postnatal day 1 to 7 and decreased from day 14 to 60, while they increased in the stroma from day 1 to 3 and decreased from day 7 to 60.

These results from the female genital organs showed that both DNA and RNA syntheses, as expressed by labeling indices and grain counting, were active in all kinds of cells, such as surface epithelial cells, stromal cells and follicular cells of the ovaries between postnatal days 1 to 7, then they decreased from day 14 to 60. However, the DNA synthesis in the epithelial cells and the stromal cells of both the uteri and the oviducts was active at postnatal day 1 and 3 and decreased from day 7 to 60. The RNA synthesis in the uteri and oviducts was active at postnatal day 1, increased from day 1 to day 14, and decreased from day 30 to 60. The unparalleled alteration of the DNA and RNA syntheses was shown between the ovary and the uterus or oviduct (Li and Nagata 1995).

We also studied PCNA/cyclin immunostaining in the ovary, oviduct and uterus (Li 1994). It was demonstrated that PCNA/cyclin positive cells were observed in the ovarian follicular epithelium, ovarian interstitial cells, tubal epithelial cells, tubal interstitial cells, uterine epithelial cells and uterine interstitial cells. The positive cells increased from postnatal 1 day to 3 and 7 days, then decreased from 14 days to senescence. These results accorded well with the results obtained from the ^{3}H-thymidine radioautography (Li 1994, Li and Nagata 1995). Moreover, the mucosubstance synthesis incorporating sulfuric acid was also carried out (Oliveira et al. 1991, 1995, Li et al. 1992).

3.2.2.4 The RNA synthesis in gametogenesis

The gametogenesis consists of both spermatogenesis in male germ cells and the oogenesis in female germ cells, leading to the implantation and further development of blastcysts. The macromolecular synthesis, DNA, RNA and protein synthesis, in both the testis and the ovary were already described in the sections of male and female reproductive systems (8.1.1. and 8.1.2.) previously.

3.2.2.5 The RNA synthesis in implantation

In order to detect the changes of DNA, RNA and protein synthesis of the developing blastcysts in mouse endometrium during activation of the implantation, ovulations of female BALB/C strain mice were controlled by pregnant mare serum gonadotropin and human chorionic gonadotropin, then pregnant female mice were ovariectomized on the 4th day of pregnancy (Yamada 1993, Yamada and Nagata 1992a,b, 1993). The delay implantation state was maintained for 48 hrs and after 0 to 18 hrs of estrogen supply ^{3}H-thymidine was injected. The three regions of the endometrium, i. e. the interinplantation site, the antimesometrial and mesometrial sides of implantation site, were taken out and processed for LM and EM RAG. It was well known that the uterus of the rodent becomes receptive to blastcyst implantation only for a restricted period. This is called the implantation window which is intercalated between refractory states of the endometrium whose cycling is regulated by ovarian hormones (Yoshinaga 1988). We studied the changes of DNA synthesis by ^{3}H-thymidine (Yamada and Nagata 1992a,b) incorporations in the endometrial cells of pregnant-ovariectomized mice after time-lapse effect of nidatory estradiol. As the results, the endometrial cells showed topographical and chronological differences in the nucleic acid synthesis. The cells labeled with ^{3}H-thymidine increased after nidatory estradiol effects in the stromal cells around the blastocyst, but not in the epithelial cells. The results suggested that the presence of the blastocysts in the uterine lumen induced selective changes in the behavior of endometrial cells after nidatory estradiol effect showing the changes of DNA synthesis.

As for a lower vertebrate, cell proliferation and migration of scleroblasts and their precursor cells during ethisterone-induced anal-fin process formation of the medaka, orizias latipes, was studied by LM RAG labeled with ^{3}H-thymidine (Uwa and Nagata 1976). The results showed that the labeling index in the posterior margin of the joint plate rapidly increased and the scleroblast population in the central portion increased simultaneously from the 3rd to 5th day of ethisterone treatment. These results indicated that the scleroblasts and their precursor cells migrated from the peripheral portion to the central portion along the proxi-distal axis of the joint plate.

3.2.3 The protein synthesis of the female genital organs

We studied the protein synthesis of female reproductive organs of aging mice after the administration of ^{3}H-leucine at various ages.

3.2.3.1 The protein synthesis in the uterus

We studied the protein synthesis of the developing blastocysts in female mouse endometrium during activation of the implantation. The ovulations of female BALB/C strain adult mice were controlled by pregnant mare serum gonadotropin and human chorionic gonadotropin, then pregnant female mice were ovariectomized on the 4th day of pregnancy (Yamada 1993, Yamada and Nagata 1992a,b, 1993). The delay implantation state was maintained for 48 hrs and after 0 to 18 hrs of estrogen supply. After the mice were injected with ^{3}H-leucine, they were sacrificed and the uteri were processed for LM and EMRAG. We studied the changes of protein synthesis by ^{3}H-leucine incorporations (Yamada 1993, Yamada and Nagata 1992a). As the results, the endometrial cells showed topographical and chronological differences in the protein synthesis. The cells labeled with ^{3}H-leucine were observed in both epithelial cells and stromal cells. Quantitative analysis revealed that the number of silver grains increased from 0 hr to 3 and 6 hr, reaching the peak at 6 hr and decreased from 12 to 18 hr. The protein synthesis in the decidual cells of pregnant mice uteri was compared to the endometrial cells of virgin mice uteri using ^{3}H-proline and ^{3}H-tryptophane incorporations. The results demonstrated that silver grains were localized over the endoplasmic reticulum and the Golgi apparatus of fibroblasts and accumulated over collagen fibrils in the extracellular matrix suggesting that the decidual cells produced collagen in the matrix. The collagen synthesis in the mouse decidual cells by ^{3}H-proline showed that silver grains were localized over the endoplasmic reticulum and Golgi apparatus of fibroblasts and accumulated over collagen fibrils in the extracellular matrix (Oliveira et al. 1991, 1995). However, analytical studies on protein synthesis in aging mice at various ages were not yet carried out.

3.2.3.2 The protein synthesis in the implantation

In order to detect the changes of DNA, RNA and protein synthesis of the developing blastocysts in female mouse endometrium during activation of the implantation, ovulations of female BALB/C strain mice were controlled by pregnant mare serum gonadotropin and human chorionic gonadotropin, then pregnant female mice were ovariectomized on the 4th day of pregnancy (Yamada 1993, Yamada and Nagata 1992a,b, 1993). The delay implantation state was maintained for 48 hrs and after 0 to 18 hrs of estrogen supply and ^{3}H-leucine was injected. The three regions of the endometrium, i. e. the interinplantation site, the antimesometrial and mesometrial sides of implantation site, were taken out and processed for LM and EM RAG. It was well known that the uterus of the rodent becomes receptive to blastocyst implantation only for a restricted period. This is called the implantation window which is intercalated between refractory states of the endometrium whose cycling is regulated by ovarian hormones (Yoshinaga 1988). We studied the changes of protein synthesis by ^{3}H-leucine (Yamada 1993, Yamada and Nagata 1992a) incorporations in the endometrial cells of pregnant-ovariectomized mice after time-lapse effect of nidatory estradiol. As the results, the endometrial cells showed topographical and chronological differences in the nucleic acid and protein synthesis. The cells labeled with ^{3}H-leucine were observed in both epithelial cells and stromal cells. Quantitative analysis revealed that the number of silver grains as expressed by grain counting per mm^2 in both the stromal and epithelial cells on the antimesometrial side with ^{3}H-leucine increased from 0 hr to 3 and 6 hr, reaching the peak at 6 hr and decreased from 12 to 18 hr. These results suggested that the presence of the blastocysts in the uterine lumen induced selective changes in the behavior of endometrial cells after nidatory estradiol effect showing the changes of DNA, RNA and protein synthesis. The time coincident peak of RNA and protein synthesis detected in the endometrial cells at the anti-mesometrial side of the implantation site, probably reflected the activation moment of the implantation window. The protein synthesis in the decidual cells of pregnant mice uteri was compared to the endometrial cells of virgin mice uteri using ^{3}H-proline and ^{3}H-tryptophane incorporations. The results demonstrated that silver grains were localized over the endoplasmic reticulum and the Golgi apparatus of fibroblasts and accumulated over collagen fibrils in the extracellular matrix suggesting that the decidual cells produced collagen in the matrix. On the other hand, collagen synthesis in the mouse decidual cells was studied by LM and EM RAG using ^{3}H-proline (Oliveira et al. 1991, 1995). Silver grains were localized over the endoplasmic reticulum and Golgi apparatus of fibroblasts and accumulated over collagen fibrils in the extracellular matrix. The results suggested that the decidual cells produced collagen into the matrix. The quantitative analysis showed that both incorporations in the decidual cells and the matrix increased in the pregnant mice than the endometrial cells in virgin mice.

3.2.4 The glucide synthesis in the reproductive system

Among the reproductive organs, only the mucosubstance synthesis with radiosulfate, $^{35}SO_4$, was studied in the ovaries of mice during the estrus cycle.

3.2.4.1 The glucide synthesis in the ovary

Litter mate groups of female ddY mice, aged 8-10 weeks, were divided into 4 groups, diestrus, proestrus, estrus and metestrus according to the vaginal smears. The ovaries were taken out, labeled with $^{35}SO_4$ in vitro and radioautographed. In all the animals, silver grains were localized over the granulosa and theca cells. Almost all compartments of the ovaries were labeled. The grain counts per cell changed according to cell cycle. From the results, it was concluded that all the cells of the ovary incorporated mucosubstances throughout the estrus cycle (Li et al. 1992).

4. References

Chen, S., Gao, F., Kotani, A., Nagata, T.: Age-related changes of male mouse submandibular gland: A morphometric and radioautographic study. Cell. Mol. Biol. 41, 117-124, 1995.

Clermont Y.: The contractime elements in the limiting membrane of the seminiferous tubules of rats. Exp. Cell Res. 15, 438-342, 1958.

Clermont, Y.: Renewal of spermatogonia in man. Amer. J. Anat. 112, 35-51, 1963.

Cui, H.: Light microscopic radioautographic study on DNA synthesis of nerve cells in the cerebella of aging mice. Cell. Mol. Biol. 41, 1139-1154, 1995.

Cui, H., Gao, F., Nagata, T.: Light microscopic radioautographic study on protein synthesis in perinatal mice corneas. Acta Histochem. Cytochem. 33, 31-37, 2000.

Duan, H., Gao, F., Li, S., Hayashi, K., Nagata, T.: Aging changes and fine structure and DNA synthesis of esophageal epithelium in neonatal, adult and old mice. J. Clin. Electron Microsc. 25, 452-453, 1992.

Duan, H., Gao, F., Li, S., Nagata, T.: Postnatal development of esophageal epithelium in mouse: a light and electron microscopic radioautographic study. Cell. Mol. Biol. 39, 309-316, 1993.

Duan, H., Gao, F., Oguchi, K., Nagata, T.: Light and electron microscopic radioautographic study on the incorporation of ^{3}H-thymidine into the lung by means of a new nebulizer. Drug Res. 44, 880-883, 1994.

Feulgen, R., Rossenbeck, H.: Mikroskopische-chemischer Nachweis einer Nukeinsaeure von Thymus der Thymonukeinsaeure Z. Physik. Chem. 135, 203-248, 1924.

Fujii, Y., Ohno, S., Yamabayashi, S., Usuda, N., Saito, H., Furuta, S., Nagata, T.: Electron microscopic radioautography of DNA synthesis in primary cultured cells from an IgG myeloma patient. J. Clin. Electr. Microsc. 13, 582-583, 1980.

Gao, F.: Study on the macromolecular synthesis in aging mouse seminiferous tubules by light and electron microscopic radioautography. Cell. Mol. Biol. 39, 659-672, 1993.

Gao, F., Toriyama, K., Nagata, T.: Light microscopic radioautographic study on the DNA synthesis of prenatal and postnatal aging mouse retina after labeled thymidine injection. Cell. Mol. Biol. 38, 661-668, 1992a.

Gao, F., Li, S., Duan, H., Ma, H., Nagata, T.: Electron microscopic radioautography on the DNA synthesis of prenatal and postnatal mice retina after labeled thymidine injection. J. Clin. Electron Microsc. 25, 721-722, 1992b.

Gao, F., Toriyama, K., Ma, H., Nagata, T.: Light microscopic radioautographic study on DNA synthesis in aging mice corneas. Cell. Mol. Biol. 39, 435-441, 1993.

Gao, F., Ma, H., Sun, L., Jin, C., Nagata, T.: Electron microscopic radioautographic study on the nucleic acids and protein synthesis in the aging mouse testis. Med. Electron Microsc. 27, 360-362, 1994.

Gao, F., Chen, S., Sun, L., Kang, W., Wang, Z., Nagata, T.: Radioautographic study of the macromolecular synthesis of Leydig cells in aging mice testis. Cell. Mol. Biol. 41, 145-150, 1995a.

Gao, F., Jin, C., Ma, H., Sun, L., Nagata, T.: Ultrastructural and radioautographic studies on DNA synthesis in Leydig cells of aging mouse testis. Cell. Mol. Biol. 41, 151-160, 1995b.

Gunarso, W.: Radioautographic studies on the nucleic acid synthesis of the retina of chick embryo. I. Light microscopic radioautography. Shinshu Med. J. 32, 231-240, 1984a.

Gunarso, W.: Radioautographic studies on the nucleic acid synthesis of the retina of chick embryo. II. Electron microscopic radioautography. Shinshu Med. J. 32, 241-248, 1984b.

Gunarso, W., Gao, F., Cui, H., Ma, H., Nagata, T.: A light and electron microscopic radioautographic study on RNA synthesis in the retina of chick embryo. Acta Histochem. 98, 309-32, 1996.

Gunarso, W., Gao, F., Nagata, T.: Development and DNA synthesis in the retina of chick embryo observed by light and electron microscopic radioautography. Cell. Mol. Biol. 43, 189-201, 1997.

Hanai, T.: Light microscopic radioautographic study of DNA synthesis in the kidneys of aging mice. Cell. Mol. Biol. 39, 81-91, 1993.

Hanai, T., Nagata, T.: Electron microscopic radioautographic study on DNA and RNA synthesis in perinatal mouse kidney. In, Radioautography in Medicine, Nagata, T., Ed., pp. 127-131, Shinshu University Press, Matsumoto, 1994a.

Hanai, T., Nagata, T.: Study on the nucleic acid synthesis in the aging mouse kidney by light and electron microscopic radioautography. In, Radioautography in Medicine, Nagata, T., Ed., pp. 209-214, Shinshu University Press, Matsumoto, 1994b.

Hanai, T., Nagata, T.: Electron microscopic study on nucleic acid synthesis in perinatal mouse kidney tissue. Med. Electron Microsc. 27, 355-357, 1994c.

Hanai, T., Usuda, N., Morita, T., Shimizu, T., Nagata, T.: Proliferative activity in the kidneys of aging mice evaluated by PCNA/cyclin immunohistochemistry. Cell. Mol. Biol. 39, 181-191, 1993.

Hayashi, K., Gao, F., Nagata, T.: Radioautographic study on ^{3}H-thymidine incorporation at different stages of muscle development in aging mice. Cell. Mol. Biol. 39, 553-560, 1993.

Ito, M.: The radioautographic studies on aging change of DNA synthesis and the ultrastructural development of mouse adrenal gland. Cell. Mol. Biol. 42, 279-292, 1996.

Ito, M., Nagata, T.: Electron microscopic radioautographic studies on DNA synthesis and ultrastructure of aging mouse adrenal gland. Med. Electron Microsc. 29, 145-152, 1996.

Izumiyama, K., Kogure, K., Kataoka, S., Nagata, T.: Quantitative analysis of glucose after transient ischemia in the gerbil hippocampus by light and electron microscope radioautography. Brain Res. 416, 175-179, 1987.

Jamieson, J. D., Palade, G. E.: Intracellular transport of secretory proteins in the pancreatic exocrine cells. J. Cell Biol. 34, 577-615, 1967.

Jin, C.: Study on DNA synthesis of aging mouse colon by light and electron microscopic radioautography. Cell. Mol. Biol. 42, 255-268, 1996.

Jin, C., Nagata, T.: Light microscopic radioautographic study on DNA synthesis in cecal epithelial cells of aging mice. J. Histochem. Cytochem. 43, 1223-1228, 1995a.

Jin, C., Nagata, T.: Electron microscopic radioautographic study on DNA synthesis in cecal epithelial cells of aging mice. Med. Electron Microsc. 28, 71-75, 1995b.

Joukura, K.: The aging changes of glycoconjugate synthesis in mouse kidney studied by ^{3}H-glucosamine radioautography. Acta Histochem. Cytochem. 29, 57-63, 1996.

Joukura, K., Nagata, T.: Aging changes of ^{3}H-glucosamine incorporation into mouse kidney observed by radioautography. Acta Histochem. Cytochem. 28, 494-494, 1995.

Joukura, K., Usuda, N., Nagata, T.: Quantitative study on the aging change of glycoconjugates synthesis in aging mouse kidney. Proc. Xth Internat. Cong. Histochem. Cytochem., Acta Histochem. Cytochem. 29, Suppl. 507-508, 1996.

Kobayashi, K., Nagata, T.: Light microscopic radioautographic studies on DNA, RNA and protein syntheses in human synovial membranes of rheumatoid arthritis patients. J. Histochem. Cytochem. 42, 982-982, 1994.

Komiyama, K., Iida, F., Furihara, R., Murata, F., Nagata, T.: Electron microscopic radioautographic study on ^{125}I-albumin in rat gastric mucosal epithelia. J. Clin. Electron Microsc. 11, 428-429, 1978.

Kong, Y.: Electron microscopic radioautographic study on DNA synthesis in perinatal mouse retina. Cell. Mol. Biol. 39, 55-64, 1993.

Kong, Y., Nagata, T.: Electron microscopic radioautographic study on nucleic acid synthesis of perinatal mouse retina. Med. Electron Microsc. 27, 366-368, 1994.

Kong, Y., Usuda, N., Nagata, T.: Radioautographic study on DNA synthesis of the retina and retinal pigment epithelium of developing mouse embryo. Cell. Mol. Biol. 38, 263-272, 1992a.

Kong, Y., Usuda, N., Morita, T., Hanai, T., Nagata, T.: Study on RNA synthesis in the retina and retinal pigment epithelium of mice by light microscopic radioautography. Cell. Mol. Biol. 38, 669-678, 1992b.

Leblond, C. P.: Localization of newly administered iodine in the thyroid gland as indicated by radioiodine. J. Anat. 77, 149-152, 1943.

Leblond, C. P.: The life history of cells in renewing systems. Am. J. Anat. 160, 113-158, 1981.

Leblond, C. P., Messier, B.: Renewal of chief cells and goblet cells in the small intestine as shown by radioautography after injection of thymidine-^{3}H into mice. Anat. Rec. 132: 247-259. 1958.

Li, S.: Relationship between cellular DNA synthesis, PCNA expression and sex steroid hormone receptor status in the developing mouse ovary, uterus and oviduct. Histochemistry 102, 405-413, 1994.

Li, S., Nagata, T.: Nucleic acid synthesis in the developing mouse ovary, uterus and oviduct studied by light and electron microscopic radioautography. Cell. Mol. Biol. 41, 185-195, 1995.

Li, S., Gao, F., Duan, H., Nagata, T.: Radioautographic study on the uptake of $^{35}SO_4$ in mouse ovary during the estrus cycle. J. Clin. Electron Microsc. 25, 709-710, 1992.

Liang, Y.: Light microscopic radioautographic study on RNA synthesis in the adrenal glands of aging mice. Acta Histochem. Cytochem. 31, 203-210, 1998.

Liang, Y., Ito, M., Nagata, T.: Light and electron microscopic radioautographic studies on RNA synthesis in aging mouse adrenal gland. Acta Anat. Nippon. 74, 291-300, 1999.

Ma, H.: Light microscopic radioautographic study on DNA synthesis of the livers in aging mice. Acta Anat. Nippon. 63, 137-147, 1988.

Ma, H., Nagata, T.: Electron microscopic radioautographic study on DNA synthesis of the livers in aging mice. J. Clin. Electron Microsc. 21, 335-343, 1988a.

Ma, H., Nagata, T.: Studies on DNA synthesis of aging mice by means of electron microscopic radioautography. J. Clin. Electron Microsc. 21, 715-716, 1988b.

Ma, H., Nagata, T.: Electron microscopic radioautographic studies on DNA synthesis in the hepatocytes of aging mice as observed by image analysis. Cell. Mol. Biol. 36, 73-84, 1990a.

Ma, H., Nagata, T.: Study on RNA synthesis in the livers of aging mice by means of electron microscopic radioautography. Cell. Mol. Biol. 36, 589-600, 1990b.

Ma, H., Nagata, T.: Collagen and protein synthesis in the livers of aging mice as studied by electron microsopic radioautography. Ann. Microsc. 1, 13-22, 2000.

Ma, H., Gao, F., Olea, M. T., Nagata, T.: Protein synthesis in the livers of aging mice studied by electron microscopic radioautography. Cell. Mol. Biol. 37, 607-615, 1991.

Matsumura, H., Kobayashi, Y., Kobayashi, K., Nagata, T.: Light microscopic radioautographic study of DNA synthesis in the lung of aging salamander, Hynobius nebulosus. J. Histochem. Cytochem. 42, 1004-1004, 1994.

Momose, Y., Nagata, T.: Radioautographic study on the intracellular localization of a hypolipidemic agent, bezafibrate, a peroxisome proliferator, in cultured rat hepatocytes. Cell. Mol. Biol. 39, 773-781, 1993a.

Momose, Y., Naito, J., Nagata, T.: Radioautographic study on the localization of an anti-allergic agent, tranilast, in the rat liver. Cell. Mol. Biol. 35, 347-355, 1989.

Momose, Y., Shibata, N., Kiyosawa, I., Naito, J., Watanabe, T., Horie, S., Yamada, J., Suga, T., Nagata, T.: Morphometric evaluation of species differences in the effects of bezafibrate, a hypolipidemic agent, on hepatic peroxisomes and mitochondria. J. Toxicol. Pathol. 6, 33-45, 1993b.

Momose, Y., Naito, J., Suzawa, H., Kanzawa, M., Nagata, T.: Radioautographic study on the intracellular localization of bezafibrate in cultured rat hepatoctyes. Acta Histochem. Cytochem. 28, 61-66, 1995.

Morita, T.: Radioautographic study on the aging change of ^{3}H-glucosamine uptake in mouse ileum. Cell. Mol. Biol. 39, 875-884, 1993.

Morita, T., Usuda, N. Hanai, T., Nagata, T.: Changes of colon epithelium proliferation due to individual aging with PCNA/cyclin immunostaining comparing with ^{3}H-thymidine radioautography. Histochemistry 101, 13-20, 1994.

Murata, F., Momose,Y. , Yoshida, K., Nagata, T.: Incorporation of ^{3}H-thymidine into the nucleus of mast cells in adult rat peritoneum. Shinshu Med. J. 25, 72-77, 1977a.

Murata, F., Momose, Y., Yoshida, K., Ohno, S., Nagata, T.: Nucleic acid and mucosubstance metabolism of mastocytoma cells by means of electron microscopic radioautography. Acta Pharmacol. Toxicol. 41, 58-59, 1977b.

Murata, F., Yoshida, K., Ohno, S., Nagata, T.: Ultrastructural and electron microscopic radioautographic studies on the mastocytoma cells and mast cells. J. Clin. Electron Microsc. 11, 561-562, 1978.

Murata, F., Yoshida, K., Ohno, S., Nagata, T.: Mucosubstances of rabbit granulocytes studied by means of electron microscopic radioautography and X-ray microanalysis. Histochemistry 61, 139-150, 1979.

Nagata, T.: On the relationship between cell division and cytochrome oxidase in the Yoshida sarcoma cells. Shinshu Med. J. 5: 383-386, 1956.

Nagata, T.: Studies on the amitosis in the Yoshida sarcoma cells. I. Observation on the smear preparation under normal conditions. Med. J. Shinshu Univ. 2: 187-198, 1957a.

Nagata, T.: Studies on the amitosis in the Yoshida sarcoma cells. II. Phase-contrast microscopic observations under normal conditions. Med. J. Shinshu Univ. 2: 199-207, 1957b.

Nagata, T.: Cell divisions in the liver of the fetal and newborn dogs. Med. J. Shinshu Univ. 4: 65-73, 1959.

Nagata, T.: A radioautographic study of the DNA synthesis in rat liver, with special reference to binucleate cells. Med. J. Shinshu Univ. 7, 17-25, 1962.

Nagata, T.: A quantitative study on the ganglion cells in the small intestine of the dog. Med. J. Shinshu Univ. 10, 1-11, 1965.

Nagata, T.: A radioautographic study on the RNA synthesis in the hepatic and the intestinal epithelial cells of mice after feeding with special reference to binuclearity. Med. J. Shinshu Univ. 11, 49-61, 1966.

Nagata, T.: On the increase of binucleate cells in the ganglion cells of dog small intestine due to experimental ischemia. Med. J. Shinshu Univ. 12, 93-113, 1967a.

Nagata, T.: A radioautographic study on the protein synthesis in the hepatic and the intestinal epithelial cells of mice, with special reference to binucleate cells. Med. J. Shinshu Univ. 12, 247-257, 1967b.

Nagata, T.: Chapter 3. Application of microspectrophotometry to various substances. In , Introduction to Microspectrophotometry. Isaka, S., Nagata, T., Inui, N., Eds., Olympus Co., Tokyo, pp. 49-155, 1972a.

Nagata, T.: Electron microscopic dry-mounting autoradiography. Proc. 4th Internat. Cong. Histochem. Cytochem. Kyoto, pp. 43-44, 1972b.

Nagata, T.: Electron microscopic radioautography of intramitochondrial RNA synthesis of HeLa cells in culture. Histochemie 32, 163-170, 1972c.

Nagata, T.: Quantitative electron microscope radioautography of intramitochondrial nucleic acid synthesis. Acta Histochem. Cytochem. 5, 201-203, 1972d.

Nagata, T.: Electron microscopic observation of target cells previously observed by phase-contrast microscopy: Electron microscopic radioautography of laser beam irradiated cultured cells. J. Clin. Electron Microsc. 17, 589-590, 1984.

Nagata, T.: Principles and techniques of radioautography. In, Histo- and Cyto-chemistry 1985, Japan Society of Histochemistry and Cytochemistry, Ed., Gakusai Kikaku Co., Tokyo, pp. 207-226, 1985.

Nagata, T.:. Electron microscopic radioautography and analytical electron microscopy. J. Clin. Electron Microsc. 24, 441-442, 1991.

Nagata, T.: Radiolabeling of soluble and insoluble compounds as demonstrated by light and electron microscopy. Recent Advances in Cellular and Molecular Biology, Wegmann, R. J., Wegmann, M. A., Eds. Peters Press, Leuven, Vol. 6, pp. 9-21, 1992.

Nagata, T.: Quantitative analysis of histochemical reactions: Image analysis of light and electron microscopic radioautograms. Acta Histochem. Cytochem. 26, 281-291, 1993a.

Nagata, T. Quantitative light and electron microscopic radioautographic studies on macromolecular synthesis in several organs of prenatal and postnatal aging mice. Chinese J. Histochem. Cytochem. 2: 106-108, 1993b.

Nagata, T.: Electron microscopic radioautography with cryo-fixation and dry-mounting procedure. Acta Histochem. Cytochem. 27: 471-489, 1994a.

Nagata, T.: Application of electron microscopic radioautography to clinical electron microscopy. Med. Electron Microsc. 27; 191-212, 1994b.

Nagata, T.: Radioautography in Medicine. Shinshu University Press, 268pp, Matsumoto, 1994c.

Nagata, T.: Radioautography, general and special. In, Histo- and Cyto-chemistry 1994, Japan Society of Histochemistry and Cytochemistry, ed, pp. 219-231, Gakusai Kikaku Co., Tokyo, 1994d.

Nagata, T., Application of electron microscopic radioautography to clinical electron microscopy. Med. Electron Microsc. 27, 191-212, 1994e.

Nagata, T.: Light and electron microscopic radioautographic study on macromolecular synthesis in digestive organs of aging mice. Cell. Mol. Biol. 41, 21-38, 1995a.

Nagata, T.: Histochemistry of the organs: Application of histochemistry to anatomy. Acta Anat. Nippon. 70, 448-471, 1995b.

Nagata, T.: Three-dimensional observation of whole mount cultured cells stained with histochemical reactions by ultrahigh voltage electron microscopy. Cell. Mol. Biol. 41, 783-792, 1995c.

Nagata, T.: Morphometry in anatomy: image analysis on fine structure and histochemical reactions with special reference to radioautography. Ital. J. Anat. 100 (Suppl. 1), 591-605, 1995d.

Nagata, T.: Technique and application of electron microscopic radioautography. J. Electron Microsc. 45, 258-274, 1996a.

Nagata, T.: Techniques of light and electron microscopic radioautography. In, Histochemistry and Cytochemistry 1996. Proc. Xth Internat. Congr. Histochem. Cytochem. Acta Histochem. Cytochem. 29 (Suppl.), 343-344, 1996b.

Nagata, T.: Remarks: Radioautographology, general and special. Cell. Mol. Biol. 42 (Suppl.), 11-12, 1996c.

Nagata, T.: On the terminology of radioautography vs. autoradiography. J. Histochem. Cytochem. 44, 1209-1209, 1996d.

Nagata, T.: Techniques and applications of microscopic radioautography. Histol. Histopathol. 12, 1091-1124, 1997a.

Nagata T.: Three-dimensional observation on whole mount cultured cells and thick sections stained with histochemical reactions by high voltage electron microscopy. In, Recent Advances in Microscopy of Cells, Tissues and Organs, Motta, P., Ed., Antonio Delfino Editore, Roma, pp. 37-44, 1997b.

Nagata, T.: Radioautographic study on collagen synthesis in the ocular tissues. J. Kaken Eye Res. 15, 1-9, 1997c.

Nagata, T.: Techniques of radioautography for medical and biological research. Braz. J. Biol. Med. Res. 31, 185-195, 1998a.

Nagata, T.: Radioautographology, the advocacy of a new concept. Braz. J. Biol. Med. Res. 31, 201-241, 1998b.

Nagata, T.: Radioautographic studies on DNA synthesis of the bone and skin of aging salamander. Bull. Nagano Women's Jr. College 6, 1-14, 1998c.

Nagata, T.: 3D observation of cell organelles by high voltage electron microscopy. Microscopy and Analysis, Asia Pacific Edition, 9, 29-32, 1999a.

Nagata, T.: Application of histochemistry to anatomy: Histochemistry of the organs, a novel concept. Proc. XV Congress of the International Federation of Associations of Anatomists, Ital. J. Anat. Embryol. 104 (Suppl. 1), 486-486, 1999b.

Nagata, T.: Aging changes of macromolecular synthesis in various organ systems as observed by microscopic radioautography after incorporation of radiolabeled precursors. Methods Find. Exp. Clin. Pharmacol. 21, 683-706, 1999c.

Nagata, T.: Radioautographic study on protein synthesis in mouse cornea. J. Kaken Eye Res. 8, 8-14, 1999d.

Nagata, T.: Radioautographology, general and special: a novel concept. Ital. J. Anat. Embryol. 104 (Suppl. 1), 487-487, 1999e.

Nagata, T.: Three-dimensional observations on thick biological specimens by high voltage electron microscopy. Image Analysis Stereolog. 19, 51-56, 2000a.

Nagata, T.: Biological microanalysis of radiolabeled and unlabeled compounds by radioautography and X-ray microanalysis. Scanning Microscopy International, 14, on line, 2000b.

Nagata, T.: Electron microscopic radioautographic study on protein synthesis in pancreatic cells of perinatal and aging mice. Bull. Nagano Women's Jr. College 8, 1-22, 2000c.

Nagata, T.: Light microscopic radioautographic study on radiosulfate incorporation into the tracheal cartilage in aging mice. Acta Histochem. Cytochem. 32, 377-383, 2000d.

Nagata, T.: Introductory remarks: Special radioautographology. Cell. Mol. Biol. 46 (Congress Suppl.), 161-161, 2000e.

Nagata, T.: Special radioautographology: the eye. J. Kaken Eye Res. 18, 1-13, 2000f.

Nagata, T.: Three-dimensional high voltage electron microscopy of thick biological specimens. Micron 32, 387-404, 2001a.

Nagata, T.: Three-dimensional and four-dimensional observation of histochemical and cytochemical specimens by high voltage electron microscopy. Acta Histochem. Cytochem. 34, 153-169, 2001b.

Nagata, T. : Special cytochemistry in cell biology. In, Internat. Rev. Cytol. Jeon, K.W., ed., Vol. 211, Chapter 2, Academic Press, New York, pp. 33-154, 2001c.

Nagata, T. : Radioautographology General and Special, In, Prog. Histochem. Cytochem., Graumann, W., Ed., Vol. 37 No. 2, Urban & Fischer, Jena, pp. 57-226, 2002.

Nagata T.: Light and electron microscopic study on macromolecular synthesis in amitotic hepatocyte mitochondria of aging mice. Cell. Mol. Biol. 49, 591-611, 2003.

Nagata, T.: X-ray microanalysis of biological specimens by high voltage electron microscopy. In, Prog. Histochem. Cytochem., Graumann, W., Ed., Vol. 39, No. 4, Urban & Fischer Verlag, Jena, pp. 185-320, 2004.

Nagata T.: Aging changes of macromolecular synthesis in the uro-genital organs as revealed by electron microscopic radioautography. Ann. Rev. Biomed. Sci. 6, 13-78, 2005.

Nagata T.: Electron microscopic radioautographic study on protein synthesis in hepatocyte mitochondria of developing mice. Ann. Microsc. 6, 43-54, 2006a.

Nagata T.: Electron microscopic radioautographic study on nucleic acids synthesis in hepatocyte mitochondria of developing mice. The Sci. World J. 6: 1583-1598, 2006b.

Nagata T.: Macromolecular synthesis in hepatocyte mitochondria of aging mice as revealed by electron microscopic radioautography. I: Nucleic acid synthesis. In, Modern Research and Educational Topics in Microscopy. Mendez-Vilas, A. and Diaz, J. Eds., Formatex Micrscopy Series No. 3, Vol. 1, Formatex, Badajoz, Spain, pp. 245-258, 2007a.

Nagata T.: Macromolecular synthesis in hepatocyte mitochondria of aging mice as revealed by electron microscopic radioautography. II: Protein synthesis. In, Modern Research and Educational Topics in Microscopy. Mendez-Vilas, A. and Diaz, J. eds., Formatex Micrscopy Series No. 3, Vol. 1, Formatex, Badajoz, Spain, pp. 259-271, 2007b.

Nagata, T.: Electron microscopic radioautographic study on macromolecular synthesis in hepatocyte mitochondria of aging mouse. J. Cell Tissue Res. 7, 1019-1029, 2007c.

Nagata, T.; Electron microscopic radioautographic study on nucleic acids synthesis in hepatocyte mitochondria of developing mice. Trends Cell Molec. Biol. 2, 19-33, 2007d.

Nagata, T.; Aging changes of macromolecular synthesis in the mitochondria of mouse hepatocytes as revealed by microscopic radioautography. Ann. Rev. Biomed. Sci. 9, 30-36, 2007e.

Nagata, T.: Radioautographology, Bull. Shinshu Institute Alternat. Med. 2, 3-32, 2007f.

Nagata, T.: Electron microscopic radioautographic study on mitochondrial DNA synthesis in adrenal cortical cells of developing mice. J. Cell. Tis. Res. 8, 1303-1312, 2008a.

Nagata T.: Electron microscopic radioautographic study on mitochondrial DNA synthesis in adrenal cortical cells of developing and aging mice. The Sci. World J. 8, 683-697. 2008b.

Nagata, T.: Sexual difference between the macromolecular synthesis of hepatocyte mitochondria in male and female mice in aging as revealed by electron microscopic radioautography. Chapter 22. In, Women and Aging: New Research, H. T. Bennninghouse, A. D. Rosset, Eds. Nova Biomed. Books, New York, pp. 461-487, 2009a

Nagata, T.: Protein synthesis in hepatocytes of mice as revealed by electron microscopic radioautography. In, Protein Biosynthesis. Esterhouse, T. E. and Petrinos, L. B., Eds., Nova Biomed. Books, New York, pp. 133-161, 2009b.

Nagata, T.: Electron microscopic radioatuographic studies on macromolecular synthesis in mitochondria of various cells. 18EMSM Conference Proc. 9th Asia-Pacific Microscopy Conference (APMC9), Kuala Lumpur, Malaysia, pp. 48-50, 2009c.

Nagata, T.: Recent studies on macromolecular synthesis labeled with ^{3}H-thymidine in various organs as revealed by electron microscopic radioautography. Current Radiopharmaceutics 2, 118-128, 2009d.

Nagata, T.: Electron microscopic radioautographic study on mitochondrial DNA synthesis in adrenal medullary cells of developing and aging mice. J. Cell Tissue Res. 9, 1793-1802, 2009e.

Nagata, T.: Applications of high voltage electron microscopy to thick biological specimens. Ann. Microsc. 9, 4-40, 2009f.

Nagata, T.: Electron microscopic radioautographic study on DNA synthesis of mitochondria in adrenal medullary cells of aging mice. Open Anat. J. 1, 14-24, 2009g.

Nagata, T.: Electron microscopic radioautographic studies on macromolecular synthesis in mitochondria of animal cells in aging. Ann. Rev. Biomed. Sci. 11, 1-17, 2009h.

Nagata, T.: Electron microscopic radioautographic studies on macromoleclular synthesis in mitochondria of some organs in aging animals. Bull. Shinshu Inst. Alternat. Med. Welfare 4, 15-38, 2009i.

Nagata, T.: Electron microscopic radioautographic study on mitochondrial DNA synthesis in adreno-cortical cells of aging ddY mice. Bull. Shinshu Inst. Alternat. Med. Welfare 4, 51-66, 2009j.

Nagata T.: Electron microscopic radioautographic study on mitochondrial RNA synthesis in adrenocortical cells of aging mice. Open Anat J. 2, 91-97, 2010a.

Nagata T. Electron microscopic radioautographic study on mitochondrial RNA synthesis in adrenal medullary cells of aging and senescent mice. J Cell Tissue Res. 10, 2213-2222, 2010b.

Nagata, T.: Macromolecular synthesis in the livers of aging mice as revealed by electron microscopic radioautography. In, Prog. Histochem. Cytochem., Sasse, D., Ed., Elsevier, Amsterdam, Boston, London, New York, Oxford, Paris, Philadelphia, San Diego, St. Louis, Vol. 45, No. 1, pp. 1-80, 2010c.

Nagata, T.: Electron microscopic radioautographic study on protein synthesis of mitochondria in adrenal medullary cells of aging mice. Bulletin Shinshu Inst Alternat Med Welfare 5, 25-37, 2010d.

Nagata, T.: Electron microscopic radioautographic study on mitochondrial RNA synthesis in adrenal cortical and medullary cells of aging mice. J. Biomed. Sci. Enginer. 4, 219-232, 2010e.

Nagata, T.: Electron microscopic radioautographic study on protein synthesis of mitochondria in adrenal cortical cells of aging mice. Bulletin Shinshu Inst. Alternat. Med. Welfare 5, 38-52, 2010f.

Nagata T.: Electron microscopic radioautographic study on mitochondrial DNA, RNA and protein synthesis in adrenal cells of aging mice. Formatex Microscopy Series No. 3, Vol. 3, Formatex, Badajoz, Spain, in press, 2010g.

Nagata, T.: Electron microscopic radioautographic studies on macromolecular synthesis in mitochondria of animal cells in aging. Ann. Rev. Biomed. Sci. 12, 1-29, 2010h.

Nagata, T., Cui, H., Gao, F.: Radioautographic study on glycoprotein synthesis in the ocular tissues. J. Kaken Eye Res. 13, 11-18, 1995.

Nagata, T., Cui, H., Kong, Y.: The localization of TGF-b1 and its mRNA in the spinal cords of prenatal and postnatal aging mice demonstrated with immunohistochemical and in situ hybridization techniques. Bull. Nagano Women's Jr. College, 7, 75-88, 1999a.

Nagata, T., Cui, H., Liang, Y.: Light microscopic radioautographic study on the protein synthesis in the cerebellum of aging mouse. Bull. Nagano Women's Jr. College, 9, 41-60 (2001).

Nagata, T., Fujii, Y., Usuda, N.: Demonstration of extranuclear nucleic acid synthesis in mammalian cells under experimental conditions by electron microscopic radioautography. Proc. 10th Internat. Congr. Electr. Microsc. 2, 305-306, 1982b.

Nagata, T., Hirano, I., Shibata, O., Nagata, T.: A radioautographic study on the DNA synthesis in the hepatic and the pancreatic acinar cells of mice during the postnatal growth, with special reference to binuclearity. Med. J. Shinshu Univ. 11, 35-42, 1966.

Nagata, T., Ito, M., Chen, S.: Aging changes of DNA synthesis in the submandibular glands of mice as observed by light and electron microscopic radioautography. Ann. Microsc. 1, 4-12, 2000a.

Nagata, T. Ito, M., Liang, Y.: Study of the effects of aging on macromolecular synthesis in mouse steroid secreting cells using microscopic radioautography. Methods Find. Exp. Clin. Pharmacol. 22, 5-18, 2000b.

Nagata, T., Iwadare, I., Murata, F.: Electron microscopic radioautography of nucleic acid synthesis in cultured cells treated with several carcinogens. Acta Pharmacol. Toxicol. 41, 64-65, 1977c.

Nagata, T., Kawahara, I.: Radioautographic study of the synthesis of sulfomucin in digestive organs of mice. J. Trace Microprobe Analysis 17, 339-355, 1999.

Nagata, T., Kawahara, I., Usuda, N., Maruyama, M., Ma, H.: Radioautographic studies on the glycoconjugate synthesis in the gastrointestinal mucosa of the mouse. In, Glycoconjugate in Medicine, Ohyama, M., Muramatsu, T., Eds, pp. 251-256, Professional Postgrad. Service, Tokyo, 1988a.

Nagata, T., Kong, Y.: Distribution and localization of TGFb1 and bFGF, and their mRNAs in aging mice. Bull. Nagano Women's Jr. College 6, 87-105, 1998.

Nagata, T., Ma, H., Electron microscopic radioautographic study on mitochondrial DNA synthesis in hepatocytes of aging mouse. Ann. Microsc. 5, 4-18, 2005a.
Nagata, T., Ma, H., Electron microscopic radioautographic study on RNA synthesis in hepatocyte mitochondria of aging mouse. Microsc. Res. Tech. 67, 55-64, 2005b.
Nagata, T., Momoze, S.: Aging changes of the amitotic and binucleate cells in dog livers. Acta Anat. Nipponica 34, 187-190, 1959.
Nagata, T., Morita, T., I. Kawahara, I.: Radioautographic studies on radiosulfate incorporation in the digestive organs of mice. Histol. Histopathol. 14, 1-8, 1999b.
Nagata, T., Murata, F.: Electron microscopic dry-mounting radioautography for diffusible compounds by means of ultracryotomy. Histochemistry 54, 75-82, 1977.
Nagata, T., Murata, F., Yoshida, K., Ohno, S., Iwadare, N.: Whole mount radioautography of cultured cells as observed by high voltage electron microscopy. Proc. Fifth Internat. Conf. High Voltage Electron Microsc. 347-350, 1977d.
Nagata, T., Nawa, T.: A modification of dry-mounting technique for radioautography of water-soluble compounds. Histochemie 7, 370-371, 1966a.
Nagata, T., Nawa, T.: A radioautographic study on the nucleic acids synthesis of binucleate cells in cultivated fibroblasts of chick embryos. Med. J. Shinshu Univ. 11, 1-5, 1966b.
Nagata, T., Nawa, T., Yokota, S.: A new technique for electron microscopic dry-mounting radioautography of soluble compounds. Histochemie 18, 241-249, 1969.
Nagata, T., Nishigaki, T., Momose, Y.: Localization of anti-allergic agent in rat mast cells demonstrated by light and electron microscopic radioautography. Acta Histochem. Cytochem. 19, 669-683, 1986b.
Nagata, T., Ohno, S., Kawahara, I., Yamabayashi, S., Fujii, Y., Murata, F.: Light and electron microscopic radioautography of nucleic acid synthesis in mitochondria and peroxisomes of rat hepatic cells during and after DEHP administration. Acta Histochem. Cytochem. 16, 610-611, 1979.
Nagata, T., Ohno, S., Murata, F.: Electron microscopic dry-mounting radioautography for soluble compounds. Acta Phamacol. Toxicol. 41, 62-63, 1977a.
Nagata, T., Ohno, S., Yoshida, K., Murata, F.: Nucleic acid synthesis in proliferating peroxisomes of rat liver as revealed by electron microscopical radioautography. Histochem. J. 14, 197-204, 1982a.
Nagata, T., Olea, M. T.: Electron microscopic radioautographic study on the protein synthesis in aging mouse spleen. Bull. Nagano Women's Jr. College 7, 1-9, 1999.
Nagata, T., Shibata, O., Omochi, S.: A new method for radioautographic osbservation on isolated cells. Histochemie 2, 255-259, 1961
Nagata, T., Shibata, O., Nawa, T.: Simplified methods for mass production of radioautograms. -Acta Anat. Nippon.42, 162-166, 1967a.
Nagata, T., Shibata, O., Nawa, T.: Incorporation of tritiated thymidine into mitochondrial DNA of the liver and kidney cells of chickens and mice in tissue culture. Histochemie 10, 305-308, 1967b.
Nagata, T., Shimamura, K., Onozawa, M., Kondo, T., Ohkubo, K., Momoze, S.: Relationship of binuclearity to cell function in some organs. I. Frequencies of binucleate cells in some organs of toads in summer and winter. Med. J. Shinshu Univ. 5, 147-152, 1960a.
Nagata, T., Shimamura, K., Kondo, T., Onozawa, M., Momoze, S., Okubo, M.: Relationship of binuclearity to cell function in some organs. II. Variation of frequencies of

binucleate cells in some organs of dogs owing to aging. Med. J. Shinshu Univ. 5, 153-158, 1960b.

Nagata, T., Steggerda, F. R.: Histological study on the deganglionated small intestine of the dog. Physiologist 6, 242-242, 1963.

Nagata, T., Steggerda, F. R.: Observations on the increase of binucleate cells in the ganglion cells of the dog's intestine due to experimental ischemia. Anat. Rec. 148, 315-315, 1964.

Nagata, T., Toriyama, K., Kong, Y., Jin, C., Gao, F.: Radioautographic study on DNA synthesis in the ciliary bodies of aging mice. J. Kaken Eye Res.12, 1-11, 1994.

Nagata, T., Usuda, N.: Image processing of electron microscopic radioautograms in clinical electron microscopy. J. Clin. Electron. Microsc. 18, 451-452, 1985.

Nagata, T., Usuda, N.: Studies on the nucleic acid synthesis in pancreatic acinar cells of aging mice by means of electron microscopic radioautography. J. Clin. Electron Microsc. 19, 486-487, 1986.

Nagata, T., Usuda, N.: Electron microscopic radioautography of protein synthesis in pancreatic acinar cells of aging mice. Acta Histochem. Cytochem. 26, 481-481, 1993a.

Nagata, T., Usuda, N.: In situ hybridization by electron microscopy using radioactive probes. J. Histochem. Cytochem. 41, 1119-1119, 1993b.

Nagata, T., Usuda, N., Ma, H.: Electron microscopic radioautography of nucleic acid synthesis in pancreatic acinar cells of prenatal and postnatal aging mice. Proc. XIth Intern. Cong. Electr. Microsc. 3, 2281-2282, 1984.

Nagata, T., Usuda, N., Ma, H.: Electron microscopic radioautography of lipid synthesis in pancreatic cells of aging mice. J. Clin. Electr. Microsc. 23, 841-842, 1990.

Nagata, T., Usuda, N., Maruyama, M., Ma, H.: Electron microscopic radioautographic study on lipid synthesis in perinatal mouse pancreas. J. Clin. Electr. Microsc. 21, 756-757, 1988b.

Nagata, T., Usuda, N., Suzawa, H., Kanzawa, M.: Incorporation of ^{3}H-glucosamine into the pancreatic cells of aging mice as demonstrated by electron microscopic radioautography. J. Clin. Electron Microsc. 25, 646-647, 1992.

Nagata, T., Yamabayashi, S.: Intracellular localization of ^{3}H-befunolol by means of electron microscopic radioautography of cryo-fixed ultrathin sections. J. Clin. Electron Microsc. 16, 737-738, 1983.

Nagata, T., Yoshida, K., Murata, F.: Demonstration of hot and cold mercury in the human thyroid tissues by means of radioautography and chemography. Acta Pharmacol. Toxicol. 41, 60-61, 1977b.

Nagata, T., Yoshida, K., Ohno, S., Murata, F.: Ultrastructural localization of soluble and insoluble ^{3}H-methyl prednisolone as revealed by electron microscopic dry-mounting radioautography. Proc. 9th Internat. Congr. Electr. Microsc. 2, 40-41, 1978b.

Nishigaki, T., Momose, Y., Nagata, T.: Light microscopic radioautographic study of the localization of anti-allergic agent, tranilast, in rat mast cells. Histochem. J. 19, 533-536, 1987.

Nishigaki, T., Momose, Y., Nagata, T.: Electron microscopic radioautographic study of the localization of an anti-allergic agent, tranilast, in rat mast cells. Cell. Mol. Biol. 36, 65-71, 1990a.

Nishigaki, T., Momose, Y., Nagata, T.: Localization of the anti-allergic agent tranilast in the urinary bladder of rat as demonstrated by light microscopic radioautography. Drug Res. 40, 272-275, 1990b.

Oguchi, K., Nagata, T.: A radioautographic study of activated satellite cells in dystrophic chicken muscle. In, Current Research in Muscular Dystrophy Japan. The Proc. Ann. Meet. Muscular Dystrophy Res. 1980, pp. 16-17, Ministry of Welfare of Japan, Tokyo, 1980.

Oguchi, K., Nagata, T.: Electron microscopic radioautographic observation on activated satellite cells in dystrophy chickens. In, Clinical Studies on the Etiology of Muscular Dystrophy. Annual Report on Neurological Diseases 1981, pp. 30-33, Ministry of Welfare of Japan, Tokyo, 1981.

Ohno, S., Fujii, Y., Usuda, N., Endo, T., Hidaka, H., Nagata, T.: Demonstration of intracellular localization of calmodulin antagonist by wet-mounting radioautography. J. Electron Microsc. 32, 1-12, 1983.

Ohno, S., Fujii, Y., Usuda, N., Nagata, T., Endo, T., Tanaka, T., Hidaka, H.: Intracellular localization of calmodulin antagonists (W-7). In, Calmodulin and intracellular Ca^{2+} receptors. Kakiuchi, S., Hidaka, H, Means, A. R., Eds., pp. 39-48, Plenum Publishing Co., New York, 1982.

Olea, M. T.: An ultrastructural localization of lysosomal acid phosphatase activity in aging mouse spleen: a quantitative X-ray microanalytical study. Acta Histochem. Cytochem. 24, 201-208, 1991.

Olea, M. T., Nagata, T.: X-ray microanalysis of cerium in mouse spleen cells demonstrating acid phosphatase activity using high voltage electron microscopy, Cell. Mol. Biol. 37, 155-163, 1991.

Olea, M. T., Nagata, T. : Simultaneous localization of ^{3}H-thymidine incorporation and acid phosphatase activity in mouse spleen: EM radioautography and cytochemistry. Cell. Mol. Biol. 38, 115-122, 1992a.

Olea, M. T., Nagata, T.: A radioautographic study on RNA synthesis in aging mouse spleen after ^{3}H-uridine labeling in vitro. Cell. Mol. Biol. 38, 399-405, 1992b.

Oliveira, S. F., Nagata, T., Abrahamsohn, P. A., Zorn, T. M. T.: Electron microscopic radioautographic study on the incorporation of ^{3}H-proline by mouse decidual cells. Cell. Mol. Biol. 37, 315-323, 1991.

Oliveira, S. F., Abrahamsohn, P. A., Nagata, T., Zorn, T. M. T.: Incorporation of ^{3}H-amino acids by endometrial stromal cells during decidualization in the mouse. A radioautographical study. Cell. Mol. Biol. 41, 107-116, 1995.

Pearse, A. G. E.: Histochemistry, Theoretical and Applied. 4th Ed. Vol. 1. 439 pp., 1980, Vol. 2. 1055 pp., 1985, Vol. 3. Ed. with P. Stoward, 728 pp. Churchill Livingstone, Edinburgh, London and New York, 1991.

Sakai, Y., Ikado, S., Nagata, T.: Electron microscopic radioautography of satellite cells in regenerating muscles. J. Clin. Electr. Microsc. 10, 508-509, 1977.

Sato, A.: Quantitative electron microscopic studies on the kinetics of secretory granules in G-cells. Cell Tissue Res. 187, 45-59, 1978.

Sato, A., Iida, F., Furihara, R., Nagata, T.: Electron microscopic raioautography of rat stomach G-cells by means of ^{3}H-amino acids. J. Clin. Electron Microsc. 10, 358-359, 1977.

Shimizu, T., Usuda, N., Yamanda, T., Sugenoya, A., Iida, F.: Proliferative activity of human thyroid tumors evaluated by proliferating cell nuclear antigen/cyclin immnohistochemical studies. Cancer 71, 2807-2812, 1993.

Sun, L.: Age related changes of RNA synthesis in the lungs of aging mice by light and electron microscopic radioautography. Cell. Mol. Biol. 41, 1061-1072, 1995.

Sun, L., Gao, F., Duan, H., Nagata, T.: Light microscopic radioautography of DNA synthesis in pulmonary cells in aging mice. In, Radioautography in Medicine, Nagata, T. Ed., pp. 201-205, Shinshu University Press, Matsumoto, 1994.

Sun, L., Gao, F., Nagata, T.: Study on the DNA synthesis of pulmonary cells in aging mice by light microscopic radioautography. Cell. Mol. Biol. 41, 851-859, 1995a.

Sun, L., Gao, F., Jin, C., Duan, H., Nagata, T.: An electron microscopic radioautographic study on the DNA synthesis of pulmonary tissue cells in aging mice. Med. Electron. Microsc. 28, 129-131, 1995b.

Sun, L., Gao, F., Jin, C., Nagata, T.: DNA synthesis in the tracheae of aging mice by means of light and electron microscopic radioautography. Acta Histochem. Cytochem. 30, 211-220, 1997a.

Sun, L., Gao, F., Nagata, T.: A Light Microscopic radioautographic study on protein synthesis in pulmonary cells of aging mice. Acta Histochem. Cytochem. 30, 463-470, 1997b.

Suzuki, K., Imada, T., Gao, F., Ma, H., Nagata, T.: Radioautographic study of benidipine hydrochloride: localization in the mesenteric artery of spontaneously hypertensive rat. Drug Res. 44, 129-133, 1994.

Terauchi, A., Mori, T., Kanda, H., Tsukada, M., Nagata, T.: Radioautographic study of 3H-taurine uptake in mouse skeletal muscle cells. J. Clin. Electron Microsc. 21, 627-628, 1988.

Terauchi, A., Nagata, T.: Observation on incorporation of ^{3}H-taurine in mouse skeletal muscle cells by light and electron microscopic radioautography. Cell. Mol. Biol. 39, 397-404, 1993.

Terauchi, A., Nagata, T.: In corporation of ^{3}H-taurine into the blood capillary cells of mouse skeletal muscle. Radioautography in Medicine, Nagata, T. ed., Shinshu University Press, Matsumoto, 1994.

Toriyama, K.: Study on the aging changes of DNA and protein synthesis of bipolar and photo-receptor cells of mouse retina by light and electron microscopic radioautography. Cell. Mol. Biol. 41, 593-601, 1995.

Tsukahara, S., Yoshida, K., Nagata, T.: A radioautographic study on the incorporation of ^{14}C-bupranolol (beta-blocking agent) into the rabbit eye. Histochemistry 68, 237-244, 1980.

Usuda, N., Nagata, T.: Electron microscopic radioautography of acyl-CoA mRNA by in situ hybridization. J. Clin. Electron Microsc. 25, 332-333, 1992.

Usuda, N., Nagata, T.: The immunohistochemical and in situ hybridization studies on hepatic peroxisomes. Acta Histochem. Cytochem. 28, 169-172, 1995.

Usuda, N., Hanai, T., Morita, T., Nagata, T.: Radioautographic demonstration of peroxisomal acyl-CoA oxidase mRNA by in situ hybridization. In, Recent advances in cellular and molecular biology, Vol. 6. Molecular biology of nucleus, peroxisomes, organelles and cell movement. Wegmann, R. J., Wegmann, M., Eds, pp.181-184, Peeters Press, Leuven, 1992.

Uwa, H., Nagata, T.: Cell population kinetics of the scleroblast during ethisterone-induced anal-fin process formation in adult females of the Medaka. Dev. Growth Different. 9, 693-694, 1976.

Watanabe, I., Makiyama, M. C. K., Nagata, T.: Electron microscopic radioautographic observation of the submandibular salivary gland of aging mouse. Acta Microscopica 6. 130-131, 1997.

Yamabayashi, S., Gunarso, W., Tsukahara, S., Nagata, T.: Incorporation of ^{3}H-befunolol (beta blocking agent) into melanin granules of ocular tissues in the pigmented rabbits. I. Light microscopic radioautography. Histochemistry 73, 371-375, 1981.

Yamada, A. T.: Timely and topologically defined protein synthesis in the periimplanting mouse endometrium revealed by light and electron microscopic radioautography. Cell. Mol. Biol. 39, 1-12, 1993.

Yamada, A., Nagata, T.: Ribonucleic acid and protein synthesis in the uterus of pregnant mouse during activation of implantation window. Med. Electron Microsc. 27, 363-365, 1992a.

Yamada, A., Nagata, T.: Light and electron microscopic raioautography of DNA synthesis in the endometria of pregnant-ovariectomized mice during activation of implantation window. Cell. Mol. Biol. 38, 763-774, 1992b.

Yamada, A., Nagata, T.: Light and electron microscopic raioautography of RNA synthesis of peri-implanting pregnant mouse during activation of receptivity for blastocyst implantation. Cell. Mol. Biol. 38, 211-233, 1993.

Yoshinaga, K.: Uterine receptivity for blastcyst implantation. Ann. N. Y. Acad. Sci. USA, 541, 424-431, 1988.

Yoshizawa, S., Nagata, A., Honma, T., Oda, M., Murata, F., Nagata, T.: Study of ethionine pancreatitis by means of electron microscopic radioautography. J. Clin. Electron Microsc. 7, 349-350, 1974.

Yoshizawa, S., Nagata, A., Honma, T., Oda, M., Murata, F., Nagata, T.: Radioautographic study of protein synthesis in pancreatic exocrine cells of alcoholic rats. J. Clin. Electron. Microsc. 10, 372-373, 1977.

7

Macromolecular Synthesis in the Digestive and Respiratory Systems

Tetsuji Nagata[1,2]
[1]*Department of Anatomy and Cell Biology, Shinshu University School of Medicine, Matsumoto*
[2]*Shinshu Institute of Alternative Medicine and Welfare, Nagano*
Japan

1. Introduction

This second chapter deals with the second parts of the application of microscopic radioautography to some of the visceral organ systems. The visceral organs can be divided into 5 organ systems according to anatomy and histology, i.e., the digestive system, the respiratory system, the urinary system, the reproductive system and the endocrine system. Among of them the digestive system consists of 2 parts, i.e., the digestive tract and the digestive glands. The former consists of simple tube structures such as the oral cavity, the esophagus, the stomach and the intestines, while the latter consists of complicated glandular structures such as the large digestive glands , i.e., the liver and the pancreas, while the respiratory system consists of 2 parts, one the respiratory tract such as the nose, the pharynx, the trachea, the bronchus, and final essential part the lungs. This chapter deals with the digestive organs and the respiratory organs, respectively.

1.1 Macromolecular synthesis in the digestive system

The digestive system consists of the digestive tract and the digestive glands. The digestive tract can be divided into several portions, from the upper part to the lower part, i.e., the oral cavity, the pharynx, the esophagus, the stomach, the small and large intestines and the anus, while the digestive glands consist of the large glands such as the salivary glands, the liver and the pancreas and the small glands affiliated to the digestive tracts in the gastrointestinal walls such as the gastric glands including the fundic gland and the pyloric gland, the intestinal glands of Lieberkühn and the duodenal glands of Brunner. We have published many papers from our laboratory dealing with the macromolecular synthesis in respective digestive organs from the oral cavity to the gastrointestinal tracts and the digestive glands (Nagata 1992, 1993a,b, 1994a,b,c,d,e, 1995a,b,c,d, 1996a,b,c, 1999a,b,c, 2002, Nagata et al. 1979, 1982a, 2000a, Chen et al. 1995). The outline of the results concerning to the macromolecular synthesis in the digestive organs should be here described in the order of systematic anatomy and special histology as follows.

1.1.1 Macromolecular synthesis in the oral cavity

The oral cavity consists of the lips, tongue, teeth, and the salivary glands. The DNA synthesis of mucosal epithelia of the 2 upper and lower lips and the tongue as well as the 3 large salivary glands and many small glands of aging mice from fetal day 19 to postnatal 2 years were studied by LM and EM RAG labeled with ^{3}H-thymidine. The glucide and glycoprotein syntheses by ^{3}H-glucosamine and radiosulfate incorporations of the submandibular and sublingual glands of aging mice were also studied.

1.1.2 The DNA synthesis in the oral cavity

We first studied the DNA synthesis of the submandibular glands in 10 groups of aging mice at various ages from embryo to postnatal 2 years (Chen et al. 1995, Nagata et al. 2000a). The submandibular gland of male mouse embryonic day 19 consisted of the glandular acini and duct system (Fig. 5A). The duct system was composed of the juxtaacinar cells (JA), the intercalated duct cells (ICD) and the striated duct cells (SD). Many labeled developing acinar cells (AC), JA and ICD cells were observed. At postnatal day 1 to 3 (Fig. 5B), there was more JA cells and secretory granules than those of former stage. JA cells were cuboidal cells in shape, characterized by small darkly stained granules in the supranuclear cytoplasm and by basophilic mitochondria mostly at the basal half of the cells. JA cells were present at the acinar-intercalated duct junction of the mouse submandibular gland. Many labeled AC, JA, ICD and SD cells were also observed by electron microscopy (Fig. 5C). At postnatal 2 weeks to 3 months, developing immature acinar cells gradually matured to acinar cells, and JA cells increased and granular convoluted duct cells (GCT) appeared.

At postnatal 6 months to 2 years, the GCT cells were very well developed and were composed of the taller cells packed with many granules and became highly convoluted, and only a few labeled cells were found. The aging changes of frequency of 5 main individual cell types in submandibular glands of male mouse from embryonic day 19 to postnatal 2 years of age were counted. On embryonic day 19 of age, the gland consisted of developing acinar cells (49%), intercalated duct cells (37%), juxta-acinar (JA) cells (3%), striated duct (SD) cells (11%). At birth, JA cells increased rapidly to 32%, thereafter decreased gradually. At 1 month of age, JA cells disappeared and granular convoluted tubule (GCT) cells appeared and increased rapidly in number with age. They reached a maximum at 6 months. Then they decreased gradually from 6-21 months. The quantity proportion of acini was relatively stable during these periods. The frequency of ICD cells (Fig. 5C) was the highest (37%) at 1 day after birth. Thereafter it gradually decreased month by month and reached 2.6% at 21 months, while the ratio of SD cells persisted in 7%-12% from embryonic day 19 to postnatal 2 weeks and it disappeared at 3 months after birth. The proliferative activity of the cell population is expressed by the labeling index which is defined as the percentage of labeled nuclei with ^{3}H-thymidine in a given cell population. The labeling index of the entire gland cells increased from 13.6% at embryonic 19 to 18.3% at neonate, when it reached the first peak (Fig. 6A, B). Then it declined to 2.2% at 1 week of age. A second small peak (2.9%) occurred at 2 weeks. Thereafter, the labeling index decreased progressively to less than 1% at 4 weeks of age and then remained low. The analysis of the labeling indices of respective cell types revealed that the first peak at neonate was due to the increased labeling indices of AC, ICD and JA cells, and the second peak at 2 weeks was due to the increase of ICD and SD cells. Thereafter, the labeling index of ICD cells decreased steadily but remained higher than those of any other cell types. Since the labeling index of ICD cells was more than the other cell types and persisted for a long time, it was suggested that ICD cells concerned with the generation of other cell types (Nagata 2002).

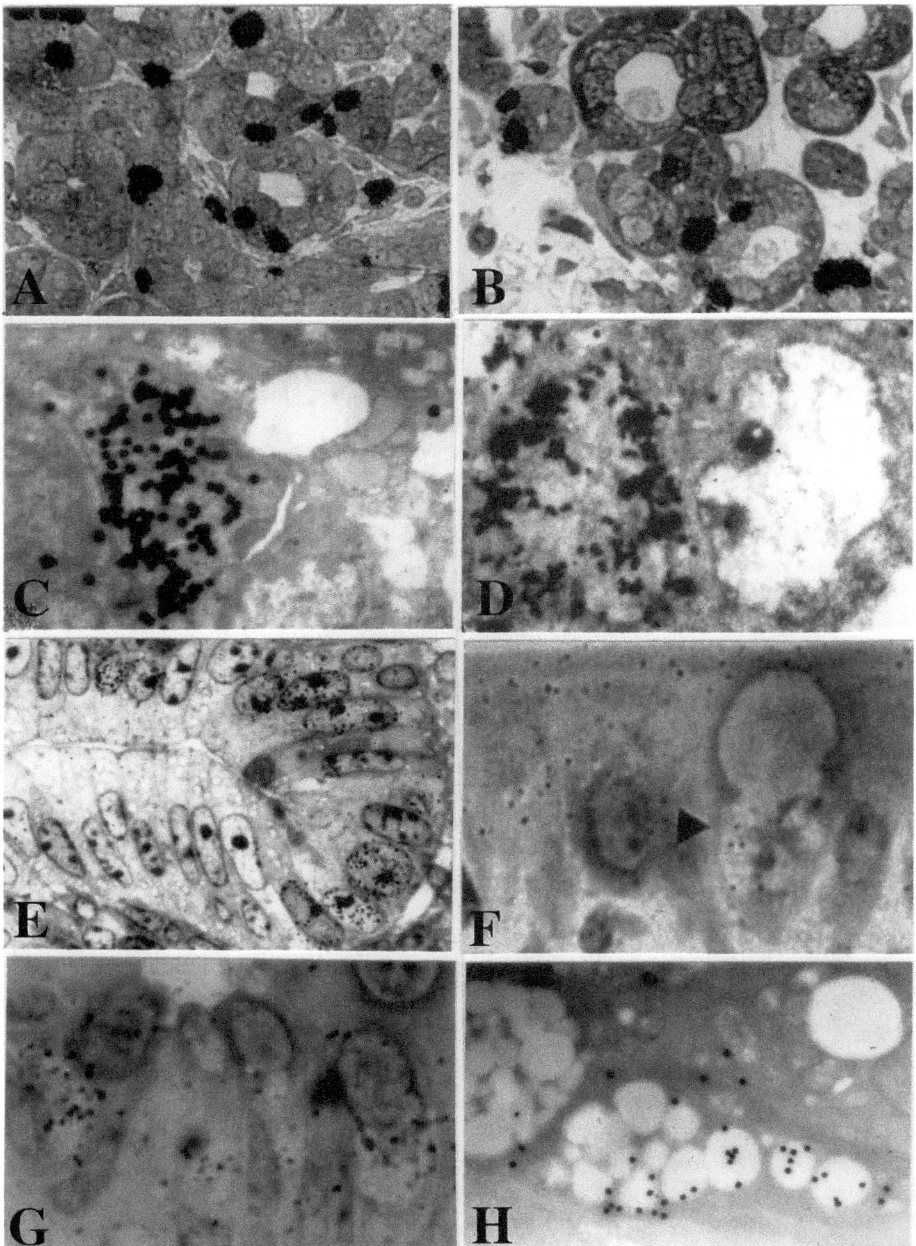

Fig. 6. LM and EM RAG of the digestive organs. From Nagata, T.: Radioautographology, General and Special. In, Prog. Histochem. Cytochem. Vol. 37, No. 2, p. 118, 2002, Urban & Fischer, Jena, Germany

Fig. 6A. LMRAG of the submandibular gland of male mouse embryonic day 19 labeled with ^{3}H-thymidine consisted with the glandular acini and duct system. The duct system was composed of juxtaacinar cells (JA), intercalated duct cells (ICD) and striated duct cells (ICD). Many labeled developing acinar cells (AC), JA and ICD cells were observed. x500.
Fig. 6B. LMRAG of the submandibular gland at postnatal day 3, labeled with ^{3}H-thymidine. There were more JA cells and secretory granules than those of former stage (Fig. 5A). x500.
Fig. 6C. EMRAG of an ICD cell of a mouse at postnatal day 3, labeled with ^{3}H-thymidine observed by electron microscopy. Many silver grains are observed over the nucleus of an ICD. x10,000.
Fig. 6D. EMRAG of the esophageal epithelial cells of a newborn mouse at postnatal day 1, labeled with ^{3}H-thymidine. Many silver grains are observed over one of the nuclei at left. x10,000.
Fig. 6E. LMRAG of the colonic epithelial cells of a mouse embryo at fetal day 19, labeled with ^{3}H-thymidine. Many silver grains are observed over the nuclei of several epithelial cells in the bottom of the crypt. x800.
Fig. 6F. LMRAG of the ileum epithelial cells labeled with ^{3}H-glucosamine of an old mouse at postnatal 6 months. Many silver gains are localized over the Golgi region of the 3 goblet cells as well as over the cytoplasm of several absorptive columnar epithelial cells. x 1,000.
Fig. 6G. LMRAG of the colonic epithelial cells of a mouse at postnatal month 1, labeled with $^{35}SO_4$ in vitro and radioautographed. x1,000.
Fig. 6H. EMRAG of a goblet cell in the deeper crypt of the colonic epithelial cells of an adult mouse after injection of $^{35}SO_4$ and radioautographed. Many silver grains are observed over the Golgi region and mucous droplets of the goblet cell, demonstrating the incorporation of radiosulfate into sulfomucins. x4,800.

1.1.3 The glucide synthesis in the oral cavity

We studied the incorporations of ^{3}H-glucosamine in the submandibular glands of 10 groups of litter mice at various ages. The animals from embryonic day 19, postnatal day 1, 3, 7, 14, and 1, 3, 6 months to 1 and 2 years were sacrificed after administration of ^{3}H-glucosamine and the submandibular glands were processed for LM and EM RAG (Watanabe et al. 1997, Nagata 2002). The results showed that the silver grains appeared over the endoplasmic reticulum, Golgi apparatus and the secretory granules of the acinar cells, demonstrating the glycoprotein synthesis in these cells. Grain counting revealed that the counts increased from the fetal stage at embryonic day 19 to postnatal day 1 to 3, 7, 14, reaching the peak at day 14, then decreased to month 1, 3, 6, to year 1 and 2, showing the aging changes, inverse proportion to DNA synthesis of these cells.

On the other hand, the sulfate uptake and accumulation in sulfomucin in several digestive organs of mice were also studied by light microscopic radioautography (Nagata and Kawahara 1999, Nagata et al. 1999b). Two litters of normal ddY mice 30 days after birth, each consisting of 3 animals, were studied. One litter of animals was sacrificed at 30 min after the intraperitoneal injections with phosphate buffered $Na_2{}^{35}SO_4$, and the other litter animals were sacrificed at 12 hr after the injections. Then the submandibular glands and the sublingual glands were taken out, fixed, embedded in epoxy resin, sectioned, radioautographed and analyzed by light microscopy. As the results, many silver grains were observed on serous cells of the salivary glands at 30 min and 12 hr after the injections

(10-20/cell). The numbers of silver grains at 30 min were less than those at 12 hr. From the results, it was concluded that glycoprotein synthesis was demonstrated in both the submandibular and sublingual glands by radiosulfate incorporation. In the salivary glands the silver grains were more observed in serous cells than mucous cells at 30 min, while in mucous cells more at 12 hr than 30 min after the injection. These results show the time difference of glycoprotein synthesis in the two salivary glands, showing inverse proportion to DNA synthesis of these cells (Watanabe et al. 1997, Nagata 2002).

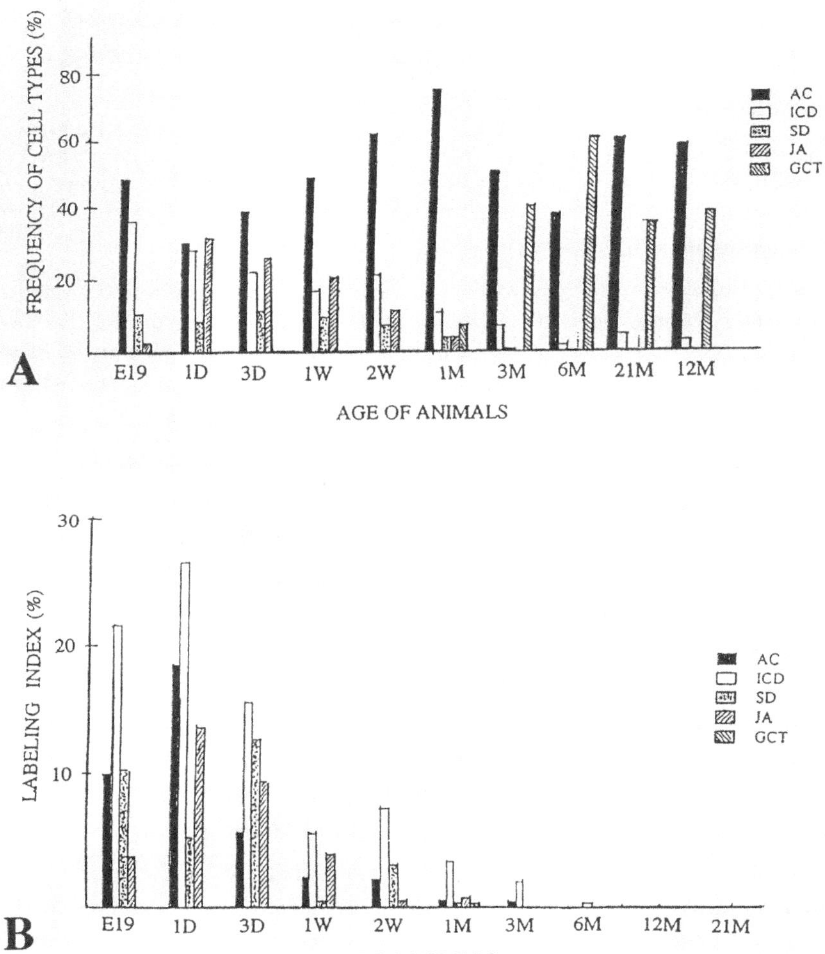

Fig. 7. Histogram showing the frequencies (A) and labeling indices (B) of the five individual cell types in the submandibular glands of male ddY mice at respective ages. From Nagata, T.: Radioautographology, General and Special. In, Prog. Histochem. Cytochem. Vol. 37, No. 2, p.118, 2002, Urban & Fischer, Jena, Germany

1.2 Macromolecular synthesis in the esophagus

The esophagus is the characteristic digestive tract including all the layers, the mucous membrane covered with the stratified squamous epithelia, the submucosa, the muscular layer and the serosa or adventitia. We studied the DNA synthesis of the esophagus of aging mice labeled with ^{3}H-thymidine by LM and EM RAG (Duan et al. 1992, 1993). The labeled cells were mainly found in the basal layer of the esophageal epithelium (Fig. 6D). By electron microscopy the nuclei and nucleoli of labeled cells were larger than those of unlabeled cells, but contained fewer cell organelles (Duan et al. 1993). The labeling indices in respective aging groups showed a peak at postnatal day 1 and decreased with aging keeping a constant level around a few % from 6 months to 2 years after birth.

1.3 Macromolecular synthesis in the stomach

The stomach consists of the mucosa covered with the surface epithelia of the columnar epithelia, including the gastric glands, the submucosa, the muscular layer and the serosa.

1.3.1 DNA synthesis in the stomach

As for the turnover of fundic glandular cells shown by ^{3}H-thymidine radioautography, it was extensively investigated with LM RAG by Leblond and co-workers (Leblond 1981, Leblond et al. 1958). They demonstrated that the DNA synthesis in the stomach increased at perinatal stages and decreased due to aging and senescence. However, the activity never reached zero but low activity continued until senescence. We studied the macromolecular synthesis including DNA, RNA, protein and glycoproteins in the gastric mucosa of both human and animal tissues by LM and EMRAG (Sato et al. 1977). As for the DNA synthesis, we obtained the same results as Leblond et al (1958, 1981). We have not carried out this study so much. Therefore, the minute details will be here omitted.

1.3.2 Protein synthesis in the stomach

We observed the secretion process in G-cells by EM RAG using ^{3}H-amino acid (Sato 1978, Sato et al. 1977, Komiyama et al. 1978). When the stomach tissues were taken out from the adult Wistar rats at postnatal month 1 and were labeled with either ^{3}H-glutamic acid or ^{3}H-glycine in vitro at varying time intervals, silver grains in the EM radioautograms appeared first over the Golgi zones, then migrated to secretory granules and were stored in the cytoplasm, suggesting the secretory kinetics. We also studied the mechanism of serum albumin passing through the gastric epithelial cells into the gastric cells by EM RAG (Sato et al. 1977). When adult Wistar rat stomach tissues were labeled with ^{132}I-albumin in vitro at varying time intervals, silver grains in the radioautograms appeared over rough endoplasmic reticulum within 3 min, then moved to the Golgi apparatus in 10 min, and on to secretory granules and into the lumen in 30 min, suggesting the pathway of serum albumin absorption from the blood vessels through the gastric mucous epithelial cells into the gastric lumen (Komiyama et al. 1978). These results demonstrated that the stomach cells of adult rats synthesized proteins and secreted. However, aging changes of these protein synthesis between the young and senescent animal were not yet completed.

1.3.3 The glucide synthesis in the stomach

When incorporation of radiosulfate into sulfated complex carbohydrate in rat stomach was studied by labeling with $^{35}SO_4$ in vivo, silver grains appeared over the glandular cells of the pyloric gland but not those of the fundic gland, demonstrating the mucous synthesis in the former glands (Nagata et al. 1988a, Nagata and Kawahara 1999). The radiosulfate uptake and accumulation in the stomach of mouse were also studied by light microscopic radioautography (Nagata et al. 1999b). Two litters of normal ddY mice 30 days after birth, each consisting of 3 animals, were studied. One litter animals were sacrificed at 30 min after the intraperitoneal injections with phosphate buffered $Na_2{}^{35}SO_4$, and the other litter animals were sacrificed 12 hr after the injections. Then the antrum and the fundus tissues of the stomachs were taken out. The tissues were fixed, dehydrated, embedded in epoxy resin, sectioned, radioautographed and analyzed. As the results, many silver grains were observed on the mucosa and submucosa of the stomach at 30 min after the injection. Then at 12 hr after the injection silver grains were observed on some of the fundic glands. The numbers of silver grains observed in the stomach especially over the pyloric glands at 30 min (a few per cell) were less than those (several per cell) at 12 hr. The results showed the time difference of glycoprotein synthesis in the stomach, showing inverse proportion to DNA synthesis (Nagata and Kawahara 1999, Nagata 2002).

1.4 Macromolecular synthesis in the intestines

The intestines of mammals are divided into 2 portions, small and large intestines, which can be further divided into several portions, the small intestines into the duodenum, the jejunum and the ileum, while the large intestines into the caecum, the appendix vermiformis, the colon and the rectum. The intestinal tracts in any portions consist of the mucosa covered with columnar epithelial cells including absorptive and secretory cells, the submucosa, the smooth muscular layer and the serosa. We studied the macromolecular synthesis, both the DNA and the proteins in the intestines by LM and EMRAG mainly in the epithelial cells (Nagata 2002).

1.4.1 The DNA synthesis in the intestines

We studied the DNA synthesis in the intestines by LM and EMRAG mainly in the epithelial cells (Nagata 2002). The DNA synthesis of small and large intestines of mice were studied by ^{3}H-thymidine RAG (Fig. 6E). The cells labeled with ^{3}H-thymidine were localized in the crypts of both small and large intestines, a region defined as the proliferative zone. In the colon of aging mice from fetal to postnatal 2 years, the labeled cells in the columnar epithelia were frequently found in the perinatal groups from embryo to postnatal day 1. However, the labeling indices became constant from the suckling period until senescence (Morita 1993, Morita et al. 1994). On the other hand, we examined the labeling indices of respective cell types in each layer of mouse colon such as columnar epithelial cells, lamina propria, lamina muscularis mucosae, tunica submucosa, inner circular muscle layer, outer longitudinal muscle layer, outer connective tissue and serous membrane of the colon and found that most labeling indices decreased after birth to 2 months except the epithelial cells which kept constant value to senescence (Jin and Nagata 1995a,b, Jin 1996) (Fig. 8). Similar results were also obtained from the cecal tissues of mouse by LM and EMRAG. We also studied immunostaining for PCNA/cyclin and compared to the results obtained from RAG (Morita et al. 1994). The colonic tissues of litter mice of six aging groups from the embryonic day 19, to newborn postnatal day 1, 5, 21, adult 2 months and senescent 12 months were fixed in methacarn solution, sectioned and immunostained for cyclin proliferating nuclear antigen (PCNA/cyclin) with the monoclonal antibody and the avidin-biotin peroxidase complex technique. The reaction appeared in the colonic epithelium from G1 to S phase of the cell cycle. The immunostaining positive cells were localized in the crypts of colons similarly to the labeled cells with ^{3}H-thymidine by radioautography, a region defined as the proliferative zone. The positive cells in the columnar epithelia were frequently found in the perinatal groups from embryo to postnatal day 1, and became constant from postnatal day 5 until senescence. Comparing the results by immunostaining with the labeling index by radioautography, it was found that the former was higher in each aging group than the latter. The reason for the difference should be due to that PCNA/cyclin positive cells included not only S-phase cells but also the late G1 cells.

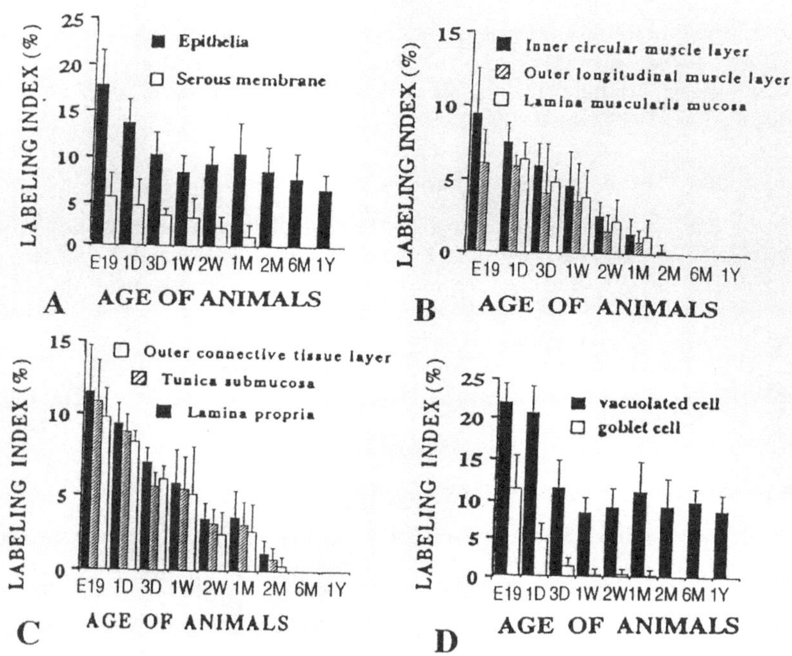

Fig. 8. Histogram showing aging changes of average labeling indices in respective tissue layers and cells of mouse colons at various ages from embryo to postnatal year 1, labeled with ^{3}H-thyidine. From Nagata, T.: Radioautographology, General and Special. In, Prog. Histochem. Cytochem. Vol. 37, No. 2, p. 123, 2002, Urban & Fischer, Jena, Germany

1.4.2 The RNA synthesis in the intestines

We studied the RNA synthesis of the small intestines of mice after feeding or refeeding under the restricted conditions (Nagata 1966). Five groups of ddY mice, each consisting of 5 individuals, total 25, were injected with ^{3}H-uridine, an RNA precursor, and sacrificed at different time intervals after feeding. The animals of the first group were injected with ^3H-uridine at 9 a.m. and fed at 10 a.m. for 30 min. and sacrificed at 11 a.m. 1 hour after the feeding and 2 hours after the injection, the 2nd group was sacrificed at 1 p.m. 3 hours after feeding and 4 hours after the injection, the 3rd group at 5 p.m., 7 and 8 hours later, the 4th group at 9 a.m. on the next day 23 and 24 hours later, and finally the 5th group at 1. p.m. on the next day 3 hours after refeeding and 28 hours after the injection. Then, the jejunums obtained from each animal were prepared for isolated cell radioautograms according to Nagata et al. (1961). The results demonstrated that the grain counts in mononucleate villus cells reached the maximum (20-30 grains per cell) 4 hours after injection and decreased (10-20/cell) after 28 hours, while the counts in mononucleate villus cells only increased gradually from 4 hours (10/cell) to 28 hours (20/cell). In contrast to this, the grain counts of binucleate cells which appeared in villus cells increased parallely to the mononucleate villus cells (10-20/cell). It was concluded that the RNA synthesis in the jejunal epithelial cells was high in the following order: mononucleate crypt cells, binucleate cells and mononucleate villus cells. These results revealed that the feeding or refeeding affected the RNA synthesis of the intestinal epithelial cells (Nagata 1966).

1.4.3 The protein synthesis in the intestines

We first studied the incorporations of ^3H-leucine and ^3H-tryptophane in mouse small intestines in connection to the binuclearity before and after feeding (Nagata 1967b). The results showed that the incorporations of both amino acids were greater in binucleate intestinal epithelial columnar cells than mononucleate villus and crypt cells at both before and after feeding. However, the aging changes of these incorporations were not studied.

1.4.4 The glucide synthesis in the intestines

We also studied the aging changes of glucide synthesis by ^3H-glucosamine uptake in the small intestines of mouse (Morita 1993), and found that the silver grains in the ileum columnar epithelial cells were mainly localized over the brush borders and the Golgi regions in these cells (Fig. 6F). The grain counting revealed that the numbers of silver grains over the brush borders and cytoplasm of the columnar epithelial cells increased in the villi (10-15/cell) than in the crypts (1-2/cell) from 6 months up to 2 years due to aging. The grain counting in other cell types also revealed that the number of silver grains in goblet cells, basal granulate cells, Paneth cells increased by aging, but did not in the undifferentiated cells.

The glycoprotein synthesis in goblet cells as well as in absorptive epithelial cells was also studied using $^{35}SO_4$ incorporation in the duodenums, the jejunums and the colons of adult mice at varying time intervals at 30, 60, and 180 min after the administration (Nagata et al. 1988a, Nagata and Kawahara 1999, Nagata et al. 1999b). Silver grains were localized over the columnar absorptive cells and the goblet cells, especially over the Golgi regions and mucous granules of the goblet cells. By EM RAG the intracellular localization of silver grains in goblet cells was clearly shown in the Golgi apparatus. The results from grain counting revealed that the average grain counts were different in the upper and deeper regions of the crypts in the 4 portions and it was shown that silver grains over goblet cells in the lower region of the crypt transferred rapidly from 30 min to 180 min, while they transferred slowly in goblet cells in the upper region of the colonic crypt, leading to the conclusion that the rates of transport and secretion of mucous products of the goblet cells at these two levels in the crypts were different. By EM RAG silver grains first appeared over the Golgi zone at 30 min. and then moved to the secretory granules at 60 and 180 min. The incorporation of $Na_2{}^{35}SO_4$ into sulfated complex carbohydrate was investigated in the mouse small and large intestines by LM and EM RAG as well as in the submandibular glands and the stomachs. Quantitative differences have been observed in the relative uptake of radiosulfate in the various labeled cells of each organ. Incorporation by the colon in goblet cells exceeded that elsewhere in the deep goblet cells of the colonic crypts migration of label progressed during the time tested from the supranuclear Golgi region to the deep position of the goblet and then extended throughout the mucosubstance in the goblet in the superficial goblet cells of the colon. The radioautographic and cytochemical staining differences between secretory cells in the deeper region compared with the upper region of the colonic crypts are considered to reflect differences in the rate of transport of secretory products in the theca and the rate of secretion at the low levels in the crypt (Figs. 6G,H). These results showed the time differences of glycoprotein synthesis in respective organs. The sulfate uptake and accumulation in several mouse digestive organs were also studied by LM RAG. Two litters of normal ddY mice 30 days after birth, each consisting of 3 animals, were studied. One litter of animals was sacrificed 30 min after the intraperitoneal injections with phosphate buffered $Na_2{}^{35}SO_4$, and the other litter animals were sacrificed 12 hr after the injections. Then several digestive organs, the parotid gland, the submandibular gland, the sublingual gland, antrum and fundus of the stomach, the duodenum, the jejunum, the ileum, the caecum, the ascending colon and the descending colon were taken out and radioautographed. As the results, many silver grains were observed on villous cells and crypt cells of the small intestines and whole mucosa of the large intestines at 30 min after the injection. Then at 12 hr after the injection silver grains were observed on mucigen granules of goblet cells in the small intestines and the large intestines. The numbers of silver grains observed in respective organs at 30 min were less than those at 12 hr. From the results, it was concluded that the time difference of the glycoprotein synthesis was demonstrated in several digestive organs by radiosulfate incorporation, in reverse proportion to DNA synthesis. The total S contents in colonic goblet cells in upper and deeper regions of colonic crypts in aging mice were also analyzed by X-ray microanalysis (Nagata et al. 2000b, Nagata 2004). The results accorded well with the results from RAG (Nagata 2002) showing increase and decrease of mucosubstances in these cells due to development and aging to senescence.

1.5 Macromolecular synthesis in the liver

The liver is the largest gland in the human and the mammalian body and consists of several types of cells (Nagata 2010c). The hepatocyte is the main component of the liver, composing the liver parenchyma which form the hepatic lobules, surrounded by other types of cells such as the connective tissue cells, sinusoidal endothelial cells, satellite cells of Kupffer, Ito's fat-storing cells and bile epithelial cells. In the livers of perinatal animals, the liver tissues include hematopoietic cells such as erythroblasts, myeloblasts and magakaryocytes. We studied macromolecular synthesis by LM and EMRAG mainly in hepatocytes of rats and mice (Nagata 1993b, 1994a,b,c,d, 1995a,b,c, 1996a, 1997a, 1999c, 2002, 2003, 2006b, 2007a, 2009a,d,h,i, 2010c,h).

1.5.1 The DNA synthesis in the liver

We first studied the DNA synthesis in the liver tissues at various ages from embryo to postnatal 2 years (Nagata 1993a,b, 1994a,b,c,d, 1995a,b,c,d, 1996a,b,c,d, 1997a,b,c, 1998a,b,c,

1999a,b,c, 2002, Nagata et al. 1977a, Nagata and Nawa 1966b). The results obtained from the tissues of 3 groups of animals injected respectively with 3 kinds of RI-labeled precursors, i.e. ^{3}H-thymidine, ^{3}H-uridine and ^{3}H-leucine, were already reported as several original articles and reviews (Ma 1988, Ma and Nagata 1988a,b, 1990a,b, 2000, Ma et al. 1991, Nagata 1996a,b, 1997a, 1998a, 1999c, 2001c, 2002, 2003, 2006a,b, 2007a,b, 2009a,b,c,d) or as a monograph in the series of Prog. Histochem. Ccytochem (Nagata 2002, 2004, 2010c). Therefore, the results from the livers in aging mice should be briefly summarized in this article.

1.5.1.1 The DNA synthesis in hepatocyte nuclei

As for the nucleic acid synthesis in hepatocytes, we first studied the difference between the mononucleate and binucleate hepatocytes of adult rats, injected with ^{3}H-thymidine and radioautographed (Nagata 1962, 1994d). The results showed that the frequency of labeled cells was greater in the mononucleate cells (Fig. 9A) than in the binucleate cells. The labeled binucleate cells were classified into two types, i.e., a hepatocyte whose one of the two nuclei was labeled and a hepatocyte whose two nuclei were labeled. The former was more frequently observed than the latter. Grain counts revealed that the amount of DNA synthesized in the binucleate cell whose one nucleus was labeled was the same as the mononucleate cell, while the total amount of DNA synthesized in the binucleate cell whose two nuclei were labeled was almost twice as that of the mononucleate cell. These results suggested that the two nuclei of binucleate hepatocytes synthesized DNA independently from each other.

On the other hand, LM and EMRAG of prenatal and postnatal normal mice at various ages labeled with ^{3}H-thymidine revealed that many silver grains were localized over the nuclei of various cell types consisting the liver, i.e., hepatocytes (Fig. 9A), sinusoidal endothelial cells (Fig. 8B), Kupffer's cells, Ito's fat-storing cells, bile ductal epithelia cells, fibroblasts and hematopoietic cells (Ma 1988, Ma and Nagata 1988a,b, 1990, Nagata 1995a). In hematopoietic cells in the livers of perinatal animals, silver grains were observed over the nuclei of erythroblasts, myeloblasts, lymphoblasts and megekaryocytes. However, most hematopoietic cells disappeared on postnatal day 14. At fetal day 19, the liver tissues were chiefly consisted of hepatocytes and haematopoietic cells and no lobular orientation was observed. At postnatal day 1 and 3, lobular formation started and finally the hepatic lobules were formed at day 9 after birth.

During the perinatal period, almost all kinds of cells were labeled with ^{3}H-thymidine. Percentage of labeled hepatocytes was the highest at fetal day 19, and rapidly decreased after birth to day 3. From day 9 to 14, percentage of labeled hepatocytes (labeling index) decreased gradually and finally to the lowest at 24 months (Fig. 10A). When the labeling indices of hepatocytes in 3 hepatic acinar zones were analyzed, the indices decreased in zone 2 (intermediate zone) and zone 3 (peripheral zone) on days 3 and 9 after birth, whereas they increased in zone 1 (central) on day 9, and then they altogether decreased from day 14 to 24 months (Fig. 10B). When the size and number of cell organelles in both labeled and unlabeled hepatocytes were estimated quantitatively by image analysis with an image analyzer, Digigramer G/A (Mutoh Kogyo Co. Ltd., Tokyo, Japan) on EMRAG, the area size of the cytoplasm, nucleus, mitochondria, endoplasmic reticulum, and the number of mitochondria in the unlabeled hepatocytes were more than the labeled cells (Ma and Nagata 1988a,b, Nagata 1995a,d).

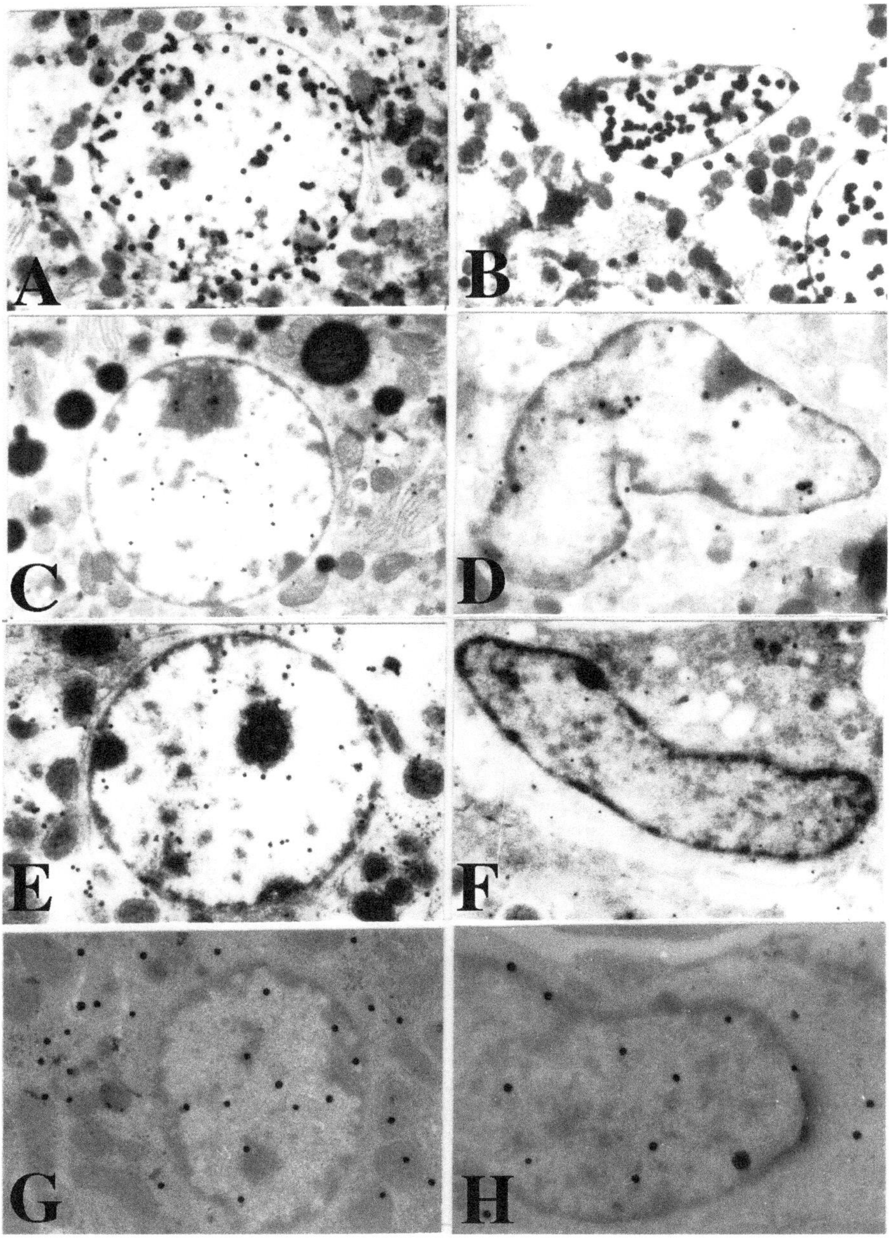

Fig. 9. EM RAG of the liver. From Nagata, T.: Radioautographology, General and Special. In, Prog. Histochem. Cytochem. Vol. 37, No. 2, p. 130, 2002, Urban & Fischer, Jena, Germany
Fig. 9A. EMRAG of a hepatocyte of the liver of a 14 day old mouse labeled with ^{3}H-thymidine. Many silver grains were observed over the nucleus and mitochondria.

Fig. 9B. EM RAG of a sinusoidal endothelial cells of the liver of a 14 day old mouse labeled with ^{3}H-thymidine. Many silver grains were observed over the nucleus and mitochondria.
Fig. 9C. EM RAG of a hepatocyte of the liver of a 14 day old mouse labeled with ^{3}H-uridine. Many silver grains were observed over the nucleus and mitochondria.
Fig. 9D. EM RAG of an Ito's fat-storing cell of the liver of a newborn 14 day old mouse labeled with ^{3}H-uridine. Many silver grains were observed over the nucleus and mitochondria.
Fig. 9E. EM RAG of a hepatocyte of the liver of a 1 month old mouse labeled with ^{3}H-leucine. Many silver grains were observed over the nucleus and mitochondria.
Fig. 9F. EM RAG of a sinusoidal endothelial cells of the liver of a newborn 14 day old mouse labeled with ^{3}H-leucine. Many silver grains were observed over the nucleus and mitochondria.
Fig. 9G. EMRAG of a hepatocyte of the liver of an adult 2 month old mouse labeled with ^{3}H-proline. Many silver grains were observed over the nucleus and mitochondria.
Fig. 9H. EMRAG of a Kupffer cell of the liver of a newborn 1 day old mouse labeled with ^{3}H-proline. Many silver grains were observed over the nucleus and mitochondria.

These data demonstrated that the cell organelles of the hepatocytes which synthesized DNA were not well developed as compared to those not synthesizing DNA during the postnatal development. In some of unlabeled hepatocytes, several silver grains were occasionally observed localizing over mitochondria and peroxisomes as was formerly reported (Nagata et al. 1967a,b, 1982b). The mitochondrial DNA synthesis was first observed in cultured hepatocytes of chickens and mice in vitro (Nagata et al. 1967a,b). The percentages of labeled cells in other cell types in the liver of aging mice such as sinusoidal endothelial cells, Kupffer's cells, Ito's fat-storing cells, bile ductal epithelia cells and fibroblasts showed also decreases from perinatal period to postnatal 24 months.

1.5.1.2 The DNA synthesis in hepatocyte mitochondria

When we observed DNA synthesis in the nuclei of mononucleate and binucleate hepatocytes, we also observed DNA synthesis in hepatocyte mitochondria (Ma 1988, Ma and Nagata 1988a,b, 1990a,b, Nagata and Ma 2005a). The results of visual grain counts on the number of mitochondria labeled with silver grains obtained from 10 mononucleate hepatocytes of each animal labeled with ^{3}H-thymidine demonstrating DNA synthesis in 8 aging groups at perinatal stages, prenatal embryo day 19, postnatal day 3, 9 and 14, month 1, 6, 12 and 24, were obtained. The number of total mitochondria per cell increased from perinatal stage (35-50/cell) to postnatal month 6 (95-105/cell), reaching the maximum, decreased to month 24 (85-90/cell), while the number of labeled mitochondria per cell increased from perinatal stage to postnatal day 14, reaching the maximum, decreased to month 6, then increased again to month 12, reaching the second peak and decreased again to month 24. Thus, the labeling indices in respective aging stages were calculated from the number of labeled mitochondria which showed an increase from perinatal stage to postnatal day 14, reaching the maximum and decreased to month 24. The results showed that the numbers of labeled mitochondria with ^{3}H-thymidine showing DNA synthesis increased from prenatal embryo day 19 (3.8/cell) to postnatal day 14 (6.2/cell), reaching the maximum, and then decreased to month 6 (3.7/cell) and again increased to year 1 (6.0/cell), while the labeling indices increased from prenatal day 19 (11.8%) to postnatal day 14 (16.9%), reaching the maximum, then decreased to month 6 (4.1%), year 1 (6.4%) and year 2 (2.3%). The increase of the total number of mitochondria in mononucleate hepatocytes was stochastically significant ($P<0.01$), while the changes of number of labeled mitochondria and labeling index in mononucleate hepatocytes were not significant ($P<0.01$).

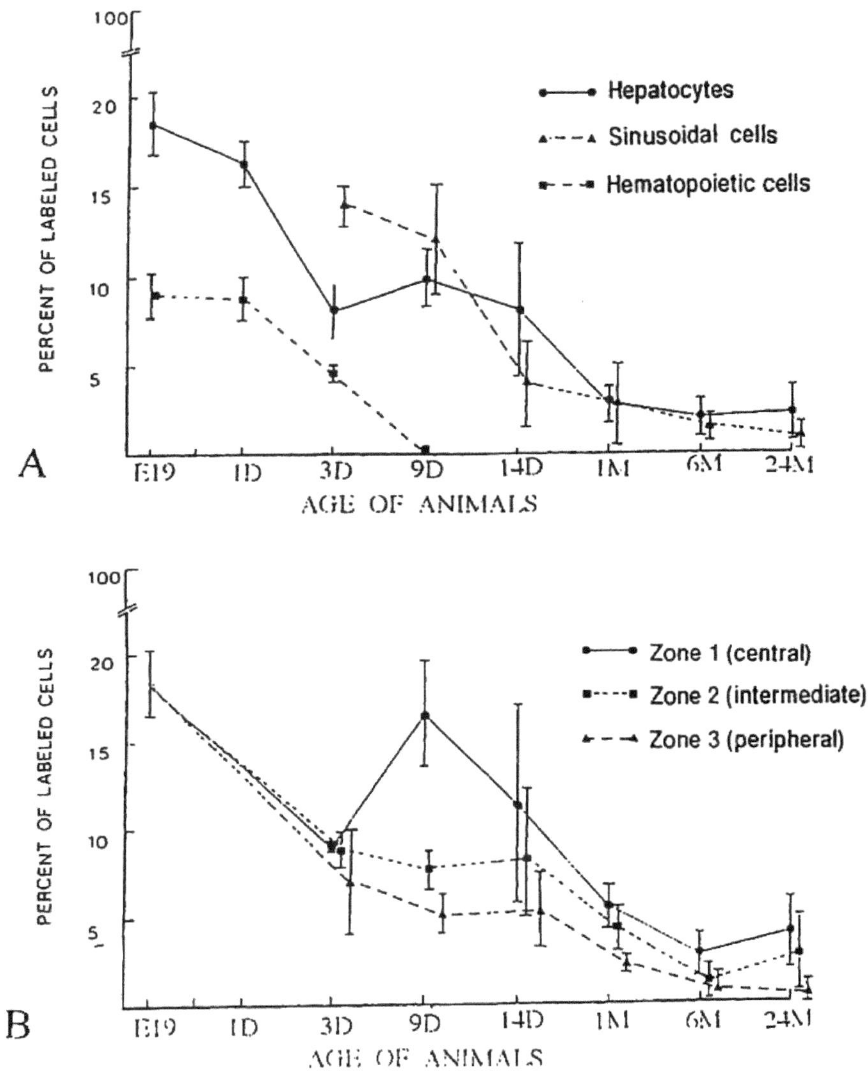

Fig. 10. Transitional curves of the labeling indices in the livers of aging mice after injection of ^{3}H-thymine. Mean ± Standard Deviation. From Nagata, T.: Special Cytochemistry in Cell Biology, In, Internat. Rev. Cytol. Vol. 211, No. 1, p. 92, 2001, Academic Press, San Diego, USA, London, UK.

Fig. 10A. Labeling indices of hepatocytes, sinusoidal endothelial cells and hematopoietic cells, respectively.

Fig. 10B. Labeling indices of hepatocytes, sinusoidal endothelial cells and hematopoietic cells, respectively.

As for the binucleate hepatocytes, on the other hand, because the appearances of binucleate hepatocytes showing silver grains in their nuclei demonstrating DNA synthesis were not so many in the adult and senescent stages from postnatal month 1 to 24, only binucleate cells at perinatal stages when reasonable numbers of labeled hepatocytes were found in respective groups were analyzed. The number of mitochondria in binucleate hepatocytes at postnatal day 1 to 14 kept around 80 (77-84/cell) which did not show such remarkable changes, neither increase nor decrease, as shown in mononucleate cells. Thus, the number of mitochondria per binucleate cell, the number of labeled mitochondria per binucleate cell and the labeling index of binucleate cell in 4 groups from postnatal day 1 to 14 were counted. The number of mitochondria and the number of labeled mitochondria were more in binucleate cells than mononucleate cells (Nagata 2007a,b,c,d,e, Nagata and Ma 2005a,b, Nagata et al. 1977a).

1.5.2 The RNA synthesis in the liver

The RNA synthesis in the liver was studied by ^{3}H-uridine RAG. Silver grains due to RNA synthesis appeared over the nuclei and cytoplasm of hepatocytes.

1.5.2.1 The RNA synthesis in hepatocyte nuclei

When the RI-labeled precursor ^{3}H-uridine was administered to experimental animals, or cultured cells were incubated in a medium containing ^{3}H-uridine in vitro and LM RAG was prepared, silver grains first appeared over the chromatin of the nucleus and nucleolus of all the cells within several minutes, then silver grains spread over the cytoplasm within 30 minutes showing messenger RNA and ribosomal RNA (Nagata 1966, Nagata and Nawa 1966a,b, Nagata et al.1969).

We studied quantitative changes of RNA synthesis in the livers of adult mice before and after feeding by incorporations of ^{3}H-uridine. Five groups of ddY mice, each consisting of 5 individuals, total 25, were injected with ^{3}H-uridine and sacrificed at different time intervals. The animals of the first group were injected with ^{3}H-uridine at 9 a.m. and fed at 10 a.m. for 30 min. and sacrificed at 11 a.m. 1 hour after the feeding and 2 hours after the injection, the 2nd group was sacrificed at 1 p.m. 3 hours after feeding and 4 hours after the injection, the 3rd group at 5 p.m., 7 and 8 hours later, the 4th group at 9 a.m. on the next day 23 and 24 hours later, and finally the 5th group at 1. p.m. on the next day 3 hours after refeeding and 28 hours after the injection. Then, the livers were taken out from each animal, prepared for isolated cell radioautograms according to Nagata et al. (1961). The results demonstrated that the grain counts in both mononucleate and binucleate hepatocytes before feeding (15-25 grains per cell) increased 4 hours after feeding (30-40 grains per cell), reached the maximum in 24 hours (40-50 grains per cell), then decreased on the next day (30-40 grains per cell). It was concluded that the RNA synthesis in the binucleate hepatocytes was a little higher than the mononucleate hepatocytes at the same stages and both increased and decreased after feeding. These results revealed that the feeding or refeeding affected the RNA synthesis of the livers (Nagata 1966).

Then, we studied aging changes of ^{3}H-uridine incorporation in the livers and pancreases of aging mice at various ages from prenatal embryos to postnatal aged mice to senescence at month 12 and 24 by LM and EMRAG (Ma and Nagata 1990b, Nagata 1999c). When aged mice were injected with ^{3}H-uridine, LM and EM RAG showed that silver grains were localized over the nucleoli, nuclear chromatin (both euchromatin and heterochromatin), mitochondria and rough surfaced endoplasmic reticulum of hepatocytes (Fig. 9C) and other types of cells such as sinusoidal endothelial cells, Kupffer's cells, Ito's fat-storing cells (Fig. 9D), ductal epithelial cells, fibroblasts and haematopoietic cells in the livers at various ages. By quantitative analysis, the total number of silver grains in nucleus, nucleolus and cytoplasm of each hepatocyte increased gradually from fetal day 19 to postnatal days, reached the maximum at postnatal day 14 (30%), then decreased to 24 months (5%). The number of silver grains in nucleolus, when classified into two compartments, grains over granular components and those over fibrillar components both increased paralelly after birth, reached the maxima on day 14 (granular 6-7, fibrillar 1-2/per cell), then decreased to 24 months with aging. However, when the ratio (%) of silver grains over euchromatin, heterochromatin of the nuclei and granular components and fibrillar components of the nucleoli are calculated, the ratio remained constant at each aging point.

1.5.2.2 The RNA synthesis in hepatocyte mitochondria

The intramitochondrial RNA synthesis was first found in the cultured HeLa cells and the cultured liver cells in vitro using EM RAG (Nagata 1972c, d). Then, it was also found in any other cells in either in vitro or vivo (Nagata et al. 1977c, Nagata 2002). Observing light microscopic radioautograms labeled with ^{3}H-uridine, the silver grains were found over both the karyoplasm and cytoplasm of almost all the cells not only at the perinatal stages from embryo day 19 to postnatal day 1, 3, 9, 14, but also at the adult and senescent stages from postnatal month 1 to 2, 6, 12 and 24 (Nagata 2007a, c, d, e, f, Nagata and Ma 2005b). By electron microscopic observation, silver grains were detected in most mononucleate hepatocytes in respective aging groups localizing not only over euchromatin and nucleoli in the nuclei but also over many cell organelles such as endoplasmic reticulum, ribosomes, and mitochondria as well as cytoplasmic matrices from perinatal stage at embryonic day 19, postnatal day 1, 3, 9, 14, to adult and senescent stages at postnatal month 1, 2, 12 and 24. The silver grains were also observed in binucleate hepatocytes at postnatal day 1, 3, 9, 14, month 1, 2, 6, 12 and 24. The localizations of silver grains over the mitochondria were mainly on the mitochondrial matrices but a few over the mitochondrial membranes and cristae when observed by high power magnification.

As the results, it was found that almost all the hepatocytes were labeled with silver grains showing RNA synthesis in their nuclei and mitochondria. Preliminary quantitative analysis on the number of mitochondria in 10 mononucleate hepatocytes whose nuclei were intensely labeled with many silver grains (more than 10 per nucleus) and other 10 mononucleate hepatocytes whose nuclei were not so intensely labeled (number of silver grains less than 9) in each aging group revealed that there was no significant difference between the number of mitochondria, number of labeled mitochondria and the labeling indices in both types of hepatocytes ($P<0.01$). Thus, the number of mitochondria and the labeling indices were calculated in 10 hepatocytes selected at random in each animal in respective aging stages regardless whether their nuclei were very intensely labeled or not. The results obtained from the number of mitochondria in mononucleate hepatocytes per cellular profile area showed an increase from the prenatal day (mean ± standard deviation 26.2± /cell) to postnatal day 1 to day 14 (38.4-51.7/cell), then to postnatal month 1-2 (53.7-89.2/cell), reaching the maximum, then decreased to year 1-2 (83.7-80.4/cell) and the increase was stochastically significant ($P<0.01$). The results of visual grain counts on the number of mitochondria labeled with silver grains obtained from 10 mononucleate hepatocytes of each animal labeled with ^{3}H-uridine demonstrating RNA synthesis in 10 aging groups at perinatal stages, prenatal embryo day 19, postnatal day 1, 3, 9 and 14, month 1, 6 and year 1 and 2, were counted. The labeling indices in respective aging stages were calculated from the number of labeled mitochondria and the number of total mitochondria per cellular profile area, respectively. The results showed that the numbers of labeled mitochondria with ^{3}H-uridine showing RNA synthesis increased from prenatal embryo day 19 (3.3/cell) to postnatal month 1 (9.2/cell), reaching the maximum, and then decreased to month 6 (3.5/cell) and again increased to year 1 (4.0/cell) and year 2 (4.3/cell), while the labeling indices increased from prenatal day 19 (12.4%) to postnatal month 1 (16.7%), reaching the maximum, then decreased to year 1 (4.8%) and year 2 (5.3%). Stochastical analysis revealed that the increases and decreases of the number of labeled mitochondria from the perinatal stage to the adult and senescent stage were significant in contrast that the increases and decreases of the labeling indices were not significant ($P<0.01$). As for the binucleate hepatocytes, on the other hand, because the appearances of binucleate hepatocytes were not so many in the embryonic stage, only several binucleate cells (5-8 at least) at respective stages when enough numbers of binucleate cells available from postnatal day 1 to year 2 were analyzed. The results of visual counts on the number of mitochondria labeled with silver grains obtained from several (5 to 8) binucleate hepatocytes labeled with ^{3}H-uridine demonstrating RNA synthesis in 8 aging groups at perinatal stages, postnatal day 1, 9, 14, and month 1, 2, 6, and year 1 and 2, were counted and the labeling indices in respective aging stages were calculated from the number of labeled mitochondria and the number of total mitochondria per cellular profile area calculated, respectively. The results showed that the number of labeled mitochondria increased from postnatal day 1 (2.3/cell) to day 9 (5.2/cell) and remained almost constant around 4-5, but the labeling indices increased from postnatal day 1 (2.1%) to postnatal day 9 (13.6%), remained almost constant around 13% (12.5-13.6%) from postnatal day 9 to month 1, then decreased to month 2 (6.1%) to month 6 (3.9%), and slightly increased to year 1 (6.3%) and 2 (5.3%). The increases and decreases of the number of labeled mitochondria and the labeling indices in binucleate hepatocytes were stochastically not significant ($P<0.01$).

We also studied intracellular localization of mRNA in adult rat hepatocytes localizing over the peroxisomes by means of in situ hybridization technique (Usuda and Nagata 1992, 1995, Usuda et al. 1992). However, its relationship to the aging of animals was not yet studied.

1.5.3 The protein synthesis in the liver

As for the protein synthesis in the liver, we first studied the incorporations of ^{3}H-leucine and ^{3}H-tryptophane in mouse hepatocytes in connection to the binuclearity before and after feeding (Nagata 1967b, Nagata et al. 1967a, Ma et al. 1991). Then, we also studied mitochondrial protein synthesis in the liver later (Nagata 2006a,b, 2007b,c,e, 2009b, 2010c).

1.5.3.1 The protein synthesis in hepatocyte nuclei

We first studied the incorporations of amino-acids, ^{3}H-leucine and ^{3}H-tryptophane, in mouse mononucleate and binucleate hepatocytes before and after feeding (Nagata 1967b, Nagata et al. 1967a, Ma et al. 1991). The results showed that the incorporations of both amino acids were greater in binucleate hepatocytes than mononucleate. When ^{3}H-leucine was injected into several groups of mice at various ages and the liver tissues were processed for LM and EM RAG, silver grains were observed over all cell types of the liver, i.e., hepatocytes (Fig. 9E), sinusoidal endothelial cells (Fig. 9F), ductal epithelial cells, Kupffer's cells, Ito's fat storing cells, fibroblasts and haematopoietic cells. In hepatocytes, number of silver grains in cytoplasm and karyoplasm increased from perinatal animals to postnatal 1 month young adult animals and decreased with aging to senescence at 24 months. Number of silver grains observed over respective cell organelles, the Golgi apparatus, mitochondria and endoplasmic reticulum, changed with aging, reaching the maxima at 1 month but the ratio remained constant at each point. When ^{3}H-proline was injected into mice at various ages from prenatal embryos to postnatal senescence, quantitative changes of collagen and protein synthesis in the livers were studied by electron microscopic radioautography (Ma and Nagata 2000, Nagata 2006a). The silver grains due to ^{3}H-proline showing collagen synthesis were localized over the nuclei, cytoplasmic matrix, endoplasmic reticulum, the Golgi apparatus, mitochondria and peroxisomes of almost all the cells such as hepatocytes (Fig. 9F), sinusoidal endothelial cells, Kupffer's cells (Fig. 9G), Ito's fat-storing cells, ductal epithelial cells, fibroblasts and haematopoietic cells at various ages. The number of silver grains in the cell bodies and nuclei, cytoplasmic matrix, endoplasmic reticulum, mitochondria, the Golgi apparatus and peroxisomes of hepatocytes gradually increased from embryo, reaching the maxima at postnatal month 1 and 6, and decreased with aging until 24 months. The grain counts of the cell bodies reached the maximum at month 6 and the nuclei at month 2, while that of endoplasmic reticulum at month 6 and mitochondria at month 1. The number of silver grains localized over the extracellular collagen fibrils and matrices was not so many in respective aging groups and did not show any remarkable changes with aging. From the results, it was concluded that ^{3}H-proline was incorporated not only into collagen but also into the structural proteins of hepatocytes and increased and decreased due to aging under normal aging conditions.

1.5.3.2 The protein synthesis in hepatocyte mitochondria

When the aging mice at various ages from embryo to senescence were injected with ^{3}H-leucine, it was found that almost all the hepatocytes, from embryonic day 19, postnatal day 1, 3, 9, 14, to adult and senescent stages at postnatal month 1, 2, 6, 12 and 24, incorporated silver grains (Fig. 9E). The silver grains were also observed in binucleate hepatocytes at postnatal day 1, 3, 9, 14, month 1, 2, 6, 12 and 24 (Nagata 2007a,b,c,d, 2006a,b, 2007b,c,e). The localizations of silver grains observed over the mitochondria were mainly on the mitochondrial matrices but a few over their nuclei, cytoplasmic matrix, endoplasmic reticulum, ribosomes, Golgi apparatus and mitochondria (Nagata 2006a, b, 2007b, c, e). In the mitochondria the silver grains were localized over the mitochondrial membranes and cristae when observed by high power magnification. Preliminary quantitative analysis on the number of mitochondria in 20 mononucleate hepatocytes whose nuclei were intensely labeled with many silver grains (more than 10 per nucleus) and other 20 mononucleate hepatocytes whose nuclei were not so intensely labeled (number of silver grains less than 9) in each aging group revealed that there was no significant difference between the number of mitochondria, number of labeled mitochondria and the labeling indices in both types of hepatocytes ($P<0.01$).

On the other hand, the numbers of mitochondria, the numbers of labeled mitochondria and the labeling indices were calculated in 10 binucleate hepatocytes selected at random in each animal in respective aging stages, regardless whether their nuclei were very intensely labeled or not, except the prenatal stage at embryonic day 19, because no binucleate cell was found at this stage, resulted in no significant difference between them. Thus, the numbers of mitochondria, the numbers of labeled mitochondria and the labeling indices were calculated in 20 hepatocytes selected at random in each animal in respective aging stages regardless whether their nuclei were very intensely labeled or not. The results obtained from the total numbers of mitochondria in mononucleate hepatocytes showed an increase from the prenatal day (34.5/cell) to postnatal days 1 (44.6/cell), 3 (45.8/cell), 9 (43.6/cell), 14 (48.5/cell), to postnatal months 1 (51.5/cell), 2 (52.3/cell), reaching the maximum at month 6 (60.7/cell), then decreased to years 1 (54.2/cell) and 2 (51.2/cell). The increase and decrease were stochastically significant ($P<0.01$). The results obtained from visual counting on the numbers of mitochondria labeled with silver grains from 20 mononucleate hepatocytes of each animal labeled with ^3H-leucine demonstrating protein synthesis in 10 aging groups at perinatal stages, prenatal embryo day 19, postnatal day 1, 3, 9 and 14, month 1, 2, 6 and year 1 and 2, were counted. The labeling indices in respective aging stages were calculated from the numbers of labeled mitochondria and the numbers of total mitochondria per cell. The results showed that the numbers of labeled mitochondria with ^3H-leucine showing protein synthesis increased from prenatal embryo day 19 (8.3/cell) to postnatal days 1 (9.6/cell), 3 (8.1/cell), 9 (8.9/cell), 14 (9.5/cell), and month 1 (11.2/cell), reaching the maximum, and then decreased to months 2 (9.1/cell), 6 (8.8/cell) to years 1 (6.7/cell) and 2 (2.2/cell), while the labeling indices increased from prenatal day 19 (20.1%) to postnatal days 1 (21.2%), 3 (21.6%), 9 (22.2%), 14 (23.1%), reaching the maximum, then decreased to month 1 (21.7%), 2 (17.4%), 6 (14.6%), and years 1 (12.4%) and 2 (4.4%). Stochastical analysis revealed that the increases and decreases of the numbers of labeled mitochondria as well as the labeling indices from the perinatal stage to the adult and senescent stages were significant ($P<0.01$).

The results obtained from the numbers of mitochondria in binucleate hepatocytes showed an increase from the postnatal days 1 (66.2/cell), to 3 (66.4/cell), 14 (81.8/cell), to postnatal months 1 (89.9/cell), 2 (95.1/cell), and 6 (102.1), reaching the maximum at month 12 (128.0/cell), then decreased to years 2 (93.9/cell). The increase and decrease were stochastically significant ($P<0.01$). The results obtained from visual counting on the numbers of mitochondria labeled with silver grains from 10 binucleate hepatocytes of each animal labeled with ^3H-leucine demonstrating protein synthesis in 10 aging groups at postnatal day 1, 3, and 14, month 1, 2, 6 and year 1 and 2, were counted. The labeling indices in respective aging stages were calculated from the numbers of labeled mitochondria and the numbers of total mitochondria per cell which showed that the numbers of labeled mitochondria with ^3H-leucine showing protein synthesis increased from postnatal day 1 (7.3/cell) to day 3 (6.8/cell), 14 (10.2/cell), and month 1 (15.0/cell), 2 (15.9/cell), reaching the maximum at month 6 (19.6/cell), then decreased to year 1 (8.3/cell) and 2 (5.1/cell), while the labeling indices increased from postnatal day 1 (11.8%) to 3 (10.2%), 14 (12.5%), month 1 (18.3%) and 2 (18.7%), reaching the maximum at month 6 (19.2%), then decreased to year 1 (6.4%) and 2 (5.5%). Stochastical analysis revealed that the increases and decreases of the numbers of labeled mitochondria as well as the labeling indices from the newborn stage to the adult and senescent stages were significant ($P<0.01$). The silver grains due to ^3H-leucine were also observed in sinusoidal endothelial cells (Fig. 8F), Kupfer's cells, Ito's fat-storing cells, ductal epithelial cells, fibroblasts and hematopoietic cells at various ages.

When ^3H-prolin was injected into aging mice at various ages, silver grains were localized over the nuclei, cytoplasmic matrix, endoplasmic reticulum, the Golgi apparatus, mitochondria and peroxisomes of all the cell such as hepatocytes (Fig. 9G), sinusoidal endothelial cells, Kupfer's cells (Fig. 9H), Ito's fat-storing cells, ductal epithelial cells, fibroblasts and hematopoietic cells at various ages. The number of silver grains in hepatocytes gradually increased from perinatal stage to postnatal month 1 to 6 due to aging, reaching the maximum and decreased to momth 24. The number of silver grains localized the extracellular collage fibrils and matrices was not so great in respective aging groups and did not show any remarkable changes with aging. From the results, it was concluded that ^{3}H-prolin was incorporated not only into collagen but also into the structural proteins of hepatocytes under normal aging conditions (Ma and Nagata 2000).

1.5.4 The glucide synthesis in the liver

We first studied ^{3}H-glucose incorporation into glycogen in the livers of adult mice, in connection to soluble compounds (Nagata and Murata 1977, Nagata et al. 1977a,d). Soluble ^{3}H-glucose, which was demonstrated by cryo-fixation (at -196°C) in combination with dry-mounting radioautography, was localized over the nuclei, nucleoli, all the cell organelles and cytoplasmic ground substance of all the hepatocytes diffusely. On the other hand, by conventional chemical fixation and wet-mounting radioautography, silver grains were localized only over glycogen granules, endoplasmic reticulum and Golgi apparatus showing glycogen synthesis. However, the relationship of glycogen synthesis to aging has not yet been fully clarified.

1.5.5 The lipids synthesis in the liver

We observed lipids synthesis in the liver using ^{3}H-glycerol in connection to soluble compounds (Nagata 1994a,d, Nagata and Murata 1977a, Nagata et al. 1977a,d). When adult mice were injected with ^{3}H-glycerol and the livers were taken out, cryo-fixed in liquid nitrogen at -196°C, then freeze-substituted, embedded in epoxy resin, dry-sectioned, and prepared for dry-mounting radioautography, many silver grains appeared over the nuclei and cytoplasm of hepatocytes diffusely. However, when the same liver tissues were fixed chemically in buffered glutaraldehyde and osmium tetroxide at 4°C, dehydrated, embedded, wet-sectioned and radioautographed by conventional wet-mounting procedures, very few silver grains were observed only over the endoplasmic reticulum and the lipid droplets, which demonstrated insoluble macromolecular lipid synthesis accumulating into the lipid droplets. However, the aging change of the lipid synthesis in the liver has not yet been fully clarified.

1.6 Macromolecular synthesis in the pancreas

The pancreas is a large gland, next to the liver in men and animals, among the digestive glands connected to the intestines. It consists of exocrine and endocrine portions and takes the shape of a compound acinous gland. The exocrine portion is composed of ductal epithelial cells, centro-acinar cells, acinar cells and connective tissue cells, while the endocrine portion, the islet of Langerhans, is composed of 3 types of endocrine cells, A, B, C cells and connective tissue cells. Intracellular transport of secretory proteins in the pancreatic exocrine cells were formerly studied by Jamieson and Palade (1967) by EMRAG. We studied the macromolecular synthesis of the aging mouse pancreas at various ages.

1.6.1 The DNA synthesis in the pancreas

We first studied the DNA synthesis of mouse pancreas by LM and EMRAG using ^{3}H-thymidine (Nagata and Usuda 1984, 1985, 1986, Nagata et al. 1986). Light and electron microscopic radioautograms of the pancreas revealed that the nuclei of pancreatic acinar cells (Fig. 11A), centro-acinar cells (Fig. 11B), ductal epithelial cells, and endocrine cells were labeled with ^{3}H-thymidine. The labeling indices of these cells in 5 groups of litter mate mice, fetal day 15, postnatal day 1, 20, 60 (2 months) and 730 (2 years) were analyzed.

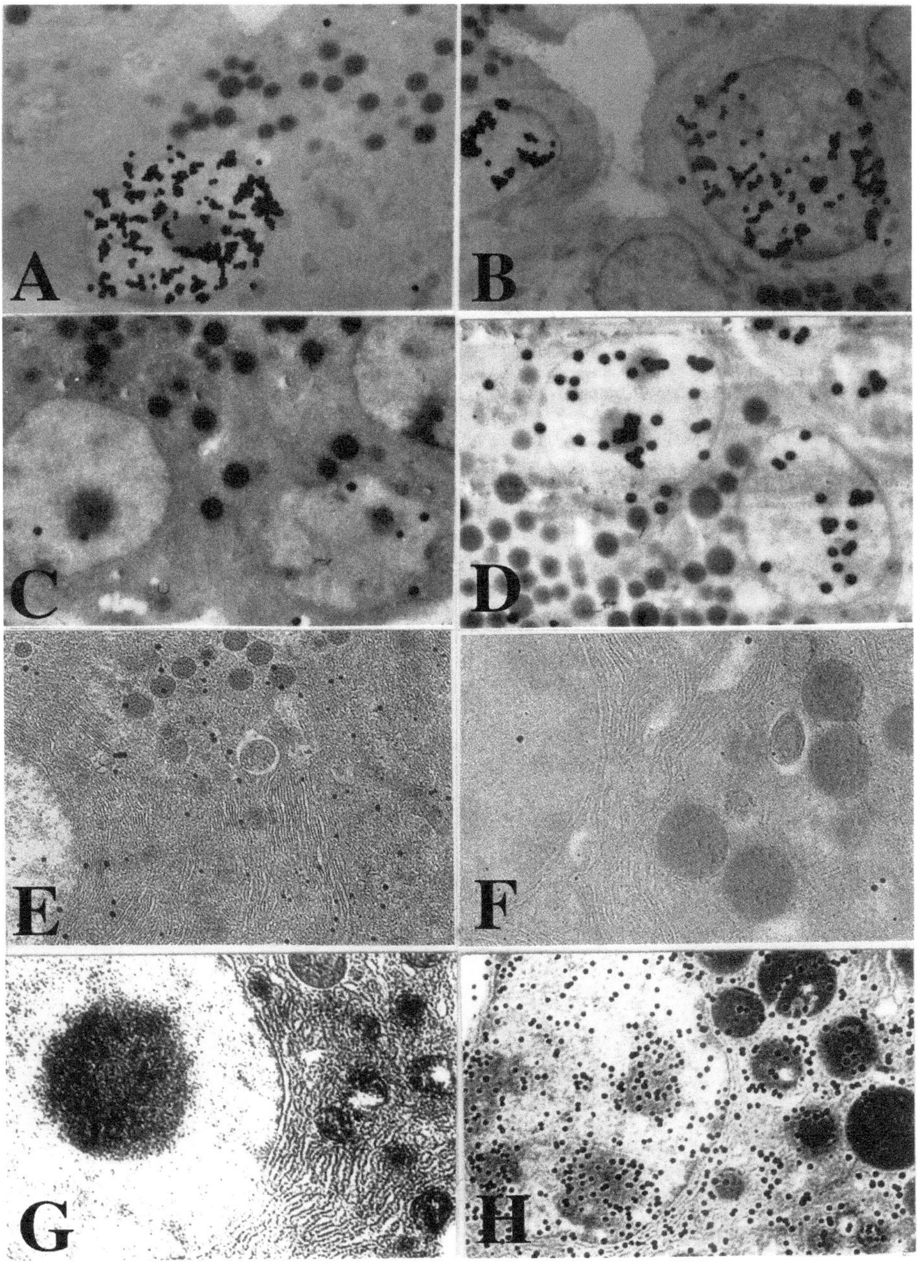

Fig. 11. EM RAG of the pancreas. From Nagata, T.: Radioautographology, General and Special. In, Prog. Histochem. Cytochem. Vol. 37, No. 2, p. 140, 2002, Urban & Fischer, Jena, Germany

Fig. 11A. EM RAG of 2 pancreatic acinar cells of a 14 day old mouse labeled with ^{3}H-thymidine, showing DNA synthesis. x10,000.
Fig. 11B. EM RAG of 2 centro-acinar cells of a 14 day old mouse labeled with ^{3}H-thymidine, showing DNA synthesis. x10,000.
Fig. 11C. EM RAG of 3 pancreatic acinar cells of a 1 day old mouse labeled with ^{3}H-uridine, showing RNA synthesis. x10,000.
Fig. 11D. EM RAG of 3 pancreatic acinar cells of a 14 day old mouse labeled with ^{3}H-uridine, showing RNA synthesis. x10,000.
Fig. 11E. EM RAG of a pancreatic acinar cell of a 30 day old mouse labeled with ^{3}H-leucine, showing protein synthesis. x10,000.
Fig. 11F. EM RAG of a pancreatic acinar cell of a 12 month old mouse labeled with ^{3}H-leucine, showing protein synthesis. x10,000.
Fig. 11G. EM RAG of a pancreatic acinar cell of a 1 day old mouse labeled with ^{3}H-glucosamine, showing glucide synthesis. x10,000.
Fig. 11H. EM RAG of a pancreatic acinar cell of a 14 day old mouse labeled with ^{3}H-glucosamine, showing glucide synthesis. x10,000.

The labeling indices of these cells reached the maxima at day 1 after birth and decreased gradually to 2 years (Fig. 12). The maximum in the acinar cells proceeded to the ductal and centro-acinar cells, suggesting that the acinar cells completed their development earlier than the ductal and centro-acinar cells (Nagata et al. 1986).

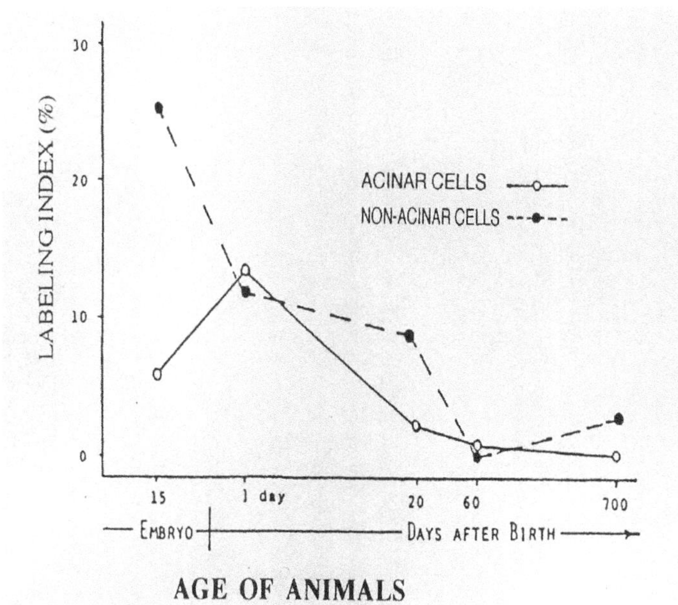

Fig. 12. Transitional curves of the labeling indices of respective cell types of the pancreas of aging mice labeled with ^{3}H-thymidine, showing DNA synthesis. Mean ± Standard Deviation. From Nagata, T.: Radioautographology, General and Special. In, Prog. Histochem. Cytochem. Vol. 37, No. 2, p. 142, 2002, Urban & Fischer, Jena, Germany

1.6.2 The RNA synthesis in the pancreas

On the other hand, LM and EMRAG of pancreas of mouse injected with ^{3}H-uridine demonstrated its incorporation into exocrine and then in endocrine cells, and more in pancreatic acinar cells (Figs. 11C,D) than in ductal or centro-acinar cells (Nagata and Usuda 1986a,b, Nagata et al. 1986). Among the acinar cells, the number of silver grains increased after birth to day 14 and then decreased with aging. Quantification of silver grains in the nucleoli, chromatin, and cell body were carried out by X-ray microanalysis (Nagata 1991, 1993, 2004, Nagata and Usuda 1985), which verified the results obtained by visual grain counting. In EMRAG obtained from the pancreas of fetal day 19 embryos, newborn day 1 and newborn day 14 mice labeled with ^{3}H-uridine, demonstrating RNA synthesis, the number of silver grains in the nucleoli, nuclear chromatin and cytoplasm increased (Nagata 1985, 1991, 1993a,b, 2004, Nagata and Usuda 1985, 1986). In order to quantify the silver contents of grains observed over the nucleoli, nuclei and cytoplasm, X-ray spectra were recorded by energy dispersive X-ray microanalysis (JEM-4000EX TN5400), demonstrating Ag-Ka peaks at higher energies. Thus, P/B ratios expressing relative silver contents were determined and compared between the two age groups. The results obtained by X-ray microanalysis in different cell compartments at postnatal day 1 and day 14 and he results obtained by visual grain counting in different cell compartments in day 1 day and day 14 animals were compared. The number of silver grains was calculated to express the counts per unit area to be compared with the XMA counts. These two results, the silver content analyzed by X-ray microanalysis and the results obtaining from visual grain counting were in good accordance with each other.

1.6.3 The protein synthesis in the pancreas

As for the protein synthesis in the pancreas, ^{3}H-leucine incorporation into endoplasmic reticulum, Golgi apparatus and to secretory granules of pancreatic acinar cells was first demonstrated by Jamieson and Palade (1967). We first studied ^{3}H-glycine incorporation into these cell organelles of mouse pancreatic acinar cells in connection with soluble compounds by EM RAG (Nagata 2000c, 2007a). It was demonstrated that soluble ^{3}H-glycine distributed not only in these cell organelles but also in the karyoplasm and cytoplasm diffusely. Then, the quantitative aspects of protein synthesis with regards the aging from fetal day 19, to postnatal day 1, 3, 7, 14 and 1, 2, 6 and 12 months were also clarified (Nagata and Usuda 1993a, Nagata 2000c). The results showed an increase of silver grain counts labeled with ^{3}H-leucine after birth, reaching a peak from postnatal 2 weeks to 1 month (Fig. 10E), and decreasing from 2 months to 1 year (Fig. 11F).

On the other hand, we also studied ^{3}H-leucine incorporations into the pancreatic acinar cells of both normal adult rats and experimentally pancreatitis induced rats with either ethionine or alcohol (Yoshizawa et al. 1974, 1977). The results showed that the incorporations as indicated silver grain counts in the pancreatitis rats were less than normal control rats. However, its relation to the aging was not yet studied.

1.6.4 The glucide synthesis in the pancreas

Concerning the glucide synthesis of the pancreas, we first studied the incorporation of ^{3}H-glucose into the pancreatic acinar cells of mouse in connection with soluble compounds by

EM RAG (Nagata et al. 1977a). It was demonstrated that soluble ^{3}H-glucose distributed not only in such cell organelles as endoplasmic reticulum, Golgi apparatus, mitochondria but also in the karyoplasm and cytoplasm diffusely. Then, the incorporation of ^{3}H-glucosamine into the pancreases of aging mice at various ages was studied by LM and EM RAG (Nagata et al. 1992). When perinatal baby mice received ^{3}H-glucosamine injections and the pancreatic tissues were radioautographed, silver grains were observed over exocrine and endocrine pancreatic cells. However, the number of silver grains was not so many (Fig. 11G). When juvenile mice at the age of 14 days after birth were examined, many silver grains appeared over the exocrine pancreatic acinar cells (Fig. 11H). Less silver grains were observed over endocrine pancreatic cells and ductal epithelial cells. The grains in the exocrine pancreatic acinar cells were localized over the nucleus, endoplasmic reticulum, Golgi apparatus and secretory granules, demonstrating glycoprotein synthesis. Adult mice at the ages of postnatal 1 month, 6 month or senile mice at the ages of 12 months or 24 months showed very few silver grains on radioautograms. Thus, the glucide synthesis in the pancreatic acinar cells of mice revealed quantitative changes, increase and decrease of ^{3}H-glucosamine incorporation with aging (Nagata 1994a,b,c,d,e, Nagata et al. 1992).

1.6.5 The lipids synthesis in the pancreas

In order to demonstrate the lipids synthesis in the pancreas, several litters of ddY mice aged fetal day 19, postnatal day 1, 3, 7, 14, and 1, 2, 6 up to 12 months, were injected with ^{3}H-glycerol and the pancreas tissues were prepared for LM and EM RAG. The silver grains were observed in both exocrine and endocrine cells of respective ages (Nagata 1995a,b, Nagata et al. 1988b, 1990). In perinatal animals from fetal day 19 to postnatal 1, 3, and 7 days, cell organelles were not well developed in exocrine and endocrine cells and number of silver grains was very few. In 14 day old juvenile animals, cell organelles such as endoplasmic reticulum, Golgi apparatus, mitochondria and secretory granules were well developed and many silver grains were observed over these organelles and nuclei in both exocrine and endocrine cells. The number of silver grains was more in exocrine cells than endocrine cells. In 1, 2, 6 month old adult animals, number of silver grains remained constant. In 12 month old senescent animals, silver grains were fewer than younger animals. It was demonstrated that the number of silver grains expressed the quantity of lipids synthesis, which increased from perinatal atages to adult and senescent stages and finally decreased to senescence.

2. Macromolecular synthesis in the respiratory system

The respiratory system consists of 2 parts, the air-conducting portion and the respiratory portion. The former are the nose, the pharynx, the larynx and the trachea, while the latter the lung. We studied the macromolecular synthesis in the pulmonary tissues as well as the tracheal tissues at various ages from embryo to postnatal 2 years.

2.1 The DNA synthesis in the respiratory system

Among the air-conducting portion and the respiratory portion we studied the macromolecular synthesis in the tracheal tissues as well as the pulmonary tissues at various ages from embryo to postnatal 2 years (Sun et al. 1994, 1995a,b, 1997a,b, Nagata 2000d).

2.1.1 The DNA synthesis in the trachea

The tracheas of mammals are composed of ciliated pseudostratified columnar epithelia, connective tissues, smooth muscles and hyalin cartilages surrounding the epithelia. The changes of DNA synthesis of tracheal cells in aging mice were studied by LM and EMRAG (Sun et al. 1997a, Nagata 2000d). The tracheae of 8 groups of mice from fetal day 18 to 2 years after birth were examined. The results demonstrated that the DNA syntheses and morphology of tracheal cells in the mouse tracheae changed due to aging. The radioautograms revealed that the DNA synthesis in the nuclei of ciliated cells was observed only in the fetal animals (Fig. 13A). However, the DNA synthesis in nonciliated cells and basal cells was observed in both prenatal and postnatal animals (Fig. 13B). The labeling indices of respective cell types were analyzed (Sun et al. 1997a). As the results, the labeling indices of the epithelial cells showed their maxima on fetal day 18, then fell down from postnatal day 3 to 2 years (Fig. 14A). The ciliated cell could not synthesize DNA and proliferate in the postnatal stage. They are supposed to be derived by the division and transformation of the basal cells. On the other hand, the DNA synthesis of chondrocytes was the highest on embryonic day 18, and rapidly declined on postnatal day 3 (Fig. 13C). The chondrocytes lost the ability of synthesizing DNA at 2 months after birth (Fig. 14B). The labeling indices of other cells (including fibroblasts, smooth muscle and glandular cells) were the highest on fetal day 18 and fell down markedly on the third day after birth and decreased progressively due to aging (Fig. 14C).

2.1.2 The DNA synthesis in the lung

We studied the pulmonary tissues at various ages from embryo to postnatal 2 years of mice (Sun et al. 1995a,b). The pulmonary tissues obtained from ddY strain mice at various ages from embryo day 19 to adult postnatal day 30 and to year 2 consisted of several types of cells, i.e., the type I epithelial cells or the small alveolar epithelial cells, type II epithelial cells or large alveolar epithelial cells, interstitial cells and endothelial cells, which incorporated macromolecular precursors respectively (Figs. 13E, F, G, H).

2.1.2.1 The DNA synthesis in the pulmonary cells

The pulmonary tissues obtained from ddY strain mice at embryonic to early postnatal stages consisted of undifferentiated cells (Fig. 13E). However, they differentiated into several types of cells due to aging, the type I epithelial cells or the small alveolar epithelial cells (Fig. 13E), the type II epithelial cells or the large alveolar epithelial cells (Fig. 13F), the interstitial cells (Fig. 13G), the endothelial cells (Fig. 13H) and alveolar phagocytes or dust cells as we had formerly observed (Sun et al. 1995a,b). At embryonic day 16 and 18, the fetal lung tissues appeared as glandular organizations consisting of many alveoli bordering undifferentiated cuboidal cells and no squamous epithelial cells were seen (Figs. 13E, 13G). Mitotic figures were frequently observed in cuboidal epithelial cells. After birth, the structure of the alveoli was characterized by further development of the alveolar-capillary networks from postnatal day 1 to 3 and 7 (Fig. 13H). During the development, the cellular composition of the alveolar epithelium resembled that of the adult lung, with a mixed population of the type I and type II epithelial cells. Up to 1 and 2 weeks after birth, the lung tissues showed complete alveolar structure and single capillary system almost the same as the adult after 1 month (Fig. 13F) to 2 to 6 months, and further to senescent stage over 12 months to 22 months. On electron microscopic radioautograms of the pulmonary tissues labeled with ^{3}H-thymidine, silver grains were observed over the nuclei of some pulmonary cells corresponding to the DNA synthesis in S-phase as observed by light microscopic radioautograpy (Sun et al. 1995a,b, 1997a).

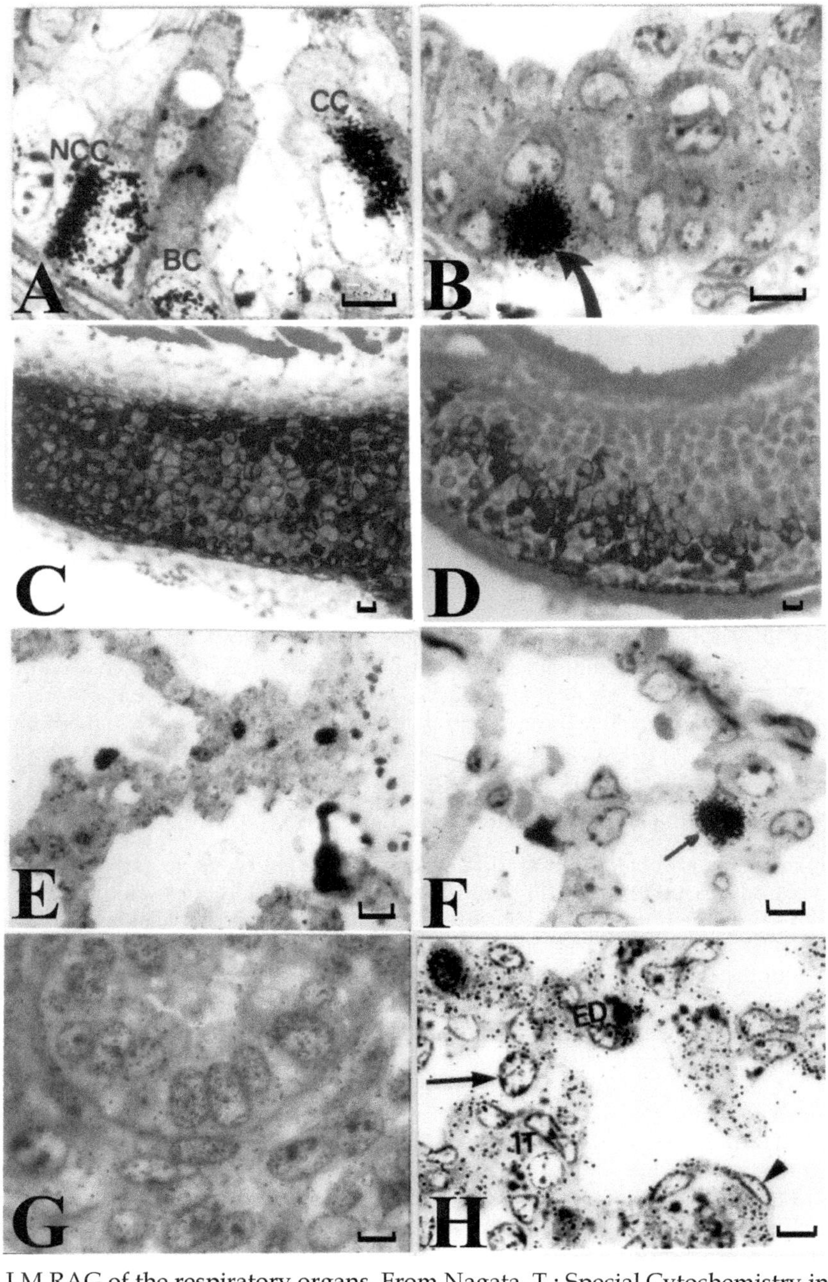

Fig. 13. LM RAG of the respiratory organs. From Nagata, T.: Special Cytochemistry in Cell Biology, In, Internat. Rev. Cytol. Vol. 211, No. 1, p. 102, 2001, Academic Press, San Diego, USA, London, UK.

Fig. 13A. LM RAG of the tracheal epithelial cells of a prenatal day 18 mouse embryo labeled with ^{3}H-thymidine, showing DNA synthesis. x1,125.
Fig. 13B. LM RAG of the tracheal epithelial cells of a postnatal 1 month old mouse labeled with ^{3}H-thymidine, showing DNA synthesis. x1,125.
Fig. 13C. LM RAG of the tracheal cartilage cells of a prenatal day 19 mouse embryo labeled with $^{35}SO_4$ showing mucosubstance synthesis. x400.
Fig. 13D. LM RAG of the tracheal cartilage cells of a postnatal day 3 mouse labeled with $^{35}SO_4$ showing mucosubstance synthesis. x400.
Fig. 13E. LM RAG of the lung of a 1 day old mouse labeled with ^{3}H-thymidine, showing DNA synthesis. x750.
Fig. 13F. LM RAG of the lung of a 1 month old mouse labeled with ^{3}H-thymidine, showing DNA synthesis. x750.
Fig. 13G. LM RAG of the lung of a prenatal day 16 mouse embryo labeled with ^{3}H-uridine, showing RNA synthesis. x750.
Fig. 13H. LM RAG of the lung of a 7 day old mouse labeled with ^{3}H-leucine, showing protein synthesis. x750.

The DNA synthetic activity of respective pulmonary cells as expressed by labeling indices demonstrated increases from perinatal stage to developmental stage and decreased due to aging. We also studied inhalation of ^{3}H-thymidine in air by means of a nebulizer into the lungs of 1 week old young mice as experimental studies and observed by LM and EM RAG (Duan et al. 1994). After 10 min. inhalation, the lung tissues were taken out and processed by either rapid freezing and freeze-substitution for dry-mounting radioautography or conventional chemical fixation for wet-mounting radioautography. By wet-mounting RAG silver grains were observed in the nuclei of a few alveolar type 2 cells and interstitial cells demonstrating DNA synthesis. By dry-mounting RAG, numerous silver grains were located diffusely over all the epithelial cells and interstitial cells demonstrating soluble compounds. The results showed that ^{3}H-thymidine inhaled into the lung distributed over all the pulmonary cells but only some of the alveolar type 2 cells and interstitial cells did synthesize DNA.

On the other hand, we studied the aging changes of DNA synthesis in the lungs of salamanders, *Hynobius nebolosus*, from larvae at 2 month after fertilization, juvenile at 1 month, adults at 10 and 12 months after metamorphosis, and finally to senescence at 5 years by LM RAG after ^{3}H-thymidine injections (Matsumura et at. 1994). The results showed that the labeling indices of in the ciliated cells and mucous cells in the superficial layer of young animals were higher than those of the basal cells and they decreased in adults, demonstrating aging changes in salamanders.

2.1.2.2 The DNA synthesis in mitochondria of mouse pulmonary cells

On electron microscopic radioautograms of the pulmonary tissues labeled with ^{3}H-thymidine, silver grains were observed over not only the nuclei of some pulmonary cells corresponding to the DNA synthesis in S-phase as observed by LM radioautograpy (Sun et al. 1995a,b) but also over the mitochondria by EMRAG (Sun et al. 1995b). Some mitochondria in both S-phase cells and interphase cells which did not show any silver grains over their nuclei were labeled with silver grains showing intramitochondrial DNA synthesis. The intramitochondrial DNA synthesis was observed in all cell types, the type I epithelial cells, the type II epithelial cells (Figs. 13F,G), the interstitial cells (Fig. 13F) and the endothelial cells. Because enough numbers of electron photographs (more than 5) were not obtained from all the cell types in respective aging groups, only some cell types and some aging groups when enough numbers of electron photographs were available were used for quantitative analysis. The numbers of mitochondria per cell profile area, the numbers of labeled mitochondria per cell and the labeling indices of the type I epithelial cells in only a few aging groups were observed and counted. The labeling indices in respective aging stages were calculated from the number of labeled mitochondria and the number of total mitochondria per cellular profile area which were calculated, respectively. These results demonstrated that the labeling indices in these cell types decreased from prenatal stages at embryo day 16 to day 18 (20-25%), and further decreased to postnatal days up to senescent stages due to aging (Fig. 15).

2.2 The RNA synthesis in the respiratory system

We studied the lungs of aging mice among the respiratory organs after administration of ^{3}H-uridine at various ages from prenatal embryonic day 16 to postnatal senescent month 22 as observed by LM and EM RAG (Fig. 13H).

2.2.1 RNA synthesis of mouse pulmonary cells

When the lung tissues of mice were labeled with ^{3}H-uridine, RNA synthesis was observed in all cells of the lungs at various ages by RAG (Sun 1995, Sun et al. 1997b, Nagata 2002). Observing the light microscopic radioautograms labeled with ^{3}H-uridine, the silver grains were found over both the karyoplasm and cytoplasm of almost all the cells not only at the perinatal stages from embryo day 16, 18, to postnatal day 1, 3, 7, 14, but also at the adult and senescent stages from postnatal month 1 to 2, 6, 12 and 22. The number of silver grains, by electron micrography, changed with aging. The grain counts in type I epithelial cells increased from the 1st day after birth and reached a peak at 1 week and decreased gradually to month 22, while the counts in type II epithelial cells (Fig. 13H), interstitial cells and endothelial cells increased from embryo day 16 and reached peaks at 1 week after birth, then decreased to senescence.

2.2.2 The RNA synthesis of mitochondria in pulmonary cells

By electron microscopic radioautography, silver grains were observed in most pulmonary cells in respective aging groups localizing not only over euchromatin and nucleoli in the nuclei but also over many cell organelles such as endoplasmic reticulum, ribosomes, and mitochondria as well as cytoplasmic matrices from perinatal stage at embryonic day 16, 18, to postnatal day 1, 3, 7, 14, to adult and senescent stages at postnatal month 1, 2, 6, 12 and 22 (Sun 1995, Sun et al. 1997b, Nagata 2002). The localizations of silver grains over the mitochondria were mainly on the mitochondrial matrices but a few over the mitochondrial membranes and cristae when observed by high power magnification. However, quantitative analyses on the number of mitochondria, the number of labeled mitochondria and the labeling index were not performed because enough number of EM RAG was not obtained.

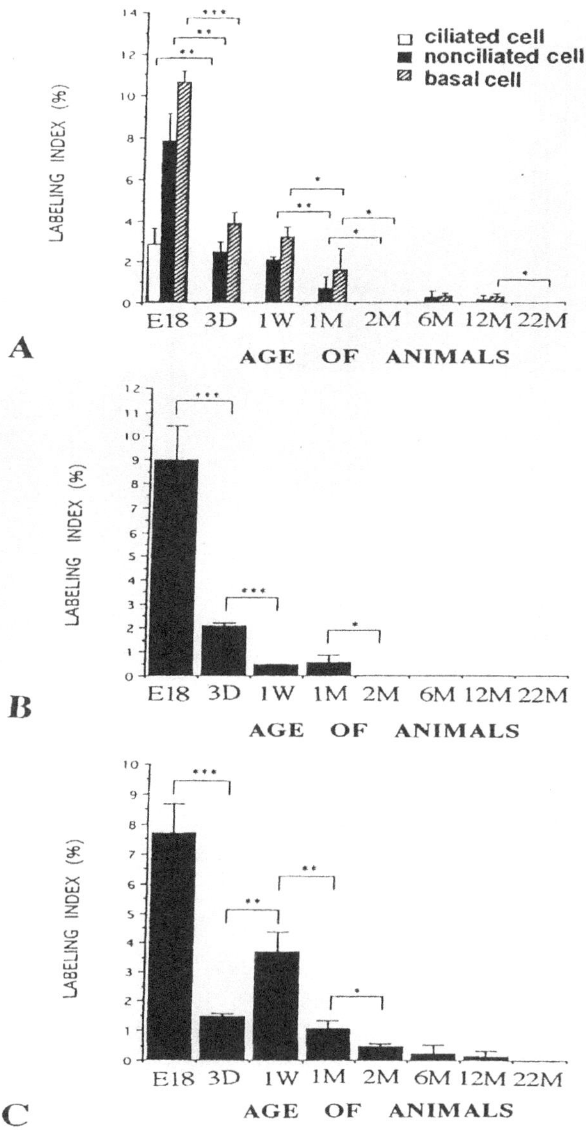

Fig. 14. Histogram showing aging changes of average labeling indices in respective cell types of the trachea of aging mice labeled with ^{3}H-thymidine. Mean ± Standard Deviation. From Nagata, T.: Radioautographology, General and Special. In, Prog. Histochem. Cytochem. Vol. 37, No. 2, p. 148, 2002, Urban & Fischer, Jena, Germany

Fig. 14A. Epithelial cells.

Fig. 14B. Chondrocytes in the cartilage.

Fig. 14C. Other cells.

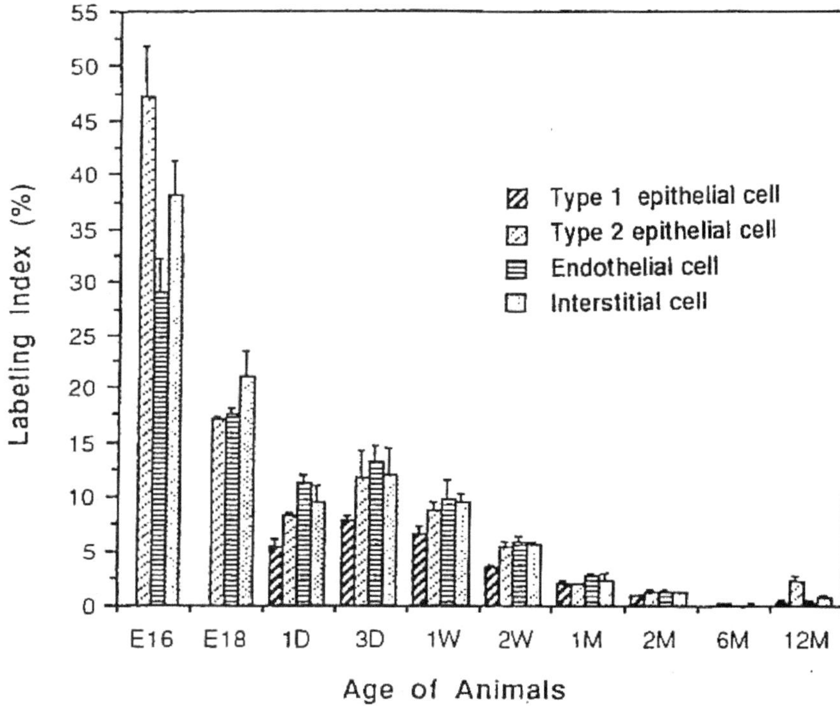

Fig. 15. Histogram showing aging changes of average labeling indices in respective cell types of the tracheal epithelial cells of aging mice labeled with ^{3}H-thymidine. Mean ± Standard Deviation. From Nagata, T.: Radioautographology, General and Special. In, Prog. Histochem. Cytochem. Vol. 37, No. 2, p. 149, 2002, Urban & Fischer, Jena, Germany

2.3 The protein synthesis in the respiratory system

We studied the protein synthesis only in the lungs of aging mice at various ages among the respiratory organs including the respiratory tract.

2.3.1 Protein synthesis of aging mouse pulmonary cells

When the lung tissues of mice at various ages were labeled with ^{3}H-leucine, protein synthesis was observed in all types of cells, type I and type II epithelial cells, interstitial cells and endothelial cells, of the lungs at various ages from embryo to senescence (Sun et al. 1997b). Observing the light microscopic radioautograms, the silver grains were detected over both the karyoplasm and cytoplasm of almost all the cells not only at the perinatal stages from embryo day 16, 18, to postnatal day 1 to 14 (Fig. 13I), but also at the adult and senescent stages from postnatal month 1 to 2, 6, 12 and 22. The number of silver grains, by light microscopic radioautography, changed with aging. The grain counts in type I epithelial cells were the highest on postnatal day 1, decreased on day 3 and increased again at 1 week, then decreased gradually to month 22, while the counts in type II epithelial cells, interstitial cells and endothelial cells reached the highest levels on fetal day 16, declined progressively with aging from fetal day 18 to postnatal day 3, then increased again from postnatal to day 7, reached peaks at 1 week (day 7) after birth, then decreased to senescence. However, grain counting on cell organelles by electron microscopy was not performed because enough numbers of EM RAG were not obtained at that time.

2.4 The glucide synthesis in the respiratory system

The incorporation of $^{35}SO_4$ in the trachea of aging mice was studied among the respiratory organs (Nagata 2000d). As the results, silver grains indicating the incorporations of radiosulfate were found over the cartilage matrices and the cartilage capsules in the hyaline cartilages of the tracheae of fetal (Fig. 13C) and postnatal newborn mice (Fig. 13D). The grain density as analyzed by grain densitometry was the maximum at the fetal day 19 (1200 grains per unit area). The grain density then decreased from the fetal day 19 to the postnatal day 1, 3, 7 (600/area), 14 (200/area) and reached 0 on day 30, and no silver grain was found in the animals aged from 1 to 12 months. The silver grains in the perinatal animals aged at postnatal day 1 and 3, disappearred from the internal layer to the external layer of the cartilage and from the interterritorial matrix to the territorial matrix and the cartilage capsule. In the juvenile animals aged at postnatal day 9 and 14, intense incorporations were observed disseminatedly over several groups of cartilage capsules in the external layer. The results indicated that the glycoproteins constituting the cartilage matrix were synthesized from prenatal to postnatal day 30. To the contrary, no incorporation of silver grains was observed in the aging animals from postnatal 1 to 12 months by both LM and EM RAG. These results demonstrated the aging changes of glycoprotein synthesis in the cartilage matrix of mice at various ages during development and aging.

3. References

Chen, S., Gao, F., Kotani, A., Nagata, T.: Age-related changes of male mouse submandibular gland: A morphometric and radioautographic study. Cell. Mol. Biol. 41, 117-124, 1995.

Clermont Y.: The contractime elements in the limiting membrane of the seminiferous tubules of rats. Exp. Cell Res. 15, 438-342, 1958.

Clermont, Y.: Renewal of spermatogonia in man. Amer. J. Anat. 112, 35-51, 1963.

Cui, H.: Light microscopic radioautographic study on DNA synthesis of nerve cells in the cerebella of aging mice. Cell. Mol. Biol. 41, 1139-1154, 1995.

Cui, H., Gao, F., Nagata, T.: Light microscopic radioautographic study on protein synthesis in perinatal mice corneas. Acta Histochem. Cytochem. 33, 31-37, 2000.

Duan, H., Gao, F., Li, S., Hayashi, K., Nagata, T.: Aging changes and fine structure and DNA synthesis of esophageal epithelium in neonatal, adult and old mice. J. Clin. Electron Microsc. 25, 452-453, 1992.

Duan, H., Gao, F., Li, S., Nagata, T.: Postnatal development of esophageal epithelium in mouse: a light and electron microscopic radioautographic study. Cell. Mol. Biol. 39, 309-316, 1993.

Duan, H., Gao, F., Oguchi, K., Nagata, T.: Light and electron microscopic radioautographic study on the incorporation of ^{3}H-thymidine into the lung by means of a new nebulizer. Drug Res. 44, 880-883, 1994.

Feulgen, R., Rossenbeck, H.: Mikroskopische-chemischer Nachweis einer Nukeinsaeure von Thymus der Thymonukeinsaeure Z. Physik. Chem. 135, 203-248, 1924.
Fujii, Y., Ohno, S., Yamabayashi, S., Usuda, N., Saito, H., Furuta, S., Nagata, T.: Electron microscopic radioautography of DNA synthesis in primary cultured cells from an IgG myeloma patient. J. Clin. Electr. Microsc. 13, 582-583, 1980.
Gao, F.: Study on the macromolecular synthesis in aging mouse seminiferous tubules by light and electron microscopic radioautography. Cell. Mol. Biol. 39, 659-672, 1993.
Gao, F., Toriyama, K., Nagata, T.: Light microscopic radioautographic study on the DNA synthesis of prenatal and postnatal aging mouse retina after labeled thymidine injection. Cell. Mol. Biol. 38, 661-668, 1992a.
Gao, F., Li, S., Duan, H., Ma, H., Nagata, T.: Electron microscopic radioautography on the DNA synthesis of prenatal and postnatal mice retina after labeled thymidine injection. J. Clin. Electron Microsc. 25, 721-722, 1992b.
Gao, F., Toriyama, K., Ma, H., Nagata, T.: Light microscopic radioautographic study on DNA synthesis in aging mice corneas. Cell. Mol. Biol. 39, 435-441, 1993.
Gao, F., Ma, H., Sun, L., Jin, C., Nagata, T.: Electron microscopic radioautographic study on the nucleic acids and protein synthesis in the aging mouse testis. Med. Electron Microsc. 27, 360-362, 1994.
Gao, F., Chen, S., Sun, L., Kang, W., Wang, Z., Nagata, T.: Radioautographic study of the macromolecular synthesis of Leydig cells in aging mice testis. Cell. Mol. Biol. 41, 145-150, 1995a.
Gao, F., Jin, C., Ma, H., Sun, L., Nagata, T.: Ultrastructural and radioautographic studies on DNA synthesis in Leydig cells of aging mouse testis. Cell. Mol. Biol. 41, 151-160, 1995b.
Gunarso, W.: Radioautographic studies on the nucleic acid synthesis of the retina of chick embryo. I. Light microscopic radioautography. Shinshu Med. J. 32, 231-240, 1984a.
Gunarso, W.: Radioautographic studies on the nucleic acid synthesis of the retina of chick embryo. II. Electron microscopic radioautography. Shinshu Med. J. 32, 241-248, 1984b.
Gunarso, W., Gao, F., Cui, H., Ma, H., Nagata, T.: A light and electron microscopic radioautographic study on RNA synthesis in the retina of chick embryo. Acta Histochem. 98, 309-32, 1996.
Gunarso, W., Gao, F., Nagata, T.: Development and DNA synthesis in the retina of chick embryo observed by light and electron microscopic radioautography. Cell. Mol. Biol. 43, 189-201, 1997.
Hanai, T.: Light microscopic radioautographic study of DNA synthesis in the kidneys of aging mice. Cell. Mol. Biol. 39, 81-91, 1993.
Hanai, T., Nagata, T.: Electron microscopic radioautographic study on DNA and RNA synthesis in perinatal mouse kidney. In, Radioautography in Medicine, Nagata, T., Ed., pp. 127-131, Shinshu University Press, Matsumoto, 1994a.
Hanai, T., Nagata, T.: Study on the nucleic acid synthesis in the aging mouse kidney by light and electron microscopic radioautography. In, Radioautography in Medicine, Nagata, T., Ed., pp. 209-214, Shinshu University Press, Matsumoto, 1994b.
Hanai, T., Nagata, T.: Electron microscopic study on nucleic acid synthesis in perinatal mouse kidney tissue. Med. Electron Microsc. 27, 355-357, 1994c.
Hanai, T., Usuda, N., Morita, T., Shimizu, T., Nagata, T.: Proliferative activity in the kidneys of aging mice evaluated by PCNA/cyclin immunohistochemistry. Cell. Mol. Biol. 39, 181-191, 1993.

Hayashi, K., Gao, F., Nagata, T.: Radioautographic study on ^{3}H-thymidine incorporation at different stages of muscle development in aging mice. Cell. Mol. Biol. 39, 553-560, 1993.
Ito, M.: The radioautographic studies on aging change of DNA synthesis and the ultrastructural development of mouse adrenal gland. Cell. Mol. Biol. 42, 279-292, 1996.
Ito, M., Nagata, T.: Electron microscopic radioautographic studies on DNA synthesis and ultrastructure of aging mouse adrenal gland. Med. Electron Microsc. 29, 145-152, 1996.
Izumiyama, K., Kogure, K., Kataoka, S., Nagata, T.: Quantitative analysis of glucose after transient ischemia in the gerbil hippocampus by light and electron microscope radioautography. Brain Res. 416, 175-179, 1987.
Jamieson, J. D., Palade, G. E.: Intracellular transport of secretory proteins in the pancreatic exocrine cells. J. Cell Biol. 34, 577-615, 1967.
Jin, C.: Study on DNA synthesis of aging mouse colon by light and electron microscopic radioautography. Cell. Mol. Biol. 42, 255-268, 1996.
Jin, C., Nagata, T.: Light microscopic radioautographic study on DNA synthesis in cecal epithelial cells of aging mice. J. Histochem. Cytochem. 43, 1223-1228, 1995a.
Jin, C., Nagata, T.: Electron microscopic radioautographic study on DNA synthesis in cecal epithelial cells of aging mice. Med. Electron Microsc. 28, 71-75, 1995b.
Joukura, K.: The aging changes of glycoconjugate synthesis in mouse kidney studied by ^{3}H-glucosamine radioautography. Acta Histochem. Cytochem. 29, 57-63, 1996.
Joukura, K., Nagata, T.: Aging changes of ^{3}H-glucosamine incorporation into mouse kidney observed by radioautography. Acta Histochem. Cytochem. 28, 494-494, 1995.
Joukura, K., Usuda, N., Nagata, T.: Quantitative study on the aging change of glycoconjugates synthesis in aging mouse kidney. Proc. Xth Internat. Cong. Histochem. Cytochem., Acta Histochem. Cytochem. 29, Suppl. 507-508, 1996.
Kobayashi, K., Nagata, T.: Light microscopic radioautographic studies on DNA, RNA and protein syntheses in human synovial membranes of rheumatoid arthritis patients. J. Histochem. Cytochem. 42, 982-982, 1994.
Komiyama, K., Iida, F., Furihara, R., Murata, F., Nagata, T.: Electron microscopic radioautographic study on ^{125}I-albumin in rat gastric mucosal epithelia. J. Clin. Electron Microsc. 11, 428-429, 1978.
Kong, Y.: Electron microscopic radioautographic study on DNA synthesis in perinatal mouse retina. Cell. Mol. Biol. 39, 55-64, 1993.
Kong, Y., Nagata, T.: Electron microscopic radioautographic study on nucleic acid synthesis of perinatal mouse retina. Med. Electron Microsc. 27, 366-368, 1994.
Kong, Y., Usuda, N., Nagata, T.: Radioautographic study on DNA synthesis of the retina and retinal pigment epithelium of developing mouse embryo. Cell. Mol. Biol. 38, 263-272, 1992a.
Kong, Y., Usuda, N., Morita, T., Hanai, T., Nagata, T.: Study on RNA synthesis in the retina and retinal pigment epithelium of mice by light microscopic radioautography. Cell. Mol. Biol. 38, 669-678, 1992b.
Leblond, C. P.: Localization of newly administered iodine in the thyroid gland as indicated by radioiodine. J. Anat. 77, 149-152, 1943.
Leblond, C. P.: The life history of cells in renewing systems. Am. J. Anat. 160, 113-158, 1981.

Leblond, C. P., Messier, B.: Renewal of chief cells and goblet cells in the small intestine as shown by radioautography after injection of thymidine-^{3}H into mice. Anat. Rec. 132: 247-259. 1958.

Li, S.: Relationship between cellular DNA synthesis, PCNA expression and sex steroid hormone receptor status in the developing mouse ovary, uterus and oviduct. Histochemistry 102, 405-413, 1994.

Li, S., Nagata, T.: Nucleic acid synthesis in the developing mouse ovary, uterus and oviduct studied by light and electron microscopic radioautography. Cell. Mol. Biol. 41, 185-195, 1995.

Li, S., Gao, F., Duan, H., Nagata, T.: Radioautographic study on the uptake of $^{35}SO_4$ in mouse ovary during the estrus cycle. J. Clin. Electron Microsc. 25, 709-710, 1992.

Liang, Y.: Light microscopic radioautographic study on RNA synthesis in the adrenal glands of aging mice. Acta Histochem. Cytochem. 31, 203-210, 1998.

Liang, Y., Ito, M., Nagata, T.: Light and electron microscopic radioautographic studies on RNA synthesis in aging mouse adrenal gland. Acta Anat. Nippon. 74, 291-300, 1999.

Ma, H.: Light microscopic radioautographic study on DNA synthesis of the livers in aging mice. Acta Anat. Nippon. 63, 137-147, 1988.

Ma, H., Nagata, T.: Electron microscopic radioautographic study on DNA synthesis of the livers in aging mice. J. Clin. Electron Microsc. 21, 335-343, 1988a.

Ma, H., Nagata, T.: Studies on DNA synthesis of aging mice by means of electron microscopic radioautography. J. Clin. Electron Microsc. 21, 715-716, 1988b.

Ma, H., Nagata, T.: Electron microscopic radioautographic studies on DNA synthesis in the hepatocytes of aging mice as observed by image analysis. Cell. Mol. Biol. 36, 73-84, 1990a.

Ma, H., Nagata, T.: Study on RNA synthesis in the livers of aging mice by means of electron microscopic radioautography. Cell. Mol. Biol. 36, 589-600, 1990b.

Ma, H., Nagata, T.: Collagen and protein synthesis in the livers of aging mice as studied by electron microsopic radioautography. Ann. Microsc. 1, 13-22, 2000.

Ma, H., Gao, F., Olea, M. T., Nagata, T.: Protein synthesis in the livers of aging mice studied by electron microscopic radioautography. Cell. Mol. Biol. 37, 607-615, 1991.

Matsumura, H., Kobayashi, Y., Kobayashi, K., Nagata, T.: Light microscopic radioautographic study of DNA synthesis in the lung of aging salamander, Hynobius nebulosus. J. Histochem. Cytochem. 42, 1004-1004, 1994.

Momose, Y., Nagata, T.: Radioautographic study on the intracellular localization of a hypolipidemic agent, bezafibrate, a peroxisome proliferator, in cultured rat hepatocytes. Cell. Mol. Biol. 39, 773-781, 1993a.

Momose, Y., Naito, J., Nagata, T.: Radioautographic study on the localization of an anti-allergic agent, tranilast, in the rat liver. Cell. Mol. Biol. 35, 347-355, 1989.

Momose, Y., Shibata, N., Kiyosawa, I., Naito, J., Watanabe, T., Horie, S., Yamada, J., Suga, T., Nagata, T.: Morphometric evaluation of species differences in the effects of bezafibrate, a hypolipidemic agent, on hepatic peroxisomes and mitochondria. J. Toxicol. Pathol. 6, 33-45, 1993b.

Momose, Y., Naito, J., Suzawa, H., Kanzawa, M., Nagata, T.: Radioautographic study on the intracellular localization of bezafibrate in cultured rat hepatoctyes. Acta Histochem. Cytochem. 28, 61-66, 1995.

Morita, T.: Radioautographic study on the aging change of ^{3}H-glucosamine uptake in mouse ileum. Cell. Mol. Biol. 39, 875-884, 1993.

Morita, T., Usuda, N. Hanai, T., Nagata, T.: Changes of colon epithelium proliferation due to individual aging with PCNA/cyclin immunostaining comparing with ^{3}H-thymidine radioautography. Histochemistry 101, 13-20, 1994.

Murata, F., Momose,Y. , Yoshida, K., Nagata, T.: Incorporation of ^{3}H-thymidine into the nucleus of mast cells in adult rat peritoneum. Shinshu Med. J. 25, 72-77, 1977a.

Murata, F., Momose, Y., Yoshida, K., Ohno, S., Nagata, T.: Nucleic acid and mucosubstance metabolism of mastocytoma cells by means of electron microscopic radioautography. Acta Pharmacol. Toxicol. 41, 58-59, 1977b.

Murata, F., Yoshida, K., Ohno, S., Nagata, T.: Ultrastructural and electron microscopic radioautographic studies on the mastocytoma cells and mast cells. J. Clin. Electron Microsc. 11, 561-562, 1978.

Murata, F., Yoshida, K., Ohno, S., Nagata, T.: Mucosubstances of rabbit granulocytes studied by means of electron microscopic radioautography and X-ray microanalysis. Histochemistry 61, 139-150, 1979.

Nagata, T.: On the relationship between cell division and cytochrome oxidase in the Yoshida sarcoma cells. Shinshu Med. J. 5: 383-386, 1956.

Nagata, T.: Studies on the amitosis in the Yoshida sarcoma cells. I. Observation on the smear preparation under normal conditions. Med. J. Shinshu Univ. 2: 187-198, 1957a.

Nagata, T.: Studies on the amitosis in the Yoshida sarcoma cells. II. Phase-contrast microscopic observations under normal conditions. Med. J. Shinshu Univ. 2: 199-207, 1957b.

Nagata, T.: Cell divisions in the liver of the fetal and newborn dogs. Med. J. Shinshu Univ. 4: 65-73, 1959.

Nagata, T.: A radioautographic study of the DNA synthesis in rat liver, with special reference to binucleate cells. Med. J. Shinshu Univ. 7, 17-25, 1962.

Nagata, T.: A quantitative study on the ganglion cells in the small intestine of the dog. Med. J. Shinshu Univ. 10, 1-11, 1965.

Nagata, T.: A radioautographic study on the RNA synthesis in the hepatic and the intestinal epithelial cells of mice after feeding with special reference to binuclearity. Med. J. Shinshu Univ. 11, 49-61, 1966.

Nagata, T.: On the increase of binucleate cells in the ganglion cells of dog small intestine due to experimental ischemia. Med. J. Shinshu Univ. 12, 93-113, 1967a.

Nagata, T.: A radioautographic study on the protein synthesis in the hepatic and the intestinal epithelial cells of mice, with special reference to binucleate cells. Med. J. Shinshu Univ. 12, 247-257, 1967b.

Nagata, T.: Chapter 3. Application of microspectrophotometry to various substances. In , Introduction to Microspectrophotometry. Isaka, S., Nagata, T., Inui, N., Eds., Olympus Co., Tokyo, pp. 49-155, 1972a.

Nagata, T.: Electron microscopic dry-mounting autoradiography. Proc. 4th Internat. Cong. Histochem. Cytochem. Kyoto, pp. 43-44, 1972b.

Nagata, T.: Electron microscopic radioautography of intramitochondrial RNA synthesis of HeLa cells in culture. Histochemie 32, 163-170, 1972c.

Nagata, T.: Quantitative electron microscope radioautography of intramitochondrial nucleic acid synthesis. Acta Histochem. Cytochem. 5, 201-203, 1972d.

Nagata, T.: Electron microscopic observation of target cells previously observed by phase-contrast microscopy: Electron microscopic radioautography of laser beam irradiated cultured cells. J. Clin. Electron Microsc. 17, 589-590, 1984.

Nagata, T.: Principles and techniques of radioautography. In, Histo- and Cyto-chemistry 1985, Japan Society of Histochemistry and Cytochemistry, Ed., Gakusai Kikaku Co., Tokyo, pp. 207-226, 1985.
Nagata, T.:. Electron microscopic radioautography and analytical electron microscopy. J. Clin. Electron Microsc. 24, 441-442, 1991.
Nagata, T.: Radiolabeling of soluble and insoluble compounds as demonstrated by light and electron microscopy. Recent Advances in Cellular and Molecular Biology, Wegmann, R. J., Wegmann, M. A., Eds. Peters Press, Leuven, Vol. 6, pp. 9-21, 1992.
Nagata, T.: Quantitative analysis of histochemical reactions: Image analysis of light and electron microscopic radioautograms. Acta Histochem. Cytochem. 26, 281-291, 1993a.
Nagata, T. Quantitative light and electron microscopic radioautographic studies on macromolecular synthesis in several organs of prenatal and postnatal aging mice. Chinese J. Histochem. Cytochem. 2: 106-108, 1993b.
Nagata, T.: Electron microscopic radioautography with cryo-fixation and dry-mounting procedure. Acta Histochem. Cytochem. 27: 471-489, 1994a.
Nagata, T.: Application of electron microscopic radioautography to clinical electron microscopy. Med. Electron Microsc. 27; 191-212, 1994b.
Nagata, T.: Radioautography in Medicine. Shinshu University Press, 268pp, Matsumoto, 1994c.
Nagata, T.: Radioautography, general and special. In, Histo- and Cyto-chemistry 1994, Japan Society of Histochemistry and Cytochemistry, ed, pp. 219-231, Gakusai Kikaku Co., Tokyo, 1994d.
Nagata, T., Application of electron microscopic radioautography to clinical electron microscopy. Med. Electron Microsc. 27, 191-212, 1994e.
Nagata, T.: Light and electron microscopic radioautographic study on macromolecular synthesis in digestive organs of aging mice. Cell. Mol. Biol. 41, 21-38, 1995a.
Nagata, T.: Histochemistry of the organs: Application of histochemistry to anatomy. Acta Anat. Nippon. 70, 448-471, 1995b.
Nagata, T.: Three-dimensional observation of whole mount cultured cells stained with histochemical reactions by ultrahigh voltage electron microscopy. Cell. Mol. Biol. 41, 783-792, 1995c.
Nagata, T.: Morphometry in anatomy: image analysis on fine structure and histochemical reactions with special reference to radioautography. Ital. J. Anat. 100 (Suppl. 1), 591-605, 1995d.
Nagata, T.: Technique and application of electron microscopic radioautography. J. Electron Microsc. 45, 258-274, 1996a.
Nagata, T.: Techniques of light and electron microscopic radioautography. In, Histochemistry and Cytochemistry 1996. Proc. Xth Internat. Congr. Histochem. Cytochem. Acta Histochem. Cytochem. 29 (Suppl.), 343-344, 1996b.
Nagata, T.: Remarks: Radioautographology, general and special. Cell. Mol. Biol. 42 (Suppl.), 11-12, 1996c.
Nagata, T.: On the terminology of radioautography vs. autoradiography. J. Histochem. Cytochem. 44, 1209-1209, 1996d.
Nagata, T.: Techniques and applications of microscopic radioautography. Histol. Histopathol. 12, 1091-1124, 1997a.
Nagata T.: Three-dimensional observation on whole mount cultured cells and thick sections stained with histochemical reactions by high voltage electron microscopy. In,

Recent Advances in Microscopy of Cells, Tissues and Organs, Motta, P., Ed., Antonio Delfino Editore, Roma, pp. 37-44, 1997b.
Nagata, T.: Radioautographic study on collagen synthesis in the ocular tissues. J. Kaken Eye Res. 15, 1-9, 1997c.
Nagata, T.: Techniques of radioautography for medical and biological research. Braz. J. Biol. Med. Res. 31, 185-195, 1998a.
Nagata, T.: Radioautographology, the advocacy of a new concept. Braz. J. Biol. Med. Res. 31, 201-241, 1998b.
Nagata, T.: Radioautographic studies on DNA synthesis of the bone and skin of aging salamander. Bull. Nagano Women's Jr. College 6, 1-14, 1998c.
Nagata, T.: 3D observation of cell organelles by high voltage electron microscopy. Microscopy and Analysis, Asia Pacific Edition, 9, 29-32, 1999a.
Nagata, T.: Application of histochemistry to anatomy: Histochemistry of the organs, a novel concept. Proc. XV Congress of the International Federation of Associations of Anatomists, Ital. J. Anat. Embryol. 104 (Suppl. 1), 486-486, 1999b.
Nagata, T.: Aging changes of macromolecular synthesis in various organ systems as observed by microscopic radioautography after incorporation of radiolabeled precursors. Methods Find. Exp. Clin. Pharmacol. 21, 683-706, 1999c.
Nagata, T.: Radioautographic study on protein synthesis in mouse cornea. J. Kaken Eye Res. 8, 8-14, 1999d.
Nagata, T.: Radioautographology, general and special: a novel concept. Ital. J. Anat. Embryol. 104 (Suppl. 1), 487-487, 1999e.
Nagata, T.: Three-dimensional observations on thick biological specimens by high voltage electron microscopy. Image Analysis Stereolog. 19, 51-56, 2000a.
Nagata, T.: Biological microanalysis of radiolabeled and unlabeled compounds by radioautography and X-ray microanalysis. Scanning Microscopy International, 14, on line, 2000b.
Nagata, T.: Electron microscopic radioautographic study on protein synthesis in pancreatic cells of perinatal and aging mice. Bull. Nagano Women's Jr. College 8, 1-22, 2000c.
Nagata, T.: Light microscopic radioautographic study on radiosulfate incorporation into the tracheal cartilage in aging mice. Acta Histochem. Cytochem. 32, 377-383, 2000d.
Nagata, T.: Introductory remarks: Special radioautographology. Cell. Mol. Biol. 46 (Congress Suppl.), 161-161, 2000e.
Nagata, T.: Special radioautographology: the eye. J. Kaken Eye Res. 18, 1-13, 2000f.
Nagata, T.: Three-dimensional high voltage electron microscopy of thick biological specimens. Micron 32, 387-404, 2001a.
Nagata, T.: Three-dimensional and four-dimensional observation of histochemical and cytochemical specimens by high voltage electron microscopy. Acta Histochem. Cytochem. 34, 153-169, 2001b.
Nagata, T. : Special cytochemistry in cell biology. In, Internat. Rev. Cytol. Jeon, K.W., ed., Vol. 211, Chapter 2, Academic Press, New York, pp. 33-154, 2001c.
Nagata, T. : Radioautographology General and Special, In, Prog. Histochem. Cytochem., Graumann, W., Ed., Vol. 37 No. 2, Urban & Fischer, Jena, pp. 57-226, 2002.
Nagata T.: Light and electron microscopic study on macromolecular synthesis in amitotic hepatocyte mitochondria of aging mice. Cell. Mol. Biol. 49, 591-611, 2003.
Nagata, T.: X-ray microanalysis of biological specimens by high voltage electron microscopy. In, Prog. Histochem. Cytochem., Graumann, W., Ed., Vol. 39, No. 4, Urban & Fischer Verlag, Jena, pp. 185-320, 2004.

Nagata T.: Aging changes of macromolecular synthesis in the uro-genital organs as revealed by electron microscopic radioautography. Ann. Rev. Biomed. Sci. 6, 13-78, 2005.

Nagata T.: Electron microscopic radioautographic study on protein synthesis in hepatocyte mitochondria of developing mice. Ann. Microsc. 6, 43-54, 2006a.

Nagata T.: Electron microscopic radioautographic study on nucleic acids synthesis in hepatocyte mitochondria of developing mice. The Sci. World J. 6: 1583-1598, 2006b.

Nagata T.: Macromolecular synthesis in hepatocyte mitochondria of aging mice as revealed by electron microscopic radioautography. I: Nucleic acid synthesis. In, Modern Research and Educational Topics in Microscopy. Mendez-Vilas, A. and Diaz, J. Eds., Formatex Micrscopy Series No. 3, Vol. 1, Formatex, Badajoz, Spain, pp. 245-258, 2007a.

Nagata T.: Macromolecular synthesis in hepatocyte mitochondria of aging mice as revealed by electron microscopic radioautography. II: Protein synthesis. In, Modern Research and Educational Topics in Microscopy. Mendez-Vilas, A. and Diaz, J. eds., Formatex Micrscopy Series No. 3, Vol. 1, Formatex, Badajoz, Spain, pp. 259-271, 2007b.

Nagata, T.: Electron microscopic radioautographic study on macromolecular synthesis in hepatocyte mitochondria of aging mouse. J. Cell Tissue Res. 7, 1019-1029, 2007c.

Nagata, T.; Electron microscopic radioautographic study on nucleic acids synthesis in hepatocyte mitochondria of developing mice. Trends Cell Molec. Biol. 2, 19-33, 2007d.

Nagata, T.; Aging changes of macromolecular synthesis in the mitochondria of mouse hepatocytes as revealed by microscopic radioautography. Ann. Rev. Biomed. Sci. 9, 30-36, 2007e.

Nagata, T.: Radioautographology, Bull. Shinshu Institute Alternat. Med. 2, 3-32, 2007f.

Nagata, T.: Electron microscopic radioautographic study on mitochondrial DNA synthesis in adrenal cortical cells of developing mice. J. Cell. Tis. Res. 8, 1303-1312, 2008a.

Nagata T.: Electron microscopic radioautographic study on mitochondrial DNA synthesis in adrenal cortical cells of developing and aging mice. The Sci. World J. 8, 683-697. 2008b.

Nagata, T.: Sexual difference between the macromolecular synthesis of hepatocyte mitochondria in male and female mice in aging as revealed by electron microscopic radioautography. Chapter 22. In, Women and Aging: New Research, H. T. Bennninghouse, A. D. Rosset, Eds. Nova Biomed. Books, New York, pp. 461-487, 2009a

Nagata, T.: Protein synthesis in hepatocytes of mice as revealed by electron microscopic radioautography. In, Protein Biosynthesis. Esterhouse, T. E. and Petrinos, L. B., Eds., Nova Biomed. Books, New York, pp. 133-161, 2009b.

Nagata, T.: Electron microscopic radioatuographic studies on macromolecular synthesis in mitochondria of various cells. 18EMSM Conference Proc. 9th Asia-Pacific Microscopy Conference (APMC9), Kuala Lumpur, Malaysia, pp. 48-50, 2009c.

Nagata, T.: Recent studies on macromolecular synthesis labeled with ^{3}H-thymidine in various organs as revealed by electron microscopic radioautography. Current Radiopharmaceutics 2, 118-128, 2009d.

Nagata, T.: Electron microscopic radioautographic study on mitochondrial DNA synthesis in adrenal medullary cells of developing and aging mice. J. Cell Tissue Res. 9, 1793-1802, 2009e.

Nagata, T.: Applications of high voltage electron microscopy to thick biological specimens. Ann. Microsc. 9, 4-40, 2009f.

Nagata, T.: Electron microscopic radioautographic study on DNA synthesis of mitochondria in adrenal medullary cells of aging mice. Open Anat. J. 1, 14-24, 2009g.

Nagata, T.: Electron microscopic radioautographic studies on macromolecular synthesis in mitochondria of animal cells in aging. Ann. Rev. Biomed. Sci. 11, 1-17, 2009h.

Nagata, T.: Electron microscopic radioautographic studies on macromoleclular synthesis in mitochondria of some organs in aging animals. Bull. Shinshu Inst. Alternat. Med. Welfare 4, 15-38, 2009i.

Nagata, T.: Electron microscopic radioautographic study on mitochondrial DNA synthesis in adreno-cortical cells of aging ddY mice. Bull. Shinshu Inst. Alternat. Med. Welfare 4, 51-66, 2009j.

Nagata T.: Electron microscopic radioautographic study on mitochondrial RNA synthesis in adrenocortical cells of aging mice. Open Anat J. 2, 91-97, 2010a.

Nagata T. Electron microscopic radioautographic study on mitochondrial RNA synthesis in adrenal medullary cells of aging and senescent mice. J Cell Tissue Res. 10, 2213-2222, 2010b.

Nagata, T.: Macromolecular synthesis in the livers of aging mice as revealed by electron microscopic radioautography. In, Prog. Histochem. Cytochem., Sasse, D., Ed., Elsevier, Amsterdam, Boston, London, New York, Oxford, Paris, Philadelphia, San Diego, St. Louis, Vol. 45, No. 1, pp. 1-80, 2010c.

Nagata, T.: Electron microscopic radioautographic study on protein synthesis of mitochondria in adrenal medullary cells of aging mice. Bulletin Shinshu Inst Alternat Med Welfare 5, 25-37, 2010d.

Nagata, T.: Electron microscopic radioautographic study on mitochondrial RNA synthesis in adrenal cortical and medullary cells of aging mice. J. Biomed. Sci. Enginer. 4, 219-232, 2010e.

Nagata, T.: Electron microscopic radioautographic study on protein synthesis of mitochondria in adrenal cortical cells of aging mice. Bulletin Shinshu Inst. Alternat. Med. Welfare 5, 38-52, 2010f.

Nagata T.: Electron microscopic radioautographic study on mitochondrial DNA, RNA and protein synthesis in adrenal cells of aging mice. Formatex Microscopy Series No. 3, Vol. 3, Formatex, Badajoz, Spain, in press, 2010g.

Nagata, T.: Electron microscopic radioautographic studies on macromolecular synthesis in mitochondria of animal cells in aging. Ann. Rev. Biomed. Sci. 12, 1-29, 2010h.

Nagata, T., Cui, H., Gao, F.: Radioautographic study on glycoprotein synthesis in the ocular tissues. J. Kaken Eye Res. 13, 11-18, 1995.

Nagata, T., Cui, H., Kong, Y.: The localization of TGF-b1 and its mRNA in the spinal cords of prenatal and postnatal aging mice demonstrated with immunohistochemical and in situ hybridization techniques. Bull. Nagano Women's Jr. College, 7, 75-88, 1999a.

Nagata, T., Cui, H., Liang, Y.: Light microscopic radioautographic study on the protein synthesis in the cerebellum of aging mouse. Bull. Nagano Women's Jr. College, 9, 41-60 (2001).

Nagata, T., Fujii, Y., Usuda, N.: Demonstration of extranuclear nucleic acid synthesis in mammalian cells under experimental conditions by electron microscopic radioautography. Proc. 10th Internat. Congr. Electr. Microsc. 2, 305-306, 1982b.

Nagata, T., Hirano, I., Shibata, O., Nagata, T.: A radioautographic study on the DNA synthesis in the hepatic and the pancreatic acinar cells of mice during the postnatal

growth, with special reference to binuclearity. Med. J. Shinshu Univ. 11, 35-42, 1966.
Nagata, T., Ito, M., Chen, S.: Aging changes of DNA synthesis in the submandibular glands of mice as observed by light and electron microscopic radioautography. Ann. Microsc. 1, 4-12, 2000a.
Nagata, T. Ito, M., Liang, Y.: Study of the effects of aging on macromolecular synthesis in mouse steroid secreting cells using microscopic radioautography. Methods Find. Exp. Clin. Pharmacol. 22, 5-18, 2000b.
Nagata, T., Iwadare, I., Murata, F.: Electron microscopic radioautography of nucleic acid synthesis in cultured cells treated with several carcinogens. Acta Pharmacol. Toxicol. 41, 64-65, 1977c.
Nagata, T., Kawahara, I.: Radioautographic study of the synthesis of sulfomucin in digestive organs of mice. J. Trace Microprobe Analysis 17, 339-355, 1999.
Nagata, T., Kawahara, I., Usuda, N., Maruyama, M., Ma, H.: Radioautographic studies on the glycoconjugate synthesis in the gastrointestinal mucosa of the mouse. In, Glycoconjugate in Medicine, Ohyama, M., Muramatsu, T., Eds, pp. 251-256, Professional Postgrad. Service, Tokyo, 1988a.
Nagata, T., Kong, Y.: Distribution and localization of TGFb1 and bFGF, and their mRNAs in aging mice. Bull. Nagano Women's Jr. College 6, 87-105, 1998.
Nagata, T., Ma, H., Electron microscopic radioautographic study on mitochondrial DNA synthesis in hepatocytes of aging mouse. Ann. Microsc. 5, 4-18, 2005a.
Nagata, T., Ma, H., Electron microscopic radioautographic study on RNA synthesis in hepatocyte mitochondria of aging mouse. Microsc. Res. Tech. 67, 55-64, 2005b.
Nagata, T., Momoze, S.: Aging changes of the amitotic and binucleate cells in dog livers. Acta Anat. Nipponica 34, 187-190, 1959.
Nagata, T., Morita, T., I. Kawahara, I.: Radioautographic studies on radiosulfate incorporation in the digestive organs of mice. Histol. Histopathol. 14, 1-8, 1999b.
Nagata, T., Murata, F.: Electron microscopic dry-mounting radioautography for diffusible compounds by means of ultracryotomy. Histochemistry 54, 75-82, 1977.
Nagata, T., Murata, F., Yoshida, K., Ohno, S., Iwadare, N.: Whole mount radioautography of cultured cells as observed by high voltage electron microscopy. Proc. Fifth Internat. Conf. High Voltage Electron Microsc. 347-350, 1977d.
Nagata, T., Nawa, T.: A modification of dry-mounting technique for radioautography of water-soluble compounds. Histochemie 7, 370-371, 1966a.
Nagata, T., Nawa, T.: A radioautographic study on the nucleic acids synthesis of binucleate cells in cultivated fibroblasts of chick embryos. Med. J. Shinshu Univ. 11, 1-5, 1966b.
Nagata, T., Nawa, T., Yokota, S.: A new technique for electron microscopic dry-mounting radioautography of soluble compounds. Histochemie 18, 241-249, 1969.
Nagata, T., Nishigaki, T., Momose, Y.: Localization of anti-allergic agent in rat mast cells demonstrated by light and electron microscopic radioautography. Acta Histochem. Cytochem. 19, 669-683, 1986b.
Nagata, T., Ohno, S., Kawahara, I., Yamabayashi, S., Fujii, Y., Murata, F.: Light and electron microscopic radioautography of nucleic acid synthesis in mitochondria and peroxisomes of rat hepatic cells during and after DEHP administration. Acta Histochem. Cytochem. 16, 610-611, 1979.
Nagata, T., Ohno, S., Murata, F.: Electron microscopic dry-mounting radioautography for soluble compounds. Acta Phamacol. Toxicol. 41, 62-63, 1977a.

Nagata, T., Ohno, S., Yoshida, K., Murata, F.: Nucleic acid synthesis in proliferating peroxisomes of rat liver as revealed by electron microscopical radioautography. Histochem. J. 14, 197-204, 1982a.

Nagata, T., Olea, M. T.: Electron microscopic radioautographic study on the protein synthesis in aging mouse spleen. Bull. Nagano Women's Jr. College 7, 1-9, 1999.

Nagata, T., Shibata, O., Omochi, S.: A new method for radioautographic osbservation on isolated cells. Histochemie 2, 255-259, 1961

Nagata, T., Shibata, O., Nawa, T.: Simplified methods for mass production of radioautograms. -Acta Anat. Nippon.42, 162-166, 1967a.

Nagata, T., Shibata, O., Nawa, T.: Incorporation of tritiated thymidine into mitochondrial DNA of the liver and kidney cells of chickens and mice in tissue culture. Histochemie 10, 305-308, 1967b.

Nagata, T., Shimamura, K., Onozawa, M., Kondo, T., Ohkubo, K., Momoze, S.: Relationship of binuclearity to cell function in some organs. I. Frequencies of binucleate cells in some organs of toads in summer and winter. Med. J. Shinshu Univ. 5, 147-152, 1960a.

Nagata, T., Shimamura, K., Kondo, T., Onozawa, M., Momoze, S., Okubo, M.: Relationship of binuclearity to cell function in some organs. II. Variation of frequencies of binucleate cells in some organs of dogs owing to aging. Med. J. Shinshu Univ. 5, 153-158, 1960b.

Nagata, T., Steggerda, F. R.: Histological study on the deganglionated small intestine of the dog. Physiologist 6, 242-242, 1963.

Nagata, T., Steggerda, F. R.: Observations on the increase of binucleate cells in the ganglion cells of the dog's intestine due to experimental ischemia. Anat. Rec. 148, 315-315, 1964.

Nagata, T., Toriyama, K., Kong, Y., Jin, C., Gao, F.: Radioautographic study on DNA synthesis in the ciliary bodies of aging mice. J. Kaken Eye Res.12, 1-11, 1994.

Nagata, T., Usuda, N.: Image processing of electron microscopic radioautograms in clinical electron microscopy. J. Clin. Electron. Microsc. 18, 451-452, 1985.

Nagata, T., Usuda, N.: Studies on the nucleic acid synthesis in pancreatic acinar cells of aging mice by means of electron microscopic radioautography. J. Clin. Electron Microsc. 19, 486-487, 1986.

Nagata, T., Usuda, N.: Electron microscopic radioautography of protein synthesis in pancreatic acinar cells of aging mice. Acta Histochem. Cytochem. 26, 481-481, 1993a.

Nagata, T., Usuda, N.: In situ hybridization by electron microscopy using radioactive probes. J. Histochem. Cytochem. 41, 1119-1119, 1993b.

Nagata, T., Usuda, N., Ma, H.: Electron microscopic radioautography of nucleic acid synthesis in pancreatic acinar cells of prenatal and postnatal aging mice. Proc. XIth Intern. Cong. Electr. Microsc. 3, 2281-2282, 1984.

Nagata, T., Usuda, N., Ma, H.: Electron microscopic radioautography of lipid synthesis in pancreatic cells of aging mice. J. Clin. Electr. Microsc. 23, 841-842, 1990.

Nagata, T., Usuda, N., Maruyama, M., Ma, H.: Electron microscopic radioautographic study on lipid synthesis in perinatal mouse pancreas. J. Clin. Electr. Microsc. 21, 756-757, 1988b.

Nagata, T., Usuda, N., Suzawa, H., Kanzawa, M.: Incorporation of ^{3}H-glucosamine into the pancreatic cells of aging mice as demonstrated by electron microscopic radioautography. J. Clin. Electron Microsc. 25, 646-647, 1992.

Nagata, T., Yamabayashi, S.: Intracellular localization of ^{3}H-befunolol by means of electron microscopic radioautography of cryo-fixed ultrathin sections. J. Clin. Electron Microsc. 16, 737-738, 1983.
Nagata, T., Yoshida, K., Murata, F.: Demonstration of hot and cold mercury in the human thyroid tissues by means of radioautography and chemography. Acta Pharmacol. Toxicol. 41, 60-61, 1977b.
Nagata, T., Yoshida, K., Ohno, S., Murata, F.: Ultrastructural localization of soluble and insoluble ^{3}H-methyl prednisolone as revealed by electron microscopic dry-mounting radioautography. Proc. 9th Internat. Congr. Electr. Microsc. 2, 40-41, 1978b.
Nishigaki, T., Momose, Y., Nagata, T.: Light microscopic radioautographic study of the localization of anti-allergic agent, tranilast, in rat mast cells. Histochem. J. 19, 533-536, 1987.
Nishigaki, T., Momose, Y., Nagata, T.: Electron microscopic radioautographic study of the localization of an anti-allergic agent, tranilast, in rat mast cells. Cell. Mol. Biol. 36, 65-71, 1990a.
Nishigaki, T., Momose, Y., Nagata, T.: Localization of the anti-allergic agent tranilast in the urinary bladder of rat as demonstrated by light microscopic radioautography. Drug Res. 40, 272-275, 1990b.
Oguchi, K., Nagata, T.: A radioautographic study of activated satellite cells in dystrophic chicken muscle. In, Current Research in Muscular Dystrophy Japan. The Proc. Ann. Meet. Muscular Dystrophy Res. 1980, pp. 16-17, Ministry of Welfare of Japan, Tokyo, 1980.
Oguchi, K., Nagata, T.: Electron microscopic radioautographic observation on activated satellite cells in dystrophy chickens. In, Clinical Studies on the Etiology of Muscular Dystrophy. Annual Report on Neurological Diseases 1981, pp. 30-33, Ministry of Welfare of Japan, Tokyo, 1981.
Ohno, S., Fujii, Y., Usuda, N., Endo, T., Hidaka, H., Nagata, T.: Demonstration of intracellular localization of calmodulin antagonist by wet-mounting radioautography. J. Electron Microsc. 32, 1-12, 1983.
Ohno, S., Fujii, Y., Usuda, N., Nagata, T., Endo, T., Tanaka, T., Hidaka, H.: Intracellular localization of calmodulin antagonists (W-7). In, Calmodulin and intracellular Ca^{2+} receptors. Kakiuchi, S., Hidaka, H, Means, A. R., Eds., pp. 39-48, Plenum Publishing Co., New York, 1982.
Olea, M. T.: An ultrastructural localization of lysosomal acid phosphatase activity in aging mouse spleen: a quantitative X-ray microanalytical study. Acta Histochem. Cytochem. 24, 201-208, 1991.
Olea, M. T., Nagata, T.: X-ray microanalysis of cerium in mouse spleen cells demonstrating acid phosphatase activity using high voltage electron microscopy, Cell. Mol. Biol. 37, 155-163, 1991.
Olea, M. T., Nagata, T. : Simultaneous localization of ^{3}H-thymidine incorporation and acid phosphatase activity in mouse spleen: EM radioautography and cytochemistry. Cell. Mol. Biol. 38, 115-122, 1992a.
Olea, M. T., Nagata, T.: A radioautographic study on RNA synthesis in aging mouse spleen after ^{3}H-uridine labeling in vitro. Cell. Mol. Biol. 38, 399-405, 1992b.
Oliveira, S. F., Nagata, T., Abrahamsohn, P. A., Zorn, T. M. T.: Electron microscopic radioautographic study on the incorporation of ^{3}H-proline by mouse decidual cells. Cell. Mol. Biol. 37, 315-323, 1991.

Oliveira, S. F., Abrahamsohn, P. A., Nagata, T., Zorn, T. M. T.: Incorporation of ^{3}H-amino acids by endometrial stromal cells during decidualization in the mouse. A radioautographical study. Cell. Mol. Biol. 41, 107-116, 1995.
Pearse, A. G. E.: Histochemistry, Theoretical and Applied. 4th Ed. Vol. 1. 439 pp., 1980, Vol. 2. 1055 pp., 1985, Vol. 3. Ed. with P. Stoward, 728 pp. Churchill Livingstone, Edinburgh, London and New York, 1991.
Sakai, Y., Ikado, S., Nagata, T.: Electron microscopic radioautography of satellite cells in regenerating muscles. J. Clin. Electr. Microsc. 10, 508-509, 1977.
Sato, A.: Quantitative electron microscopic studies on the kinetics of secretory granules in G-cells. Cell Tissue Res. 187, 45-59, 1978.
Sato, A., Iida, F., Furihara, R., Nagata, T.: Electron microscopic raioautography of rat stomach G-cells by means of ^{3}H-amino acids. J. Clin. Electron Microsc. 10, 358-359, 1977.
Shimizu, T., Usuda, N., Yamanda, T., Sugenoya, A., Iida, F.: Proliferative activity of human thyroid tumors evaluated by proliferating cell nuclear antigen/cyclin immnohistochemical studies. Cancer 71, 2807-2812, 1993.
Sun, L.: Age related changes of RNA synthesis in the lungs of aging mice by light and electron microscopic radioautography. Cell. Mol. Biol. 41, 1061-1072, 1995.
Sun, L., Gao, F., Duan, H., Nagata, T.: Light microscopic radioautography of DNA synthesis in pulmonary cells in aging mice. In, Radioautography in Medicine, Nagata, T. Ed., pp. 201-205, Shinshu University Press, Matsumoto, 1994.
Sun, L., Gao, F., Nagata, T.: Study on the DNA synthesis of pulmonary cells in aging mice by light microscopic radioautography. Cell. Mol. Biol. 41, 851-859, 1995a.
Sun, L., Gao, F., Jin, C., Duan, H., Nagata, T.: An electron microscopic radioautographic study on the DNA synthesis of pulmonary tissue cells in aging mice. Med. Electron. Microsc. 28, 129-131, 1995b.
Sun, L., Gao, F., Jin, C., Nagata, T.: DNA synthesis in the tracheae of aging mice by means of light and electron microscopic radioautography. Acta Histochem. Cytochem. 30, 211-220, 1997a.
Sun, L., Gao, F., Nagata, T.: A Light Microscopic radioautographic study on protein synthesis in pulmonary cells of aging mice. Acta Histochem. Cytochem. 30, 463-470, 1997b.
Suzuki, K., Imada, T., Gao, F., Ma, H., Nagata, T.: Radioautographic study of benidipine hydrochloride: localization in the mesenteric artery of spontaneously hypertensive rat. Drug Res. 44, 129-133, 1994.
Terauchi, A., Mori, T., Kanda, H., Tsukada, M., Nagata, T.: Radioautographic study of 3H-taurine uptake in mouse skeletal muscle cells. J. Clin. Electron Microsc. 21, 627-628, 1988.
Terauchi, A., Nagata, T.: Observation on incorporation of ^{3}H-taurine in mouse skeletal muscle cells by light and electron microscopic radioautography. Cell. Mol. Biol. 39, 397-404, 1993.
Terauchi, A., Nagata, T.: In corporation of ^{3}H-taurine into the blood capillary cells of mouse skeletal muscle. Radioautography in Medicine, Nagata, T. ed., Shinshu University Press, Matsumoto, 1994.
Toriyama, K.: Study on the aging changes of DNA and protein synthesis of bipolar and photo-receptor cells of mouse retina by light and electron microscopic radioautography. Cell. Mol. Biol. 41, 593-601, 1995.

Tsukahara, S., Yoshida, K., Nagata, T.: A radioautographic study on the incorporation of ^{14}C-bupranolol (beta-blocking agent) into the rabbit eye. Histochemistry 68, 237-244, 1980.
Usuda, N., Nagata, T.: Electron microscopic radioautography of acyl-CoA mRNA by in situ hybridization. J. Clin. Electron Microsc. 25, 332-333, 1992.
Usuda, N., Nagata, T.: The immunohistochemical and in situ hybridization studies on hepatic peroxisomes. Acta Histochem. Cytochem. 28, 169-172, 1995.
Usuda, N., Hanai, T., Morita, T., Nagata, T.: Radioautographic demonstration of peroxisomal acyl-CoA oxidase mRNA by in situ hybridization. In, Recent advances in cellular and molecular biology, Vol. 6. Molecular biology of nucleus, peroxisomes, organelles and cell movement. Wegmann, R. J., Wegmann, M., Eds, pp.181-184, Peeters Press, Leuven, 1992.
Uwa, H., Nagata, T.: Cell population kinetics of the scleroblast during ethisterone-induced anal-fin process formation in adult females of the Medaka. Dev. Growth Different. 9, 693-694, 1976.
Watanabe, I., Makiyama, M. C. K., Nagata, T.: Electron microscopic radioautographic observation of the submandibular salivary gland of aging mouse. Acta Microscopica 6. 130-131, 1997.
Yamabayashi, S., Gunarso, W., Tsukahara, S., Nagata, T.: Incorporation of ^{3}H-befunolol (beta blocking agent) into melanin granules of ocular tissues in the pigmented rabbits. I. Light microscopic radioautography. Histochemistry 73, 371-375, 1981.
Yamada, A. T.: Timely and topologically defined protein synthesis in the periimplanting mouse endometrium revealed by light and electron microscopic radioautography. Cell. Mol. Biol. 39, 1-12, 1993.
Yamada, A., Nagata, T.: Ribonucleic acid and protein synthesis in the uterus of pregnant mouse during activation of implantation window. Med. Electron Microsc. 27, 363-365, 1992a.
Yamada, A., Nagata, T.: Light and electron microscopic raioautography of DNA synthesis in the endometria of pregnant-ovariectomized mice during activation of implantation window. Cell. Mol. Biol. 38, 763-774, 1992b.
Yamada, A., Nagata, T.: Light and electron microscopic raioautography of RNA synthesis of peri-implanting pregnant mouse during activation of receptivity for blastocyst implantation. Cell. Mol. Biol. 38, 211-233, 1993.
Yoshinaga, K.: Uterine receptivity for blastcyst implantation. Ann. N. Y. Acad. Sci. USA, 541, 424-431, 1988.
Yoshizawa, S., Nagata, A., Honma, T., Oda, M., Murata, F., Nagata, T.: Study of ethionine pancreatitis by means of electron microscopic radioautography. J. Clin. Electron Microsc. 7, 349-350, 1974.
Yoshizawa, S., Nagata, A., Honma, T., Oda, M., Murata, F., Nagata, T.: Radioautographic study of protein synthesis in pancreatic exocrine cells of alcoholic rats. J. Clin. Electron. Microsc. 10, 372-373, 1977.

8

Macromolecular Synthesis in the Endocrine, Nervous and Sensory Systems

Tetsuji Nagata[1,2]
[1]*Department of Anatomy and Cell Biology, Shinshu University School of Medicine, Matsumoto*
[2]*Shinshu Institute of Alternative Medicine and Welfare, Nagano*
Japan

1. Introduction

This chapter deals with the last forth part of the application of microscopic radioautography to the organ systems, including the endocrine system, the nervous system and the sensory organs. The endocrine system includes the hypophysis, the pineal body, the thyroid gland, the parathyroid gland, the thymus, the adrenal gland, the islet of Langerhans and the reproductive glands, i. e. the testis and the ovary. The nervous system includes the central nervous system, i.e. the brains and spinal cord and the peripheral nerves, i.e. the ganglion and nerves, while the sensory system includes, the skin, the visual, the stato-acoustic, the olfactory and the gustatory organs. We have studied some of these organs, not all of them yet.

2. Macromolecular synthesis in the endocrine system

Among the endocrine organs, we studied macromolecular synthesis in the adrenal gland and the steroid secreting cells of both sexes, the Leydig cells of the testis and the ovarian follicular cells in mice. On the other hand, incorporation of mercury chloride into the human thyroid tissues was also studied (Nagata 2002).

2.1 The DNA synthesis in the endocrine system

Among the endocrine organs, we studied DNA synthesis in the adrenal glands and steroid secreting cells of both sexes, the Leydig cells of the testis and the ovarian follicular cells in mice.

2.1.1 The DNA synthesis in the adrenal gland

We studied the adrenal tissues of aging mice, both the adrenal cortex and the medulla, from embryo to postnatal 2 years in senescence. Some of the results were already published in several original articles (Ito 1996, Ito and Nagata 1996, Liang 1998, Liang et al. 1999, Nagata 1994, 1999c, 2000a,b, 2008a,b, 2009c,d,e,f,g,h,i,j, 2010a, Nagata et al. 2000b). The results shall be summarized in this review.

2.1.1.1 The DNA synthesis in the adrenal cortex

We studied the adrenal tissues of mice at various ages from embryo to postnatal 2 years (Ito 1996, Ito and Nagata 1996, Nagata 2008a,b, 2009c,d,e,f,g,h,i,j). The adrenal tissues obtained from ddY strain mice at various ages from embryo day 19 to postnatal day 30 of both sexes, consisted of the adrenal cortex and the adrenal medulla. The former consisted of 3 layers, zona glomerulosa (Fig. 20A), zona fasciculata (Fig. 20B) and zona reticularis (Fig. 20C), developing gradually with aging from perinatal stage at embryonic day 19 to postnatal stages as day 1, 3, 9, 14, month 1, 2, 6, 12, 24 as observed by light microscopy.

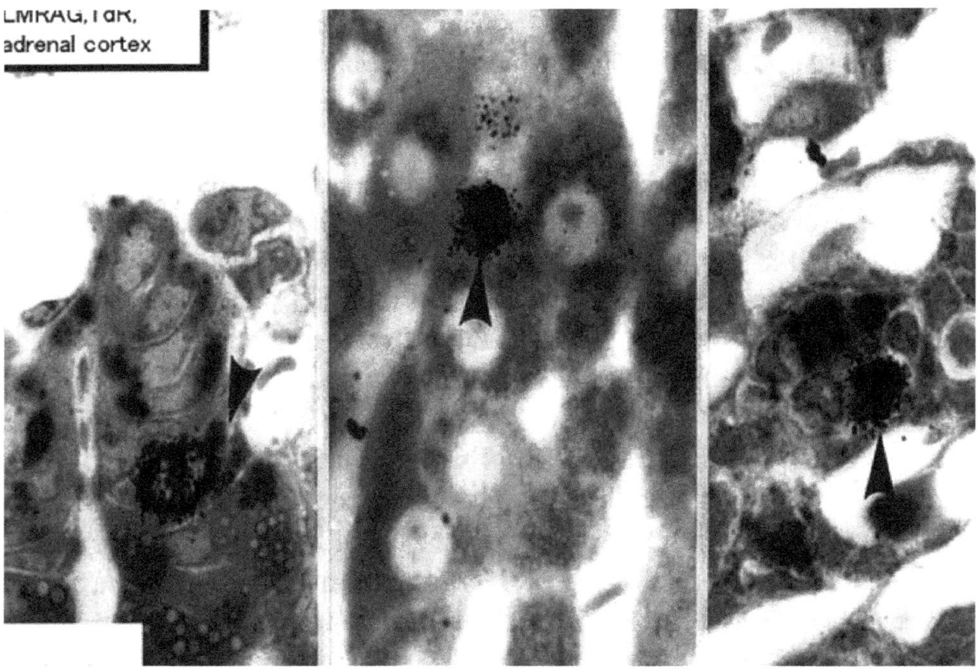

Fig. 20. LM RAG of a young mouse adrenal cortex, labeled with ^{3}H-thymine, showing DNA synthesis (arrow) in 3 layers, zona glomerulosa (left), zona fasciculate (middle), zona reticularis (right). From Nagata, T.: Annals Microsc. Vol. 10, p. 61, 2010, Microsc. Soc. Singapore.

At embryonic day 19 and postnatal day 1, the 3 layers of the adreno-cortical cells, zona glomerulosa (Fig. 21D), zona fasciculate (Fig. 21E) and zona reticularis were composed mainly of polygonal cells, while the specific orientation of the 3 layers was not yet well established as observed by EMRAG. At postnatal day 3, orientation of 3 layers, especially the zona glomerulosa became evident. At postnatal day 9 and 14, the specific structure of the 3 layers was completely formed and the arrangements of the cells in respective layers became typical especially at day 14 (Fig. 21D) and month 1 (Fig. 21E) to 24. Observing the ultrastructure of the adreno-cortical cells by electron microscopy, cell organelles including mitochondria were not so well developed at perinatal and early postnatal stages from embryonic day 19 to postnatal day 9. However, these cell organelles, mitochondria, endoplasmic reticulum, Golgi apparatus, appeared well developed similarly to the adult stages at postnatal day 14. The zona glomerulosa is the thinnest layer found at the outer zone, covered by the capsule, consisted of closely packed groups of columnar or pyramidal cells forming arcades of cell columns. The cells contained many spherical mitochondria and well developed smooth surfaced endoplasmic reticulum but a compact Golgi apparatus in day 14 animals. The zona fasciculate was the thickest layer, consisted of polygonal cells that were larger than the glomerulosa cells, arranged in long cords disposed radially to the medulla containing many lipid droplets. The mitochondria were less numerous and were more variable in size and shape than those of the glomerulosa cells, while the smooth surfaced endoplasmic reticulum were more developed and the Golgi apparatus was larger than the glomerulosa. In the zona reticularis, the parallel arrangement of cell cords were anastomosed showing networks continued to the medullar cells. The mitochondria were less numerous and were more variable in size and shape than those of the glomerulosa cells like the fasciculata cells, as well as the smooth surfaced endoplasmic reticulum were developed and the Golgi apparatus was large like the fasciculata cells. However, the structure of the adrenal cortex tissues showed changes due to development and aging at respective developmental stages.

Observing both LM and EM RAG of the adrenal cortex labeled with ^{3}H-thymidine, demonstrating DNA synthesis, the silver grains were found over the nuclei of some adreno-cortical cells in S-phase of cell cycle mainly in perinatal stages at embryonic day 19, postnatal day 1 and day 3, while less at day 9 and day 14 to month 1-24 (Ito 1996, Ito and Nagata 1996, Nagata 2008a,b,c,d, 2009c,d,e). Those labeled cells were found in all the 3 layers (Fig. 20), the zona glomerulosa (Fig. 20 left), the zona fasciculata (Fig. 20 middle) and the zona reticularis (Fig. 20 right), at respective aging stages. In labeled adreno-cortical cells in the 3 layers the silver grains were mainly localized over the euchromatin of the nuclei and only a few or several silver grains were found over the mitochondria of these cells as observed by EM RAG (Fig. 21D,E).

To the contrary, most adreno-cortical cells were not labeled with any silver grains in their nuclei nor cytoplasm, showing no DNA synthesis after labeling with ^{3}H-thymidine. The labeling indices in respective 3 zones in the cortex as well as the medulla showed the maxima at perinatal stages and decreased due to aging (Fig. 22A,B). Among many unlabeled adreno-cortical cells, however, most cells in the 3 layers were observed to be labeled with several silver grains over their mitochondria due to the incorporations of ^{3}H-thymidine especially at the perinatal stages from embryonic day 19 to postnatal day 1, day 3, day 9 and 14 (Fig. 21D). The ultrastructural localizations of silver grains over the mitochondria were mainly on the mitochondrial matrices and some over the cristae or membranes.

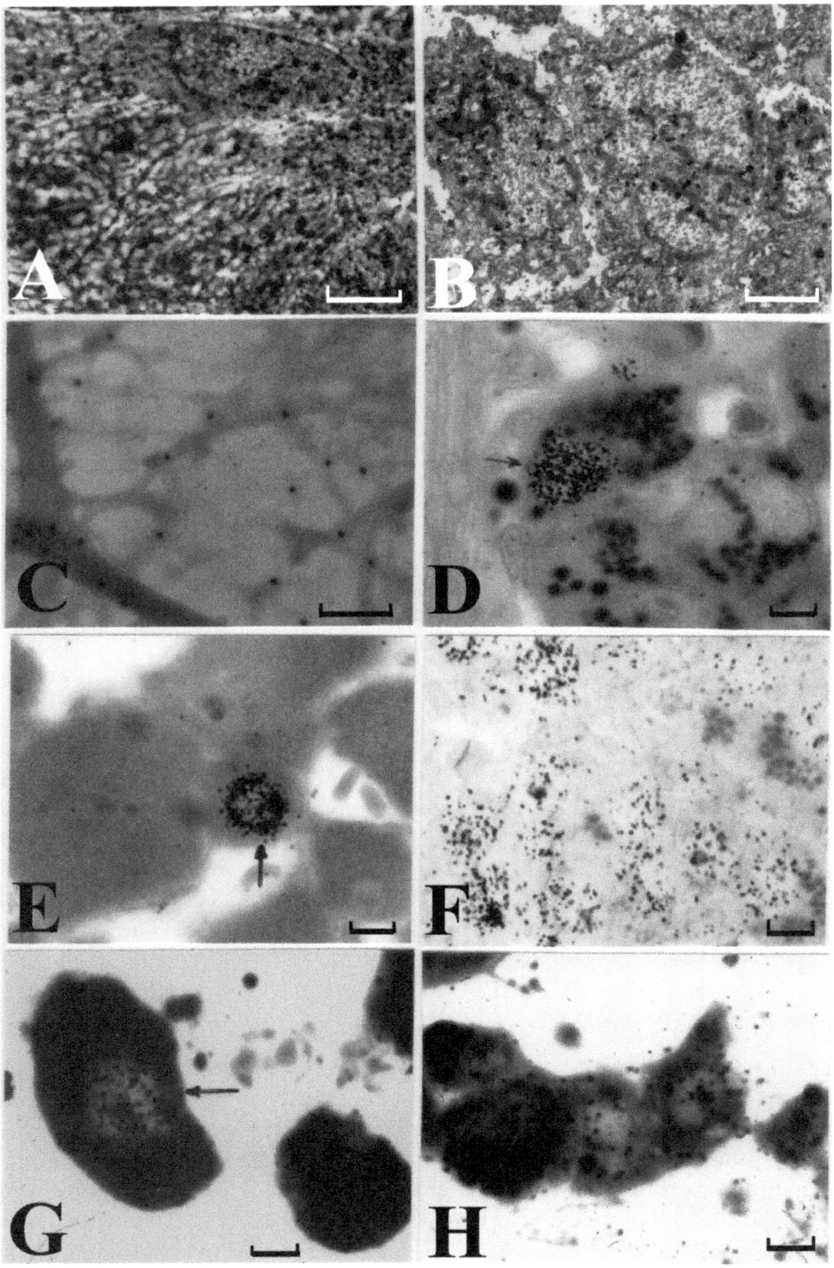

Fig. 21. LM and EM RAG of the endocrine organs. From Nagata, T.: Special Cytochemistry in Cell Biology, In, Internat. Rev. Cytol. Vol. 211, No. 1, p. 115, 2001, Academic Press, San Diego, USA, London, UK.

Fig. 21A. EM RAG of human thyroid cancer cells, labeled with $^{205}HgCl_2$ in vitro, quick frozen, freeze-dried, embedded in Epoxy resin, dry-sectioned, and radioautographed by dry-mounting procedure for demonstrating soluble compounds, many silver grains showing soluble $^{205}HgCl_2$ incorporations. x15,000.
Fig. 21B. EM RAG of human thyroid cancer cells, labeled with $^{205}HgCl_2$ in vitro, chemically fixed doubly in buffered glutaraldehyde and osmium tetroxide, dehydrated, embedded in Epoxy resin, wet-sectioned, and radioautographed by wet-mounting procedure for demonstrating insoluble compounds, less silver grains showing insoluble $^{205}HgCl_2$ incorporations. x15,000.
Fig. 21C. EM photo of a human thyroid cancer cell, fixed in paraformaldehyde and glutaraldehyde mixture, embedded in Lowicryl K4M, sectioned and immuno-stained with anti-keratin antibody by the protein-A gold technique, demonstrating keratin filaments. x15,000.
Fig. 21D. LM RAG of the zona glomelurosa of the adrenal cortex of a postnatal day 14 mouse labeled with ^3H-thymidine, showing DNA synthesis. x900.
Fig. 21E. LM RAG of the zona fasciculata of the adrenal cortex of a postnatal month 6 mouse labeled with ^3H-thymidine, showing DNA synthesis. x900.
Fig. 21F. LM RAG of the zona gromeulosa of the adrenal cortex of a prenatal day 19 mouse labeled with ^3H-uridine, showing RNA synthesis. x1,000.
Fig. 21G. LM RAG of the interstitial tissues of a postnatal month 12 male mouse labeled with ^3H-thymidine, showing DNA synthesis in Leydig cells. x1,000.
Fig. 21H. LM RAG of the interstitial tissues of a postnatal day 3 male mouse labeled with ^3H-uridine, showing RNA synthesis in Leydig cells. x1,000.

2.1.1.1.1 The number of mitochondria of mouse adreno-corticl cells

Preliminary quantitative analysis on the number of mitochondria in 10 adreno-cortical cells whose nuclei were labeled with silver grains and other 10 cells whose nuclei were not labeled in each aging group revealed that there was no significant difference between the number of mitochondria and the labeling indices ($P<0.01$). Thus, the number of mitochondria and the labeling indices were later calculated regardless whether their nuclei were labeled or not (Ito 1996, Ito and Nagata 1996, Nagata 2008a,b,c,d, 2009c,d,e). The results obtained from the number of mitochondria in adreno-cortical cells in the 3 layers of respective animals in 10 aging groups at perinatal stages, prenatal embryo day 19, postnatal day 1, 3, 9 and 14, showed an gradual increase from the prenatal day 19 (glomerulosa 12.5, fasciculata 14.9, reticularis 15/2/cell) to postnatal day 14 and month 1, 2 (glomerulosa 62.2, fasciculata 64.0, reticularis 68.2/cell) to month 6 and 12. The increase from embryo day 19 to postnatal month 2 was stochastically significant ($P <0.01$). Then, they did not change significanctly from month 12 to 24 (Fig. 22A).

2.1.1.1.2 The DNA synthesis in the mitochondria of mouse adreno-cortical cells

The results of visual grain counts on the number of mitochondria labeled with silver grains obtained from 10 adreno-cortical cells in the 3 layers of each animal labeled with ^3H-thymidine demonstrating DNA synthesis in 10 aging groups at perinatal stages, prenatal embryo day 19, postnatal day 1, 3, 9 and 14, and month 1, 2, 6, 12 and 24 were reported previously (Ito 1996, Ito and Nagata 1996, Nagata 2008a,b, 2009c,d,e). The results demonstrated that the numbers of labeled mitochondria with ^3H-thymidine showing DNA synthesis gradually increased from prenatal embryo day 19 (glomerulosa 0.3, fasciculata 0.5, reticularis 0.4/cell) to postnatal day 14, month 1 and 2 (glomerulosa 5.3, fasciculata 5.0, reticularis 6.2/cell), reaching the maximum, then decreased to month 6, 12 and 24 (Fig. 22A). The increase and decrease were stochastically significant (P <0.01).

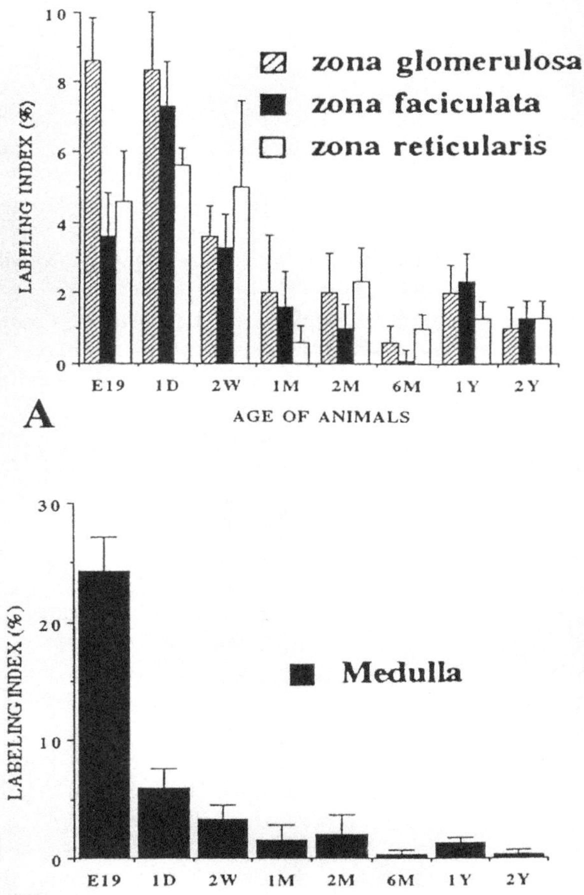

Fig. 22. Histogram showing aging changes of average labeling indices in respective cell types of the adrenal glands of aging mice labeled with ^{3}H-thymidine showing DNA synthesis. Mean ± Standard Deviation. From Nagata, T.: Radioautographology, General and Special. In, Prog. Histochem. Cytochem. Vol. 37, No. 2, p. 171, 2002, Urban & Fischer, Jena, Germany

Fig. 22A. The adrenal cortex.

Fig. 22B. The adrenal medulla.

2.1.1.1.3 The labeling index of DNA synthesis in mouse adreno-cortical mitochondria

On the other hand, the labeling indices in respective aging stages were calculated from the number of labeled mitochondria, dividing by the number of total mitochondria per cell which were mentioned previously (Ito 1996, Ito and Nagata 1996, Nagata 2008a,b, 2009c,d,e,j, 2010e). The results showed that the labeling indices gradually increased from prenatal day 19 (glomerulosa 2.4, fasciculata 2.7, reticularis 2.6%) to postnatal day 14, month 1 and 2 (glomerulosa 8.5, fasciculata 7.8, reticularis 8.8%), reaching the maximum and decreased to month 6 (glomerulosa 4.1, fasciculata 4.2, reticularis 3.8%), 12 and 24 (Fig. 23C). The increase and decrease were stochastically significant ($P < 0.01$).

2.1.1.2 The DNA synthesis in mouse adreno-medullary cells

We studied the adrenal tissues of mice at various ages from embryo to postnatal 2 years. The adrenal tissues obtained from ddY strain mice at various ages from embryo day 19 to postnatal day 30 of both sexes, consisted of the adrenal cortex and the adrenal medulla (Ito 1996, Ito and Nagata 1996, Nagata 2008a,b, 2009c,d,e,g, 2010d,e,f,g). The former consited of 3 layers, zona glomerulosa, zona fasciculata and zona reticularis, developing gradually with aging as observed by light microscopy (Fig. 20), while the latter consisted of 2 cell types in one layer when observed by electron microscopy (Nagata 2009c,d,e, 2010d,e,f,g). At embryonic day 19 and postnatal day 1, the 3 layers of the adreno-cortical cells, zona glomerulosa, zona fasciculate and zona reticularis were composed mainly of polygonal cells, while the specific orientation of the 3 layers was not yet well established. However, the orientation of 3 layers became evident at day 3 and completely formed at day 14 (Fig. 20) and to month 1-24 (Fig. 21D,E,F). On the other hand, the medulla consisted of only one layer containing 2 types of cells, adrenalin cell and noradrenalin cell. The former contains adrenalin granlules with low electron density, while the latter contains noradrenalin granules with high electron density.

The adrenal medulla is the deepest layer in the adrenal glands, surrounded by the 3 layers of the adrenal cortex as observed by light and electron microscopy (Ito 1996, Ito and Nagata 1996, Nagata 2008a,b,c,d, 2009c,d,e, 2010d,e,f,g), containing either adrenalin granules or noradrenalin granules. Quantitative analysis revealed that the numbers of mitochondria in both adrenalin and noradrenalin cells at various ages increased from fetal day 19 to postnatal month 1 due to aging of animals, respectively, but did not decrease to month 24, while the number of labeled mitochondria and the labeling indices of intramitochondrial DNA synthesis changed due to aging. When they were labeled with ^{3}H-thymidine silver grains appeared over some nuclei of both cell types at perinatal stages, but they appeared almost all the cell bodies containing mitochondria. Quantitative analysis revealed that the numbers of mitochondria in both adrenalin and noradrenalin cells at various ages increased from fetal day 19 to postnatal month 1 due to aging of animals, respectively, while the number of labeled mitochondria and the labeling indices of intramitochondrial DNA synthesis incorporating ^{3}H-thymidine increased from fetal day 19 to postnatal day 14 (2 weeks), reaching the maxima, and decreased to month 24. It was shown that the activity of intramitochnodrial DNA synthesis in the adrenal medullary cells in aging mice increased and decreased due to aging of animals.

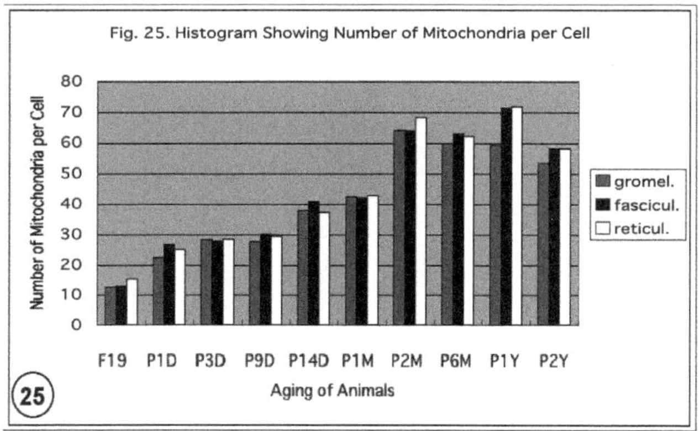

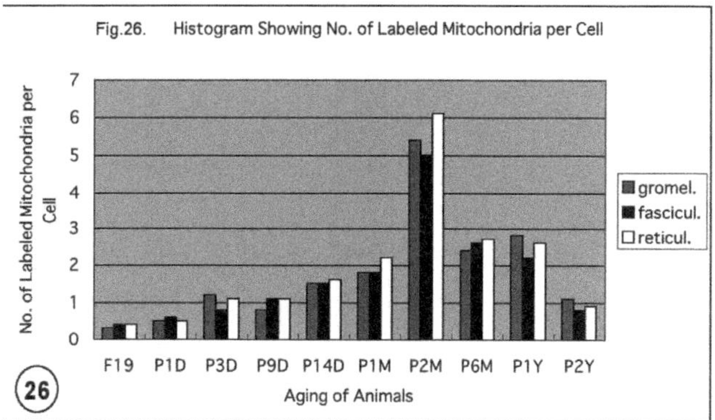

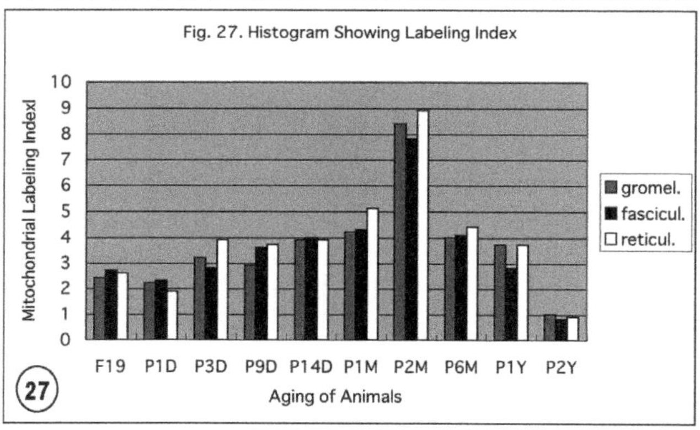

Fig. 23. Histogram showing aging changes of the respective cell types of the adrenal glands of aging mice labeled with ^{3}H-thymidine showing DNA synthesis. Mean ± Standard Deviation. From Nagata, T.: Current Radiopharmaceutics, Vol. 2, p. 173, 2002.
Fig. 23A(25). The number of mitochondria per cell.
Fig. 23B(26). The number of labeled mitochondria per cell.
Fig. 23C(27). The mitochondrial labeling index.

2.1.2 The DNA synthesis in the islets of Langerhans

When we studied macromolecular synthesis in the exocrine pancreatic cells of aging mice by LM and EMRAG we also studied the islet cells of Langerhans together with the exocrine cells, using RI labeled precursors such as ^{3}H-thymidine for DNA (Nagata and Usuda 1985, 1986, Nagata et al. 1986a,b), ^{3}H-uridine for RNA (Nagata and Usuda 1985, 1993b, Nagata et al. 1986a,b), ^{3}H-leucine for proteins (Nagata 2000, Nagata and Usuda 1993a, 1995), ^{3}H-glucosamine for glucides (Nagata et al. 1992), ^{3}H-glycerol for lipids (Nagata et al. 1988b, 1990). The results showed that the islets cells, A, B and C cells, incorporated those precursors to synthesize DNA, RNA, proteins and glucides. The labeling index of DNA synthesis and the densities of silver grains showing RNA, proteins and glucides syntheses were high at prenatal and earlier postnatal stages from day 1 to day 14, then decreased from 1 month to 1 years due to aging. However, the labeling indices by ^{3}H-thymidine and the grain counts by ^{3}H-uridine and ^{3}H-leucine in the endocrine cells were less than those in the exocrine cells at the same ages.

2.1.3 The DNA synthesis in the Leydig cells of the testis

The cells of Leydig can be found in the interstitial tissues between the seminiferous tubules of the testis of mammals (Gao 1993, Gao et al. 1994, 1995a,b, Nagata et al. 2000b). They are identified as spherical, oval, or irregular in shape and their cytoplasms contain lipid droplets. We studied the macromolecular synthesis of the cells in the testis of several groups of litter ddY mice at various ages from fetal day 19 to postnatal aging stages up to 2 years senescence by LM and EMRAG using ^{3}H-thymidine, ^{3}H-uridine and ^{3}H-leucine incorporations.

The Leydig cells from embryonic stage to senescent stages were labeled with ^{3}H-thymidine as observed by LMRAG (Fig. 21G). The changes of the numbers of labeled Leydig cells with the ^{3}H-thymidine incorporation into the nuclei showing the DNA synthesis were found in these cells at different aging stages. Only a few cells were labeled after ^{3}H-thymidine at embryonic day 19. At early postnatal stages, there was a slight increase of the number of labeled cells. The number of labeled cells from perinatal stage to postnatal 14 days and 1, 2, 6 months were similar to the values found at prenatal and early postnatal stages. The notable increases in the number of labeled cells of Leydig were found from 9 months to 2 years in senescence. The labeling indices with ^{3}H-thymidine in perinatal stages to postnatal 6 months were low (5-10%) but increased at 9 months and maintained high level (50-60%) to 2 years (Gao 1993, Gao et al. 1994, 1995a, Nagata et al. 2000b). The labeling indices at senescent stages still maintained a relatively high level and they were obviously higher than those of young animals. By electron microscopy, typical Leydig cells contained abundant cell organelles such as smooth surfaced endoplasmic reticulum, Golgi apparatus and mitochondria with tubular cristae. The silver grains were mainly localized over the euchromatin of labeled nucleus. Some of the grains were also localized over some of the mitochondria in both the nuclei labeled and unlabeled cells.

2.1.4 The DNA synthesis in the ovarian follicles

The ovarian follicles in the ovaries of mature mice are one of the steroid secreting organs in female animals. We studied the DNA and RNA synthesis of the follicular cells in the developing ddY mice ovaries in several aging groups at postnatal day 1, 3, 7, 14, 30 and 60 by LM and EMRAG using ^{3}H-thymidine and ^{3}H-uridine (Li 1994, Li and Nagata 1995). From the results it was shown that both DNA and RNA synthesis in the ovarian follicular cells were observed (Fig. 16G). Quantitative analysis, as expressed with labeling indices and grain counts, revealed that both increased significantly from postnatal day 1 to 3, then decreased from day 7 to 60 (Fig. 19A). Comparing the results to other female genital cells, a paralleled alteration of both DNA and RNA synthesis was revealed between the ovarian follicular cells and other uterine or oviductal cells (Fig. 19B,C). On the other hand, the glycoconjugate synthesis as shown by the uptake of $^{35}SO_4$ in mouse ovary during the estrus cycle was also demonstrated (Li et al. 1992).

2.2 The RNA synthesis in the endocrine system

We studied the RNA synthesis of the adrenal glands and the cells of Leydig in the testis of aging mice among the endocrine organs after the administration of ^{3}H-uridine at various ages.

2.2.1 The RNA synthesis in the adrenal glands

The RNA synthesis in the adrenal glands in aging mice was studied in both the adrenal cortex and the adrenal medulla after administration of ^{3}H-uridine in many groups of mice at various ages from perinatal stages to senescence at year 2.

2.2.1.1 The RNA synthesis in aging mouse adreno-cortical cells

Observing both LM and EM RAG labeled with ^{3}H-uridine, demonstrating RNA synthesis, the silver grains were found over the nuclei and cytoplasm of almost all the adreno-cortical cells from perinatal stages to postnatal month 1-24 (Liang 1998, Liang et al. 1999, Nagata et al. 2000b, Nagata 2010a). Those labeled cells were found in all the 3 layers, the zona glomerulosa (Fig. 21F), the zona fasciculata and the zona reticularis, at respective aging stages. In labeled adreno-cortical cells in the 3 layers the silver grains were mainly localized over the euchromatin of the nuclei and several silver grains were found over the endoplasmic reticulum, ribosomes and mitochondria of these cells. The ultrastructural localizations of silver grains over the mitochondria were mainly on the mitochondrial matrices and some over the cristae or membranes.

2.2.1.2 The RNA synthesis in the mitochondria of mouse adreno-cortical cells

The results of visual grain counts on the number of mitochondria labeled with silver grains obtained from 10 adreno-cortical cells in the 3 layers of each animal labeled with ^{3}H-uridine demonstrating RNA synthesis in 10 aging groups at perinatal stages, prenatal embryo day 19, postnatal day 1, 3, 9 and 14, and month 1, 2, 6, 12 and 24 were reported previously (Liang 1998, Liang et al. 1999, Nagata et al. 2000b, Nagata 2010a). The results demonstrated that the numbers of labeled mitochondria with ^{3}H-uridine showing RNA synthesis gradually increased from prenatal embryo day 19 (glomerulosa 0.3, fasciculata 0.5, reticularis 0.4/cell) to postnatal day 14, month 1 and 2 (glomerulosa 5.3, fasciculata 5.0, reticularis 6.2/cell), reaching the maximum, then decreased to month 6, 12 and 24 (Fig. 23B). The increase and decrease were stochastically significant ($P < 0.01$).

2.2.1.3 The labeling index of RNA synthesis in mouse adreno-cortical mitochondria

On the other hand, the labeling indices in respective aging stages were calculated from the number of labeled mitochondria (Fig. 14B), dividing by the number of total mitochondria per cell (Fig. 24A) which were mentioned previously (Liang 1998, Liang et al. 1999, Nagata et al. 2000b, Nagata 2010a). The results showed that the labeling indices gradually increased from prenatal day 19 (glomerulosa 10.4, fasciculata 12.1, reticularis 13.1%) to postnatal day 1 (glomerulosa 12.6, fasciculata 11.4, reticularis 11.1%), 3, 9 (glomerulosa 16.6, fasciculata 18.0, reticularis 18.0%), reaching the maximum and decreased to day 14, month 1 (glomerulosa 11.4, fasciculata 11.0, reticularis 10.7%) 2 (glomerulosa 8.5, fasciculata 7.8, reticularis 8.8%), month 6 (glomerulosa 4.1, fasciculata 4.2, reticularis 3.8%), 12 and 24 (Fig. 24C). The increase and decrease were stochastically significant (P <0.01).

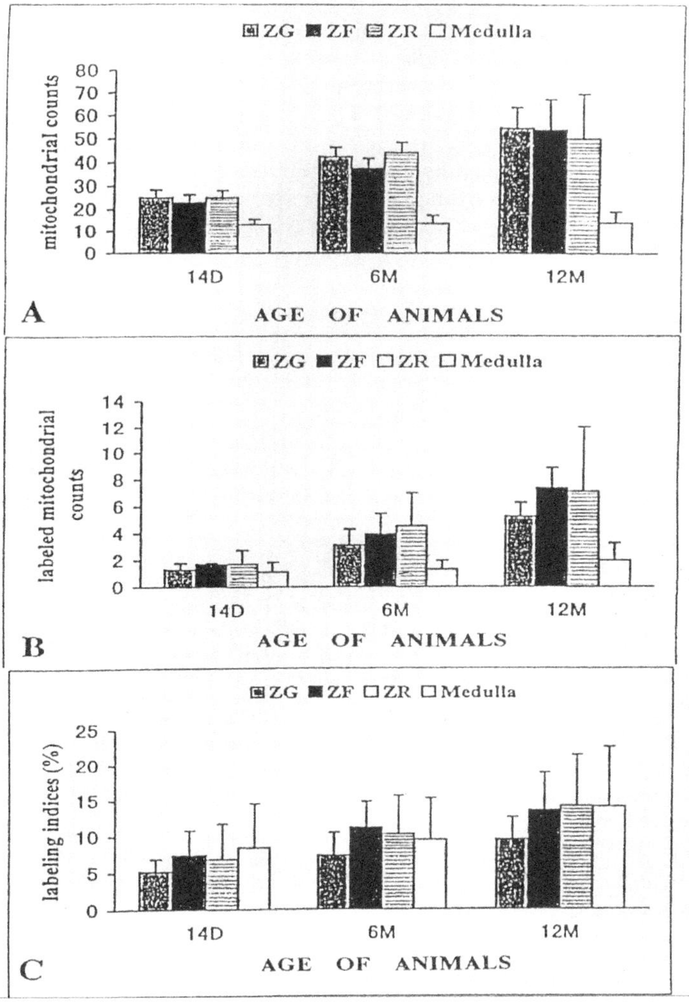

Fig. 24. Histogram showing aging changes of the respective cell types of the adrenal glands of aging mice labeled with ^{3}H-uridine showing RNA synthesis. Mean ± Standard Deviation. From Nagata, T.: Radioautographology, General and Special. In, Prog. Histochem. Cytochem. Vol. 37, No. 2, p. 173, 2002, Urban & Fischer, Jena, Germany

Fig. 24A. The number of mitochondria per cell.
Fig. 24B. The number of labeled mitochondria per cell.
Fig. 24C. The mitochondrial labeling index.

2.2.2 The RNA synthesis in aging mouse adreno-medullary cells

The adrenal medulla consists of 2 cell types, the adrenalin cells and noradrenalin cells. When they were labeled with ^{3}H-uridine, an RNA precursor, silver grains appeared over almost all the cells, both nuclei and cytoplasm containing mitochondria (Liang et al. 1999, Nagata et al. 2000b, 2010b). Quantitative analysis revealed that the numbers of mitochondria in both adrenalin and noradrenalin cells at various ages increased from fetal day 19 to postnatal month 1 due to aging of animals, respectively, but did not decrease to month 24 (Fig. 24A), while the number of labeled mitochondria (Fig. 24B) and the labeling indices of intramitochondrial RNA synthesis incorporating ^{3}H-uridine increased from fetal day 19 to postnatal month 1, reaching the maxima, but did not decrease to month 24 (Fig. 24C). It was shown that the activity of intramitochnodrial RNA synthesis in the adrenal medullary cells in aging mice increased but did not decrease due to aging of animals in contrast to DNA synthesis (Nagata 2010b).

2.2.3 The RNA synthesis in the Leydig cells of the testis

The cells of Leydig can be found in the interstitial tissues between the seminiferous tubules of the testis of mammals (Gao 1993, Gao et al. 1994, 1995a, Nagata et al. 2000b). They are identified as spherical, oval, or irregular in shape and their cytoplasms contain lipid droplets. We studied the macromolecular synthesis of the cells in the testis of several groups of litter ddY mice at various ages from fetal day 19 to postnatal aging stages up to 2 years in senescence by LM and EMRAG using ^{3}H-thymidine, ^{3}H-uridine and ^{3}H-leucine incorporations.

The incorporation of ^{3}H-uridine into RNA was observed in almost all the Leydig cells in the interstitial tissues of the testis from embryonic day 19 to 2 years after birth. A few silver grains over the nuclei and cytoplasm of the Leydig cells labeled with ^{3}H-uridine were observed at embryonic day 19. The silver grains over those cells slightly decreased at postnatal day 1, 3 (Fig. 21H), 7 and 14. The number of the silver grains over the nuclei increased from postnatal 1 months onwards. The average number of silver grains over the cytoplasm increased gradually and reached the maximum at 12 months after birth. At each stage, the activity of RNA synthesis was specifically localized over the euchromatin in the nucleus and nucleolus as observed by EMRAG. From adult to senescent stages, the activity of RNA synthesis maintained a high level in their nuclei as compared to the cytoplasm. In the cytoplasm of Leydig cells in respective aging groups some of the mitochondria and endoplasmic reticulum were also labeled with silver grains. It is noteworthy that the average grain counts increased prominently in the senescent aging groups at 1 and 2 years after birth.

2.3 The protein synthesis in the endocrine system

We studied the protein synthesis of the adrenal glands and the cells of Leydig in the testis of aging mice among the endocrine organs after the administration of ^{3}H-leucine at various ages.

2.3.1 The protein synthesis in the adrenal gland

Observing both LM and EM RAG labeled with ^{3}H-leucine, demonstrating protein synthesis, the silver grains were found over the nuclei and cytoplasm of almost all the adreno-cortical and adreno-medullary cells from perinatal stages to postnatal month 1-24 (Nagata 2010c,d,e,f,g). Those labeled cells were found in all the 3 layers, the zona glomerulosa, the zona fasciculata and the zona reticularis, as well as in all the cells in the adreno-medullae at respective aging stages.

2.3.1.1 Protein synthesis in mitochondria of mouse adrenal cells

In order to study the aging changes of intramitochondrial protein synthesis of mouse adrenal cells, 10 groups of developing and aging mice, each consisting of 3 individuals, total 30, from fetal day 19 to postnatal newborn at day 1, 3, 9, 14, adult at month 1, 2, 6 and senescent animals at month 12 (year 1) and 24 (year 2) were injected with ^{3}H-leucine, an protein precursor, sacrificed 1 hr later and the adrenal tissues were fixed and processed for electron microscopic radioautography. On electron microscopic radioautograms obtained from each animal, the number of mitochondria per cell, the number of labeled mitochondria with ^{3}H-leucine showing protein synthesis per cell and the mitochondrial labeling index in each adreno-cortical cells, in 3 zones, as well as in each adreno-medullary cells, 2 types of cells in the medulla, the adrenalin cells and the noradrenalin cells, were calculated and the results in respective aging groups were compared with each others (Nagata 2010c,d,e,f,g). Preliminary quantitative analysis on the number of mitochondria in either 10 adreno-cortical cells or adreno-medullary cells whose nuclei and cytoplasm were labeled with silver grains and other 10 cells whose nuclei and cytoplasm were not labeled in each aging group revealed that there was no significant difference between the number of mitochondria and the labeling indices ($P<0.01$). Thus, the number of mitochondria and the labeling indices were calculated regardless whether their nuclei were labeled or not (Nagata 2010d,e,f,g). The results demonstrated that the number of mitochondria in adreno-cortical cells in 3 zones, the zona glomerulosa, fasciculata and reticularis of respective mice at various ages increased from fetal day 19 to postnatal month 1 reaching the plateau from month 1 to 24 due to development and aging of animals, respectively, while the number of labeled mitochondria per cell and the labeling index of intramitochondrial protein synthesis incorporating ^{3}H-leucine increased from fetal day 19 to postnatal day 3 to month 2 and decreased to month 24. We carried out the quantitative analysis of these incorporations into nuclei and cell organelles of adrenal cells, both adrenal cortical cells and medullary cells, in aging mice from prenatal to postnatal newborn, juvenile, adult and senescent individuals (Nagata 2010a,b).

Observing EM radioautograms, the silver grains were found over the nuclei of some adreno-cortical cells labeled with ^{3}H-leucine demonstrating protein synthesis in all aging stages from perinatal stages at embryonic day 19, postnatal day 1 and day 3, day 9 and day 14 and adults at month 1, month 2, month 6, month 12 and month 24. Those labeled cells were found in all the 3 layers, the zona glomerulosa (Fig. 25A), the zona fasciculata and the zona reticularis (Fig. 25B, C), at respective aging stages. In the labeled adreno-cortical cells in 3 layers the silver grains were mainly localized over the euchromatin of the nuclei or a few or several silver grains were found over cytoplasmic organelles, especially over some of the mitochondria showing protein synthesis incorporating ^{3}H-leucine. The localizations of silver grains over the mitochondria were mainly on the mitochondrial matrices and some over the mitochondrial membranes as observed by high power magnification (Fig. 25C).

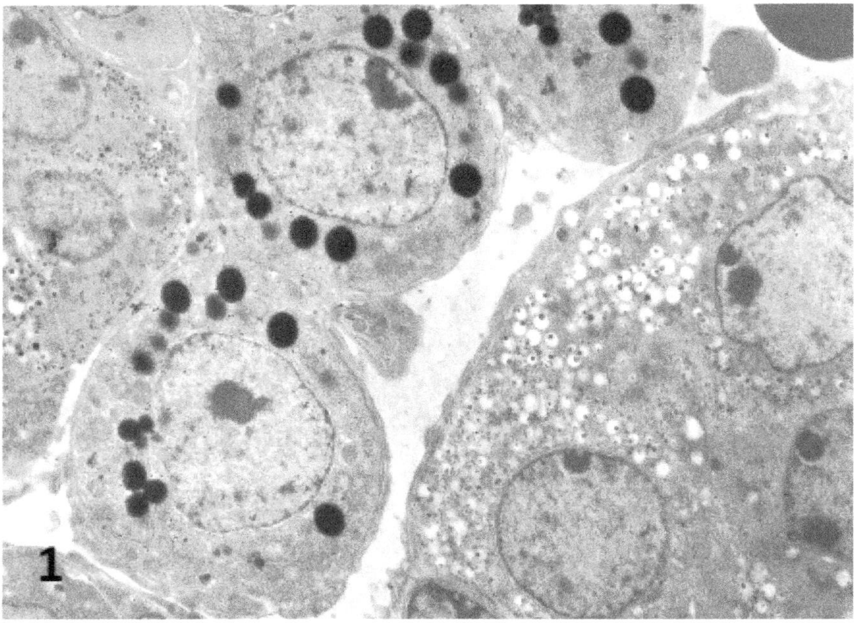

25A

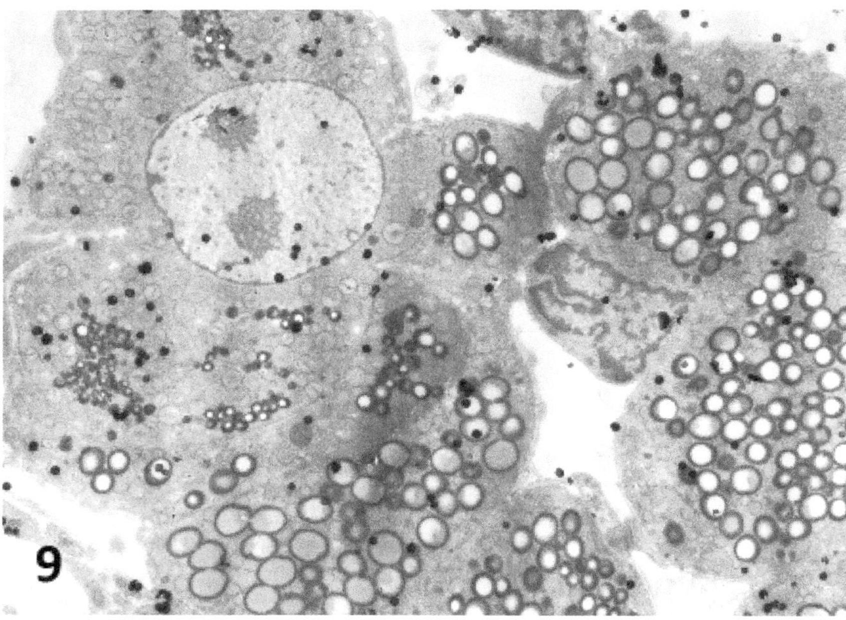

25B

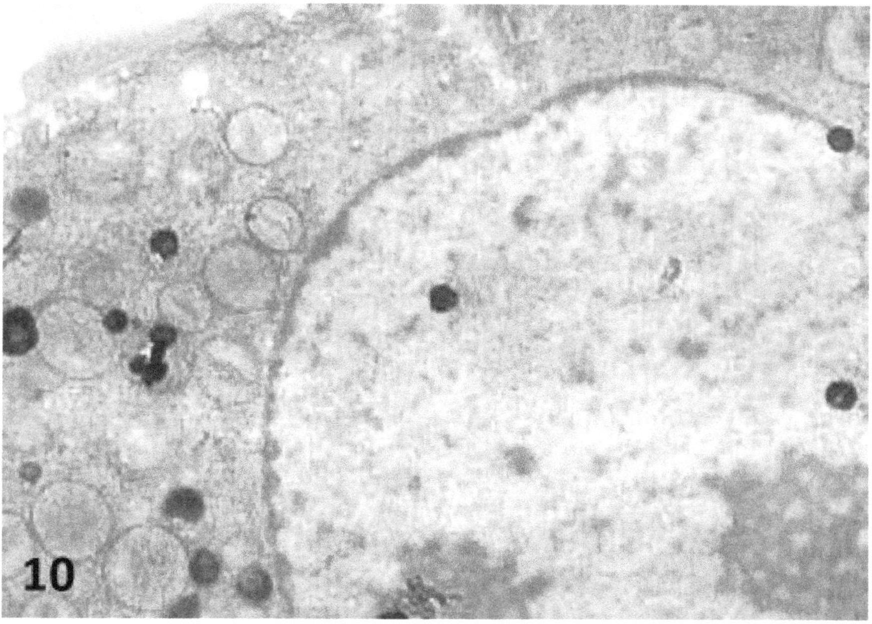

25C

Fig. 25. EMRAG of the adrenal cortical cells aging mice labeled with ^{3}H-leucine showing protein synthesis in the nucleus as well as in a few mitochondria. From Nagata, T.: Annals of Microscopy Vol. 4, p. 54-71, 2011.

Fig. 25A. EMRAG of the zona glomerulosa of a juvenile mouse at postnatal day 14, labeled with ^{3}H-leucine showing protein synthesis (several silver grains) in the nucleus as well as in a few mitochondria. x 3,000.

Fig. 25B. EMRAG of the zona reticularis of an old adult mouse aged at postnatal month 12, labeled with ^{3}H-leucine showing protein synthesis in the nucleus and a few mitochondria. x 3,000

Fig. 25C. High power magnification EMRAG of Fig. 24B, the zona reticularis of an old adult mouse aged at postnatal month 12, showing protein synthesis in a few mitochondria at upper left corner. x15,000.

2.3.1.1.1 Number of mitochondria of adreno-cortical cells in aging mice labeled with ^{3}H-leucine

Preliminary quantitative analysis on the number of mitochondria in either 10 adreno-cortical cells whose nuclei and cytoplasm were labeled with ^{3}H-leucine showing silver grains and other 10 cells whose nuclei were not labeled in each aging group revealed that there was no significant difference between the number of mitochondria and the labeling indices ($P<0.01$). Thus, the number of mitochondria and the labeling indices were calculated regardless whether their nuclei were labeled or not (Fig. 26A). The results obtained from the number of mitochondria in adreno-cortical cells in the 3 layers of respective animals in 10 aging groups at perinatal stages, prenatal embryo day 19, postnatal day 1, 3, 9 and 14, showed an gradual increase from the prenatal day 19 (glomerulosa 13.5, fasciculata 14.9, reticularis 15.2/cell) to postnatal day 14 (glomerulosa 37.7, fasciculata 37.8, reticularis 39.8/cell), and to adult stages at postnatal month 1 (glomerulosa 41.5, fasciculata 42.3, reticularis 42.9/cell), then increased at month 2 (glomerulosa 64.2, fasciculata 65.1, reticularis 67.2/cell), but kept plateau from month 6 (glomerulosa 61.7, fasciculata 62.9, reticularis 62.1/cell), to month 12 (glomerulosa 59.4, fasciculata 70.5, reticularis 71.4/cell) and month 24 (glomerulosa 59.5, fasciculata 62.2, reticularis 63.3/cell). The increase from embryo day 19 to postnatal month 1 was stochastically significant ($P < 0.01$).

2.3.1.1.2 Number of labeled mitochondria of adreno-cortical cells in aging mice labeled with ^{3}H-leucine

The results of visual grain counting on the number of mitochondria labeled with silver grains obtained from 10 adreno-cortical cells in the 3 layers of each animal labeled with ^{3}H-leucine demonstrating protein synthesis in 10 aging groups at perinatal stages, prenatal embryo day 19, postnatal day 1, 3, 9 and 14, month 1, 3, 6, 12 and 24, showed that the numbers of labeled mitochondria with ^{3}H-leucine showing protein synthesis per cell gradually increased from prenatal embryo day 19 (glomerulosa 0.3, fasciculata 0.4, reticularis 0.4/cell) to postnatal day 1 (glomerulosa 0.5, fasciculata 0.6, reticularis 0.5/cell), day 3 (glomerulosa 1.2, fasciculata 0.8, reticularis 1.1/cell), day 9 (glomerulosa 0.8, fasciculata 1.1, reticularis 1.1/cell), day 14 (glomerulosa 1.5, fasciculata 1.5, reticularis 1.6/cell), and month 1 (glomerulosa 1.8, fasciculata 1.8, reticularis 2.2/cell) and month 2 (glomerulosa 5.4, fasciculata 5.3, reticularis 5.8/cell), reaching the maximum, then decreased to month 6 (glomerulosa 4.5, fasciculata 4.8, reticularis 5.1/cell), month 12 (glomerulosa 5.2, fasciculata 5.8, reticularis 6.0/cell) and 24 (glomerulosa 3.8, fasciculata 4.1, reticularis 4.3/cell), as demonstrated in Fig. 26B.

2.3.1.1.3 Labeling index of mitochondria of adrenal cortical cells in aging mice labeled with ^{3}H-leucine

Finally, the labeling indices of adreno-cortical cells showing protein synthesis in respective aging stages were calculated from the number of labeled mitochondria (Fig. 26B) dividing by the number of total mitochondria per cell (Fig.26A), which were plotted in Fig. 26C, respectively.

The results showed that the labeling indices gradually increased from prenatal day 19 (glomerulosa 2.2, fasciculata 2.7, reticularis 2.6%) to postnatal newborn stage at postnatal day 1 (glomerulosa 2.2, fasciculate 2.4, reticularis 2.0%) and day 3 (glomerulosa 4.5 fasciculata 2.9, reticularis 3.9%), and to juvenile stage at postnatal day 9 (glomerulosa 2.8, fasciculate 3.7, reticularis 3.7%), day 14 (glomerulosa 3.9, fasciculate 3.9, reticularis 4.0%) and to the adult stage at month 1 (glomerulosa 4.3, fasciculate 4.2 reticularis 5.1%) and month 2 (glomerulosa 8.5, fasciculate 8.1, reticularis 8.6%), reaching the maximum, and decreased to month 6 (glomerulosa 7.3, fasciculate 7.6, reticularis 8.1%) to month 12 (glomerulosa 8.4, fasciculate 8.2, reticularis 8.4%) and finally to senescence at month 24 (glomerulosa 6.1, fasciculate 6.6 reticularis 6.8%), as is shown in the histogram (Fig. 26C).

From the results obtained it was shown that the labeling indices of the adreno-cortical cells in 3 layers of each animal labeled with ^{3}H-leucine demonstrating protein synthesis in 10 groups gradually increased from prenatal embryo day 19 to postnatal day 1, 3, 9, and 14, month 1, 2, reaching the maximum, and decreased to 6, 12 and 24, due to aging and senescence.

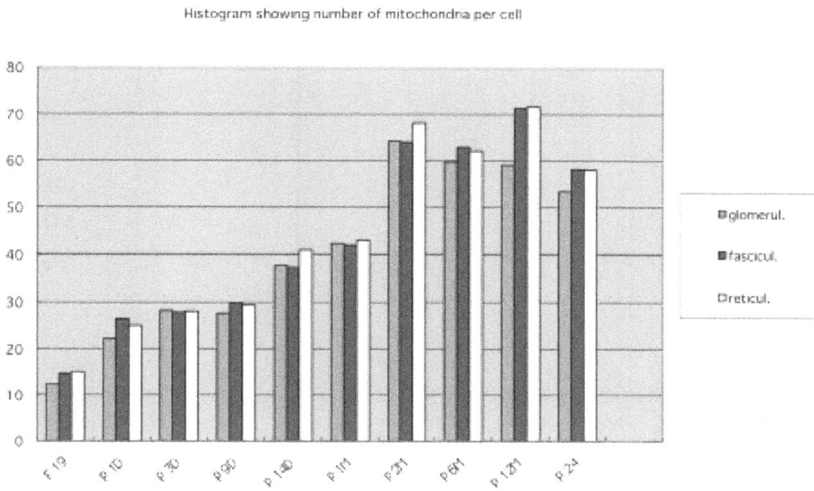

11

Fig. 26A.

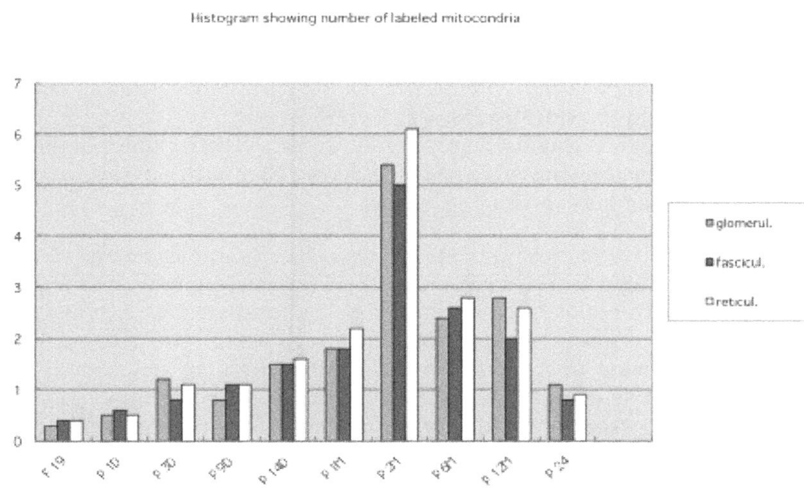

12

Fig. 26B.

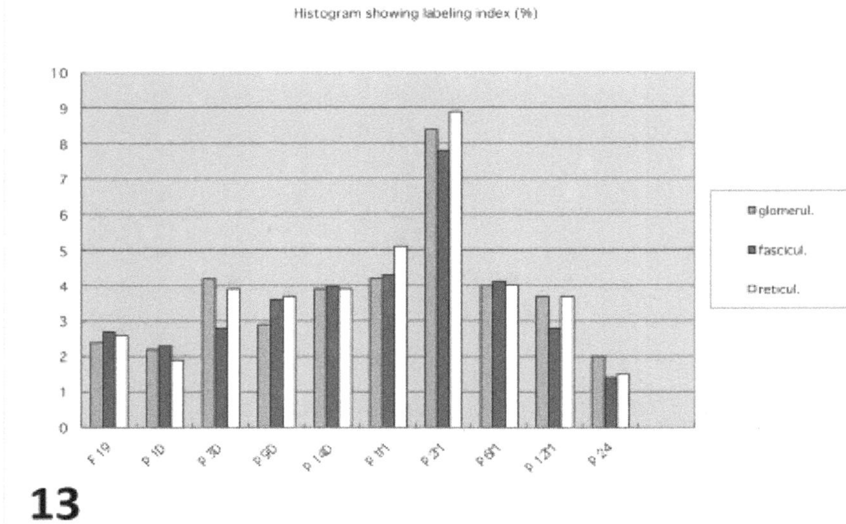

Fig. 26C.

Fig. 26. Histogram showing aging changes of the mitochondria in each adreno-cortical cell in the 3 layers of respective animals in 10 aging groups. From Nagata, T.: Annals of Microscopy Vol. 4, p. 54-71, 2011.
Fig. 26A. Histogram showing aging changes of the average numbers of mitochondria per cell in each adreno-cortical cell in the 3 layers of respective animals in 10 aging groups
Fig. 26B. Histogram showing aging changes of the average numbers of labeled mitochondria with ^{3}H-leucine showing protein synthesis per cell in each adrenocortical cell in the 3 layers of respective animals in 10 aging groups.
Fig. 26C. Histogram showing aging changes of the average labeling index of mitochondria labeled with ^{3}H-leucine showing protein synthesis per cell in each adreno-cortical cell in the 3 layers of respective animals in 10 aging groups.

As for the macromolecular synthesis in various cells in various organs of experimental animals observed by light and electron microscopic radioautography, it is well known that the silver grains due to radiolabeled ^{3}H-thymidine demonstrated DNA synthesis, while the grains due to ^{3}H-uridine demonstrated RNA synthesis. On the other hand, the radioautograms showing incorporations of ^{3}H-leucine into mitochondria indicating mitochondrial protein synthesis (Nagata 2002, 2009c,d,e, 2010a,b,c). It was shown from the results that the silver grains localized over the mitochondria independently from the nuclei whether the nuclei were labeled with silver grains or not in almost all the cells in the 3 layers of the adreno-cortical cells from prenatal embryo day 19 to postnatal month 24, administered with ^{3}H-leucine during the development and aging. The numbers of labeled mitochondria showing protein synthesis increased from perinatal day to postnatal adult stage at month 2, then kept plateau, while the labeled mitochondria with ^{3}H-leucine showing protein synthesis increased from perinatal stage to postnatal adult stage at month

2, then decreased at month 24, while the labeling indices increased from perinatal embryonic day to postnatal newborn and juvenile stages at day 9, then decreased from day 14 to senescence at month 24, then decreased to the adult stages at month 1 and 2, to month 6, 12 and 24. These changes demonstrate the chronological aging changes. The results obtained previously (Nagata 2008a,b,c,d, 2009a, Nagata & Ito 1996) indicated that mitochondria in the adreno-cortical cells proliferated from newborn to adult stages around month 1 and 2, showing mitochondrial DNA synthesis, while the mitochondrial RNA synthesis increased from newborn stage to postnatal day 9, then decreased from day 14, reaching the maximum, then decreased to month 24, but the RNA synthetic activity was kept plateau from day 14 to month 12, then decreased to senescent at month 24 due to aging (Nagata 2010e).

2.3.1.2 Protein synthesis in mitochondria of mouse adreno-medullary cells

The adreno-medullary tissues obtained from ddY strain mice at various ages from embryo day 19 to postnatal day 30, consisted mainly of 2 cell types, as observed by electron microscopy, adrenalin cells and noradrenalin cells, developing gradually (Nagata 2010a,b). At embryonic day 19 and postnatal day 1, the adreno-medullary cells were composed mainly of polygonal epitheloid cells, surrounded by blood capillaries and fibroblasts. The medullary cells can be divided into 2 types by the ultrastructure of granules. Some of the medullary cells possessed many granules of medium electron density which were believed to correspond to the adrenalin granules, while some other cells possessed many granules of very high electron density which were believed to correspond to the noradrenalin granules. However, the numbers of mitochondria found in their cytoplasm were not so many. At postnatal day 1, day 3 and day 9, the 2 types of cells differentiated and the numbers of granules, both adrenalin and noradrenalin granules, increased respectively. Likewise, the numbers of mitochondria also increased from prenatal day to postnatal days. At postnatal day 14 to month 1, month 2, month 6, month 12 and month 24 the numbers of adrenalin and noradrenalin granules as well as mitochondria increased. At postnatal month 1 and 2, the ultrastructures of 2 cell types were completely developed and the arrangements of the cells in the medulla became typical as adult tissues. Thus, the ultrastructure of the adrenal medullary cells showed changes due to development and aging at respective developmental stages. The number of mitochondria per cell increased from perinatal stage to postnatal stages due to aging. The data were stochastically analyzed using variance and Student's t-test. The increases of mitochondrial numbers in both adrenalin and noradrenalin cells from embryonic day 19 to postnatal month 6 were considered to be significant at P value <0.01.

Observing electron microscopic radioautograms, the silver grains were found over the nuclei of some adreno-medullary cells labeled with ^{3}H-leucine, demonstrating protein synthesis less in perinatal stages at embryonic day 19, postnatal day 1, day 3, day 9, while more day 14 and adults at month 1, month 2, month 6, month 12 and month 24.

However, those labeled cells were found in all the 2 cell types, adrenalin cells and noradrenalin cells, at respective aging stages. In the labeled adreno-medullary cells the silver grains were mainly localized over the euchromatin of the nuclei and only a few or several silver grains were found over the mitochondria of adrenalin cells. Likewise, most noradrenalin cells were also labeled with several silver grains in their nuclei as well as in their cytoplasm not only over the mitochondria but also over ribosomes, showing protein synthesis due to the incorporations of ^{3}H-leucine especially at the senescent stages from postnatal month 12 when observed at high power magnifications by high voltage electron microscopy. The localizations of silver grains over the mitochondria were not only on the mitochondrial matrices but also over mitochondrial membranes (Nagata 2010a,b).

2.3.1.2.1 Number of mitochondria of adreno-medullary cells in aging mice labeled with ^{3}H-leucine

Preliminary quantitative analysis on the number of mitochondria in 10 adreno-medullary cells whose nuclei were labeled with silver grains and other 10 cells whose nuclei were not labeled in each aging group revealed that there was no significant difference between the number of mitochondria and the labeling indices (P<0.01). Thus, the number of mitochondria and the labeling indices were calculated in 10 adreno-medullary cells regardless whether their nuclei were labeled or not. The results obtained from the number of mitochondria in adreno-medullary cells of respective animals in 10 aging groups at perinatal stages, from prenatal embryo day 19 to postnatal day 1, 3, 9, 14, and month 1, 2, 6, 12, showed an gradual increase from the prenatal day 19 (adrenalin 17.8, noradrenalin 18.2/cell) to postnatal day 1 (adrenalin 21.5, noradrenalin 22.4/cell), day 3 (adrenalin 22.5, noradrenalin 22.9/cell), day 9 (adrenalin 23.5, noradrenalin 23.9/cell), day 14 (adrenalin 24.1, noradrenalin 24.4/cell), and to adult stages at postnatal month 1 (adrenalin 24.7, noradrenalin 23.9/cell), month 2 (adrenalin 25.1, noradrenalin 24.5/cell) and month 6 (adrenalin 24.8, noradrenalin 24.3/cell), month 12 (adrenalin 24.5, noradrenalin 24.1/cell), and month 24 (adrenalin 23.5, noradrenalin 23.3/cell). The increases of mitochondrial numbers in both adrenalin and noradrenalin cells from embryonic day 19 to postnatal month 2 were considered to be significant at P value <0.01. However, the slight decrease from month 6 to 24 was not significant (Nagata 2010a or b).

2.3.1.2.2 Number of labeled mitochondria of adrenal medullary cells in aging mice labeled with ^{3}H-leucine

We counted the number of mitochondria labeled with silver grains obtained from 10 adreno-medullary cells in the 3 layers of each animal labeled with ^{3}H-leucine demonstrating protein synthesis in 10 aging groups at perinatal stages, prenatal embryo day 19, postnatal day 1, 3, 9 and 14, month 1, 3, 6, 12 and 24, as well as the number of all the mitochondria and calculated the labeling index.

The results demonstrated that the numbers of labeled mitochondria with ^{3}H-leucine showing protein synthesis per cell gradually increased from prenatal embryo day 19 (adrenalin 0.5, noradrenalin 0.5/cell), to postnatal day 1 (adrenalin 0.65, noradrenalin 0.6/cell), day 3 (adrenalin 0.7, noradrenalin 0.75/cell), day 9 (adrenalin 0.8, noradrenalin 0.8/cell), day 14 (adrenalin 0.8, noradrenalin 0.9/cell), reaching the maxima at month 1 (adrenalin 0.9, noradrenalin 0.85/cell), and decreased to month 2 (adrenalin 0.8, noradrenalin 0.82/cell), month 6 (adrenalin 0.81, noradrenalin 0.8/cell) month 12 (adrenalin 0.7, noradrenalin 0.75/cell) and month 24 (adrenalin 0.7, noradrenalin 0.65/cell). The data were stochastically analyzed using variance and Student's t-test. The increases of the numbers of labeled mitochondria in both adrenalin and noradrenalin cells from embryo day 19 to postnatal day 14 and month 1, as well as the decreases from month 1 to month 24 were stochastically significant (P <0.01).

2.3.1.2.3 Labeling index of mitochondria of adreno-medullary cells in aging mice labeled with ^{3}H-leucine

From the results, the labeling indices of mitochondrial protein synthesis in 2 cell types in respective aging stages were calculated from the number of labeled mitochondria dividing by the number of total mitochondria per cell.

The results showed that the labeling indices gradually increased from prenatal day 19 (adrenalin 2.8, noradrenalin 2.6%) to postnatal newborn day 1 (adrenalin 2.8, noradrenalin 2.4%), day 3 (adrenalin 3.3, noradrenalin 2.9%), day 9 (adrenalin 3.4, noradrenalin 3.3%) to juvenile stage at day 14 (adrenalin 3.6, noradrenalin 3.8%), reaching the maximum, and decreased to adult stages at month 1 (adrenalin 3.6, noradrenalin 3.5%), month 2 (adrenalin 3.1, noradrenalin 3.3%), month 6 (adrenalin 3.2, noradrenalin 3.3%), month 12 (adrenalin 2.8, noradrenalin 3.1%) and month 24 (adrenalin 2.9, noradrenalin 2.8%). From the results, the increases of the mitochondrial labeling indices in both the adrenalin and noradrenalin cells from embryo day 19 to postnatal day 14, as well as the decreases from day 14 to month 24 were stochastically significant ($P < 0.01$).

From the results obtained it was shown that intramitochondrial protein synthesis was observed in adreno-medullary cells, both the adrenalin cells and noradrenalin cells of developing and aging mice at various ages from prenatal embryos to postnatal newborn, young juvenile, adult and senescent stages and the number of mitochondria per cell showed increases due to aging, while the number of labeled mitochondria per cell and the labeling indices showed increases and slight decreases due to aging.

As for the macromolecular synthesis in various cells in various organs of experimental animals observed by light and electron microscopic radioautography, it is well known that the silver grains due to radiolabeled ^{3}H-thymidine demonstrate DNA synthesis, while the grains due to ^{3}H-uridine demonstrate RNA synthesis and ^{3}H-leucine protein synthesis (Nagata 1996, 1997, 2002, 2010). We formerly observed the intramitochondrial DNA synthesis as well as RNA synthesis in various cells in mice and rats (Nagata 1972a,b,c,d, 1974, 1984, 2001, 2001, Nagata et al. 1975, 1976, 1977). Lately we observed in the intramitochondrial DNA synthesis as well as RNA synthesis in the adrenal cortical and medullary cells, both the adrenalin cells and the noradrenalin cells at various ages from fetal day 19 to postnatal newborn day 1, 3, 9, juvenile day 14 and to adult month 1, 2, 6, 12 and 24 (Nagata 2008a,b, 2009a, 2010b,e), as well as the intramitochondrial protein synthesis during the aging and senescence (Nagata 2010d,f,g,h).

These results demonstrated that the numbers of silver grains showing nuclear protein synthesis did not give any significant difference between the 2 cells types in the same aging groups from perinatal to senescent stages. The radioautograms showing incorporations of ^{3}H-leucine into mitochondria indicating mitochondrial protein synthesis resulted in silver grain localization over the mitochondria independently from the nuclei whether the nuclei were labeled with silver grains or not in both cell types, adrenalin and noradrenalin cells, in the medullae from prenatal embryo day 19 to postnatal day 1, 3, 9 and 14, to postnatal month 1, 2, 6, 12 and 24, during the development and aging. The numbers of labeled mitochondria showing protein synthesis as well as the labeling indices increased from perinatal embryonic day to postnatal newborn and juvenile stages at day 14 to month 12, reaching the maxima, and then did not decrease to the senescent stages at month 24 (year 2).

From the results obtained, it was shown that intramitochondrial protein synthesis was observed in adreno-medullary cells, both the adrenalin cells and noradrenalin cells of developing and aging mice at various ages from prenatal embryos to postnatal newborn, young juvenile, adult and senescent stages and the number of mitochondria per cell showed increases due to aging, while the number of labeled mitochondria per cell and the labeling indices showed increases and slight decreases due to aging (Nagata 2010a,b,c,d,e,f,g,h).

As for the macromolecular synthesis in various cells in various organs of experimental animals observed by light and electron microscopic radioautography, it is well known that the silver grains due to radiolabeled ^{3}H-thymidine demonstrate DNA synthesis, while the grains due to ^{3}H-uridine demonstrate RNA synthesis and ^{3}H-leucine protein synthesis (Nagata et al. 1967a, b, Nagata 1972a, b, c, 2001, 2002). The previous results obtained from the studies on the adreno-cortical cells of aging mice by light microscopic radioautography revealed that silver grains indicating DNA synthesis incorporating ^{3}H-thymidine were observed over the nuclei of some adreno-cortical cells at perinatal stages from postnatal day 1 to day 14 (Ito 1996, Ito & Nagata 1996). However, they did not observe the intramitochondrial RNA synthesis. We formerly observed the intramitochondrial DNA synthesis (Nagata et al. 1967a, b, Nagata 1972a, b, c, 2001, 2002) as well as RNA synthesis (Nagata 1972a,b,c,d, 1974, 1984, 2001, 2002, Nagata et al. 1975, 1976, 1977, Nagata & Murata 1977) in the adrenal medullary cells, both the adrenalin cells and the noradrenalin cells at various ages from fetal day 19 to postnatal newborn day 1, 3, 9, juvenile day 14 and to adult month 1, 2, 6, 12 and 24. In the present study, we also observed the intramitochondrial protein synthesis during the aging and senescence. The numbers of silver grains showing nuclear protein synthesis did not give any significant difference between the 2 cells types in the same aging groups from perinatal to senescent stages. The radioautograms showing incorporations of ^{3}H-leucine into mitochondria indicating mitochondrial protein synthesis resulted in silver grain localization over the mitochondria independently from the nuclei whether the nuclei were labeled with silver grains or not in both cell types, adrenalin and noradrenalin cells, in the medullae from prenatal embryo day 19 to postnatal day 1, 3, 9 and 14, to postnatal month 1, 2, 6, 12 and 24, during the development and aging. The numbers of labeled mitochondria showing protein synthesis as well as the labeling indices increased from perinatal embryonic day to postnatal newborn and juvenile stages at day 14 to month 12, reaching the maxima, and then did not decrease to the senescent stages at month 24 (year 2).

With regards to DNA in mitochondria in animal cells or plastids in plant cells, many studies have been reported in various cells of various plants and animals since 1960s (Nass 1966, Nass and Nass 1963, Gibor and Granick 1964, Gahan and Chayen 1965). Most of these authors observed DNA fibrils in mitochondria which were histochemically extracted by DN'ase. Electron microscopic observation of the DNA molecules isolated from the mitochondria revealed that they were circular in shape, with a circumference of 5-6 μm. It was calculated that such a single molecule had a molecular weight of about 10^{7} daltons (van Bruggen et al. 1966). Mitochondria of various cells also contained a DNA polymerase, which was supposed to function in the replication of the mitochondrial DNA. On the other hand, the incorporations of ^{3}H-thymidine into mitochondria demonstrating DNA synthesis were observed by means of electron microscopic radioautography in lower organism such as slime mold (Guttes and Guttes 1964, Schuster 1965), tetrahymena (Stone and Miller Jr.

1965) or chicken fibroblasts in tissue culture under abnormal conditions (Chévremont 1963). However, these authors used old-fashioned developers consisting of methol and hydroquinone (MQ-developer) which produced coarse spiral silver grains resulting in inaccurate localization over cell organelles when observed by electron microscopy. All of these authors showed photographs of electron radioautographs with large spiral-formed silver grains (2-3 μm in diameter) localizing not only over the mitochondria but also outside the mitochondria. In order to obtain smaller silver grains, we first used elon-ascorbic acid developer after gold latensification, which produced comma-shaped smaller silver grains (0.4-0.8 μm in diameter), then later we used phenidon developer after gold latensification, producing dot-like smaller silver grains (0.2-0.4 μm in diameter) localizing only inside the mitochondria showing ultrahigh resolution of radioautograms. These papers were the first which demonstrated intramitochondrial DNA synthesis incorporating ^{3}H-thymidine with accurate intramitochondrial localization in avian and mammalian cells. With regards the resolution of electron microscopic radioautography, on the other hand, many authors discussed the sizes of silver grains under various conditions and calculated various values of resolutions (Salpeter et al. 1969, Nadler 1971, Uchida & Mizuhira 1971, Nagata 1972b,c). Those authors who used the M-Q developers maintained the resolution to be 100-160 nm (Salpeter et al. 1969, Nadler 1971), while those authors who used the elon-ascorbic acid developer (Nagata 1972b, , Uchida & Mizuhira 1971) calculated it to be 25-50 nm. When we used phenidon developer at 16°C for 1 min after gold latensification, we could produce very fine dot-shaped silver grains and obtained the resolution around 25 nm (Nagata 1992, 1996, 1997, 2001, 2002, Murata et al. 1979). For the analysis of electron radioautographs, Salpeter et al. (1969) proposed to use the half-distance and very complicated calculations through which respective coarse spiral-shaped silver grains were judged to be attributable to the radioactive source in a certain territory within a resolution boundary circle. However, since we used phenidon developer after gold latensification to produce very fine dot-shaped silver grains, we judged only the silver grains which were located in the mitochondria which were dot-shaped very fine ones to be attributable to the mitochondria without any problem as was formerly discussed (Nagata 1972a,b, c, 1996, 1997, 2001, 2002).

Then we also demonstrated intramitochondrial DNA synthesis incorporating ^{3}H-thymidine in some other established cell lines originated from human being such as HeLa cells (Nagata 1972a,b,c,d) or mitochondrial fractions prepared from in vivo mammalian cells such as rat and mouse (Nagata 1974, Nagata et al. 1975, 1976). It was later commonly found in various cells and tissues not only in vitro obtained from various organs in vivo such as the cultured human HeLa cells (Nagata et al. 1966, 1986, Nagata 1984), cultured rat sarcoma cells (Nagata et al. 1977), mouse liver and pancreas cells in vitro (Nagata & Murata 1977, Nagata et al. 1977, 1986), but also in vivo cells obtained from various organs such as the salivary glands (Nagata et al. 2000), the liver (Nagata 2003, 2006a,b, 2007a,b,c,d,e, Nagata & Ma 2005, Nagata et al. 1979, 1982a,b, Ma & Nagata 1988a,b, Ma et al. 1994), the pancreas (Nagata 1992, Nagata et al. 1986), the trachea (Sun et al. 1997), the lung (Sun et al. 1994, 1995a,b, Nagata 2007), the kidneys (Hanai & Nagata 1994, Nagata 2005), the testis (Gao et al. 1994, 1995), the uterus (Yamada et al. 1993, 1994), the adrenal glands (Ito 1996, Ito et al. 1996, Nagata 2008a,b, 2009g,j, 2010a,b), the brains (Cui et al. 1996), and the retina (Gunarso 1984, Gunarso et al. 1996, 1997, Kong & Nagata 1994, Nagata 1996) of mice, rats and chickens. Thus, it is clear that all the cells in various organs of various animals synthesize DNA not only in their nuclei but also in their mitochondria.

The relationship between the intramitochondrial DNA synthesis and cell cycle was formerly studied in synchronized cells and it was clarified that the intramitochondrial DNA synthesis was performed without nuclear involvement (Nagata 1972b). However, the relationship between the DNA synthesis and the aging of individual animals and men has not yet been clarified except a few papers recently published by Korr and associates on mouse brain (Korr et al. 1997, 1998, Schmitz et al. 1999a,b,). They reported both nuclear DNA repair, measured as nuclear unscheduled DNA synthesis, and cytoplasmic DNA synthesis labeled with ^{3}H-thymidine in several types of cells in brains such as pyramidal cells, Purkinje cells, granular cells, glial cells, endothelial cells, ependymal cells, epithelial cells as observed by light microscopic radioautography using paraffin sections observed by LMRAG. They observed silver grains over cytoplasm of these cells by light microscopy and maintained that it was reasonable to interpret these labeling as ^{3}H-DNA outside the nuclei, which theoretically belonged to mitochondrial DNA without observing the mitochondria by electron microscopy. From the results, they concluded that distinct types of neuronal cells showed a decline of both unscheduled DNA and mitochondrial DNA syntheses with age in contrast that other cell types, glial and endothelial cells, did not show such age-related changes neither counting the number of mitochondria in respective cells nor counting the labeling indices at respective aging stages. Thus, their results from the statistics obtained from the cytoplasmic grain counting seems to be not accurate without observing mitochondria directly. To the contrary, we had studied DNA synthesis in the livers of aging mice (Nagata et al. 1979, Nagata 1982a,b, 2003, 2005, 2006a,b,c,d,e, 2010c, Ma et al. 1988, 1994, Ma & Nagata 1988) and clearly demonstrated that the number of mitochondria in each hepatocytes, especially mononucleate hepatocytes, increased with the ages of animals from the perinatal stages to adult and senescent stages, while the number of labeled mitochondria and the labeling indices increased from the perinatal stages, reaching a maximum at postnatal day 14, then decreased. We also demonstrated that the number of mitochondria and labeled mitochondria with ^{3}H-thymidine, ^{3}H-uridine and ^{3}H-leucine in the adrenal glands in aging mice increased due to aging (Nagata 2009j, 2010a,b,d,e,f,g,h).

2.3.2 The protein synthesis in the Leydig cells of the testis

We studied the macromolecular synthesis of the cells in the testis of several groups of litter ddY mice at various ages from fetal day 19 to postnatal aging stages up to 2 years senescence by LM and EM RAG using ^{3}H-thymidine, ^{3}H-uridine and ^{3}H-leucine incorporations (Gao 1993, Gao et al. 1994, 1995a, Nagata 2000b). The incorporation of ^{3}H-leucine into proteins was observed in almost all the Leydig cells in the interstitial tissues of the testis. The silver grains were located over the nuclei and cytoplasm of respective Leydig cells. The aging change of protein synthesis of Leydig cells among different aging groups was also found (Nagata 2001c, 2002). At embryonic day 19, the silver grains of Leydig cells labeled with ^{3}H-leucine was observed in both nucleus and cytoplasm and there was no obvious difference between the number of silver grains on the cytoplasm and the nucleus. The number of silver grains decreased at postnatal day 1 and then increased at day 3 and 7. However, the number of silver grains on both the nucleus and cytoplasm decreased from 1 month to 3 months and increased again from 6 months onwards maintaining a high level from adult to senescent stages. Some of the silver grains were also localized over some of the mitochondria in respective aging groups as observed by EM RAG. These results indicate that the DNA, RNA and protein syntheses in Leydig cells are maintained at rather high level even at senescent stages at postnatal 1 and 2 years when the animals survived for longer lives.

2.4 The localization of mercury in the thyroid gland

We studied incorporations of mercury chloride into human thyroid tissues of both normal and cancer cells obtained from human patients (Nagata et al. 1977b, Nagata 1994a,b,c,d,e). The tissues were obtained surgically from human patients of both sexes in various ages suffering from the cancer of thyroids and the both normal and cancer cells were cut into small pieces (3x3 mm) aseptically which were incubated in a medium (Eagle's MEM) containing RI-mercury chloride ($^{203}HgCl_2$) and fixed either cryo-fixation at -196°C and freeze-dried or chemically fixed with buffered glutaraldehyde and osmium tetroxide. The former tissue blocks were processed for dry-mounting radioautography, while the latter were processed for conventional wet-mounting radioautography. The results revealed that the silver grains appeared much more in the cancer cells processed for freeze-fixation and dry-mounting radioatuography (Fig. 21A) than the cancer cells processed for chemical fixation and wet-mounting radioatuography (Fig. 21B), as well as much more in the cancer cells than the normal cells under the same conditions. On the other hand, PCNA/cycline and both keratin kinase C and vimentin were immunostained in connection to DNA synthetic activity. It was found that PCNA/cycline, keratin kinase C and vimentin antibodeis were localized around the filaments in the thyroid cancer cells (Fig. 21C), demonstrating the relation between those antigens and DNA synthetic activity in cell cycle (Shimizu et al. 1993, Nagata 1994b,c, Gao et al. 1994).

3. Macromolecular synthesis in the nervous system

The nervous system consists of the central nervous system and the peripheral nervous system. The former is divided into the brains and the spinal cord, while the latter into the cerebrospinal system and the autonomous system. We studied macromolecular synthesis of the brains, the spinal cord in the cerebrospinal system and the autonomic peripheral nerves in the autonomous system by LM and EM RAG.

3.1 The DNA synthesis in the nervous system

We studied the DNA synthesis of the brains, the spinal cord in the cerebrospinal system and the autonomic peripheral nerves in the autonomous system by LM and EM RAG (Cui 1995, Cui et al. 1996, Izumiyama et al. 1987, Nagata 1965, 1967a, Nagata and Stegerda 1963, 1964, Nagata et al. 1999a).

3.1.1 The DNA synthesis in the brains

The brains of mammals consist of the cerebrum, the cerebellum and the brain stem. We studied on DNA synthesis and protein synthesis in the cerebellum of aging mouse (Cui 1995, Cui et al. 1996) as well as the glucose incorporation in the cerebrum of adult gerbils (Izumiyama et al. 1987). The DNA synthesis was examined in the cerebella of 9 groups of aging ddY strain mice from fetal day 19, to postnatal day 1, 3, 8, 14 and month 1, 2, 6, 12, each consisting of 3 litter animals, using ^{3}H-thymidine, a DNA precursor, by LM and EM RAG (Cui 1995, Cui et al. 1996). The labeled nuclei, by the precursor, in both the neurons and glias, i.e., neuroblasts and glioblasts, were observed in the external granular layers of the cerebella of perinatal mice from embryonic day 19 (Fig. 27A) to postnatal day 1, 3, 7 and day 14 by LMRAG and EMRAG. The labeled nuclei disappeared at postnatal 1 month. The peak of labeling index was at postnatal day 3 in both neuroblasts and glioblasts (Fig. 28A, B). The glioblasts of the external granular layer migrated inward, some of them formed the Bergmann glia cells located between Purkinje cells. Labeled nuclei of neuroblasts and glioblasts in the internal granular layers were observed at perinatal stages. The maximum of the labeling index in the internal granular layer was at postnatal day 3, similarly to the external granular layer. The endothelial cells of the cerebellar vessels were progressively labeled from embryos to neonates, reaching the peak at 1 week after birth and decreasing thereafter.

3.1.2 The DNA synthesis in the peripheral nerves

We first studied the degeneration and regeneration of autonomous nerve cells in the plexuses of Auerbach and Meissner of the jejunums of 15 dogs which were operated upon to produce experimental ischemia of the jejunal loops by perfusing with Tyrode's solution via the mesenteric arteries for 1, 2, 3 and 4 hours (Nagata 1965, 1967, Nagata and Steggerda 1963, 1964). Tissue blocks were obtained from the deganglionated portions and the adjoining normal portions, which were fixed in Carnoy's fluid, embedded in paraffin, sectioned and stained with buffered thionine, methyl-green and pyronine and PAS. Some animals were injected with either ^{3}H-thymidine or ^{3}H-cytidine and the intestinal tissues obtained from ischemic portions and normal portions were processed for LM RAG. The results revealed that the ganglion cells in Auerbach's plexus showed various degenerative changes in accordance with the duration of ischemia. After 4 hours ischemia, most of the ganglion cells in Auerbach's plexus were completely destroyed. The degenerative changes in Auerbach's plexus after 4 hours ischemia were irreversible after 1 week recovery. The ganglion cells in the Meissner's plexus, on the other hand, were less sensitive to the ischemia. They recovered completely even after 4 hours ischemia. The PAS positive substances in degenerative ganglion cells in both plexuses decreased immediately after 4 hours ischemia. The DNA contents of ganglion cells in both Auerbach's and Meissner's plexus did not show any change before and after ischemia. The RNA contents decreased immediately after the ischemia (Nagata and Steggerda 1963, 1964). The number of binucleate cells in ganglion cells in both Auerbach's and Meissner's plexuses after 4 hours ischemia increased to 4.6% and 5.7% respectively. In contrast, in the non-ischemic normal control preparations, the binucleate cells occurred only 0.5% and 1.8% in Auerbach's and Meissner's plexus respectively. The high frequency of binucleate cells in the ganglion cells persisted for more than 100 days after the ischemia, indicating a possible regeneration of ganglion cells. The radioautographic study revealed that there was no evidence for DNA synthesis in both Auerbach's and Meissner's plexus from either ischemic or normal loops. The RNA synthesis was observed to be higher in gaglion cells in normal loop than ischemic loop and higher in Auerbach's than in Meissner's as expressed by grain counting. It was higher in binucleate ganglion cells than in mononucleate cells.

3.2 The RNA synthesis in the nervous system

We studied only messenger RNA in the spinal cords of aging mice from perinatal to postnatal adult stages by means of in situ hybridizaion.

3.2.1 The RNA synthesis in the spinal cord

The localization of TGF-β1 mRNA in the segments of the spinal cords of mice was investigated by means of in situ hybridization techniques together with immunohistochemical staining (Nagata et al. 1999). The tissues of lower cervical segments of the spinal cords of BALB/c mice, from embryonic day 12, 14, 16, 19 and postnatal day 1, 3, 7, 14, 21, 28, 42 and 70, were used. For in situ hybridization, ^{35}S-labeled oligonucleotide probes for TGF-β1 were used to detect their messenger RNA. Cryosections were incubated under silicon cover slides with 100 μl of pre-incubation solution plus final concentration of 2.4 x 10^6 cpm/ml probes and 100 mM DTT for 16 hours. After washing with SSC and DTT, the slides were dried and processed for radioautography by dipping in Konica NR-M2 emulsion, which were exposed and developed. The results showed that TGF-β1 mRNA was detectable in the meninges surrounding the spinal cord, but scarcely detected in spinal cord parenchyma (Fig. 27B). The localization of TGF-β1 mRNA in the spinal cord suggested that TGF- β1 acted through paracrine mechanism in the morphogenesis of the spinal cord in mice. The localization of TGF-β1 and its mRNA in the segments of the spinal cords of mice was also investigated with immunocytochemical techniques (Nagata et al. 1999). The tissues of lower cervical segments of the spinal cords of BALB/c mice, from embryonic day 12, 14, 16, 19 and postnatal day 1, 3, 7, 14, 21, 28, 42 and 70, the same as in situ hybridization were used. For immunocytochemistry, transverse cryosections of the spinal cords were cut and stained with rabbit anti-TGF-β1 polyclonal antibody followed with ABC method. The results showed that positive immunoreactivities arose in the ventral horn motoneurons from the embryonic stage to postnatal neonates (Fig. 27B) up to the adults. The extracellular matrix of the white matter, however, showed positive immunocytochemical staining from postnatal day 14, and thereafter, and the immunoreactivity remained with aging. The whole white matter showed only background level of staining before postnatal day 14. The results indicated that TGF-β1 regulates motoneuron growth and differentiation as well as they were probably correlated with formation, differentiation and regeneration of myelin of nerve tracts. The immunostaining with βFGF antibody presented the same basal pattern as shown in TGFβ1 immunohistochemistry (Nagata and Kong 1988). The positive immunoreactivities were detected in ganglion cell layer, inner and outer plexiform layers, retinal pigment epithelial layer, choroidal and scleral layers. Since TGF-β1 mRNA was detectable in the meninges surrounding the spinal cord by in situ hybridization but scarcely detected in spinal cord parenchyma, the disparate localization of TGF-β1 polypeptide and TGF-β1 mRNA in the spinal cord suggest that TGF-β1 acts through paracrine mechanism in the morphogenesis of the spinal cord in mice. The negative control abolished virtually all reactivity when using the normal rabbit serum instead of primary antibody or using avidin-biotin-peroxidase complex solution only.

3.3 The protein synthesis in the cerebellum

When 10 groups of aging ddY mice from fetal day 19, to postnatal day 1, 3, 7, 14 and month 1, 2, 6, 12 and 24, each consisting of 3 litter mates, using ^3H-leucine, protein syntheses of both neuroblasts and glioblasts were observed by LM and EM RAG in the extragranular layers of perinatal animals (Cui 1995, Cui et al. 1996, 2000, Nagata et al. 2001). The silver grains due to ^3H-leucine demonstrating protein synthesis were localized over the nuclei and cytoplasm of neuroblasts and glioblasts of embryos at fetal day 19 and the number of silver grains increased after birth from postnatal day 1, 3 to day 7 and onward. On day 3, some Purkinje cells were recognized incorporating silver grains. The number of silver grains in these cells increased from neonatal stages to mature adult stage at postnatal day 14 and 30, then decreased from month 1, 2, 6, 12 to 24. The increase and decrease of the silver grains were due to the aging changes of protein synthesis in the cerebella due to development and senescence of individual animals.

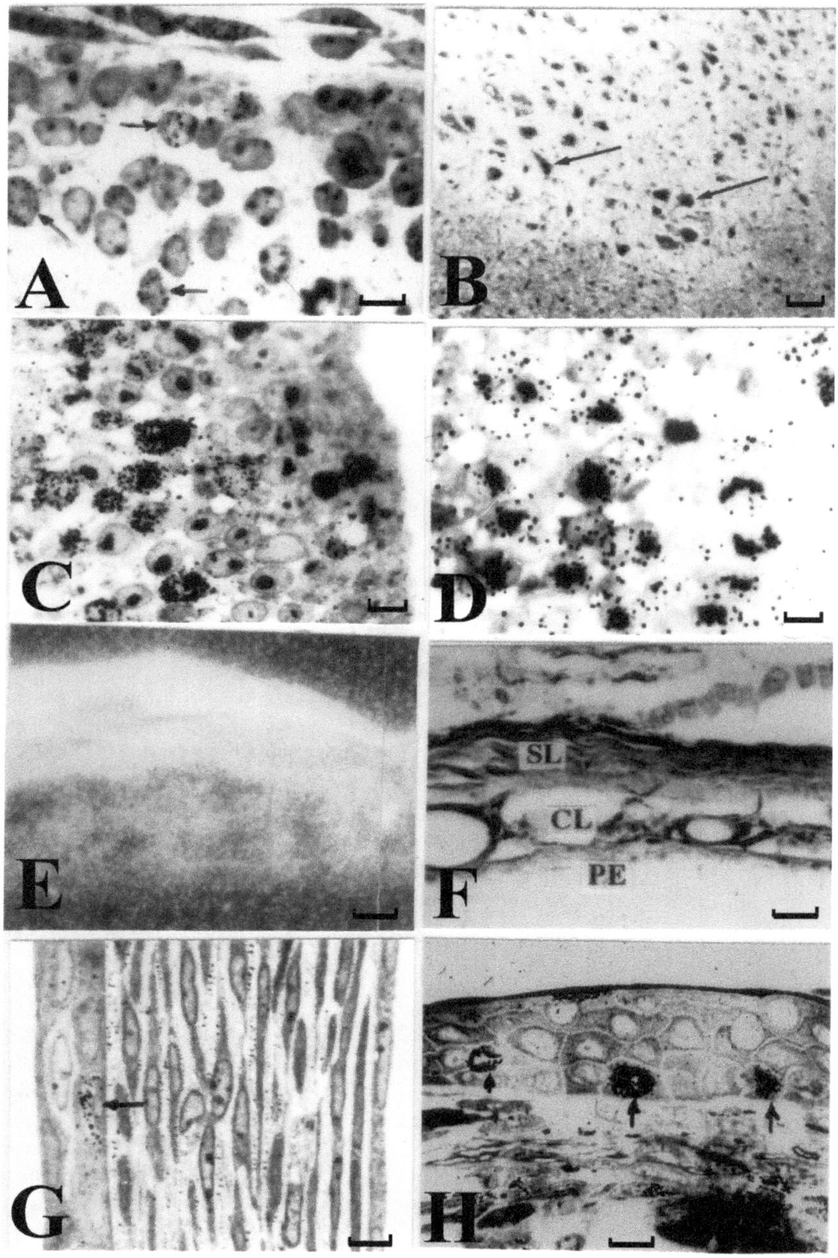

Fig. 27. LM RAG of the neuro-sensory cells. From Nagata, T., Special Cytochemistry in Cell Biology, In, Internat. Rev. Cytol. Jeon, K. W. Ed., Academic Press, San Diego, USA, London, UK, Vol. 211, No. 1, p. 122, 2001.

Fig. 27A. LM RAG of a prenatal day 19 mouse cerebellum labeled with ^{3}H-thymidine, showing DNA synthesis. x900.
Fig. 27B. LM RAG of the spinal cord of a postnatal day 14 mouse immunostained with rabbit anti-TGF-β1 polyclonal IgG followed by ABC method, showing that the ventral horn motoneurons are strongly positive. x70.
Fig. 27C. LM RAG of the optic vesicle of a day 2 chick embryo labeled with ^{3}H-thymidine, showing DNA synthesis. x750.
Fig. 27D. LM RAG of the optic vesicle of a day 2 chick embryo labeled with ^{3}H-uridine, showing RNA synthesis. x750.
Fig. 27E. Dark-field LM photo of the scleral layer (top), choroid layer (middle) and pigment epithelium (bottom) of an adult 1 month old mouse demonstrating intense silver grains by in situ hybridization for TGF-β1 mRNA. x450.
Fig. 27F. Bright-field LM photo of the scleral layer (top), choroid layer (middle) and pigment epithelium (bottom) of an adult 1 month old mouse demonstrating intense silver grains by in situ hybridization for TGF-β1 mRNA. x450.
Fig. 27G. LM RAG of the cornea of a postnatal day 14 mouse labeled with ^{3}H-thymidine, showing DNA synthesis in the epithelial nucleus (arrow) as well as in the stroma. x900.
Fig. 27H. LM RAG of the skin of the fore-limb of a salamander at 6weeks after hatching labeled with ^{3}H-thymidine, showing DNA synthesis. x900.

3.4 The glucide synthesis in the brains

The incorporation of ^{3}H-deoxyglucose was studied in the adult gerbil brains among the nervous system of experimental animals (Izumiyama et al. 1987). The changes of soluble deoxyglucose uptake in the hippocampus were studied after ^{3}H-deoxyglucose injections by means of cryo-fixation, freeze-substitution and dry-mounting radioautography to demonstrate soluble compounds under normal and post-ischemic conditions. The results demonstrated that the neurons in the hippocampus subjected to ischemia revealed higher uptake of soluble glucose than normal control. The concentration of soluble ^{3}H-deoxyglucose was higher than the chemically fixed and wet-mounted radioautograms that demonstrated only insoluble compounds. However, the relation of glycogen synthesis to aging has not yet been fully clarified.

4. Macromolecular synthesis in the sensory system

The sensory system consists of five organs, i.e., the visual organ or the eye, the stato-acoustic organ or the ear, the gustatory organ or the tongue, the olfactory organ or the nose, and the dermis or the skin. Among these sensory organs, we mainly studied the visual organ and the skin (Gunarso 1983a,b, Gunarso et al. 1996, 1997, Gao et al. 1992a,b, 1993, Kong 1993, Kong and Nagata 1994, Kong et al. 1992a,b, Nagata 1998, 1999, 2000, Nagata et al. 1994, Toriyama 1995).

4.1 The DNA synthesis in the sensory organs

Among the sensory organs, we mainly studied the DNA synthesis of the visual organ and the skin. They should be described separately.

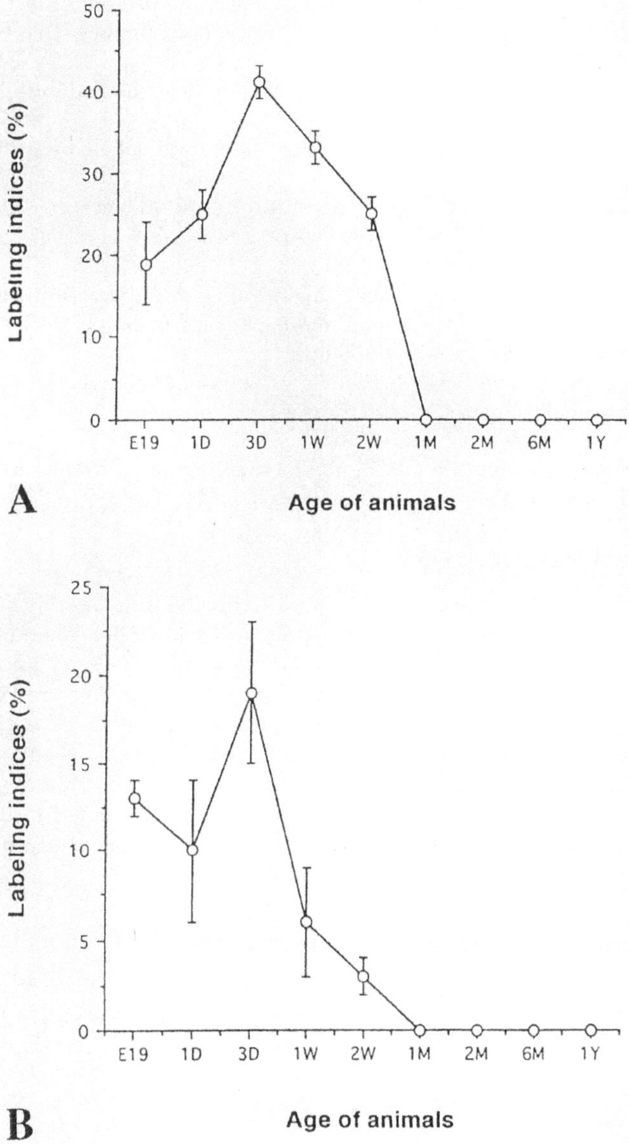

Fig. 28. Transitional curves of the labeling indices of respective cell types in the cerebella of aging mice labeled with ^{3}H-thymidine, showing DNA synthesis. Mean ± Standard Deviation. From Nagata, T.: Radioautographology, General and Special. In, Prog. Histochem. Cytochem. Vol. 37, No. 2, p. 186, 2002, Urban & Fischer, Jena, Germany

Fig. 28A. The neuroblasts in the extragranular layer of the cerebella.
Fig. 28B. The glioblasts in the extragranular layer of the cerebella.

4.1.1 The DNA synthesis in the visual organ

The visual organ consists of the eye and its accessory organs. The eye of mammals consists of the cornea, iris, cilliary body, lens, retina, choroid and sclera. We studied mainly the macromolecular synthesis in the retina of chickens and mice (Nagata 2000f). The nucleic acid syntheses, both DNA and RNA, were first studied in the ocular tissues of white Leghorn chick embryos from day 1 to day 14 incubations by LM and EMRAG (Gunarso 1984a,b, Gunarso et al. 1997, Gao et al. 1992a,b, 1993, Kong 1993, Kong and Nagata 1994, Kong et al. 1992a,b, Nagata et al. 1994). It was shown that the labeled cells with silver grains due to ^{3}H-thymidine were most frequently observed in the nuclei of the retinal cells in the posterior region of the day 2 chick embryo optic vesicle (Fig. 27C) and the labeled cells moved from anterior to posterior regions due to aging by incubation in vitro. The number of labeled cells as expressed by labeling index (%), was more in the posterior regions than the anterior and the equatorial regions and more in the outer portions than in the inner portions at day 2, but the labeling index became more in the anterior regions than the equatorial and posterior regions at day 3, 4 and 7 and it became more in the inner portions than in the outer portions at day 7, decreasing from day 2 to 3, 4 and 7 in each regions (Fig. 28). On the other hand, the silver grains due to ^{3}H-uridine were observed over the nuclei and cytoplasm of all retinal cells from day 2 to 7 (Fig. 27D) and the number of silver grains incorporating ^{3}H-uridine increased from day 1 to day 7 and it was more in the anterior regions than in the posterior regions at the same stage (Gunarso et al. 1996). On the other hand, DNA and RNA syntheses in the ocular tissues of aging ddY mice were also studied (Gao et al. 1993, Kong and Nagata 1994, Kong et al. 1992a,b). The ocular tissues taken out from several groups of litter ddY mice at ages varying from fetal day 9, 12, 14, 16, 19 to postnatal day 1, 3, 7, 14 were labeled with ^{3}H-thymidine in vitro and radioautographed (Gao et al. 1992a,b, 1993, Kong 1993, Kong et al. 1992a,b, Toriyama 1995). Silver grains showing DNA synthesis were localized over the nuclei of retinal cells and pigment epithelial cells in the anterior, equatorial and posterior regions of perinatal animals (Fig. 27A). The labeling indices of the retina and pigment epithelium were higher in earlier stages than in later stages, during which they steadily declined (Fig. 28A,B). However, the retina and the pigment epithelium followed different courses in their changes of labeling indices during embryonic development. In the retina, the labeling indices in the vitreal portions were more than those in the scleral portions during the earlier stages. However, the indices of scleral portions were more than those in the vitreal portions in the later stages. Comparing the three regions of the retinae of mice, the anterior, equatorial and posterior regions, the labeling indices of the anterior region were generally higher than those of the equatorial and posterior regions (Fig. 28A). In the pigment epithelium (Fig. 28B), the labeling indices gradually increased in the anterior region, but decreased in the equatorial and the posterior regions through all developmental stages. These results suggest that the proliferation of both the retina and pigment epithelium in the central region occurred earlier than those of the peripheral regions (Nagata 1999a, Gao et al. 1992a,b, Kong 1993, Kong and Nagata 1994, Kong et al. 1992a, b). In the juvenile and adult stages, however, the labeled cells were localized at the middle of the bipolar-photoreceptor layer of the retina, where was supposed to be the undifferentiated zone.

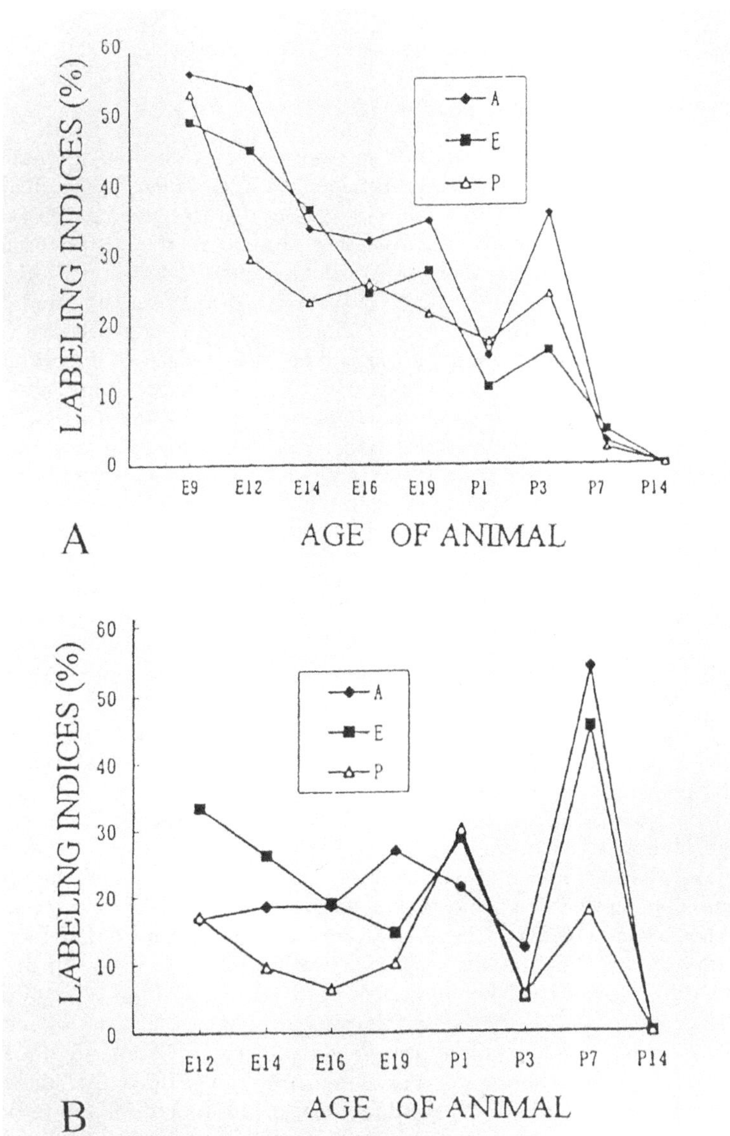

Fig. 29. Transitional curves of the labeling indices in the three regions (A: anterior, E: eqauator, P: posterior) of the retina and the pigment epithelium of aging mice labeled with ^{3}H-thymidine, showing DNA synthesis. Mean ± Standard Deviation. From Nagata, T.: Radioautographology, General and Special. In, Prog. Histochem. Cytochem. Vol. 37, No. 2, p. 191, 2002, Urban & Fischer, Jena, Germany

Fig. 29A. The labeling index in the retina.

Fig. 29B. The labeling index in the pigment epithelium.

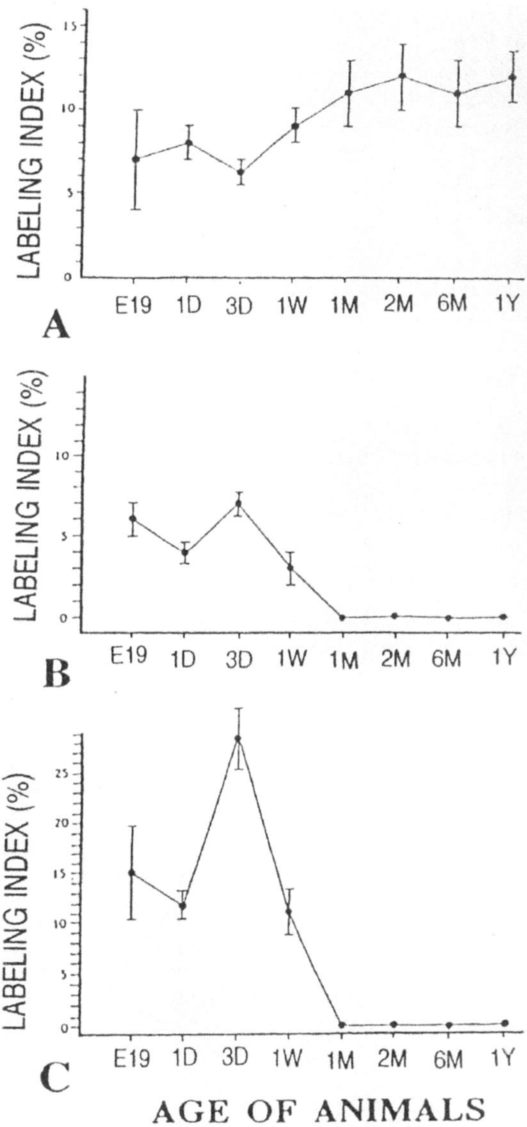

Fig. 30. Transitional curves of the labeling indices in the three layers of the central area of the cornea of aging mice labeled with ^{3}H-thymidine, showing DNA synthesis. Mean ± Standard Deviation. From Nagata, T.: Radioautographology, General and Special. In, Prog. Histochem. Cytochem. Vol. 37, No. 2, p. 192, 2002, Urban & Fischer, Jena, Germany
Fig. 30A. The labeling index in the epithelium.
Fig. 30B. The labeling index in the stroma.
Fig. 30C. The labeling index in the endothelium.

In the corneas of aging mice, DNA synthesis was observed in all 3 layers, i.e., the epithelial, stromal and endothelial layers, at perinatal stages (Gao et al. 1993). The labeled cells with ^{3}H-thymidine were localized in the epithelial cells at prenatal day 19, postnatal day 1, 14 (Fig. 27G) to 1 year, while the labeled cells in the stromal and endothelial layers were less. The labeling index of the corneal epithelial cells reached a peak at 1 month after birth and decreased to 1 year, (Fig. 30A), while the indices of the stromal (Fig. 30B) and endothelial (Fig. 30C) cells were low and reached a peak at 3 days after birth and disappeared completely from postnatal 1 month to 1 year (Nagata 1999c).

In the cilliary body, the labeled cells were located in the cilliary and pigment epithelial cells, stromal cells and smooth muscle cells from prenatal day 19 to postnatal 1 week, but no labeled cells were observed in any cell types from postnatal day 14 to 1 year (Nagata et al. 1994). The labeling indices of all the cell types in the cilliary body were at the maximum at prenatal day 19 and decreased gradually after birth reaching 0 at postnatal day 14. On the other hand, when the ocular tissues were labeled with ^{3}H-uridine, silver grains appeared over all cell types at all stages of development and aging (Toriyama 1995, Nagata 2000f). The grain counts in the retina and the pigment epithelium increased from prenatal day 9 to postnatal day 1 in the retinal cells, while they increased from prenatal day 12 to postnatal day 7 in the pigment epithelial cells (Nagata 1999a,b, Nagata et al. 1994)

4.2 The DNA synthesis in the skin

The skin which covers the surface of the animal body can histologically be divided into 3 layers, the epidermis, the dermis and the hypodermis. We studied only the epidermal cells of young salamanders after hatching to 24 months during the aging by radioautography (Nagata 1998c). The fore-limbs and hind-limbs of salamanders were composed of skeletons consisting of bones and cartilages which were covered with skeletal muscles, connective tissues and epidermis consisting of stratified squamous epithelial cells in the outermost layer. We observed both the cartilage cells in the bone and the epithelial cells in the epidermis to compare the two cell populations. The skin of a salamander consisted of epidermis and dermis or corium which was lined with connective tissue layers designated as the subcutaneous layer. The former consisted of stratified squamous epithelium, while the latter consisted of dense connective tissues. The epithelial cells in the juvenile animals at 4 weeks after hatching were cuboidal in shape and not keratinized. Radioautograms labeled with ^{3}H-thymidine at this stage showed that many cells were labeled demonstrating DNA synthesis at both the superficial and deeper layers (Fig. 27H), resulting very high labeling index. At 6 weeks after hatching, the superficial cells changed their shape from cuboidal to flattened squamous, while the deeper and basal cells remained cuboidal. The numbers of labeled cells were almost the same as the previous stage at 4 weeks, but they were localized at the basal layer. The shape of epithelial cells in juvenile animals at 8, 9, 10, and 11 weeks differentiated gradually forming the superficial corneum layer which appeared keratinized and the deeper basal layer. Radioautograms at these stages showed that the labeled cells remarkably reduced as compared with that of 4 and 6 weeks. In the adult salamanders at 8 months up to 12 months, the dermal and epidermal cells showed complete mature structure and examination of radioautograms revealed that the labeled cells were localized at only the basal cell layer and their number reached very low but at constant level. No difference was found on the morphology and labeling between the fore-limbs and hind-limbs at any stages. Comparing the labeling indices of both epidermal cells and the cartilage cells in the limbs, the labeling index of the epidermal cells was higher than the cartilage cells. The index of the dermal cells in the hind-limbs was at its maximum about 25% at 4 weeks, and fell down markedly with time from 6 weeks to 9 weeks. The labeling index of epidermal cells of the hind-limbs, on the other hand, had its maximum about 23% at 6 weeks, increasing from 20% at 4 weeks, and decreasing to about 18% at 8 weeks, then fell progressively with time, dropped to 5% at 9 weeks (Fig. 31). The labeling indices of the epidermal cells of both fore-limbs and hind-limbs were almost the same from 9 weeks to 12 months, keeping low constant level about 4-5%, but never reaching 0. These results indicated that the cutaneous cells belonged to the renewing cell population (Nagata 1998c).

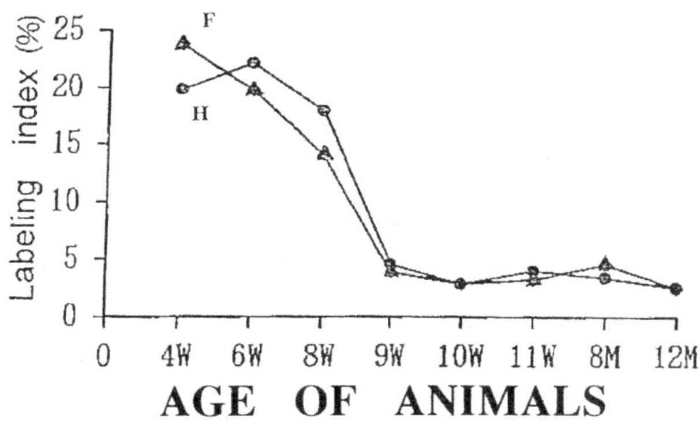

Fig. 31. Transitional curves of the labeling indices of the epithelial cells in the epidermis of the fore-limbs (F) and the hind-limbs (H) of the salamanders at various ages from 4 weeks to 12 months after hatching labeled with ^{3}H-thymidine. From Nagata, T.: Radioautographology, General and Special. In, Prog. Histochem. Cytochem. Vol. 37, No. 2, p.197, 2002, Urban & Fischer, Jena, Germany

4.3 The RNA synthesis in the sensory organs

We studied only the RNA synthesis in the chicken and mouse eyes among of the sensory organs.

4.3.1 The RNA synthesis in the eye

The RNA synthesis in the chicken eyes was studied with the ocular tissues of chicken embryos in incubation (Fig. 27D). Silver grains due to the incorporations of ^{3}H-uridine were observed over all the nuclei, cell organelles, cytoplasm of all the cells in the optic cups in development showing the RNA synthesis (Gunarso 1984a,b, Gunarso et al. 1969). Grain counting revealed that the counts gradually increased from day 2 to 7 and the numbers of silver grains were the most in the nuclei, while the numbers between the 3 portions of the optic cups, the anterior, equator and the posterior portions decreased from the anterior to the posterior at the same developmental stages (Gunarso 1984a).

On the other hand, the ocular tissues of aging mice were also labeled with ^{3}H-uridine. The silver grains demonstrating RNA synthesis appeared over all the cell types at all the stages of development and aging. The grain counts in the retina and the pigment epithelium increased from prenatal day 9 to postnatal day 1 in the retinal cells, while they increased from prenatal day 12 to postnatal day 7 in the pigment epithelial cells (Kong et al. 1992b).

On the other hand, the distribution and localization of TGF-β1 and βFGF and their mRNA in the ocular tissues of aging mice were also studied (Nagata and Kong 1998). The posterior segment of BALB/c mouse eyes from embryonic day 14, 16, 19 and postnatal 1, 3, 5, 7, 14, 28, 42 and 70 were used. For in situ hybridization, ^{35}S-labeled oligonucleotide probes for TGF-β1 and βFGF were used to detect their mRNA. Cryo-sections were picked up on glass slides which were processed for in situ hybridization and for radioautography. As the results, silver grains mainly located in the scleral layers and some in the choroidal and pigment epithelial layers, but only background level of grains were found in the whole retina. In the radioautograms from embryonic day 14 to adult mice at week 10 (day 70), the significant distribution of silver grains representing TGF-β1 mRNA was not detected in the whole retina. However, the significant silver grains were detected in scleral and chorioidal layers and mesenchymal cells at embryonic day 14, then the number of grains increased in these layers particularly in sclera from prenatal to postnatal neonate until adult (Fig. 27E). These results suggest that mRNA for TGF-β1 and βFGF were synthesized in scleral, choroidal and pigment epithelial layers, but their proteins were transferred to the target cells of the retina and elsewhere. Furthermore, it is suggested that TGF-β1 and βFGF may play important roles on retinal differentiation, development and aging, particularly during the late embryonic and newborn stages (Nagata and Kong 1998).

These results showed that RNA synthetic activities in the ocular cells changed due to the aging of individual animals.

4.4 The protein synthesis in the sensory organs

We studied only the protein synthesis in the mouse eyes among of the sensory organs.

4.4.1 The protein synthesis of the eye

The protein synthesis in the ocular tissues of aging mouse were studied in all the 3 layers of the eye, the tunica fibrosa, the tunica vasculosa and the tunica intima, or the cornea, cilliary bodies and the retina of the aging mouse at various stages after the administration of several precursors (Toriyama 1995, Nagata 1997c, 1999b,c,d, 2000e,f, 2001c, Nagata and Kong 1998, Cui et al. 2000).

The protein synthesis of the retina in aging mouse as revealed by ^{3}H-leucine incorporation demonstrated that number of silver grains in bipolar cells and photoreceptor cells was the most intense at embryonic stage and early postnatal days. The peak was 1 day after birth and decreased from 14 days to 1 year after birth. (Toriyama 1995). The protein synthesis of the cornea as revealed by ^{3}H-leucine incorporation (Nagata 1997c, 1999d, 2000f, 2001c, Nagata and Kong 1998, Cui et al. 2000) and the glycoprotein synthesis demonstrated by ^{3}H-glucosamine (Nagata et al. 1995) were also studied in several groups of aging ddY mice. Silver grains of both ^{3}H-leucine and ^{3}H-glucosamine incorporations were located in the epithelial cells, the stromal fibroblasts and the endothelial cells from prenatal day 19 to postnatal 6 months. No silver grains were observed in the lamina limitans anterior (Bowman's membrane) and the lamina limitans posterior (Descemet's membrane). The grain densities by ^{3}H-leucine incorporation in 3 layers, i. e., epithelial, stromal and endothelial layers, increased from embryonic stage to postnatal day 3 and 7, then decreased to 2 weeks and 1 year. The grain densities due to the glycoprotein synthesis with ^{3}H-glucosamine were more observed in the endothelial cells of prenatal day 19 animals, but more in the epithelial cells of postnatal day 1, 3 and 7 animals. From the results, it was shown that the glycoprotein synthetic activity in respective cell types in the cornea of mouse changed with aging of the animals.

The collagen synthesis in the ocular tissues was also demonstrated by the incorporation of ^{3}H-proline in 4 groups of mice at various ages, from prenatal day 20, postnatal day 3, 7 and 30. The results showed that the sites of ^{3}H-proline incorporation were located in the stromal fibroblasts in both cornea and the trabecular meshworks in the iridocorneal angle in prenatal and postnatal newborn mice. No silver grains were observed in the epithelial and endothelial cells. On EM RAG, silver grains were localized over the endoplasmic reticulum and Golgi apparatus of fibroblasts and over intercellular matrices consisting of collagen fibrils. From the quantitative analysis, the grain densities were more observed in the fibroblasts in postnatal day 7 animals than younger animals at fetal day 20 and postnatal day 3, 7 and 30. In the same aging groups, the grain densities were more in the cornea than the iridocorneal angle. It was concluded that the collagen synthetic activity was localized in the fibroblasts in the cornea and the trabecular meshworks in the iridocorneal angle and the activity changed with aging, reaching the maximum at postnatal day 7.

On the other hand, the distributions of some of the ophthalmological drugs used for the treatment of human glaucoma patients were examined in the ocular tissues by LM and EM RAG (Nagata 2000f). However, its relationship to the aging was not studied.

4.5 The glucide synthesis in the sensory organs

We studied the aging changes of glucide synthesis by ^{3}H-glucosamine uptake in the ocular tissues of aging mice.

4.5.1 The glucide synthesis in the eye

The glycoprotein synthesis of the cornea in aging mouse as revealed by ^{3}H-glucosamine incorporation was studied in several groups of aging mice at various ages from prenatal stages to senescence (Nagata et al. 1995). Silver grains were located in the epithelial cells, the stromal fibroblasts and the endothelial cells from prenatal day 19 to postnatal 6 months. No silver grains were observed in the lamina limitans anterior (Bowman's membrane) and the lamina limitans posterior (Descemet's membrane). On the other hand, the grain densities by ^{3}H-leucine incorporation in 3 layers, i.e., epithelial, stromal and endothelial layers, increased from embryonic stage to postnatal day 3 and 7, then decreased to week 2 and year 1. The grain densities due to the glycoprotein synthesis with ^{3}H-glucosamine were more observed in the endothelial cells of prenatal day 19 animals, but more in the epithelial cells of postnatal day 1, 3 and 7 animals. From the results, it was shown that the glycoprotein synthetic activity in respective cell types in the cornea of mouse changed with aging of the animals.

5. Macromolecular synthesis in the tumor cells

We carried out several experiments dealing with the nucleic acid synthesis in some malignant tumor cells by means of LM and EM RAG.

5.1 The DNA synthesis in the tumor cells

The DNA synthesis in nuclei and mitochondria of cultured HeLa cells, an established cell line obtained from the carcinoma of the human uterus, or IgG myeloma cells from a human patient, labeled with ^{3}H-thymidine were demonstrated (Nagata 1972b,c,d, Fujii et al. 1980). The incorporations of these precursors increased or decreased depending upon the aging of isolated cells in vitro, i.e., the days of incubation in vitro. The incorporations of ^{3}H-thymidine demonstrating DNA synthesis increased immediately after the incubation and then gradually decreased due to incubation days, reaching zero within several days.

5.2 The RNA synthesis in the tumor cells

The RNA synthesis in nuclei and mitochondria of cultured HeLa cells, or IgG myeloma cells from a human patient, labeled with ^{3}H-uridine were demonstrated (Nagata 1972b,c,d, Fujii et al. 1980). The incorporations of these precursors increased or decreased depending upon the aging of isolated cells in vitro. The incorporations of ^{3}H-uridine demonstrating RNA synthesis increased immediately after the incubation and then gradually decreased due to incubation days.

6. Conclusions

From the results obtained, it was concluded that almost all the cells in various organs of all the organ systems of experimental animals at various ages from prenatal to postnatal development and senescence during the aging of cells and individual animals demonstrated to incorporate various macromolecular precursors such as ^{3}H-thymidine, ^{3}H-uridine, ^{3}H-leucine, ^{3}H-glucose or glucosamine, ^{3}H-glycerol and others localizing in the nuclei, cytoplasmic cell organelles showing silver grains due to DNA, RNA, proteins, glucides, lipids and others those which the cells synthesized during the cell aging. Quantitative analysis carried out on the numbers of silver grains in respective cell organelles demonstrated quantitative changes, increases and decreases, of these macromolecular synthesis in connection to cell aging of respective organs. In general, DNA synthesis with ^{3}H-thymidine incorporations in most organs showed maxima at perinatal stages and gradually decreased due to aging. To the contrary, the other syntheses such as RNA, proteins, glucides and lipids increased due to aging and did not remarkably decrease until senescence. Anyway, these results indicated that macromolecular synthetic activities of respective compounds in various cells were affected from the aging of the individual animals.

Thus, the results obtained from the various cells of various organs should form a part of special radioautographology that I had formerly proposed (Nagata 1999e, 2002), i.e., application of radioautography to the aging of cells, as well as a part of special cytochemistry (Nagata 2001), as was formerly reviewed. We expect that such special radioautographology and special cytochemistry should be further developed in all the organs in the future.

7. Acknowledgments

This study was supported in part by Grant-in-Aids for Scientific Research from the Ministry of Education, Science and Culture of Japan (No. 02454564) while the author worked at Shinshu University School of Medicine by 1996 as well as Grants for Promotion of Characteristic Research and Education from the Japan Foundation for Promotion of Private Schools (No. 1997, 1998 1999, 2000) while the author worked at Nagano Women's Jr. College from 1996 to 2002. The author is also grateful to Grant-in-Aids for Scientific Research from the Japan Society for Promotion of Sciences (No. 18924034, 19924204, 20929003) while the author has been working at Shinshu Institute of Alternative Medicine and Welfare since 2005 up to the present time. The author thanks Dr. Kiyokazu Kametani, Technical Official, Department of Instrumental Analysis, Research Center for Human and Environmental Sciences, Shinshu University, for his technical assistance in electron microscopy during the course of this study.

8. References

Chen, S., Gao, F., Kotani, A., Nagata, T.: Age-related changes of male mouse submandibular gland: A morphometric and radioautographic study. Cell. Mol. Biol. 41, 117-124, 1995.

Clermont Y.: The contractime elements in the limiting membrane of the seminiferous tubules of rats. Exp. Cell Res. 15, 438-342, 1958.

Clermont, Y.: Renewal of spermatogonia in man. Amer. J. Anat. 112, 35-51, 1963.

Cui, H.: Light microscopic radioautographic study on DNA synthesis of nerve cells in the cerebella of aging mice. Cell. Mol. Biol. 41, 1139-1154, 1995.

Cui, H., Gao, F., Nagata, T.: Light microscopic radioautographic study on protein synthesis in perinatal mice corneas. Acta Histochem. Cytochem. 33, 31-37, 2000.

Duan, H., Gao, F., Li, S., Hayashi, K., Nagata, T.: Aging changes and fine structure and DNA synthesis of esophageal epithelium in neonatal, adult and old mice. J. Clin. Electron Microsc. 25, 452-453, 1992.

Duan, H., Gao, F., Li, S., Nagata, T.: Postnatal development of esophageal epithelium in mouse: a light and electron microscopic radioautographic study. Cell. Mol. Biol. 39, 309-316, 1993.

Duan, H., Gao, F., Oguchi, K., Nagata, T.: Light and electron microscopic radioautographic study on the incorporation of ^{3}H-thymidine into the lung by means of a new nebulizer. Drug Res. 44, 880-883, 1994.

Feulgen, R., Rossenbeck, H.: Mikroskopische-chemischer Nachweis einer Nukeinsaeure von Thymus der Thymonukeinsaeure Z. Physik. Chem. 135, 203-248, 1924.

Fujii, Y., Ohno, S., Yamabayashi, S., Usuda, N., Saito, H., Furuta, S., Nagata, T.: Electron microscopic radioautography of DNA synthesis in primary cultured cells from an IgG myeloma patient. J. Clin. Electr. Microsc. 13, 582-583, 1980.

Gao, F.: Study on the macromolecular synthesis in aging mouse seminiferous tubules by light and electron microscopic radioautography. Cell. Mol. Biol. 39, 659-672, 1993.

Gao, F., Toriyama, K., Nagata, T.: Light microscopic radioautographic study on the DNA synthesis of prenatal and postnatal aging mouse retina after labeled thymidine injection. Cell. Mol. Biol. 38, 661-668, 1992a.

Gao, F., Li, S., Duan, H., Ma, H., Nagata, T.: Electron microscopic radioautography on the DNA synthesis of prenatal and postnatal mice retina after labeled thymidine injection. J. Clin. Electron Microsc. 25, 721-722, 1992b.

Gao, F., Toriyama, K., Ma, H., Nagata, T.: Light microscopic radioautographic study on DNA synthesis in aging mice corneas. Cell. Mol. Biol. 39, 435-441, 1993.

Gao, F., Ma, H., Sun, L., Jin, C., Nagata, T.: Electron microscopic radioautographic study on the nucleic acids and protein synthesis in the aging mouse testis. Med. Electron Microsc. 27, 360-362, 1994.

Gao, F., Chen, S., Sun, L., Kang, W., Wang, Z., Nagata, T.: Radioautographic study of the macromolecular synthesis of Leydig cells in aging mice testis. Cell. Mol. Biol. 41, 145-150, 1995a.

Gao, F., Jin, C., Ma, H., Sun, L., Nagata, T.: Ultrastructural and radioautographic studies on DNA synthesis in Leydig cells of aging mouse testis. Cell. Mol. Biol. 41, 151-160, 1995b.

Gunarso, W.: Radioautographic studies on the nucleic acid synthesis of the retina of chick embryo. I. Light microscopic radioautography. Shinshu Med. J. 32, 231-240, 1984a.

Gunarso, W.: Radioautographic studies on the nucleic acid synthesis of the retina of chick embryo. II. Electron microscopic radioautography. Shinshu Med. J. 32, 241-248, 1984b.

Gunarso, W., Gao, F., Cui, H., Ma, H., Nagata, T.: A light and electron microscopic radioautographic study on RNA synthesis in the retina of chick embryo. Acta Histochem. 98, 309-32, 1996.

Gunarso, W., Gao, F., Nagata, T.: Development and DNA synthesis in the retina of chick embryo observed by light and electron microscopic radioautography. Cell. Mol. Biol. 43, 189-201, 1997.

Hanai, T.: Light microscopic radioautographic study of DNA synthesis in the kidneys of aging mice. Cell. Mol. Biol. 39, 81-91, 1993.

Hanai, T., Nagata, T.: Electron microscopic radioautographic study on DNA and RNA synthesis in perinatal mouse kidney. In, Radioautography in Medicine, Nagata, T., Ed., pp. 127-131, Shinshu University Press, Matsumoto, 1994a.

Hanai, T., Nagata, T.: Study on the nucleic acid synthesis in the aging mouse kidney by light and electron microscopic radioautography. In, Radioautography in Medicine, Nagata, T., Ed., pp. 209-214, Shinshu University Press, Matsumoto, 1994b.

Hanai, T., Nagata, T.: Electron microscopic study on nucleic acid synthesis in perinatal mouse kidney tissue. Med. Electron Microsc. 27, 355-357, 1994c.

Hanai, T., Usuda, N., Morita, T., Shimizu, T., Nagata, T.: Proliferative activity in the kidneys of aging mice evaluated by PCNA/cyclin immunohistochemistry. Cell. Mol. Biol. 39, 181-191, 1993.

Hayashi, K., Gao, F., Nagata, T.: Radioautographic study on ^{3}H-thymidine incorporation at different stages of muscle development in aging mice. Cell. Mol. Biol. 39, 553-560, 1993.

Ito, M.: The radioautographic studies on aging change of DNA synthesis and the ultrastructural development of mouse adrenal gland. Cell. Mol. Biol. 42, 279-292, 1996.

Ito, M., Nagata, T.: Electron microscopic radioautographic studies on DNA synthesis and ultrastructure of aging mouse adrenal gland. Med. Electron Microsc. 29, 145-152, 1996.

Izumiyama, K., Kogure, K., Kataoka, S., Nagata, T.: Quantitative analysis of glucose after transient ischemia in the gerbil hippocampus by light and electron microscope radioautography. Brain Res. 416, 175-179, 1987.

Jamieson, J. D., Palade, G. E.: Intracellular transport of secretory proteins in the pancreatic exocrine cells. J. Cell Biol. 34, 577-615, 1967.

Jin, C.: Study on DNA synthesis of aging mouse colon by light and electron microscopic radioautography. Cell. Mol. Biol. 42, 255-268, 1996.

Jin, C., Nagata, T.: Light microscopic radioautographic study on DNA synthesis in cecal epithelial cells of aging mice. J. Histochem. Cytochem. 43, 1223-1228, 1995a.

Jin, C., Nagata, T.: Electron microscopic radioautographic study on DNA synthesis in cecal epithelial cells of aging mice. Med. Electron Microsc. 28, 71-75, 1995b.

Joukura, K.: The aging changes of glycoconjugate synthesis in mouse kidney studied by ^{3}H-glucosamine radioautography. Acta Histochem. Cytochem. 29, 57-63, 1996.

Joukura, K., Nagata, T.: Aging changes of ^{3}H-glucosamine incorporation into mouse kidney observed by radioautography. Acta Histochem. Cytochem. 28, 494-494, 1995.

Joukura, K., Usuda, N., Nagata, T.: Quantitative study on the aging change of glycoconjugates synthesis in aging mouse kidney. Proc. Xth Internat. Cong. Histochem. Cytochem., Acta Histochem. Cytochem. 29, Suppl. 507-508, 1996.

Kobayashi, K., Nagata, T.: Light microscopic radioautographic studies on DNA, RNA and protein syntheses in human synovial membranes of rheumatoid arthritis patients. J. Histochem. Cytochem. 42, 982-982, 1994.

Komiyama, K., Iida, F., Furihara, R., Murata, F., Nagata, T.: Electron microscopic radioautographic study on ^{125}I-albumin in rat gastric mucosal epithelia. J. Clin. Electron Microsc. 11, 428-429, 1978.

Kong, Y.: Electron microscopic radioautographic study on DNA synthesis in perinatal mouse retina. Cell. Mol. Biol. 39, 55-64, 1993.

Kong, Y., Nagata, T.: Electron microscopic radioautographic study on nucleic acid synthesis of perinatal mouse retina. Med. Electron Microsc. 27, 366-368, 1994.

Kong, Y., Usuda, N., Nagata, T.: Radioautographic study on DNA synthesis of the retina and retinal pigment epithelium of developing mouse embryo. Cell. Mol. Biol. 38, 263-272, 1992a.

Kong, Y., Usuda, N., Morita, T., Hanai, T., Nagata, T.: Study on RNA synthesis in the retina and retinal pigment epithelium of mice by light microscopic radioautography. Cell. Mol. Biol. 38, 669-678, 1992b.

Leblond, C. P.: Localization of newly administered iodine in the thyroid gland as indicated by radioiodine. J. Anat. 77, 149-152, 1943.

Leblond, C. P.: The life history of cells in renewing systems. Am. J. Anat. 160, 113-158, 1981.

Leblond, C. P., Messier, B.: Renewal of chief cells and goblet cells in the small intestine as shown by radioautography after injection of thymidine-^{3}H into mice. Anat. Rec. 132: 247-259. 1958.

Li, S.: Relationship between cellular DNA synthesis, PCNA expression and sex steroid hormone receptor status in the developing mouse ovary, uterus and oviduct. Histochemistry 102, 405-413, 1994.

Li, S., Nagata, T.: Nucleic acid synthesis in the developing mouse ovary, uterus and oviduct studied by light and electron microscopic radioautography. Cell. Mol. Biol. 41, 185-195, 1995.

Li, S., Gao, F., Duan, H., Nagata, T.: Radioautographic study on the uptake of $^{35}SO_4$ in mouse ovary during the estrus cycle. J. Clin. Electron Microsc. 25, 709-710, 1992.
Liang, Y.: Light microscopic radioautographic study on RNA synthesis in the adrenal glands of aging mice. Acta Histochem. Cytochem. 31, 203-210, 1998.
Liang, Y., Ito, M., Nagata, T.: Light and electron microscopic radioautographic studies on RNA synthesis in aging mouse adrenal gland. Acta Anat. Nippon. 74, 291-300, 1999.
Ma, H.: Light microscopic radioautographic study on DNA synthesis of the livers in aging mice. Acta Anat. Nippon. 63, 137-147, 1988.
Ma, H., Nagata, T.: Electron microscopic radioautographic study on DNA synthesis of the livers in aging mice. J. Clin. Electron Microsc. 21, 335-343, 1988a.
Ma, H., Nagata, T.: Studies on DNA synthesis of aging mice by means of electron microscopic radioautography. J. Clin. Electron Microsc. 21, 715-716, 1988b.
Ma, H., Nagata, T.: Electron microscopic radioautographic studies on DNA synthesis in the hepatocytes of aging mice as observed by image analysis. Cell. Mol. Biol. 36, 73-84, 1990a.
Ma, H., Nagata, T.: Study on RNA synthesis in the livers of aging mice by means of electron microscopic radioautography. Cell. Mol. Biol. 36, 589-600, 1990b.
Ma, H., Nagata, T.: Collagen and protein synthesis in the livers of aging mice as studied by electron microsopic radioautography. Ann. Microsc. 1, 13-22, 2000.
Ma, H., Gao, F., Olea, M. T., Nagata, T.: Protein synthesis in the livers of aging mice studied by electron microscopic radioautography. Cell. Mol. Biol. 37, 607-615, 1991.
Matsumura, H., Kobayashi, Y., Kobayashi, K., Nagata, T.: Light microscopic radioautographic study of DNA synthesis in the lung of aging salamander, Hynobius nebulosus. J. Histochem. Cytochem. 42, 1004-1004, 1994.
Momose, Y., Nagata, T.: Radioautographic study on the intracellular localization of a hypolipidemic agent, bezafibrate, a peroxisome proliferator, in cultured rat hepatocytes. Cell. Mol. Biol. 39, 773-781, 1993a.
Momose, Y., Naito, J., Nagata, T.: Radioautographic study on the localization of an anti-allergic agent, tranilast, in the rat liver. Cell. Mol. Biol. 35, 347-355, 1989.
Momose, Y., Shibata, N., Kiyosawa, I., Naito, J., Watanabe, T., Horie, S., Yamada, J., Suga, T., Nagata, T.: Morphometric evaluation of species differences in the effects of bezafibrate, a hypolipidemic agent, on hepatic peroxisomes and mitochondria. J. Toxicol. Pathol. 6, 33-45, 1993b.
Momose, Y., Naito, J., Suzawa, H., Kanzawa, M., Nagata, T.: Radioautographic study on the intracellular localization of bezafibrate in cultured rat hepatoctyes. Acta Histochem. Cytochem. 28, 61-66, 1995.
Morita, T.: Radioautographic study on the aging change of ^{3}H-glucosamine uptake in mouse ileum. Cell. Mol. Biol. 39, 875-884, 1993.
Morita, T., Usuda, N. Hanai, T., Nagata, T.: Changes of colon epithelium proliferation due to individual aging with PCNA/cyclin immunostaining comparing with ^{3}H-thymidine radioautography. Histochemistry 101, 13-20, 1994.
Murata, F., Momose,Y. , Yoshida, K., Nagata, T.: Incorporation of ^{3}H-thymidine into the nucleus of mast cells in adult rat peritoneum. Shinshu Med. J. 25, 72-77, 1977a.
Murata, F., Momose, Y., Yoshida, K., Ohno, S., Nagata, T.: Nucleic acid and mucosubstance metabolism of mastocytoma cells by means of electron microscopic radioautography. Acta Pharmacol. Toxicol. 41, 58-59, 1977b.

Murata, F., Yoshida, K., Ohno, S., Nagata, T.: Ultrastructural and electron microscopic radioautographic studies on the mastocytoma cells and mast cells. J. Clin. Electron Microsc. 11, 561-562, 1978.
Murata, F., Yoshida, K., Ohno, S., Nagata, T.: Mucosubstances of rabbit granulocytes studied by means of electron microscopic radioautography and X-ray microanalysis. Histochemistry 61, 139-150, 1979.
Nagata, T.: On the relationship between cell division and cytochrome oxidase in the Yoshida sarcoma cells. Shinshu Med. J. 5: 383-386, 1956.
Nagata, T.: Studies on the amitosis in the Yoshida sarcoma cells. I. Observation on the smear preparation under normal conditions. Med. J. Shinshu Univ. 2: 187-198, 1957a.
Nagata, T.: Studies on the amitosis in the Yoshida sarcoma cells. II. Phase-contrast microscopic observations under normal conditions. Med. J. Shinshu Univ. 2: 199-207, 1957b.
Nagata, T.: Cell divisions in the liver of the fetal and newborn dogs. Med. J. Shinshu Univ. 4: 65-73, 1959.
Nagata, T.: A radioautographic study of the DNA synthesis in rat liver, with special reference to binucleate cells. Med. J. Shinshu Univ. 7, 17-25, 1962.
Nagata, T.: A quantitative study on the ganglion cells in the small intestine of the dog. Med. J. Shinshu Univ. 10, 1-11, 1965.
Nagata, T.: A radioautographic study on the RNA synthesis in the hepatic and the intestinal epithelial cells of mice after feeding with special reference to binuclearity. Med. J. Shinshu Univ. 11, 49-61, 1966.
Nagata, T.: On the increase of binucleate cells in the ganglion cells of dog small intestine due to experimental ischemia. Med. J. Shinshu Univ. 12, 93-113, 1967a.
Nagata, T.: A radioautographic study on the protein synthesis in the hepatic and the intestinal epithelial cells of mice, with special reference to binucleate cells. Med. J. Shinshu Univ. 12, 247-257, 1967b.
Nagata, T.: Chapter 3. Application of microspectrophotometry to various substances. In , Introduction to Microspectrophotometry. Isaka, S., Nagata, T., Inui, N., Eds., Olympus Co., Tokyo, pp. 49-155, 1972a.
Nagata, T.: Electron microscopic dry-mounting autoradiography. Proc. 4th Internat. Cong. Histochem. Cytochem. Kyoto, pp. 43-44, 1972b.
Nagata, T.: Electron microscopic radioautography of intramitochondrial RNA synthesis of HeLa cells in culture. Histochemie 32, 163-170, 1972c.
Nagata, T.: Quantitative electron microscope radioautography of intramitochondrial nucleic acid synthesis. Acta Histochem. Cytochem. 5, 201-203, 1972d.
Nagata, T.: Electron microscopic observation of target cells previously observed by phase-contrast microscopy: Electron microscopic radioautography of laser beam irradiated cultured cells. J. Clin. Electron Microsc. 17, 589-590, 1984.
Nagata, T.: Principles and techniques of radioautography. In, Histo- and Cyto-chemistry 1985, Japan Society of Histochemistry and Cytochemistry, Ed., Gakusai Kikaku Co., Tokyo, pp. 207-226, 1985.
Nagata, T.:. Electron microscopic radioautography and analytical electron microscopy. J. Clin. Electron Microsc. 24, 441-442, 1991.

Nagata, T.: Radiolabeling of soluble and insoluble compounds as demonstrated by light and electron microscopy. Recent Advances in Cellular and Molecular Biology, Wegmann, R. J., Wegmann, M. A., Eds. Peters Press, Leuven, Vol. 6, pp. 9-21, 1992.
Nagata, T.: Quantitative analysis of histochemical reactions: Image analysis of light and electron microscopic radioautograms. Acta Histochem. Cytochem. 26, 281-291, 1993a.
Nagata, T. Quantitative light and electron microscopic radioautographic studies on macromolecular synthesis in several organs of prenatal and postnatal aging mice. Chinese J. Histochem. Cytochem. 2: 106-108, 1993b.
Nagata, T.: Electron microscopic radioautography with cryo-fixation and dry-mounting procedure. Acta Histochem. Cytochem. 27: 471-489, 1994a.
Nagata, T.: Application of electron microscopic radioautography to clinical electron microscopy. Med. Electron Microsc. 27; 191-212, 1994b.
Nagata, T.: Radioautography in Medicine. Shinshu University Press, 268pp, Matsumoto, 1994c.
Nagata, T.: Radioautography, general and special. In, Histo- and Cyto-chemistry 1994, Japan Society of Histochemistry and Cytochemistry, ed, pp. 219-231, Gakusai Kikaku Co., Tokyo, 1994d.
Nagata, T., Application of electron microscopic radioautography to clinical electron microscopy. Med. Electron Microsc. 27, 191-212, 1994e.
Nagata, T.: Light and electron microscopic radioautographic study on macromolecular synthesis in digestive organs of aging mice. Cell. Mol. Biol. 41, 21-38, 1995a.
Nagata, T.: Histochemistry of the organs: Application of histochemistry to anatomy. Acta Anat. Nippon. 70, 448-471, 1995b.
Nagata, T.: Three-dimensional observation of whole mount cultured cells stained with histochemical reactions by ultrahigh voltage electron microscopy. Cell. Mol. Biol. 41, 783-792, 1995c.
Nagata, T.: Morphometry in anatomy: image analysis on fine structure and histochemical reactions with special reference to radioautography. Ital. J. Anat. 100 (Suppl. 1), 591-605, 1995d.
Nagata, T.: Technique and application of electron microscopic radioautography. J. Electron Microsc. 45, 258-274, 1996a.
Nagata, T.: Techniques of light and electron microscopic radioautography. In, Histochemistry and Cytochemistry 1996. Proc. Xth Internat. Congr. Histochem. Cytochem. Acta Histochem. Cytochem. 29 (Suppl.), 343-344, 1996b.
Nagata, T.: Remarks: Radioautographology, general and special. Cell. Mol. Biol. 42 (Suppl.), 11-12, 1996c.
Nagata, T.: On the terminology of radioautography vs. autoradiography. J. Histochem. Cytochem. 44, 1209-1209, 1996d.
Nagata, T.: Techniques and applications of microscopic radioautography. Histol. Histopathol. 12, 1091-1124, 1997a.
Nagata T.: Three-dimensional observation on whole mount cultured cells and thick sections stained with histochemical reactions by high voltage electron microscopy. In, Recent Advances in Microscopy of Cells, Tissues and Organs, Motta, P., Ed., Antonio Delfino Editore, Roma, pp. 37-44, 1997b.
Nagata, T.: Radioautographic study on collagen synthesis in the ocular tissues. J. Kaken Eye Res. 15, 1-9, 1997c.

Nagata, T.: Techniques of radioautography for medical and biological research. Braz. J. Biol. Med. Res. 31, 185-195, 1998a.
Nagata, T.: Radioautographology, the advocacy of a new concept. Braz. J. Biol. Med. Res. 31, 201-241, 1998b.
Nagata, T.: Radioautographic studies on DNA synthesis of the bone and skin of aging salamander. Bull. Nagano Women's Jr. College 6, 1-14, 1998c.
Nagata, T.: 3D observation of cell organelles by high voltage electron microscopy. Microscopy and Analysis, Asia Pacific Edition, 9, 29-32, 1999a.
Nagata, T.: Application of histochemistry to anatomy: Histochemistry of the organs, a novel concept. Proc. XV Congress of the International Federation of Associations of Anatomists, Ital. J. Anat. Embryol. 104 (Suppl. 1), 486-486, 1999b.
Nagata, T.: Aging changes of macromolecular synthesis in various organ systems as observed by microscopic radioautography after incorporation of radiolabeled precursors. Methods Find. Exp. Clin. Pharmacol. 21, 683-706, 1999c.
Nagata, T.: Radioautographic study on protein synthesis in mouse cornea. J. Kaken Eye Res. 8, 8-14, 1999d.
Nagata, T.: Radioautographology, general and special: a novel concept. Ital. J. Anat. Embryol. 104 (Suppl. 1), 487-487, 1999e.
Nagata, T.: Three-dimensional observations on thick biological specimens by high voltage electron microscopy. Image Analysis Stereolog. 19, 51-56, 2000a.
Nagata, T.: Biological microanalysis of radiolabeled and unlabeled compounds by radioautography and X-ray microanalysis. Scanning Microscopy International, 14, on line, 2000b.
Nagata, T.: Electron microscopic radioautographic study on protein synthesis in pancreatic cells of perinatal and aging mice. Bull. Nagano Women's Jr. College 8, 1-22, 2000c.
Nagata, T.: Light microscopic radioautographic study on radiosulfate incorporation into the tracheal cartilage in aging mice. Acta Histochem. Cytochem. 32, 377-383, 2000d.
Nagata, T.: Introductory remarks: Special radioautographology. Cell. Mol. Biol. 46 (Congress Suppl.), 161-161, 2000e.
Nagata, T.: Special radioautographology: the eye. J. Kaken Eye Res. 18, 1-13, 2000f.
Nagata, T.: Three-dimensional high voltage electron microscopy of thick biological specimens. Micron 32, 387-404, 2001a.
Nagata, T.: Three-dimensional and four-dimensional observation of histochemical and cytochemical specimens by high voltage electron microscopy. Acta Histochem. Cytochem. 34, 153-169, 2001b.
Nagata, T. : Special cytochemistry in cell biology. In, Internat. Rev. Cytol. Jeon, K.W., ed., Vol. 211, Chapter 2, Academic Press, New York, pp. 33-154, 2001c.
Nagata, T. : Radioautographology General and Special, In, Prog. Histochem. Cytochem., Graumann, W., Ed., Vol. 37 No. 2, Urban & Fischer, Jena, pp. 57-226, 2002.
Nagata T.: Light and electron microscopic study on macromolecular synthesis in amitotic hepatocyte mitochondria of aging mice. Cell. Mol. Biol. 49, 591-611, 2003.
Nagata, T.: X-ray microanalysis of biological specimens by high voltage electron microscopy. In, Prog. Histochem. Cytochem., Graumann, W., Ed., Vol. 39, No. 4, Urban & Fischer Verlag, Jena, pp. 185-320, 2004.
Nagata T.: Aging changes of macromolecular synthesis in the uro-genital organs as revealed by electron microscopic radioautography. Ann. Rev. Biomed. Sci. 6, 13-78, 2005.

Nagata T.: Electron microscopic radioautographic study on protein synthesis in hepatocyte mitochondria of developing mice. Ann. Microsc. 6, 43-54, 2006a.
Nagata T.: Electron microscopic radioautographic study on nucleic acids synthesis in hepatocyte mitochondria of developing mice. The Sci. World J. 6: 1583-1598, 2006b.
Nagata T.: Macromolecular synthesis in hepatocyte mitochondria of aging mice as revealed by electron microscopic radioautography. I: Nucleic acid synthesis. In, Modern Research and Educational Topics in Microscopy. Mendez-Vilas, A. and Diaz, J. Eds., Formatex Micrscopy Series No. 3, Vol. 1, Formatex, Badajoz, Spain, pp. 245-258, 2007a.
Nagata T.: Macromolecular synthesis in hepatocyte mitochondria of aging mice as revealed by electron microscopic radioautography. II: Protein synthesis. In, Modern Research and Educational Topics in Microscopy. Mendez-Vilas, A. and Diaz, J. eds., Formatex Micrscopy Series No. 3, Vol. 1, Formatex, Badajoz, Spain, pp. 259-271, 2007b.
Nagata, T.: Electron microscopic radioautographic study on macromolecular synthesis in hepatocyte mitochondria of aging mouse. J. Cell Tissue Res. 7, 1019-1029, 2007c.
Nagata, T.; Electron microscopic radioautographic study on nucleic acids synthesis in hepatocyte mitochondria of developing mice. Trends Cell Molec. Biol. 2, 19-33, 2007d.
Nagata, T.; Aging changes of macromolecular synthesis in the mitochondria of mouse hepatocytes as revealed by microscopic radioautography. Ann. Rev. Biomed. Sci. 9, 30-36, 2007e.
Nagata, T.: Radioautographology, Bull. Shinshu Institute Alternat. Med. 2, 3-32, 2007f.
Nagata, T.: Electron microscopic radioautographic study on mitochondrial DNA synthesis in adrenal cortical cells of developing mice. J. Cell. Tis. Res. 8, 1303-1312, 2008a.
Nagata T.: Electron microscopic radioautographic study on mitochondrial DNA synthesis in adrenal cortical cells of developing and aging mice. The Sci. World J. 8, 683-697. 2008b.
Nagata, T.: Sexual difference between the macromolecular synthesis of hepatocyte mitochondria in male and female mice in aging as revealed by electron microscopic radioautography. Chapter 22. In, Women and Aging: New Research, H. T. Bennninghouse, A. D. Rosset, Eds. Nova Biomed. Books, New York, pp. 461-487, 2009a
Nagata, T.: Protein synthesis in hepatocytes of mice as revealed by electron microscopic radioautography. In, Protein Biosynthesis. Esterhouse, T. E. and Petrinos, L. B., Eds., Nova Biomed. Books, New York, pp. 133-161, 2009b.
Nagata, T.: Electron microscopic radioatuographic studies on macromolecular synthesis in mitochondria of various cells. 18EMSM Conference Proc. 9th Asia-Pacific Microscopy Conference (APMC9), Kuala Lumpur, Malaysia, pp. 48-50, 2009c.
Nagata, T.: Recent studies on macromolecular synthesis labeled with ^{3}H-thymidine in various organs as revealed by electron microscopic radioautography. Current Radiopharmaceutics 2, 118-128, 2009d.
Nagata, T.: Electron microscopic radioautographic study on mitochondrial DNA synthesis in adrenal medullary cells of developing and aging mice. J. Cell Tissue Res. 9, 1793-1802, 2009e.
Nagata, T.: Applications of high voltage electron microscopy to thick biological specimens. Ann. Microsc. 9, 4-40, 2009f.

Nagata, T.: Electron microscopic radioautographic study on DNA synthesis of mitochondria in adrenal medullary cells of aging mice. Open Anat. J. 1, 14-24, 2009g.

Nagata, T.: Electron microscopic radioautographic studies on macromolecular synthesis in mitochondria of animal cells in aging. Ann. Rev. Biomed. Sci. 11, 1-17, 2009h.

Nagata, T.: Electron microscopic radioautographic studies on macromoleclular synthesis in mitochondria of some organs in aging animals. Bull. Shinshu Inst. Alternat. Med. Welfare 4, 15-38, 2009i.

Nagata, T.: Electron microscopic radioautographic study on mitochondrial DNA synthesis in adreno-cortical cells of aging ddY mice. Bull. Shinshu Inst. Alternat. Med. Welfare 4, 51-66, 2009j.

Nagata T.: Electron microscopic radioautographic study on mitochondrial RNA synthesis in adrenocortical cells of aging mice. Open Anat J. 2, 91-97, 2010a.

Nagata T. Electron microscopic radioautographic study on mitochondrial RNA synthesis in adrenal medullary cells of aging and senescent mice. J Cell Tissue Res. 10, 2213-2222, 2010b.

Nagata, T.: Macromolecular synthesis in the livers of aging mice as revealed by electron microscopic radioautography. In, Prog. Histochem. Cytochem., Sasse, D., Ed., Elsevier, Amsterdam, Boston, London, New York, Oxford, Paris, Philadelphia, San Diego, St. Louis, Vol. 45, No. 1, pp. 1-80, 2010c.

Nagata, T.: Electron microscopic radioautographic study on protein synthesis of mitochondria in adrenal medullary cells of aging mice. Bulletin Shinshu Inst Alternat Med Welfare 5, 25-37, 2010d.

Nagata, T.: Electron microscopic radioautographic study on mitochondrial RNA synthesis in adrenal cortical and medullary cells of aging mice. J. Biomed. Sci. Enginer. 4, 219-232, 2010e.

Nagata, T.: Electron microscopic radioautographic study on protein synthesis of mitochondria in adrenal cortical cells of aging mice. Bulletin Shinshu Inst. Alternat. Med. Welfare 5, 38-52, 2010f.

Nagata T.: Electron microscopic radioautographic study on mitochondrial DNA, RNA and protein synthesis in adrenal cells of aging mice. Formatex Microscopy Series No. 3, Vol. 3, Formatex, Badajoz, Spain, in press, 2010g.

Nagata, T.: Electron microscopic radioautographic studies on macromolecular synthesis in mitochondria of animal cells in aging. Ann. Rev. Biomed. Sci. 12, 1-29, 2010h.

Nagata, T., Cui, H., Gao, F.: Radioautographic study on glycoprotein synthesis in the ocular tissues. J. Kaken Eye Res. 13, 11-18, 1995.

Nagata, T., Cui, H., Kong, Y.: The localization of TGF-b1 and its mRNA in the spinal cords of prenatal and postnatal aging mice demonstrated with immunohistochemical and in situ hybridization techniques. Bull. Nagano Women's Jr. College, 7, 75-88, 1999a.

Nagata, T., Cui, H., Liang, Y.: Light microscopic radioautographic study on the protein synthesis in the cerebellum of aging mouse. Bull. Nagano Women's Jr. College, 9, 41-60 (2001).

Nagata, T., Fujii, Y., Usuda, N.: Demonstration of extranuclear nucleic acid synthesis in mammalian cells under experimental conditions by electron microscopic radioautography. Proc. 10th Internat. Congr. Electr. Microsc. 2, 305-306, 1982b.

Nagata, T., Hirano, I., Shibata, O., Nagata, T.: A radioautographic study on the DNA synthesis in the hepatic and the pancreatic acinar cells of mice during the postnatal growth, with special reference to binuclearity. Med. J. Shinshu Univ. 11, 35-42, 1966.

Nagata, T., Ito, M., Chen, S.: Aging changes of DNA synthesis in the submandibular glands of mice as observed by light and electron microscopic radioautography. Ann. Microsc. 1, 4-12, 2000a.

Nagata, T. Ito, M., Liang, Y.: Study of the effects of aging on macromolecular synthesis in mouse steroid secreting cells using microscopic radioautography. Methods Find. Exp. Clin. Pharmacol. 22, 5-18, 2000b.

Nagata, T., Iwadare, I., Murata, F.: Electron microscopic radioautography of nucleic acid synthesis in cultured cells treated with several carcinogens. Acta Pharmacol. Toxicol. 41, 64-65, 1977c.

Nagata, T., Kawahara, I.: Radioautographic study of the synthesis of sulfomucin in digestive organs of mice. J. Trace Microprobe Analysis 17, 339-355, 1999.

Nagata, T., Kawahara, I., Usuda, N., Maruyama, M., Ma, H.: Radioautographic studies on the glycoconjugate synthesis in the gastrointestinal mucosa of the mouse. In, Glycoconjugate in Medicine, Ohyama, M., Muramatsu, T., Eds, pp. 251-256, Professional Postgrad. Service, Tokyo, 1988a.

Nagata, T., Kong, Y.: Distribution and localization of TGFb1 and bFGF, and their mRNAs in aging mice. Bull. Nagano Women's Jr. College 6, 87-105, 1998.

Nagata, T., Ma, H., Electron microscopic radioautographic study on mitochondrial DNA synthesis in hepatocytes of aging mouse. Ann. Microsc. 5, 4-18, 2005a.

Nagata, T., Ma, H., Electron microscopic radioautographic study on RNA synthesis in hepatocyte mitochondria of aging mouse. Microsc. Res. Tech. 67, 55-64, 2005b.

Nagata, T., Momoze, S.: Aging changes of the amitotic and binucleate cells in dog livers. Acta Anat. Nipponica 34, 187-190, 1959.

Nagata, T., Morita, T., I. Kawahara, I.: Radioautographic studies on radiosulfate incorporation in the digestive organs of mice. Histol. Histopathol. 14, 1-8, 1999b.

Nagata, T., Murata, F.: Electron microscopic dry-mounting radioautography for diffusible compounds by means of ultracryotomy. Histochemistry 54, 75-82, 1977.

Nagata, T., Murata, F., Yoshida, K., Ohno, S., Iwadare, N.: Whole mount radioautography of cultured cells as observed by high voltage electron microscopy. Proc. Fifth Internat. Conf. High Voltage Electron Microsc. 347-350, 1977d.

Nagata, T., Nawa, T.: A modification of dry-mounting technique for radioautography of water-soluble compounds. Histochemie 7, 370-371, 1966a.

Nagata, T., Nawa, T.: A radioautographic study on the nucleic acids synthesis of binucleate cells in cultivated fibroblasts of chick embryos. Med. J. Shinshu Univ. 11, 1-5, 1966b.

Nagata, T., Nawa, T., Yokota, S.: A new technique for electron microscopic dry-mounting radioautography of soluble compounds. Histochemie 18, 241-249, 1969.

Nagata, T., Nishigaki, T., Momose, Y.: Localization of anti-allergic agent in rat mast cells demonstrated by light and electron microscopic radioautography. Acta Histochem. Cytochem. 19, 669-683, 1986b.

Nagata, T., Ohno, S., Kawahara, I., Yamabayashi, S., Fujii, Y., Murata, F.: Light and electron microscopic radioautography of nucleic acid synthesis in mitochondria and peroxisomes of rat hepatic cells during and after DEHP administration. Acta Histochem. Cytochem. 16, 610-611, 1979.

Nagata, T., Ohno, S., Murata, F.: Electron microscopic dry-mounting radioautography for soluble compounds. Acta Phamacol. Toxicol. 41, 62-63, 1977a.

Nagata, T., Ohno, S., Yoshida, K., Murata, F.: Nucleic acid synthesis in proliferating peroxisomes of rat liver as revealed by electron microscopical radioautography. Histochem. J. 14, 197-204, 1982a.
Nagata, T., Olea, M. T.: Electron microscopic radioautographic study on the protein synthesis in aging mouse spleen. Bull. Nagano Women's Jr. College 7, 1-9, 1999.
Nagata, T., Shibata, O., Omochi, S.: A new method for radioautographic osbservation on isolated cells. Histochemie 2, 255-259, 1961
Nagata, T., Shibata, O., Nawa, T.: Simplified methods for mass production of radioautograms. -Acta Anat. Nippon.42, 162-166, 1967a.
Nagata, T., Shibata, O., Nawa, T.: Incorporation of tritiated thymidine into mitochondrial DNA of the liver and kidney cells of chickens and mice in tissue culture. Histochemie 10, 305-308, 1967b.
Nagata, T., Shimamura, K., Onozawa, M., Kondo, T., Ohkubo, K., Momoze, S.: Relationship of binuclearity to cell function in some organs. I. Frequencies of binucleate cells in some organs of toads in summer and winter. Med. J. Shinshu Univ. 5, 147-152, 1960a.
Nagata, T., Shimamura, K., Kondo, T., Onozawa, M., Momoze, S., Okubo, M.: Relationship of binuclearity to cell function in some organs. II. Variation of frequencies of binucleate cells in some organs of dogs owing to aging. Med. J. Shinshu Univ. 5, 153-158, 1960b.
Nagata, T., Steggerda, F. R.: Histological study on the deganglionated small intestine of the dog. Physiologist 6, 242-242, 1963.
Nagata, T., Steggerda, F. R.: Observations on the increase of binucleate cells in the ganglion cells of the dog's intestine due to experimental ischemia. Anat. Rec. 148, 315-315, 1964.
Nagata, T., Toriyama, K., Kong, Y., Jin, C., Gao, F.: Radioautographic study on DNA synthesis in the ciliary bodies of aging mice. J. Kaken Eye Res.12, 1-11, 1994.
Nagata, T., Usuda, N.: Image processing of electron microscopic radioautograms in clinical electron microscopy. J. Clin. Electron. Microsc. 18, 451-452, 1985.
Nagata, T., Usuda, N.: Studies on the nucleic acid synthesis in pancreatic acinar cells of aging mice by means of electron microscopic radioautography. J. Clin. Electron Microsc. 19, 486-487, 1986.
Nagata, T., Usuda, N.: Electron microscopic radioautography of protein synthesis in pancreatic acinar cells of aging mice. Acta Histochem. Cytochem. 26, 481-481, 1993a.
Nagata, T., Usuda, N.: In situ hybridization by electron microscopy using radioactive probes. J. Histochem. Cytochem. 41, 1119-1119, 1993b.
Nagata, T., Usuda, N., Ma, H.: Electron microscopic radioautography of nucleic acid synthesis in pancreatic acinar cells of prenatal and postnatal aging mice. Proc. XIth Intern. Cong. Electr. Microsc. 3, 2281-2282, 1984.
Nagata, T., Usuda, N., Ma, H.: Electron microscopic radioautography of lipid synthesis in pancreatic cells of aging mice. J. Clin. Electr. Microsc. 23, 841-842, 1990.
Nagata, T., Usuda, N., Maruyama, M., Ma, H.: Electron microscopic radioautographic study on lipid synthesis in perinatal mouse pancreas. J. Clin. Electr. Microsc. 21, 756-757, 1988b.

Nagata, T., Usuda, N., Suzawa, H., Kanzawa, M.: Incorporation of ^{3}H-glucosamine into the pancreatic cells of aging mice as demonstrated by electron microscopic radioautography. J. Clin. Electron Microsc. 25, 646-647, 1992.

Nagata, T., Yamabayashi, S.: Intracellular localization of ^{3}H-befunolol by means of electron microscopic radioautography of cryo-fixed ultrathin sections. J. Clin. Electron Microsc. 16, 737-738, 1983.

Nagata, T., Yoshida, K., Murata, F.: Demonstration of hot and cold mercury in the human thyroid tissues by means of radioautography and chemography. Acta Pharmacol. Toxicol. 41, 60-61, 1977b.

Nagata, T., Yoshida, K., Ohno, S., Murata, F.: Ultrastructural localization of soluble and insoluble ^{3}H-methyl prednisolone as revealed by electron microscopic dry-mounting radioautography. Proc. 9th Internat. Congr. Electr. Microsc. 2, 40-41, 1978b.

Nishigaki, T., Momose, Y., Nagata, T.: Light microscopic radioautographic study of the localization of anti-allergic agent, tranilast, in rat mast cells. Histochem. J. 19, 533-536, 1987.

Nishigaki, T., Momose, Y., Nagata, T.: Electron microscopic radioautographic study of the localization of an anti-allergic agent, tranilast, in rat mast cells. Cell. Mol. Biol. 36, 65-71, 1990a.

Nishigaki, T., Momose, Y., Nagata, T.: Localization of the anti-allergic agent tranilast in the urinary bladder of rat as demonstrated by light microscopic radioautography. Drug Res. 40, 272-275, 1990b.

Oguchi, K., Nagata, T.: A radioautographic study of activated satellite cells in dystrophic chicken muscle. In, Current Research in Muscular Dystrophy Japan. The Proc. Ann. Meet. Muscular Dystrophy Res. 1980, pp. 16-17, Ministry of Welfare of Japan, Tokyo, 1980.

Oguchi, K., Nagata, T.: Electron microscopic radioautographic observation on activated satellite cells in dystrophy chickens. In, Clinical Studies on the Etiology of Muscular Dystrophy. Annual Report on Neurological Diseases 1981, pp. 30-33, Ministry of Welfare of Japan, Tokyo, 1981.

Ohno, S., Fujii, Y., Usuda, N., Endo, T., Hidaka, H., Nagata, T.: Demonstration of intracellular localization of calmodulin antagonist by wet-mounting radioautography. J. Electron Microsc. 32, 1-12, 1983.

Ohno, S., Fujii, Y., Usuda, N., Nagata, T., Endo, T., Tanaka, T., Hidaka, H.: Intracellular localization of calmodulin antagonists (W-7). In, Calmodulin and intracellular Ca^{2+} receptors. Kakiuchi, S., Hidaka, H, Means, A. R., Eds., pp. 39-48, Plenum Publishing Co., New York, 1982.

Olea, M. T.: An ultrastructural localization of lysosomal acid phosphatase activity in aging mouse spleen: a quantitative X-ray microanalytical study. Acta Histochem. Cytochem. 24, 201-208, 1991.

Olea, M. T., Nagata, T.: X-ray microanalysis of cerium in mouse spleen cells demonstrating acid phosphatase activity using high voltage electron microscopy, Cell. Mol. Biol. 37, 155-163, 1991.

Olea, M. T., Nagata, T. : Simultaneous localization of ^{3}H-thymidine incorporation and acid phosphatase activity in mouse spleen: EM radioautography and cytochemistry. Cell. Mol. Biol. 38, 115-122, 1992a.

Olea, M. T., Nagata, T.: A radioautographic study on RNA synthesis in aging mouse spleen after ^{3}H-uridine labeling in vitro. Cell. Mol. Biol. 38, 399-405, 1992b.

Oliveira, S. F., Nagata, T., Abrahamsohn, P. A., Zorn, T. M. T.: Electron microscopic radioautographic study on the incorporation of ^{3}H-proline by mouse decidual cells. Cell. Mol. Biol. 37, 315-323, 1991.

Oliveira, S. F., Abrahamsohn, P. A., Nagata, T., Zorn, T. M. T.: Incorporation of ^{3}H-amino acids by endometrial stromal cells during decidualization in the mouse. A radioautographical study. Cell. Mol. Biol. 41, 107-116, 1995.

Pearse, A. G. E.: Histochemistry, Theoretical and Applied. 4th Ed. Vol. 1. 439 pp., 1980, Vol. 2. 1055 pp., 1985, Vol. 3. Ed. with P. Stoward, 728 pp. Churchill Livingstone, Edinburgh, London and New York, 1991.

Sakai, Y., Ikado, S., Nagata, T.: Electron microscopic radioautography of satellite cells in regenerating muscles. J. Clin. Electr. Microsc. 10, 508-509, 1977.

Sato, A.: Quantitative electron microscopic studies on the kinetics of secretory granules in G-cells. Cell Tissue Res. 187, 45-59, 1978.

Sato, A., Iida, F., Furihara, R., Nagata, T.: Electron microscopic raioautography of rat stomach G-cells by means of ^{3}H-amino acids. J. Clin. Electron Microsc. 10, 358-359, 1977.

Shimizu, T., Usuda, N., Yamanda, T., Sugenoya, A., Iida, F.: Proliferative activity of human thyroid tumors evaluated by proliferating cell nuclear antigen/cyclin immnohistochemical studies. Cancer 71, 2807-2812, 1993.

Sun, L.: Age related changes of RNA synthesis in the lungs of aging mice by light and electron microscopic radioautography. Cell. Mol. Biol. 41, 1061-1072, 1995.

Sun, L., Gao, F., Duan, H., Nagata, T.: Light microscopic radioautography of DNA synthesis in pulmonary cells in aging mice. In, Radioautography in Medicine, Nagata, T. Ed., pp. 201-205, Shinshu University Press, Matsumoto, 1994.

Sun, L., Gao, F., Nagata, T.: Study on the DNA synthesis of pulmonary cells in aging mice by light microscopic radioautography. Cell. Mol. Biol. 41, 851-859, 1995a.

Sun, L., Gao, F., Jin, C., Duan, H., Nagata, T.: An electron microscopic radioautographic study on the DNA synthesis of pulmonary tissue cells in aging mice. Med. Electron. Microsc. 28, 129-131, 1995b.

Sun, L., Gao, F., Jin, C., Nagata, T.: DNA synthesis in the tracheae of aging mice by means of light and electron microscopic radioautography. Acta Histochem. Cytochem. 30, 211-220, 1997a.

Sun, L., Gao, F., Nagata, T.: A Light Microscopic radioautographic study on protein synthesis in pulmonary cells of aging mice. Acta Histochem. Cytochem. 30, 463-470, 1997b.

Suzuki, K., Imada, T., Gao, F., Ma, H., Nagata, T.: Radioautographic study of benidipine hydrochloride: localization in the mesenteric artery of spontaneously hypertensive rat. Drug Res. 44, 129-133, 1994.

Terauchi, A., Mori, T., Kanda, H., Tsukada, M., Nagata, T.: Radioautographic study of 3H-taurine uptake in mouse skeletal muscle cells. J. Clin. Electron Microsc. 21, 627-628, 1988.

Terauchi, A., Nagata, T.: Observation on incorporation of ^{3}H-taurine in mouse skeletal muscle cells by light and electron microscopic radioautography. Cell. Mol. Biol. 39, 397-404, 1993.

Terauchi, A., Nagata, T.: In corporation of ^{3}H-taurine into the blood capillary cells of mouse skeletal muscle. Radioautography in Medicine, Nagata, T. ed., Shinshu University Press, Matsumoto, 1994.

Toriyama, K.: Study on the aging changes of DNA and protein synthesis of bipolar and photo-receptor cells of mouse retina by light and electron microscopic radioautography. Cell. Mol. Biol. 41, 593-601, 1995.

Tsukahara, S., Yoshida, K., Nagata, T.: A radioautographic study on the incorporation of ^{14}C-bupranolol (beta-blocking agent) into the rabbit eye. Histochemistry 68, 237-244, 1980.

Usuda, N., Nagata, T.: Electron microscopic radioautography of acyl-CoA mRNA by in situ hybridization. J. Clin. Electron Microsc. 25, 332-333, 1992.

Usuda, N., Nagata, T.: The immunohistochemical and in situ hybridization studies on hepatic peroxisomes. Acta Histochem. Cytochem. 28, 169-172, 1995.

Usuda, N., Hanai, T., Morita, T., Nagata, T.: Radioautographic demonstration of peroxisomal acyl-CoA oxidase mRNA by in situ hybridization. In, Recent advances in cellular and molecular biology, Vol. 6. Molecular biology of nucleus, peroxisomes, organelles and cell movement. Wegmann, R. J., Wegmann, M., Eds, pp.181-184, Peeters Press, Leuven, 1992.

Uwa, H., Nagata, T.: Cell population kinetics of the scleroblast during ethisterone-induced anal-fin process formation in adult females of the Medaka. Dev. Growth Different. 9, 693-694, 1976.

Watanabe, I., Makiyama, M. C. K., Nagata, T.: Electron microscopic radioautographic observation of the submandibular salivary gland of aging mouse. Acta Microscopica 6. 130-131, 1997.

Yamabayashi, S., Gunarso, W., Tsukahara, S., Nagata, T.: Incorporation of ^{3}H-befunolol (beta blocking agent) into melanin granules of ocular tissues in the pigmented rabbits. I. Light microscopic radioautography. Histochemistry 73, 371-375, 1981.

Yamada, A. T.: Timely and topologically defined protein synthesis in the periimplanting mouse endometrium revealed by light and electron microscopic radioautography. Cell. Mol. Biol. 39, 1-12, 1993.

Yamada, A., Nagata, T.: Ribonucleic acid and protein synthesis in the uterus of pregnant mouse during activation of implantation window. Med. Electron Microsc. 27, 363-365, 1992a.

Yamada, A., Nagata, T.: Light and electron microscopic raioautography of DNA synthesis in the endometria of pregnant-ovariectomized mice during activation of implantation window. Cell. Mol. Biol. 38, 763-774, 1992b.

Yamada, A., Nagata, T.: Light and electron microscopic raioautography of RNA synthesis of peri-implanting pregnant mouse during activation of receptivity for blastocyst implantation. Cell. Mol. Biol. 38, 211-233, 1993.

Yoshinaga, K.: Uterine receptivity for blastcyst implantation. Ann. N. Y. Acad. Sci. USA, 541, 424-431, 1988.

Yoshizawa, S., Nagata, A., Honma, T., Oda, M., Murata, F., Nagata, T.: Study of ethionine pancreatitis by means of electron microscopic radioautography. J. Clin. Electron Microsc. 7, 349-350, 1974.

Yoshizawa, S., Nagata, A., Honma, T., Oda, M., Murata, F., Nagata, T.: Radioautographic study of protein synthesis in pancreatic exocrine cells of alcoholic rats. J. Clin. Electron. Microsc. 10, 372-373, 1977.

9

Cell Senescence as Observed by Electron Microscopic Radioautography

Tetsuji Nagata[1,2]
[1]*Department of Anatomy and Cell Biology, Shinshu University School of Medicine, Matsumoto,*
[2]*Shinshu Institute of Alternative Medicine and Welfare, Nagano, Japan*

1. Introduction

The term "cell senescence" initially means how the cells change when they get old due to their aging. It contains 2 meanings, one how a cell changes when it is isolated from in vivo original animals or plants such as in vitro cells in cell culture, while the other means how all the cells of an animal or a plant change in vivo due to the aging of the individual animal or plant. In order to study the cell senescence, we have 2 research techniques to clarify how cells get old, i.e., morphological technique and functional technique. The former employs microscopy either light microscope or electron microscope to observe the structure of cells and tissues, while the latter employs functional techniques such as physiological or biochemical to observe either the electric activities or chemical components. Since I am anatomist and had learned morphological technique, I employed to observe cells by light and electron microscopy. I had first studied the meaning of cell senescence many years ago (more than 50 years) how a cell changed when it was isolated from original experimental animals such as mice and rats by cell culture (Nagata 1956, 1957a,b), and then moved to the study on the latter cell senescence, i.e., how all the cells of an experimental animal change in vivo due to the senescence of the individual animal bodies (Nagata 1959, 1962, Nagata and Momoze 1959, Nagata et al. 1960a,b).

Recently, I have been studying the senescent changes from the viewpoint of the cell nutrients which were incorporated and synthesized into various cells in individual animals during their senescence (Nagata 2010c). Therefore, this article deals with the cell senescence of animal cells in vivo, how the metabolism, i.e., incorporations and syntheses of respective nutrients, the macromolecular precursors, in various kinds of cells change due to the senescence of individual experimental animals such as mice and rats by means of microscopic radioautography. The incorporations and syntheses of various nutrients such as DNA, RNA, proteins, glucides, lipids and others in various kinds of cells of various organ in respective organ systems such as skeletal, muscular, circulatory, digestive, respiratory, urinary, reproductive, endocrine, nervous and sensory systems should be reviewed referring many original papers already published from our laboratory.
When I was asked from the publisher to contribute this article and to edit the articles submitted

from around 40 authors throughout the world, I initially intended to write only one chapter entitled "Senescence as Analyzed by Microscopic Radioautography." However, when I was almost finishing this article in one chapter which consisted of the text around 140 pages and around 30 figures which might become over 170 pages altogether, the publisher requested me to submit only one chapter within exactly 16, 18, 20, 22, 24 or 26 pages including both text and figures. Since I could not agree with them in submitting such a short chapter dealing with all the research results of myself and my associates, I insisted to submit the large article in one chapter including so much information as the results of my research on senescence for more 50 years since 1955 up to the present time 2011, they requested me to divide the original one chapter into 3 or 4 short chapters.

Thus, I tried to divide the original draft into 4 chapters, including the foundations of radioautography and the results of its application to all the organ systems, i.e., skeletal, muscular, circulatory, digestive, respiratory, urinary, reproductive, endocrine, nervous and sensory organs altogether. As the results of dividing the initial one chapter into 4, the final chapters were consisted of more than 26 pages as the publisher requested. However, I am going to submit the longer chapters to the publisher and insist to publish those chapters as they are, otherwise, I would rather prefer to withdraw them from this book and would like to contribute them to any other suitable publishers in the world.

This first chapter deals with the methodology of microscopic radioautography as well as the first parts of the applications of radioautography to the organ systems, i.e. the organ of movement (skeletal and muscular system) and the circulatory system.

1.1 Method in microscopic radioautography

For the purpose of observing the localizations of the incorporations and syntheses of various nutrients synthesizing macromolecules in the human and animal bodies such as DNA, RNA, proteins, glucides and lipids in various kinds of cells of various organ in respective organ systems such as skeletal, muscular, circulatory, digestive, respiratory, urinary, reproductive, endocrine, nervous and sensory systems, we employed the specific techniques developed in our laboratory during these 50 years (Nagata 2002). The technique is designated as radioautography using RI-labeled compounds. Some scientists use another term autoradiography which is used as the synonym to radioautogarphy. However, the author prefers the term radioautography because of the etymological reason (Nagata 1996b). To demonstrate the localizations of macromolecular synthesis by using such RI-labeled precursors as ^{3}H-thymidine for DNA, ^{3}H-uridine for RNA, ^{3}H-leucine for proteins, ^{3}H-glucosamine or $^{35}SO_4$ for glucides and ^{3}H-glycerol for lipids are divided into macroscopic radioautography and microscopic radioautography. The techniques employ both the physical techniques using RI-labeled compounds and the histochemical techniques treating tissue sections by coating sections containing RI-labeled precursors with photographic emulsions and processing for exposure and development. Such techniques can demonstrate both the soluble compounds diffusible in the cells and tissues and the insoluble compounds bound to the macromolecules (Nagata 1972b). As the results, specimens prepared for EM RAG (electron microscopic radioautography) are very thick than conventional EM specimens and should be observed with high voltage electron microscopes in order to obtain better transmittance and resolution (Nagata 2001a,b). Such radioautographic techniques in details should be referred to other literature (Nagata 2002). On the other hand, the systematic results obtained by radioautography should be designated as radioauographology which means the science of radioautography (Nagata 1998b, 1999e, 2000e). This article deals with the results dealing with the radioautographic changes of individual cell by aging that should be included in radioautographology.

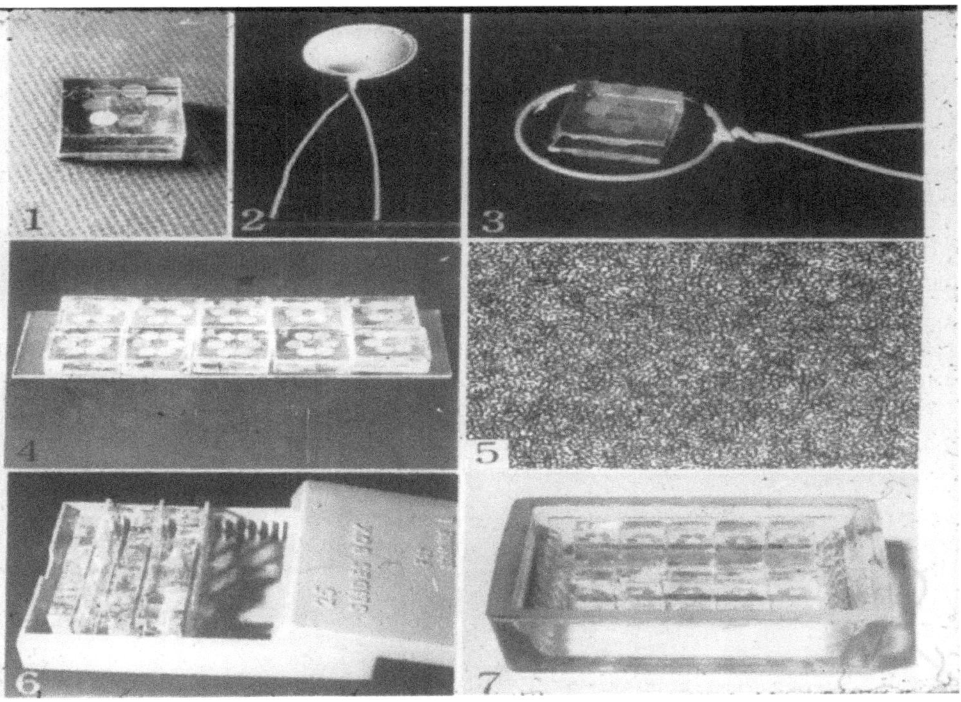

Explanation of Figures. From Nagata, T.: Acta Microsc. Vol. 6: Suppl. B. p. 42, 1997a. Brazil. Soc. Electron Microsc., San Paulo, Brazil

Fig. 1. Photographs showing the standard procedure for preparing EMRAG (electron microscopic radioautograms) by the wire-loop method (Nagata 1982, 1985).
Fig. 1-1. Six grid meshes carrying sections are placed on a square glass block.
Fig. 1-2. A large wire-loop is dipped into the melted radioautographic emulsion and a thin film of the emulsion is obtained.
Fig. 1-3. The emulsion film with the wire-loop (Fig. 1-2) is applied horizontally to the glass block on which 6 grid meshes were placed (Fig. 1-1).
Fig. 1-4. Ten glass blocks carrying 6 grid meshes each (Fig. 1-3) are attached on a glass slide with Scotch tape.
Fig. 1-5. The emulsion film picked up at random is checked by transmission electron microscopy before exposure. Note the monolayer arrangement of the silver bromide crystals in Konica NR-H2 emulsion in this figure. x6,000.
Fig. 1-6. Several glass slides, each carrying 10 glass blocks with grid meshes, are stored in a light tight slide box kept in a refrigerator at 4°C for exposure.
Fig. 1-7. All the grid meshes on glass blocks are developed, fixed and stained simultaneously.

1.2 Macromolecular synthesis

The human body as well as the bodies of any experimental animals such as mice and rats consist of various macromolecules. They are classified into nucleic acids (both DNA and RNA), proteins, glucides and lipids, according to their chemical structures. These macromolecules can be demonstrated by specific histochemical staining for respective molecules such as Feulgen reaction (Feulgen and Rossenbeck 1924) which stains all the DNA contained in the cells. Each compounds of macromolecules such as DNA, RNA, proteins, glucides, lipids can be demonstrated by respective specific histochemical stainings (Pearse 1991) and such reactions can be quantified by microscpectrophotometry using specific wave-lengths demonstrating the total amount of respective compounds (Nagata 1972a). To the contrary, radioautography can only demonstrate the newly synthesized macromolecules such as synthetic DNA or RNA or proteins depending upon the RI-labeled precursors incorporated specifically into these macromolecules such as ^{3}H-thymidine into DNA or ^{3}H-uridine into RNA or ^{3}H-amino acid into proteins (Nagata 2002).

Concerning to the newly synthesized macromolecules, the results of recent studies in our laboratory by the present author and co-workers should be reviewed in this article according to the classification of macromolecules as follows.

1.2.1 The DNA synthesis

The DNA (deoxyribonucleic acid) contained in cells can be demonstrated either by morphological histochemical techniques staining tissue sections such as Feulgen reaction (Feulgen and Rossenbeck 1924) or by biochemical techniques homogenizing tissues and cells. To the contrary, the synthetic DNA or newly synthesized DNA but not all the DNA can be detected as macromolecular synthesis together with other macromolecules such as RNA or proteins in various organs of experimental animals by either morphological or biochemical procedures employing RI-labeled precursors. We have studied the sites of macromolecular synthesis in almost all the organs of mice during their aging from prenatal to postnatal development to senescence by means of microscopic radioautography, one of the morphological methods (Nagata 1992, 1994b,c,d, 1996a,b,c,d, 1997a, 2002, 2010c). The results should be here described according to the order of organ systems in anatomy or histology.

1.2.2 The RNA synthesis

The RNA (ribonucleic acid) contained in cells can be demonstrated either by morphological histochemical techniques staining tissue sections such as methyl green-pyronin staining or by biochemical techniques homogenizing tissues and cells. To the contrary, the synthetic RNA or newly synthesized RNA but not all the RNA in the cells can be detected as macromolecular synthesis together with other macromolecules such as DNA or proteins in various organs of experimental animals by either morphological or biochemical procedures employing RI-labeled precursors. We have studied the sites of macromolecular synthesis in almost all the organs of mice during their aging from prenatal to postnatal development to senescence by means of microscopic radioautography, one of the morphological methods (Nagata 1992, 1994b,c, 1996a,b,c, 1997a, 2002, 2010c). The results obtained from RNA synthesis should be here described according to the order of organ systems in anatomy or histology. In contrast to the results obtained from DNA synthesis of almost all the organs, we have studied only several parts of the organ systems with regards RNA. The skeletal system, the muscular system or the circulatory system were not so much studied.

1.2.3 The protein synthesis

The proteins found in animal cells are composed of various amino-acids which initially form low molecular polypeptides and finally macromolecular compounds designated as proteins. They are chemically classified into two, simple proteins and conjugated proteins. Therefore, the proteins can be demonstrated by showing specific reactions to respective amino-acids composing any proteins. Thus, the proteins contained in cells can be demonstrated either by morphological histochemical techniques staining tissue sections such as Millon reaction (Millon 1849) or tetrazonium reaction or otherwise by biochemical techniques homogenizing tissues and cells. To the contrary, the newly synthesized proteins but not all the proteins in the cells can be detected as macromolecular synthesis together with other macromolecules such as DNA or RNA in various organs of experimental animals by either morphological or biochemical procedures employing RI-labeled precursors. We have studied the sites of macromolecular synthesis in almost all the organs of mice during their aging from prenatal to postnatal development to senescence by means of microscopic radioautography (Nagata 1992, 1994b,c, 1996a,b, 1997a, 2002, 2010c). The results obtained from protein synthesis should be described according to the order of organ systems in anatomy or histology. In contrast to the results obtained from DNA synthesis of almost all the organs, we have studied only several parts of the organ systems.

1.2.4 The glucide synthesis

The glucides found in animal cells and tissues are composed of various low molecular sugars such as glucose or fructose called monosaccharides which form compounds of polysaccharides or complex mucopolysaccharides connecting to sulfated compounds. The former are called simple polysaccharides, while the latter mucopolysubstances. Thus, the glucides are chemically classified into 3 groups, monosaccharides such as glucose or fructose, disaccharides such as sucrose and polysaccharides such as mucosubstances. However, in most animal cells polysaccharides are much more found than monosaccharides or disaccharides. The polysaccharides can be classified into 2, i.e. simple polyscaccharides and mucosubstances. Anyway, they are composed of various low molecular sugars that can be demonstrated by either histochemical reactions or biochemical techniques. To the contrary, the newly synthesized glucides but not all the glucides in the cells and tissues can be detected as macromolecular synthesis together with other macromolecules such as DNA, RNA or proteins in various organs of experimental animals by either morphological or biochemical procedures employing RI-labeled precursors. We have studied the sites of macromolecular synthesis in almost all the organs of mice during their aging from prenatal to postnatal development to senescence by means of microscopic radioautography (Nagata 1992, 1994a,b,c, 1996a, 1997a, 2002, 2010c). The results obtained from glucides synthesis are described according to the order of organ systems in anatomy or histology. In contrast to the results obtained from DNA synthesis of almost all the organs, we have studied only several parts of the organ systems. The skeletal system, the muscular system and the circulatory system were not yet studied.

1.2.5 The lipid synthesis

The lipids found in animal cells are chemically composed of various low molecular fatty acids. They are esters of high fatty acids and glycerol that can biochemically be classified into simple lipids and compound lipids such as phospholipids, glycolipids or proteolipids. The simple lipids are composed of only fatty acids and glycerol, while the latter composed of lipids and other components such as phosphates, glucides or proteins. In order to demonstrate intracellular localization of total lipids, we can employ either histochemical reactions or biochemical techniques. To the contrary, the newly synthesized lipids but not all of the lipids in the cells can be detected as macromolecular synthesis similarly to the other macromolecules such as DNA, RNA, proteins or glucides in various organs of experimental animals by either morphological or biochemical procedures employing RI-labeled precursors. We have studied the sites of macromolecular synthesis in almost all the organs of mice during their aging from prenatal to postnatal development to senescence by means of microscopic radioautography (Nagata 1992, 1994a,b,c,d,e, 1996a, 1997a, 2002, 2010c). However, we have not studied the lipids synthesis so much as compared to other compounds. We have studied only a few organs of the digestive system.

1.3 The intracellular localization of the other substances

The other substances than macromolecules that can also be demonstrated by radioautography are target tracers not the precursors for the macromolecular synthesis. They are hormones such as ^{3}H-methyl prednisolone (Nagata et al. 1978b), neurotransmitters and inhibitors such as ^{14}C-bupranolol, a beta-blocking agent (Tsukahara et al. 1980) or ^{3}H-befunolol (Nagata and Yamabayashi 1983, Yamabayashi et al. 1981), vitamins, drugs such as synthetic anti-allergic agent ^{3}H-tranilast (Nagata et al. 1986b, Nishigaki et al. 1987, 1990a.b, Momose et al. 1989), hypolipidemic agent bezafibrate (Momose et al. 1993a,b, 1995), calmodulin antagonist (Ohno et al. 1982, 1983) or anti-hypertensive agent ^{3}H-benidipine hydrochloride (Suzuki et al. 1994), toxins, inorganic substances such as mercury (Nagata et al. 1977b) and others such as laser beam irradiation (Nagata 1984). The details are referred to the previous publication on the radioatuographology (Nagata 2002). However, their relationships to the cell aging and senescence were not studied.

1.4 Macromolecular synthesis in the normal organ systems

With regards to the macromolecular synthesis such as DNA, RNA, proteins, glucides, lipids etc in various cells and tissues, we have studied various cells and tissues in almost all the organ systems in the experimental animals such as mice and rats. Therefore, the results are classified into the organ systems in anatomy and histology, i.e. the organ of movement including the skeletal system and the muscular system, the circulatory organs, the digestive organs, the respiratory organs, the urinary organs, the reproductive organs, the endocrine organs, the nervous system, and the sensory organs. Thus, the results should be described according to this order in the following chapters divided into 4 chapters.

1.5 Macromolecular synthesis in the tumor cells

As for the tumor cells, on the other hand, which do not belong to any organ systems of the normal organs but grow in any organ systems, the macromolecular synthesis in the tumor cells should be described separately from the normal organ system at the end of this book in chapter 4.

2. Macromolecular synthesis in the organ of movement

The Organ of Movement or locomotive organ of men and experimental animals consists of both the skeletal system and the muscular system. The former consists of many bones, around 200 in case of men while the latter consists of many skeletal muscles around 600 in case of men. We studied the macromolecular synthesis in a part of these locomotive organs in the experimental animals, but not all of them. The results should be described in the following 2 sections, the skeletal system and the muscular system.

2.1 Macromolecular synthesis in the skeletal system

The skeletal system of men and experimental animals consists of bones, joints and ligaments. We studied the DNA synthetic activities of the bones and joints of experimental animals in development and aging (Kobayashi and Nagata 1994, Nagata 1998c).

2.1.1 The DNA synthesis in the skeletal system

We studied the DNA synthetic activities of the bones and joints of experimental animals in development and aging to senescence (Kobayashi and Nagata 1994, Nagata 1998c).

2.1.1.1 The DNA synthesis in the bone

We studied the ossifications of salamander skeletons from hatching to senescence (Nagata 1998c). The fore-limbs (Fig. 2A) and hind-limbs (Fig. 2B) of salamanders were composed of skeletons consisting of bones and cartilages which were covered with skeletal muscles, connective tissues and epidermis consisting of stratified squamous epithelial cells in the outermost layer. The bones of juvenile salamanders at 4 weeks consisted of the hyaline cartilage (Figs. 2A, 2B). The hyaline cartilage consisted of spherical or polygonal cartilage cells or chondrocytes at the center. They were surrounded by rich interstitial ground substance which stained deep blue with toluidine blue staining. The spherical cartilage cells at the center of the bone changed their shapes to flattened shape under the perichondrium or free joint surfaces. Some of the nuclei of the chondrocytes were covered with silver grains when labeled with ^{3}H-thymidine (Figs. 2A, 2B). Mitotic figures were frequently seen in spherical cartilage cells in young animals. Examination of radioautograms at the young stages such as 4 weeks after hatching showed that many spherical cartilage cells and flattened cartilage cells were predominantly labeled. At 6 weeks after hatching, the size of bones enlarged and the number of cartilage cells increased. At this stage, however, the number of labeled cells in the cartilage cells in both fore-limbs (Fig. 2C) and hind-limbs (Fig. 2D) decreased as compared with the previous stage (Figs. 2A, 2B). The size of bones in juvenile animals at 8, 9, 10, and 11 weeks enlarged gradually (Fig. 2E, 2F). Radioautograms at these stages showed that the number of the labeled cells remarkably reduced as compared with those of 4 and 6 weeks. In the adult salamanders at 8 months up to 12 months, the bones showed complete mature structure and examination of radioautograms revealed that the number of labeled cells reached almost zero (Kobayashi and Nagata 1994, Nagata 2006c). No difference was found on the morphology and labeling between the fore-limbs and hind-limbs at any stage.

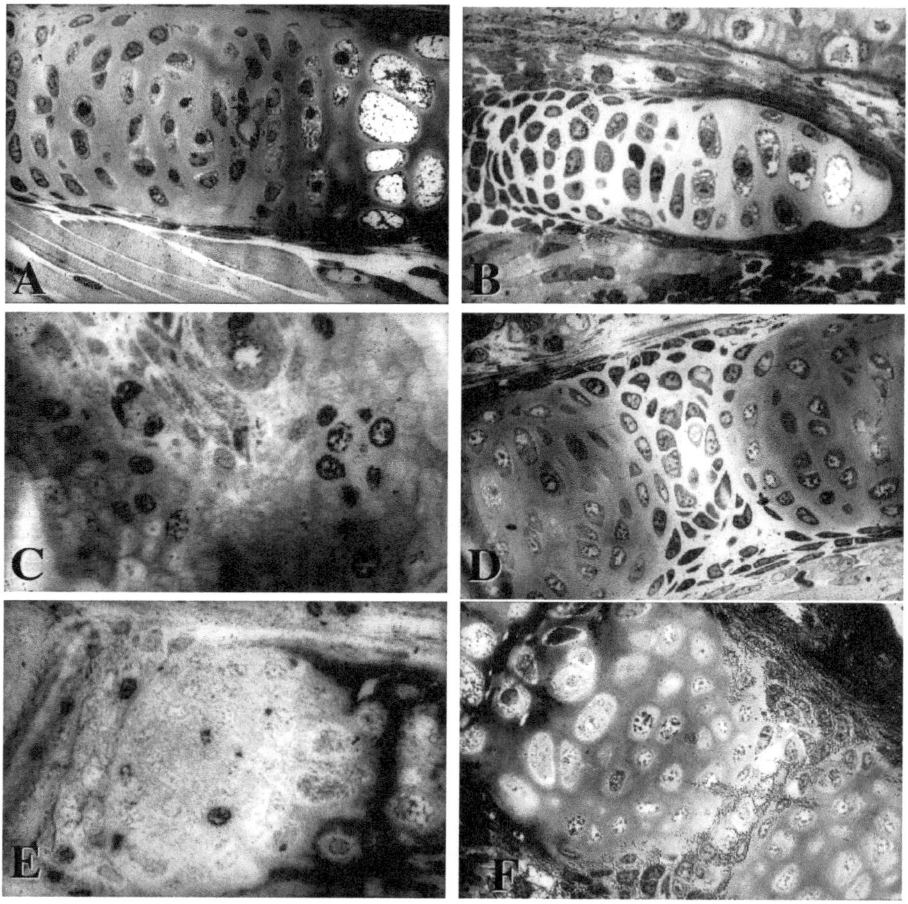

Fig. 2. Light microscopic radioautograms of the bones of either fore-limbs or hind-limbs of salamanders at various ages from 4 weeks to 8 weeks after hatching, injected with ^{3}H-thymidine, fixed and processed for radioautography. Some of the cartilage cells (arrows) are labeled with silver grains due to ^{3}H-thymidine incorporation demonstrating DNA synthesis. From Nagata, T.: Bulletin Shinshu Inst. Alternat. Med. Vol. 2, p. 53, 2006, Nagano, Japan
Fig. 2A. Light microscopic radioautogram of the bone of a fore-limb of a salamander at 4 weeks after hatching. Many cartilage cells (arrows) are labeled with silver grains due to ^{3}H-thymidine. Magnification. x 1200.
Fig. 2B. Light microscopic radioautogram of the bone of a hind-limb of a salamander at 4 weeks after hatching. Many cartilage cells (arrows) are labeled with silver grains due to ^{3}H-thymidine. Magnification. x 1200.
Fig. 2C. Light microscopic radioautogram of the bone of a fore-limb of a salamander at 6 weeks after hatching. Only a few cartilage cells (arrow) are labeled. The numbers of labeled cells are fewer than the bone of a younger salamander at 4 weeks after hatching (Fig. 1A). Magnification x 1200.

Fig. 2D. Light microscopic radioautogram of the bone of a hind-limb of a salamander at 6 weeks after hatching. Only a few cartilage cells (arrow) are labeled. Magnification x 1200.
Fig. 2E. Light microscopic radioautogram of the bone of a fore-limb of a salamander at 8 weeks after hatching. Only a few cartilage cells (arrow) are labeled. Magnification x 1200.
Fig. 2F. Light microscopic radioautogram of the bone of a hind-limb of a salamander at 8 weeks after hatching. Only a few cartilage cells (arrow) are labeled. Magnification x 1200.

The labeling indices of respective cell types changed with aging as expressed by mean in each group. The labeling index of the cartilage cells was lower than the epithelial cells. The peak of the labeling index of the cartilage cells in both fore-limbs and hind-limbs was found about 15-18% at 4 weeks after hatching (Fig. 2). The labeling index of the cartilage cells in both limbs at 6 weeks rapidly decreased to about 4-6%, then increased at 8 weeks to about 7-8% and finally decreased to 2-3% gradually from 8 weeks to 9 weeks with aging and fell down to 0-1% at 10 weeks. The labeling index of cartilage cells from 10 weeks to 12 months kept very low around 0-1% (Fig. 3). Thus, the cartilages and bones of fore-limbs and hind-limbs of salamanders are demonstrated to complete the development by 10 weeks after hatching (Kobayashi and Nagata 1994, Nagata 2006c).

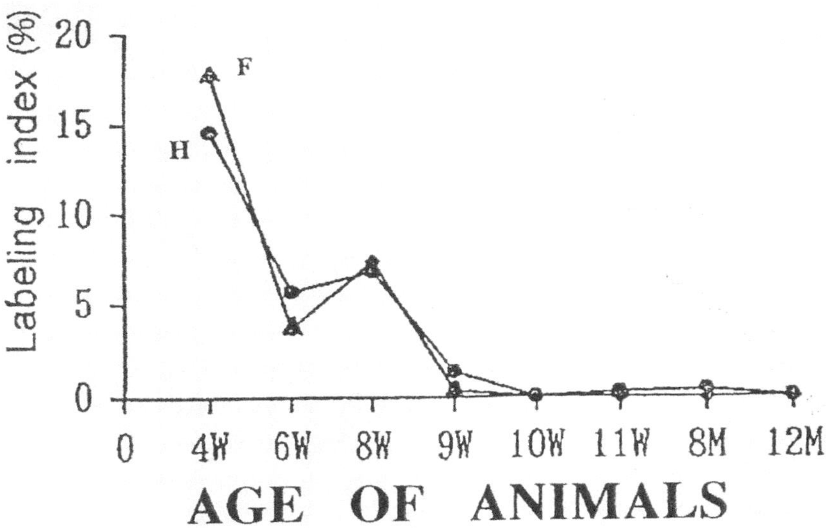

Fig. 3. Transitional curves of the labeling indices of the cartilage cells in the bones of the fore-limbs and the hind-limbs of salamanders labeled with ^{3}H-thymidine at various ages from 4 weeks to 60 months (5 years) after hatching. Mean ± S.D. From Nagata, T.: Bulletin Shinshu Inst. Alternat. Med. Vol. 2, p. 54, 2006, Nagano, Japan,

2.1.1.2 The DNA synthesis in the joint

The joints of an experimental animal such as mouse or human being are consisted of either 2 or 3 bones and the synovial membranes covering the ends of the bones. The synovial membranes are composed of the collagenous fibers interspersed with the synovial cells which are fibroblasts and lining cells. We studied macromolecular synthesis, both DNA and

RNA syntheses of the synovial cells of the joints surgically obtained from 15 elderly human patients of both sexes aged from 50 to 70, suffering from rheumatoid arthritis (Kobayashi and Nagata 1994). Both the normal and rheumatoid cells were cultured and labeled in vitro with media containing precursors such as ^{3}H-thymidine or ^{3}H-uridine, fixed and radioautographed. DNA synthetic cells labeled with silver grains were observed by LM RAG (light microscopic radioautography) in both normal and rheumatoid cells. As the results, some labeled synovial cells with ^{3}H-thymidine were found. However, no significant difference was observed between the labeling indices of normal and rheumatoid cells labeled with ^{3}H-thymidine. From the results, it was concluded that the synovial cells synthesized DNA in both normal and rheumatoid conditions. However, the quantities of these macromolecules synthesized in these synovial cells varied in respective individuals and no significant difference was found between the labeling indices and grain counts in both normal and rheumatoid cells (Kobayashi and Nagata 1994).

2.2 Macromolecular synthesis in the muscular system

The muscular system consists of various skeletal muscles amounting to around 600 in number in men and less in experimental animals such as rats and mice. We studied the aging changes of DNA synthesis in the intercostal muscles of aging ddY mice from prenatal day 13 through postnatal 24 months by ^{3}H-thymidine RAG (Hayashi et al. 1993).

2.2.1 The DNA synthesis in the muscular system

We studied the aging changes of DNA synthesis in the intercostal muscles of aging ddY mice from prenatal day 13 through postnatal 24 months in senescence by ^{3}H-thymidine RAG (Hayashi et al. 1993). Many nuclei were labeled in myotubes at embryonic day 13-17 (Fig. 4C) during development, then the number of labeled nuclei decreased to embryonic day 18-19 (Fig. 4D), and less due to aging after birth.

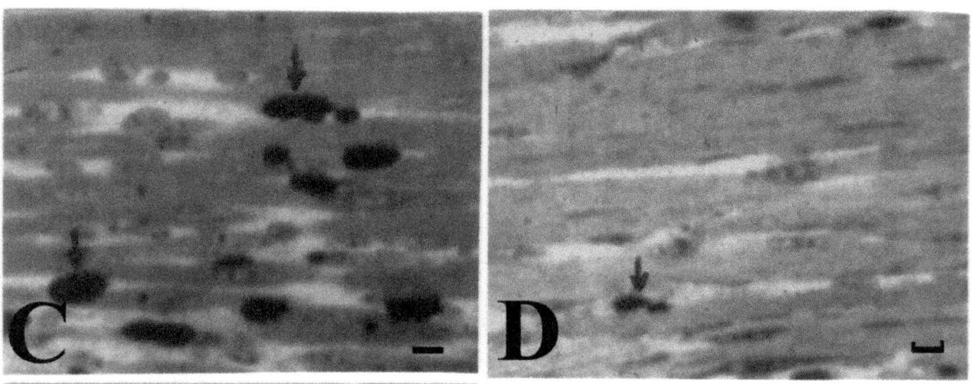

Fig. 4. Light microscopic radioautograms of the skeletal muscle cells in the myotubes labeled with ^{3}H-thymidine at embryonic day 13-17 (Fig. 3C), then the number of labeled nuclei decreased to embryonic day 18-19 (Fig. 3D), and less after birth. x260. From Nagata, T.: Special Cytochemistry in Cell Biology, In, Internat. Rev. Cytol. Vol. 211, No. 1, p. 62, 2001, Academic Press, San Diego, USA, London, UK.

The labeling indices revealed chronological changes, showing a peak at embryonic day 13 and decreasing gradually to 0% at 3 months after birth to month 24 (Fig. 5). We classified the graduation of the embryonic muscle development into 5 stages. Among them, the labeling index (LI) at stage I was the highest, while the LI at stage II was significantly lower than stage I, the LI at stage IV was significantly higher than stage II, and the LI at stage V was significantly lower than stage IV (Fig. 5). These changes accorded well with the primary and secondary myotube formation during the embryonic muscle development. We also studied the DNA synthesis of rat thigh muscles during the muscle regeneration after injury in rats (Sakai et al. 1977). When the skeletal muscles, i.e., the diaphragma, the rectus abdominis muscles and the gastrocnemius femoris muscle of adult Wistar rats were mechanically injured and labeled with ^{3}H-thymidine, satellite cells were labeled during their regeneration. The satellite cells in the muscles of dystrophy chickens and normal control chickens were also labeled with ^{3}H-thymidine, demonstrating DNA synthesis (Oguchi and Nagata 1980, 1981), which was later described in details in the review (Nagata 2002). Briefly, 2 groups of chickens, 4 dystrophy chickens and normal control chickens of both sexes aged 1 day and 21 days after hatching were used. All the animals received every 6 hrs intraperitoneal injections of ^{3}H-thymidine 4 times successively and sacrificed. The superficial pectoral muscles were taken out, fixed, embedded in Epoxy resin and processed for LM and EMRAG. The results demonstrated that many nuclei of the satellite cells in all the experimental groups were labeled but none of the nuclei in the muscle fibers were labeled. The labeling indices of normal chickens at 1 day and 21 days were 4.59 and 3.86%, respectively. These results showed that the LI decreased after hatching due to aging.

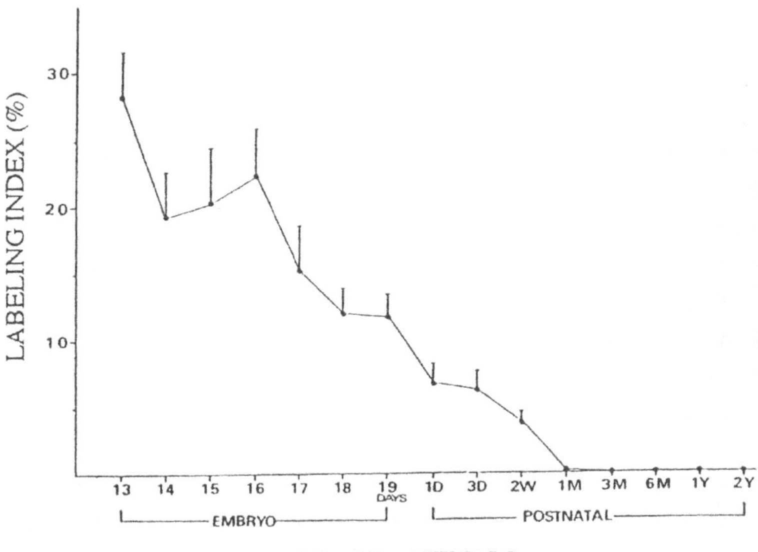

Fig. 5. The labeling indices of the muscular cells labeled with ^{3}H-thymidine revealed chronological changes, showing a peak at embryonic day 13 and decreasing gradually to 0% at 3 months after birth. From Nagata, T.: Radioautographology, General and Special. In, Prog. Histochem. Cytochem. Vol. 37, No. 2, p. 114, 2002, Urban & Fischer, Jena, Germany

2.2.2 The protein synthesis in the muscular system

We studied only ^{3}H-taurin incorporation in the skeletal muscles of both normal and dystrophy adult mice incubated in Eagle's medium containing ^{3}H-taurin in vitro at varying time intervals from 1 min to 5, 10, 30 and 60 min (1 hour). The silver grains were observed over the skeletal muscle cells as well as over the smooth muscle cells and the endothelial cells in the arteries, both the nuclei and cytoplalsm, by LM and EM RAG, showing taurin incorporation into the proteins (Terauch et al. 1988, Terauch and Nagata 1993, 1994). However, the aging changes were not studied yet.

3. Macromolecular synthesis in the circulatory system

The circulatory system or cardiovascular organs consists of the heart, the arteries, the veins, the capillaries, the blood, the lymphatic organs and the spleen. Among these cardiovascular organs, we studied the heart, the artery, some blood cells and the spleen.

3.1 The nucleic acid synthesis in the heart

Among the macromolecular synthesis, the nucleic acid synthesis, both DNA and RNA, in cultured cells from the hearts of chick embryos was studied by LM RAG (Nagata and Nawa 1966a,b). The fibroblasts of chick hearts in culture proliferated extensively and produced many binucleate cells. We compared the nucleic acid synthesis in mononucleate cells and binucleate cells in the heart fibroblasts. The incorporation of ^{3}H-thymidine into each nucleus of a binucleate cell was a little less than that of a mononucleate cell, but the total of the two nuclei of a binucleate cell was almost twice as that of a mononucleate cell. The incorporation of ^{3}H-uridine in the two nuclei of a binucleate cell was almost twice as that of a mononucleate cell, while the incorporation of ^{3}H-uridine in the cytoplasm of a binucleate cell was not so much as twice as a mononucleate cell. From these results, it was concluded that the nucleic acid synthesis both DNA and RNA increased in binucleate cells than mononucleate cells of chick embryo heart fibroblasts (Nagata and Nawa 1966a,b).

3.2 Localization of drugs in the artery

The structure of the blood vessels, both arteries and veins consist of 3 layers, from inside to outside, the tunica intima, the media and the adventitia. Those layers are formed with connective tissues and the smooth muscles. We studied the localization of anti-hypertensive drugs in the supramesenteric arteries of the spontaneous hypertensive rats (Suzuki et al. 1994). Two kinds of anti-hypertensive drugs, labeled with RI, ^{3}H-benidipine hydrochloride (Kyowa Hakko Kogyo Co., Shizuoka, Japan) and ^{3}H-nitrendipine (New England Nuclear, Boston, MA, USA) were used. Both intravenous administrations into rats and in vitro incubation for 10 to 30 min were employed. For light and electron microscopic radioautography, both the conventional wet-mounting radioautograms after chemical fixation for insoluble compounds and the dry-mounting radioautograms after cryo-fixation and freeze-substitution for soluble compounds were prepared. The silver grains due to the anti-hypertensive drugs were localized over the plasma membranes and the cytoplasm of the fibrocytes in the intima and the smooth muscle cells in the media, suggesting the pharmacological active sites. However, the localization of synthetic DNA was not studied.

3.3 The DNA synthesis in the blood cells

The mature blood cells circulating in the blood vessels of mammals are classified into 3 types, the erythrocytes, the leukocytes and the blood platelets. Those mature cells are formed either in the lymphatic tissues in the lymphatic organs or the myeloid tissues in the bone marrow, where various immature cells, lymphoblasts, erythroblasts, myeloblasts, meylocytes, and megakaryocytes can be observed. Among these blood cells, we studied macromolecular synthesis and cytochemical localization in leukocytes, megakaryocytes and blood platelets. As for the granulocytes, normal rabbit granulocytes were shown by EM RAG and X-ray microanalysis to incorporate $^{35}SO_4$ into the Golgi apparatus and to the granules demonstrating glucosaminoglycan synthesis (Murata et al. 1978, 1979).

On the other hand, the DNA, RNA and mucosubstance synthesis of mast cells from Wistar strain rats were studied by ^{3}H-thymidine, ^{3}H-uridine and $^{35}SO_4$ radioautography, demonstrating incorporation changes of those normal mast cells from abnormal mastocytoma cells (Murata et al. 1977a). Mast cells were widely found distributing in the loose connective tissues of most mammals, as well as in the serous exudate in the peritoneal cavity as one of the free cells. We studied the fine structure and nucleic acid and mucosubstance syntheses of normal mast cells and Dunn and Potter's mastocytoma cells in mice and rats by electron microscopic radioautography (Murata et al. 1977b, 1979). As the results, some of the normal mast cells and mastocytoma cells incorporated ^{3}H-thymidine, ^{3}H-uridine and $^{35}SO_4$, demonstrating DNA, RNA and mucosubstance syntheses. The incorporation of ^{3}H-thymidine was observed in the nuclei and mitochondria. The labeling index of ^{3}H-thymidine incorporation in the nuclei and mitochondria of normal mast cells was very low (0.37%) while that of mastocytoma cells was high (2-5%). These results suggested that the macromolecular synthesis such as nucleic acids (DNA, RNA) and mucosubstances were higher in tumor cells than normal blood cells.

3.4 Macromolecular synthesis in the spleen

The spleen is one of the blood cell forming organs and is composed of the lymphatic tissues. Among the macromolecular synthesis, both the DNA, RNA and protein syntheses in the spleen were studied.

3.4.1 The DNA synthesis in the spleen

We studied ^{3}H-thymidine incorporation into the splenic cells of aging mice from newborn to adult and senescence in connection with the lysosomal acid phosphatase activity (Olea 1991, Olea and Nagata 1991, 1992a). The acid phosphatase activity as demonstrated by means of cerium substrate method was observed in the splenic tissues at various ages from postnatal day 1, week 1 and 2, month 1, 2 and 10. Electron dense deposits were localized in the lysosomes of macrophages, reticular cells and littoral cells in all the aging groups. The intensity of the reaction products as visually observed increased from day 1 to week 1, reaching the peak at 1 week, and decreased from week 2 to month 10 due to aging. The incorporation of ^{3}H-thymidine, on the other hand, demonstrating DNA synthesis, was mainly observed in the hematopoietic cells in the spleens from postnatal day 1 to month 10 animals (Olea and Nagata 1991, 1992a). The labeling index was the maximum at day 1 and decreased to week 1, 2, 4, 8 and 40. A correlation between DNA synthesis and AcPase activity was examined by comparing two cell populations in the cell cycle, the S-phase cells which were labeled with ^{3}H-thymidine and the non-S-phase cells or the interphase cells which were not labeled. It was demonstrated that the former showed an increase and decrease of much more AcPase activity with the aging while the latter less activity and no change.

3.4.2 The RNA synthesis in the spleen

On the other hand, the number of labeled cells and the grain counts in the hematopoietic cells in the spleens labeled with ^{3}H-uridine, demonstrating RNA synthesis, from postnatal day 1 increased to 1 and 2 weeks, reaching the maximum, and decreased to 4, 8 and 40 weeks, different from the DNA synthesis (Olea and Nagata 1992b). These results demonstrated that AcPase activity, DNA and RNA synthetic activity changed due to aging.

3.4.3 The protein synthesis in the spleen

Among the circulatory organs, we first studied the protein synthesis in the spleens of aging mice at various ages. Several groups of litter mates, each 3, from fetal day 19 to postnatal day 1, 14, and month 6 to 12 (year 1) were administered with ^{3}H-leucine and sacrificed, the spleens were taken out and processed for LM and EM RAG (Nagata and Olea 1999). The results demonstrated that the sites of incorporations were hematopoietic cells, i.e., lymphoblasts, myeloblasts, erythroblasts and littoral cells in the splenic tissues at every aging stage. In most labeled cells silver grains were observed over the nuclei, nucleoli, endoplasmic reticulum, ribosomes, Golgi apparatus and mitochondria. Quantitative analysis revealed that grain counts in respective cells were higher in young animals than adult aged animals. The grain counts and the labeling index increased from prenatal to postnatal day 14, reaching the maximum, then decreased to month 12. These results showed the increase and decrease due to aging of animals.

4. Conclusion

This chapter deals with the introductory remarks describing the method and procedure of microscopic radioautography as well as the first parts of its application to the organ systems such as the skeletal, muscular and the circulatory organs. The method and the procedure of microscopic radioautography described the detailed technology which was developed by the present author and associates since 1955 in our laboratory. The application of radaioautography to the skeletal, muscular and circulatory systems demonstrated the sites of macromolecular syntheses such as DNA, RNA, proteins, glucides and lipids in various organs as well as the quantitative changes due to aging of the experimental animals. These results should be very important to understand the fundamental changes in the respective bones, the skeletal muscles, and the cardiovascular organs as well as the contributions to the experimental biology and medicine throughout the world.

5. References

Chen, S., Gao, F., Kotani, A., Nagata, T.: Age-related changes of male mouse submandibular gland: A morphometric and radioautographic study. Cell. Mol. Biol. 41, 117-124, 1995.

Clermont Y.: The contractime elements in the limiting membrane of the seminiferous tubules of rats. Exp. Cell Res. 15, 438-342, 1958.

Clermont, Y.: Renewal of spermatogonia in man. Amer. J. Anat. 112, 35-51, 1963.

Cui, H.: Light microscopic radioautographic study on DNA synthesis of nerve cells in the cerebella of aging mice. Cell. Mol. Biol. 41, 1139-1154, 1995.

Cui, H., Gao, F., Nagata, T.: Light microscopic radioautographic study on protein synthesis in perinatal mice corneas. Acta Histochem. Cytochem. 33, 31-37, 2000.

Duan, H., Gao, F., Li, S., Hayashi, K., Nagata, T.: Aging changes and fine structure and DNA synthesis of esophageal epithelium in neonatal, adult and old mice. J. Clin. Electron Microsc. 25, 452-453, 1992.

Duan, H., Gao, F., Li, S., Nagata, T.: Postnatal development of esophageal epithelium in mouse: a light and electron microscopic radioautographic study. Cell. Mol. Biol. 39, 309-316, 1993.

Duan, H., Gao, F., Oguchi, K., Nagata, T.: Light and electron microscopic radioautographic study on the incorporation of ^{3}H-thymidine into the lung by means of a new nebulizer. Drug Res. 44, 880-883, 1994.

Feulgen, R., Rossenbeck, H.: Mikroskopische-chemischer Nachweis einer Nukeinsaeure von Thymus der Thymonukeinsaeure Z. Physik. Chem. 135, 203-248, 1924.

Fujii, Y., Ohno, S., Yamabayashi, S., Usuda, N., Saito, H., Furuta, S., Nagata, T.: Electron microscopic radioautography of DNA synthesis in primary cultured cells from an IgG myeloma patient. J. Clin. Electr. Microsc. 13, 582-583, 1980.

Gao, F.: Study on the macromolecular synthesis in aging mouse seminiferous tubules by light and electron microscopic radioautography. Cell. Mol. Biol. 39, 659-672, 1993.

Gao, F., Toriyama, K., Nagata, T.: Light microscopic radioautographic study on the DNA synthesis of prenatal and postnatal aging mouse retina after labeled thymidine injection. Cell. Mol. Biol. 38, 661-668, 1992a.

Gao, F., Li, S., Duan, H., Ma, H., Nagata, T.: Electron microscopic radioautography on the DNA synthesis of prenatal and postnatal mice retina after labeled thymidine injection. J. Clin. Electron Microsc. 25, 721-722, 1992b.

Gao, F., Toriyama, K., Ma, H., Nagata, T.: Light microscopic radioautographic study on DNA synthesis in aging mice corneas. Cell. Mol. Biol. 39, 435-441, 1993.

Gao, F., Ma, H., Sun, L., Jin, C., Nagata, T.: Electron microscopic radioautographic study on the nucleic acids and protein synthesis in the aging mouse testis. Med. Electron Microsc. 27, 360-362, 1994.

Gao, F., Chen, S., Sun, L., Kang, W., Wang, Z., Nagata, T.: Radioautographic study of the macromolecular synthesis of Leydig cells in aging mice testis. Cell. Mol. Biol. 41, 145-150, 1995a.

Gao, F., Jin, C., Ma, H., Sun, L., Nagata, T.: Ultrastructural and radioautographic studies on DNA synthesis in Leydig cells of aging mouse testis. Cell. Mol. Biol. 41, 151-160, 1995b.

Gunarso, W.: Radioautographic studies on the nucleic acid synthesis of the retina of chick embryo. I. Light microscopic radioautography. Shinshu Med. J. 32, 231-240, 1984a.

Gunarso, W.: Radioautographic studies on the nucleic acid synthesis of the retina of chick embryo. II. Electron microscopic radioautography. Shinshu Med. J. 32, 241-248, 1984b.

Gunarso, W., Gao, F., Cui, H., Ma, H., Nagata, T.: A light and electron microscopic radioautographic study on RNA synthesis in the retina of chick embryo. Acta Histochem. 98, 309-32, 1996.

Gunarso, W., Gao, F., Nagata, T.: Development and DNA synthesis in the retina of chick embryo observed by light and electron microscopic radioautography. Cell. Mol. Biol. 43, 189-201, 1997.

Hanai, T.: Light microscopic radioautographic study of DNA synthesis in the kidneys of aging mice. Cell. Mol. Biol. 39, 81-91, 1993.

Hanai, T., Nagata, T.: Electron microscopic radioautographic study on DNA and RNA synthesis in perinatal mouse kidney. In, Radioautography in Medicine, Nagata, T., Ed., pp. 127-131, Shinshu University Press, Matsumoto, 1994a.

Hanai, T., Nagata, T.: Study on the nucleic acid synthesis in the aging mouse kidney by light and electron microscopic radioautography. In, Radioautography in Medicine, Nagata, T., Ed., pp. 209-214, Shinshu University Press, Matsumoto, 1994b.

Hanai, T., Nagata, T.: Electron microscopic study on nucleic acid synthesis in perinatal mouse kidney tissue. Med. Electron Microsc. 27, 355-357, 1994c.

Hanai, T., Usuda, N., Morita, T., Shimizu, T., Nagata, T.: Proliferative activity in the kidneys of aging mice evaluated by PCNA/cyclin immunohistochemistry. Cell. Mol. Biol. 39, 181-191, 1993.

Hayashi, K., Gao, F., Nagata, T.: Radioautographic study on ^{3}H-thymidine incorporation at different stages of muscle development in aging mice. Cell. Mol. Biol. 39, 553-560, 1993.

Ito, M.: The radioautographic studies on aging change of DNA synthesis and the ultrastructural development of mouse adrenal gland. Cell. Mol. Biol. 42, 279-292, 1996.

Ito, M., Nagata, T.: Electron microscopic radioautographic studies on DNA synthesis and ultrastructure of aging mouse adrenal gland. Med. Electron Microsc. 29, 145-152, 1996.

Izumiyama, K., Kogure, K., Kataoka, S., Nagata, T.: Quantitative analysis of glucose after transient ischemia in the gerbil hippocampus by light and electron microscope radioautography. Brain Res. 416, 175-179, 1987.

Jamieson, J. D., Palade, G. E.: Intracellular transport of secretory proteins in the pancreatic exocrine cells. J. Cell Biol. 34, 577-615, 1967.

Jin, C.: Study on DNA synthesis of aging mouse colon by light and electron microscopic radioautography. Cell. Mol. Biol. 42, 255-268, 1996.

Jin, C., Nagata, T.: Light microscopic radioautographic study on DNA synthesis in cecal epithelial cells of aging mice. J. Histochem. Cytochem. 43, 1223-1228, 1995a.

Jin, C., Nagata, T.: Electron microscopic radioautographic study on DNA synthesis in cecal epithelial cells of aging mice. Med. Electron Microsc. 28, 71-75, 1995b.

Joukura, K.: The aging changes of glycoconjugate synthesis in mouse kidney studied by ^{3}H-glucosamine radioautography. Acta Histochem. Cytochem. 29, 57-63, 1996.

Joukura, K., Nagata, T.: Aging changes of ^{3}H-glucosamine incorporation into mouse kidney observed by radioautography. Acta Histochem. Cytochem. 28, 494-494, 1995.

Joukura, K., Usuda, N., Nagata, T.: Quantitative study on the aging change of glycoconjugates synthesis in aging mouse kidney. Proc. Xth Internat. Cong. Histochem. Cytochem., Acta Histochem. Cytochem. 29, Suppl. 507-508, 1996.

Kobayashi, K., Nagata, T.: Light microscopic radioautographic studies on DNA, RNA and protein syntheses in human synovial membranes of rheumatoid arthritis patients. J. Histochem. Cytochem. 42, 982-982, 1994.

Komiyama, K., Iida, F., Furihara, R., Murata, F., Nagata, T.: Electron microscopic radioautographic study on ^{125}I-albumin in rat gastric mucosal epithelia. J. Clin. Electron Microsc. 11, 428-429, 1978.

Kong, Y.: Electron microscopic radioautographic study on DNA synthesis in perinatal mouse retina. Cell. Mol. Biol. 39, 55-64, 1993.

Kong, Y., Nagata, T.: Electron microscopic radioautographic study on nucleic acid synthesis of perinatal mouse retina. Med. Electron Microsc. 27, 366-368, 1994.

Kong, Y., Usuda, N., Nagata, T.: Radioautographic study on DNA synthesis of the retina and retinal pigment epithelium of developing mouse embryo. Cell. Mol. Biol. 38, 263-272, 1992a.

Kong, Y., Usuda, N., Morita, T., Hanai, T., Nagata, T.: Study on RNA synthesis in the retina and retinal pigment epithelium of mice by light microscopic radioautography. Cell. Mol. Biol. 38, 669-678, 1992b.

Leblond, C. P.: Localization of newly administered iodine in the thyroid gland as indicated by radioiodine. J. Anat. 77, 149-152, 1943.

Leblond, C. P.: The life history of cells in renewing systems. Am. J. Anat. 160, 113-158, 1981.

Leblond, C. P., Messier, B.: Renewal of chief cells and goblet cells in the small intestine as shown by radioautography after injection of thymidine-^{3}H into mice. Anat. Rec. 132: 247-259. 1958.

Li, S.: Relationship between cellular DNA synthesis, PCNA expression and sex steroid hormone receptor status in the developing mouse ovary, uterus and oviduct. Histochemistry 102, 405-413, 1994.

Li, S., Nagata, T.: Nucleic acid synthesis in the developing mouse ovary, uterus and oviduct studied by light and electron microscopic radioautography. Cell. Mol. Biol. 41, 185-195, 1995.

Li, S., Gao, F., Duan, H., Nagata, T.: Radioautographic study on the uptake of $^{35}SO_4$ in mouse ovary during the estrus cycle. J. Clin. Electron Microsc. 25, 709-710, 1992.

Liang, Y.: Light microscopic radioautographic study on RNA synthesis in the adrenal glands of aging mice. Acta Histochem. Cytochem. 31, 203-210, 1998.

Liang, Y., Ito, M., Nagata, T.: Light and electron microscopic radioautographic studies on RNA synthesis in aging mouse adrenal gland. Acta Anat. Nippon. 74, 291-300, 1999.

Ma, H.: Light microscopic radioautographic study on DNA synthesis of the livers in aging mice. Acta Anat. Nippon. 63, 137-147, 1988.

Ma, H., Nagata, T.: Electron microscopic radioautographic study on DNA synthesis of the livers in aging mice. J. Clin. Electron Microsc. 21, 335-343, 1988a.

Ma, H., Nagata, T.: Studies on DNA synthesis of aging mice by means of electron microscopic radioautography. J. Clin. Electron Microsc. 21, 715-716, 1988b.

Ma, H., Nagata, T.: Electron microscopic radioautographic studies on DNA synthesis in the hepatocytes of aging mice as observed by image analysis. Cell. Mol. Biol. 36, 73-84, 1990a.

Ma, H., Nagata, T.: Study on RNA synthesis in the livers of aging mice by means of electron microscopic radioautography. Cell. Mol. Biol. 36, 589-600, 1990b.

Ma, H., Nagata, T.: Collagen and protein synthesis in the livers of aging mice as studied by electron microsopic radioautography. Ann. Microsc. 1, 13-22, 2000.

Ma, H., Gao, F., Olea, M. T., Nagata, T.: Protein synthesis in the livers of aging mice studied by electron microscopic radioautography. Cell. Mol. Biol. 37, 607-615, 1991.

Matsumura, H., Kobayashi, Y., Kobayashi, K., Nagata, T.: Light microscopic radioautographic study of DNA synthesis in the lung of aging salamander, Hynobius nebulosus. J. Histochem. Cytochem. 42, 1004-1004, 1994.

Momose, Y., Nagata, T.: Radioautographic study on the intracellular localization of a hypolipidemic agent, bezafibrate, a peroxisome proliferator, in cultured rat hepatocytes. Cell. Mol. Biol. 39, 773-781, 1993a.

Momose, Y., Naito, J., Nagata, T.: Radioautographic study on the localization of an anti-allergic agent, tranilast, in the rat liver. Cell. Mol. Biol. 35, 347-355, 1989.

Momose, Y., Shibata, N., Kiyosawa, I., Naito, J., Watanabe, T., Horie, S., Yamada, J., Suga, T., Nagata, T.: Morphometric evaluation of species differences in the effects of bezafibrate, a hypolipidemic agent, on hepatic peroxisomes and mitochondria. J. Toxicol. Pathol. 6, 33-45, 1993b.

Momose, Y., Naito, J., Suzawa, H., Kanzawa, M., Nagata, T.: Radioautographic study on the intracellular localization of bezafibrate in cultured rat hepatoctyes. Acta Histochem. Cytochem. 28, 61-66, 1995.

Morita, T.: Radioautographic study on the aging change of ^{3}H-glucosamine uptake in mouse ileum. Cell. Mol. Biol. 39, 875-884, 1993.

Morita, T., Usuda, N. Hanai, T., Nagata, T.: Changes of colon epithelium proliferation due to individual aging with PCNA/cyclin immunostaining comparing with ^{3}H-thymidine radioautography. Histochemistry 101, 13-20, 1994.

Murata, F., Momose,Y. , Yoshida, K., Nagata, T.: Incorporation of ^{3}H-thymidine into the nucleus of mast cells in adult rat peritoneum. Shinshu Med. J. 25, 72-77, 1977a.

Murata, F., Momose, Y., Yoshida, K., Ohno, S., Nagata, T.: Nucleic acid and mucosubstance metabolism of mastocytoma cells by means of electron microscopic radioautography. Acta Pharmacol. Toxicol. 41, 58-59, 1977b.

Murata, F., Yoshida, K., Ohno, S., Nagata, T.: Ultrastructural and electron microscopic radioautographic studies on the mastocytoma cells and mast cells. J. Clin. Electron Microsc. 11, 561-562, 1978.

Murata, F., Yoshida, K., Ohno, S., Nagata, T.: Mucosubstances of rabbit granulocytes studied by means of electron microscopic radioautography and X-ray microanalysis. Histochemistry 61, 139-150, 1979.

Nagata, T.: On the relationship between cell division and cytochrome oxidase in the Yoshida sarcoma cells. Shinshu Med. J. 5: 383-386, 1956.

Nagata, T.: Studies on the amitosis in the Yoshida sarcoma cells. I. Observation on the smear preparation under normal conditions. Med. J. Shinshu Univ. 2: 187-198, 1957a.

Nagata, T.: Studies on the amitosis in the Yoshida sarcoma cells. II. Phase-contrast microscopic observations under normal conditions. Med. J. Shinshu Univ. 2: 199-207, 1957b.

Nagata, T.: Cell divisions in the liver of the fetal and newborn dogs. Med. J. Shinshu Univ. 4: 65-73, 1959.

Nagata, T.: A radioautographic study of the DNA synthesis in rat liver, with special reference to binucleate cells. Med. J. Shinshu Univ. 7, 17-25, 1962.

Nagata, T.: A quantitative study on the ganglion cells in the small intestine of the dog. Med. J. Shinshu Univ. 10, 1-11, 1965.

Nagata, T.: A radioautographic study on the RNA synthesis in the hepatic and the intestinal epithelial cells of mice after feeding with special reference to binuclearity. Med. J. Shinshu Univ. 11, 49-61, 1966.

Nagata, T.: On the increase of binucleate cells in the ganglion cells of dog small intestine due to experimental ischemia. Med. J. Shinshu Univ. 12, 93-113, 1967a.

Nagata, T.: A radioautographic study on the protein synthesis in the hepatic and the intestinal epithelial cells of mice, with special reference to binucleate cells. Med. J. Shinshu Univ. 12, 247-257, 1967b.

Nagata, T.: Chapter 3. Application of microspectrophotometry to various substances. In , Introduction to Microspectrophotometry. Isaka, S., Nagata, T., Inui, N., Eds., Olympus Co., Tokyo, pp. 49-155, 1972a.

Nagata, T.: Electron microscopic dry-mounting autoradiography. Proc. 4th Internat. Cong. Histochem. Cytochem. Kyoto, pp. 43-44, 1972b.

Nagata, T.: Electron microscopic radioautography of intramitochondrial RNA synthesis of HeLa cells in culture. Histochemie 32, 163-170, 1972c.

Nagata, T.: Quantitative electron microscope radioautography of intramitochondrial nucleic acid synthesis. Acta Histochem. Cytochem. 5, 201-203, 1972d.

Nagata, T.: Electron microscopic observation of target cells previously observed by phase-contrast microscopy: Electron microscopic radioautography of laser beam irradiated cultured cells. J. Clin. Electron Microsc. 17, 589-590, 1984.

Nagata, T.: Principles and techniques of radioautography. In, Histo- and Cyto-chemistry 1985, Japan Society of Histochemistry and Cytochemistry, Ed., Gakusai Kikaku Co., Tokyo, pp. 207-226, 1985.

Nagata, T.:. Electron microscopic radioautography and analytical electron microscopy. J. Clin. Electron Microsc. 24, 441-442, 1991.

Nagata, T.: Radiolabeling of soluble and insoluble compounds as demonstrated by light and electron microscopy. Recent Advances in Cellular and Molecular Biology, Wegmann, R. J., Wegmann, M. A., Eds. Peters Press, Leuven, Vol. 6, pp. 9-21, 1992.

Nagata, T.: Quantitative analysis of histochemical reactions: Image analysis of light and electron microscopic radioautograms. Acta Histochem. Cytochem. 26, 281-291, 1993a.

Nagata, T. Quantitative light and electron microscopic radioautographic studies on macromolecular synthesis in several organs of prenatal and postnatal aging mice. Chinese J. Histochem. Cytochem. 2: 106-108, 1993b.

Nagata, T.: Electron microscopic radioautography with cryo-fixation and dry-mounting procedure. Acta Histochem. Cytochem. 27: 471-489, 1994a.

Nagata, T.: Application of electron microscopic radioautography to clinical electron microscopy. Med. Electron Microsc. 27; 191-212, 1994b.

Nagata, T.: Radioautography in Medicine. Shinshu University Press, 268pp, Matsumoto, 1994c.

Nagata, T.: Radioautography, general and special. In, Histo- and Cyto-chemistry 1994, Japan Society of Histochemistry and Cytochemistry, ed, pp. 219-231, Gakusai Kikaku Co., Tokyo, 1994d.

Nagata, T., Application of electron microscopic radioautography to clinical electron microscopy. Med. Electron Microsc. 27, 191-212, 1994e.

Nagata, T.: Light and electron microscopic radioautographic study on macromolecular synthesis in digestive organs of aging mice. Cell. Mol. Biol. 41, 21-38, 1995a.

Nagata, T.: Histochemistry of the organs: Application of histochemistry to anatomy. Acta Anat. Nippon. 70, 448-471, 1995b.

Nagata, T.: Three-dimensional observation of whole mount cultured cells stained with histochemical reactions by ultrahigh voltage electron microscopy. Cell. Mol. Biol. 41, 783-792, 1995c.

Nagata, T.: Morphometry in anatomy: image analysis on fine structure and histochemical reactions with special reference to radioautography. Ital. J. Anat. 100 (Suppl. 1), 591-605, 1995d.

Nagata, T.: Technique and application of electron microscopic radioautography. J. Electron Microsc. 45, 258-274, 1996a.

Nagata, T.: Techniques of light and electron microscopic radioautography. In, Histochemistry and Cytochemistry 1996. Proc. Xth Internat. Congr. Histochem. Cytochem. Acta Histochem. Cytochem. 29 (Suppl.), 343-344, 1996b.

Nagata, T.: Remarks: Radioautographology, general and special. Cell. Mol. Biol. 42 (Suppl.), 11-12, 1996c.

Nagata, T.: On the terminology of radioautography vs. autoradiography. J. Histochem. Cytochem. 44, 1209-1209, 1996d.

Nagata, T.: Techniques and applications of microscopic radioautography. Acta Microsc. Vol. 6: Suppl. B. p. 19-42, 1997a.

Nagata T.: Three-dimensional observation on whole mount cultured cells and thick sections stained with histochemical reactions by high voltage electron microscopy. In, Recent Advances in Microscopy of Cells, Tissues and Organs, Motta, P., Ed., Antonio Delfino Editore, Roma, pp. 37-44, 1997b.

Nagata, T.: Radioautographic study on collagen synthesis in the ocular tissues. J. Kaken Eye Res. 15, 1-9, 1997c.

Nagata, T.: Techniques of radioautography for medical and biological research. Braz. J. Biol. Med. Res. 31, 185-195, 1998a.

Nagata, T.: Radioautographology, the advocacy of a new concept. Braz. J. Biol. Med. Res. 31, 201-241, 1998b.

Nagata, T.: Radioautographic studies on DNA synthesis of the bone and skin of aging salamander. Bull. Nagano Women's Jr. College 6, 1-14, 1998c.

Nagata, T.: 3D observation of cell organelles by high voltage electron microscopy. Microscopy and Analysis, Asia Pacific Edition, 9, 29-32, 1999a.

Nagata, T.: Application of histochemistry to anatomy: Histochemistry of the organs, a novel concept. Proc. XV Congress of the International Federation of Associations of Anatomists, Ital. J. Anat. Embryol. 104 (Suppl. 1), 486-486, 1999b.

Nagata, T.: Aging changes of macromolecular synthesis in various organ systems as observed by microscopic radioautography after incorporation of radiolabeled precursors. Methods Find. Exp. Clin. Pharmacol. 21, 683-706, 1999c.

Nagata, T.: Radioautographic study on protein synthesis in mouse cornea. J. Kaken Eye Res. 8, 8-14, 1999d.

Nagata, T.: Radioautographology, general and special: a novel concept. Ital. J. Anat. Embryol. 104 (Suppl. 1), 487-487, 1999e.
Nagata, T.: Three-dimensional observations on thick biological specimens by high voltage electron microscopy. Image Analysis Stereolog. 19, 51-56, 2000a.
Nagata, T.: Biological microanalysis of radiolabeled and unlabeled compounds by radioautography and X-ray microanalysis. Scanning Microscopy International, 14, on line, 2000b.
Nagata, T.: Electron microscopic radioautographic study on protein synthesis in pancreatic cells of perinatal and aging mice. Bull. Nagano Women's Jr. College 8, 1-22, 2000c.
Nagata, T.: Light microscopic radioautographic study on radiosulfate incorporation into the tracheal cartilage in aging mice. Acta Histochem. Cytochem. 32, 377-383, 2000d.
Nagata, T.: Introductory remarks: Special radioautographology. Cell. Mol. Biol. 46 (Congress Suppl.), 161-161, 2000e.
Nagata, T.: Special radioautographology: the eye. J. Kaken Eye Res. 18, 1-13, 2000f.
Nagata, T.: Three-dimensional high voltage electron microscopy of thick biological specimens. Micron 32, 387-404, 2001a.
Nagata, T.: Three-dimensional and four-dimensional observation of histochemical and cytochemical specimens by high voltage electron microscopy. Acta Histochem. Cytochem. 34, 153-169, 2001b.
Nagata, T. : Special cytochemistry in cell biology. In, Internat. Rev. Cytol. Jeon, K.W., ed., Vol. 211, Chapter 2, Academic Press, New York, pp. 33-154, 2001c.
Nagata, T. : Radioautographology General and Special, In, Prog. Histochem. Cytochem., Graumann, W., Ed., Vol. 37 No. 2, Urban & Fischer, Jena, pp. 57-226, 2002.
Nagata T.: Light and electron microscopic study on macromolecular synthesis in amitotic hepatocyte mitochondria of aging mice. Cell. Mol. Biol. 49, 591-611, 2003.
Nagata, T.: X-ray microanalysis of biological specimens by high voltage electron microscopy. In, Prog. Histochem. Cytochem., Graumann, W., Ed., Vol. 39, No. 4, Urban & Fischer Verlag, Jena, pp. 185-320, 2004.
Nagata T.: Aging changes of macromolecular synthesis in the uro-genital organs as revealed by electron microscopic radioautography. Ann. Rev. Biomed. Sci. 6, 13-78, 2005.
Nagata T.: Electron microscopic radioautographic study on protein synthesis in hepatocyte mitochondria of developing mice. Ann. Microsc. 6, 43-54, 2006a.
Nagata T.: Electron microscopic radioautographic study on nucleic acids synthesis in hepatocyte mitochondria of developing mice. The Sci. World J. 6: 1583-1598, 2006b.
Nagata T.: Macromolecular synthesis in hepatocyte mitochondria of aging mice as revealed by electron microscopic radioautography. I: Nucleic acid synthesis. In, Modern Research and Educational Topics in Microscopy. Mendez-Vilas, A. and Diaz, J. Eds., Formatex Micrscopy Series No. 3, Vol. 1, Formatex, Badajoz, Spain, pp. 245-258, 2007a.
Nagata T.: Macromolecular synthesis in hepatocyte mitochondria of aging mice as revealed by electron microscopic radioautography. II: Protein synthesis. In, Modern Research and Educational Topics in Microscopy. Mendez-Vilas, A. and Diaz, J. eds., Formatex Micrscopy Series No. 3, Vol. 1, Formatex, Badajoz, Spain, pp. 259-271, 2007b.
Nagata, T.: Electron microscopic radioautographic study on macromolecular synthesis in hepatocyte mitochondria of aging mouse. J. Cell Tissue Res. 7, 1019-1029, 2007c.

Nagata, T.; Electron microscopic radioautographic study on nucleic acids synthesis in hepatocyte mitochondria of developing mice. Trends Cell Molec. Biol. 2, 19-33, 2007d.

Nagata, T.; Aging changes of macromolecular synthesis in the mitochondria of mouse hepatocytes as revealed by microscopic radioautography. Ann. Rev. Biomed. Sci. 9, 30-36, 2007e.

Nagata, T.: Radioautographology, Bull. Shinshu Institute Alternat. Med. 2, 3-32, 2007f.

Nagata, T.: Electron microscopic radioautographic study on mitochondrial DNA synthesis in adrenal cortical cells of developing mice. J. Cell. Tis. Res. 8, 1303-1312, 2008a.

Nagata T.: Electron microscopic radioautographic study on mitochondrial DNA synthesis in adrenal cortical cells of developing and aging mice. The Sci. World J. 8, 683-697. 2008b.

Nagata, T.: Sexual difference between the macromolecular synthesis of hepatocyte mitochondria in male and female mice in aging as revealed by electron microscopic radioautography. Chapter 22. In, Women and Aging: New Research, H. T. Bennninghouse, A. D. Rosset, Eds. Nova Biomed. Books, New York, pp. 461-487, 2009a

Nagata, T.: Protein synthesis in hepatocytes of mice as revealed by electron microscopic radioautography. In, Protein Biosynthesis. Esterhouse, T. E. and Petrinos, L. B., Eds., Nova Biomed. Books, New York, pp. 133-161, 2009b.

Nagata, T.: Electron microscopic radioatuographic studies on macromolecular synthesis in mitochondria of various cells. 18EMSM Conference Proc. 9th Asia-Pacific Microscopy Conference (APMC9), Kuala Lumpur, Malaysia, pp. 48-50, 2009c.

Nagata, T.: Recent studies on macromolecular synthesis labeled with ^{3}H-thymidine in various organs as revealed by electron microscopic radioautography. Current Radiopharmaceutics 2, 118-128, 2009d.

Nagata, T.: Electron microscopic radioautographic study on mitochondrial DNA synthesis in adrenal medullary cells of developing and aging mice. J. Cell Tissue Res. 9, 1793-1802, 2009e.

Nagata, T.: Applications of high voltage electron microscopy to thick biological specimens. Ann. Microsc. 9, 4-40, 2009f.

Nagata, T.: Electron microscopic radioautographic study on DNA synthesis of mitochondria in adrenal medullary cells of aging mice. Open Anat. J. 1, 14-24, 2009g.

Nagata, T.: Electron microscopic radioautographic studies on macromolecular synthesis in mitochondria of animal cells in aging. Ann. Rev. Biomed. Sci. 11, 1-17, 2009h.

Nagata, T.: Electron microscopic radioautographic studies on macromoleclular synthesis in mitochondria of some organs in aging animals. Bull. Shinshu Inst. Alternat. Med. Welfare 4, 15-38, 2009i.

Nagata, T.: Electron microscopic radioautographic study on mitochondrial DNA synthesis in adreno-cortical cells of aging ddY mice. Bull. Shinshu Inst. Alternat. Med. Welfare 4, 51-66, 2009j.

Nagata T.: Electron microscopic radioautographic study on mitochondrial RNA synthesis in adrenocortical cells of aging mice. Open Anat J. 2, 91-97, 2010a.

Nagata T. Electron microscopic radioautographic study on mitochondrial RNA synthesis in adrenal medullary cells of aging and senescent mice. J Cell Tissue Res. 10, 2213-2222, 2010b.

Nagata, T.: Macromolecular synthesis in the livers of aging mice as revealed by electron microscopic radioautography. In, Prog. Histochem. Cytochem., Sasse, D., Ed., Elsevier, Amsterdam, Boston, London, New York, Oxford, Paris, Philadelphia, San Diego, St. Louis, Vol. 45, No. 1, pp. 1-80, 2010c.

Nagata, T.: Electron microscopic radioautographic study on protein synthesis of mitochondria in adrenal medullary cells of aging mice. Bulletin Shinshu Inst Alternat Med Welfare 5, 25-37, 2010d.

Nagata, T.: Electron microscopic radioautographic study on mitochondrial RNA synthesis in adrenal cortical and medullary cells of aging mice. J. Biomed. Sci. Enginer. 4, 219-232, 2010e.

Nagata, T.: Electron microscopic radioautographic study on protein synthesis of mitochondria in adrenal cortical cells of aging mice. Bulletin Shinshu Inst. Alternat. Med. Welfare 5, 38-52, 2010f.

Nagata T.: Electron microscopic radioautographic study on mitochondrial DNA, RNA and protein synthesis in adrenal cells of aging mice. Formatex Microscopy Series No. 3, Vol. 3, Formatex, Badajoz, Spain, in press, 2010g.

Nagata, T.: Electron microscopic radioautographic studies on macromolecular synthesis in mitochondria of animal cells in aging. Ann. Rev. Biomed. Sci. 12, 1-29, 2010h.

Nagata, T., Cui, H., Gao, F.: Radioautographic study on glycoprotein synthesis in the ocular tissues. J. Kaken Eye Res. 13, 11-18, 1995.

Nagata, T., Cui, H., Kong, Y.: The localization of TGF-b1 and its mRNA in the spinal cords of prenatal and postnatal aging mice demonstrated with immunohistochemical and in situ hybridization techniques. Bull. Nagano Women's Jr. College, 7, 75-88, 1999a.

Nagata, T., Cui, H., Liang, Y.: Light microscopic radioautographic study on the protein synthesis in the cerebellum of aging mouse. Bull. Nagano Women's Jr. College, 9, 41-60 (2001).

Nagata, T., Fujii, Y., Usuda, N.: Demonstration of extranuclear nucleic acid synthesis in mammalian cells under experimental conditions by electron microscopic radioautography. Proc. 10th Internat. Congr. Electr. Microsc. 2, 305-306, 1982b.

Nagata, T., Hirano, I., Shibata, O., Nagata, T.: A radioautographic study on the DNA synthesis in the hepatic and the pancreatic acinar cells of mice during the postnatal growth, with special reference to binuclearity. Med. J. Shinshu Univ. 11, 35-42, 1966.

Nagata, T., Ito, M., Chen, S.: Aging changes of DNA synthesis in the submandibular glands of mice as observed by light and electron microscopic radioautography. Ann. Microsc. 1, 4-12, 2000a.

Nagata, T. Ito, M., Liang, Y.: Study of the effects of aging on macromolecular synthesis in mouse steroid secreting cells using microscopic radioautography. Methods Find. Exp. Clin. Pharmacol. 22, 5-18, 2000b.

Nagata, T., Iwadare, I., Murata, F.: Electron microscopic radioautography of nucleic acid synthesis in cultured cells treated with several carcinogens. Acta Pharmacol. Toxicol. 41, 64-65, 1977c.

Nagata, T., Kawahara, I.: Radioautographic study of the synthesis of sulfomucin in digestive organs of mice. J. Trace Microprobe Analysis 17, 339-355, 1999.

Nagata, T., Kawahara, I., Usuda, N., Maruyama, M., Ma, H.: Radioautographic studies on the glycoconjugate synthesis in the gastrointestinal mucosa of the mouse. In,

Glycoconjugate in Medicine, Ohyama, M., Muramatsu, T., Eds, pp. 251-256, Professional Postgrad. Service, Tokyo, 1988a.

Nagata, T., Kong, Y.: Distribution and localization of TGFb1 and bFGF, and their mRNAs in aging mice. Bull. Nagano Women's Jr. College 6, 87-105, 1998.

Nagata, T., Ma, H., Electron microscopic radioautographic study on mitochondrial DNA synthesis in hepatocytes of aging mouse. Ann. Microsc. 5, 4-18, 2005a.

Nagata, T., Ma, H., Electron microscopic radioautographic study on RNA synthesis in hepatocyte mitochondria of aging mouse. Microsc. Res. Tech. 67, 55-64, 2005b.

Nagata, T., Momoze, S.: Aging changes of the amitotic and binucleate cells in dog livers. Acta Anat. Nipponica 34, 187-190, 1959.

Nagata, T., Morita, T., I. Kawahara, I.: Radioautographic studies on radiosulfate incorporation in the digestive organs of mice. Histol. Histopathol. 14, 1-8, 1999b.

Nagata, T., Murata, F.: Electron microscopic dry-mounting radioautography for diffusible compounds by means of ultracryotomy. Histochemistry 54, 75-82, 1977.

Nagata, T., Murata, F., Yoshida, K., Ohno, S., Iwadare, N.: Whole mount radioautography of cultured cells as observed by high voltage electron microscopy. Proc. Fifth Internat. Conf. High Voltage Electron Microsc. 347-350, 1977d.

Nagata, T., Nawa, T.: A modification of dry-mounting technique for radioautography of water-soluble compounds. Histochemie 7, 370-371, 1966a.

Nagata, T., Nawa, T.: A radioautographic study on the nucleic acids synthesis of binucleate cells in cultivated fibroblasts of chick embryos. Med. J. Shinshu Univ. 11, 1-5, 1966b.

Nagata, T., Nawa, T., Yokota, S.: A new technique for electron microscopic dry-mounting radioautography of soluble compounds. Histochemie 18, 241-249, 1969.

Nagata, T., Nishigaki, T., Momose, Y.: Localization of anti-allergic agent in rat mast cells demonstrated by light and electron microscopic radioautography. Acta Histochem. Cytochem. 19, 669-683, 1986b.

Nagata, T., Ohno, S., Kawahara, I., Yamabayashi, S., Fujii, Y., Murata, F.: Light and electron microscopic radioautography of nucleic acid synthesis in mitochondria and peroxisomes of rat hepatic cells during and after DEHP administration. Acta Histochem. Cytochem. 16, 610-611, 1979.

Nagata, T., Ohno, S., Murata, F.: Electron microscopic dry-mounting radioautography for soluble compounds. Acta Phamacol. Toxicol. 41, 62-63, 1977a.

Nagata, T., Ohno, S., Yoshida, K., Murata, F.: Nucleic acid synthesis in proliferating peroxisomes of rat liver as revealed by electron microscopical radioautography. Histochem. J. 14, 197-204, 1982a.

Nagata, T., Olea, M. T.: Electron microscopic radioautographic study on the protein synthesis in aging mouse spleen. Bull. Nagano Women's Jr. College 7, 1-9, 1999.

Nagata, T., Shibata, O., Omochi, S.: A new method for radioautographic osbservation on isolated cells. Histochemie 2, 255-259, 1961

Nagata, T., Shibata, O., Nawa, T.: Simplified methods for mass production of radioautograms. -Acta Anat. Nippon.42, 162-166, 1967a.

Nagata, T., Shibata, O., Nawa, T.: Incorporation of tritiated thymidine into mitochondrial DNA of the liver and kidney cells of chickens and mice in tissue culture. Histochemie 10, 305-308, 1967b.

Nagata, T., Shimamura, K., Onozawa, M., Kondo, T., Ohkubo, K., Momoze, S.: Relationship of binuclearity to cell function in some organs. I. Frequencies of binucleate cells in some organs of toads in summer and winter. Med. J. Shinshu Univ. 5, 147-152, 1960a.

Nagata, T., Shimamura, K., Kondo, T., Onozawa, M., Momoze, S., Okubo, M.: Relationship of binuclearity to cell function in some organs. II. Variation of frequencies of binucleate cells in some organs of dogs owing to aging. Med. J. Shinshu Univ. 5, 153-158, 1960b.

Nagata, T., Steggerda, F. R.: Histological study on the deganglionated small intestine of the dog. Physiologist 6, 242-242, 1963.

Nagata, T., Steggerda, F. R.: Observations on the increase of binucleate cells in the ganglion cells of the dog's intestine due to experimental ischemia. Anat. Rec. 148, 315-315, 1964.

Nagata, T., Toriyama, K., Kong, Y., Jin, C., Gao, F.: Radioautographic study on DNA synthesis in the ciliary bodies of aging mice. J. Kaken Eye Res.12, 1-11, 1994.

Nagata, T., Usuda, N.: Image processing of electron microscopic radioautograms in clinical electron microscopy. J. Clin. Electron. Microsc. 18, 451-452, 1985.

Nagata, T., Usuda, N.: Studies on the nucleic acid synthesis in pancreatic acinar cells of aging mice by means of electron microscopic radioautography. J. Clin. Electron Microsc. 19, 486-487, 1986.

Nagata, T., Usuda, N.: Electron microscopic radioautography of protein synthesis in pancreatic acinar cells of aging mice. Acta Histochem. Cytochem. 26, 481-481, 1993a.

Nagata, T., Usuda, N.: In situ hybridization by electron microscopy using radioactive probes. J. Histochem. Cytochem. 41, 1119-1119, 1993b.

Nagata, T., Usuda, N., Ma, H.: Electron microscopic radioautography of nucleic acid synthesis in pancreatic acinar cells of prenatal and postnatal aging mice. Proc. XIth Intern. Cong. Electr. Microsc. 3, 2281-2282, 1984.

Nagata, T., Usuda, N., Ma, H.: Electron microscopic radioautography of lipid synthesis in pancreatic cells of aging mice. J. Clin. Electr. Microsc. 23, 841-842, 1990.

Nagata, T., Usuda, N., Maruyama, M., Ma, H.: Electron microscopic radioautographic study on lipid synthesis in perinatal mouse pancreas. J. Clin. Electr. Microsc. 21, 756-757, 1988b.

Nagata, T., Usuda, N., Suzawa, H., Kanzawa, M.: Incorporation of ^{3}H-glucosamine into the pancreatic cells of aging mice as demonstrated by electron microscopic radioautography. J. Clin. Electron Microsc. 25, 646-647, 1992.

Nagata, T., Yamabayashi, S.: Intracellular localization of ^{3}H-befunolol by means of electron microscopic radioautography of cryo-fixed ultrathin sections. J. Clin. Electron Microsc. 16, 737-738, 1983.

Nagata, T., Yoshida, K., Murata, F.: Demonstration of hot and cold mercury in the human thyroid tissues by means of radioautography and chemography. Acta Pharmacol. Toxicol. 41, 60-61, 1977b.

Nagata, T., Yoshida, K., Ohno, S., Murata, F.: Ultrastructural localization of soluble and insoluble ^{3}H-methyl prednisolone as revealed by electron microscopic dry-mounting radioautography. Proc. 9th Internat. Congr. Electr. Microsc. 2, 40-41, 1978b.

Nishigaki, T., Momose, Y., Nagata, T.: Light microscopic radioautographic study of the localization of anti-allergic agent, tranilast, in rat mast cells. Histochem. J. 19, 533-536, 1987.

Nishigaki, T., Momose, Y., Nagata, T.: Electron microscopic radioautographic study of the localization of an anti-allergic agent, tranilast, in rat mast cells. Cell. Mol. Biol. 36, 65-71, 1990a.

Nishigaki, T., Momose, Y., Nagata, T.: Localization of the anti-allergic agent tranilast in the urinary bladder of rat as demonstrated by light microscopic radioautography. Drug Res. 40, 272-275, 1990b.

Oguchi, K., Nagata, T.: A radioautographic study of activated satellite cells in dystrophic chicken muscle. In, Current Research in Muscular Dystrophy Japan. The Proc. Ann. Meet. Muscular Dystrophy Res. 1980, pp. 16-17, Ministry of Welfare of Japan, Tokyo, 1980.

Oguchi, K., Nagata, T.: Electron microscopic radioautographic observation on activated satellite cells in dystrophy chickens. In, Clinical Studies on the Etiology of Muscular Dystrophy. Annual Report on Neurological Diseases 1981, pp. 30-33, Ministry of Welfare of Japan, Tokyo, 1981.

Ohno, S., Fujii, Y., Usuda, N., Endo, T., Hidaka, H., Nagata, T.: Demonstration of intracellular localization of calmodulin antagonist by wet-mounting radioautography. J. Electron Microsc. 32, 1-12, 1983.

Ohno, S., Fujii, Y., Usuda, N., Nagata, T., Endo, T., Tanaka, T., Hidaka, H.: Intracellular localization of calmodulin antagonists (W-7). In, Calmodulin and intracellular Ca^{2+} receptors. Kakiuchi, S., Hidaka, H, Means, A. R., Eds., pp. 39-48, Plenum Publishing Co., New York, 1982.

Olea, M. T.: An ultrastructural localization of lysosomal acid phosphatase activity in aging mouse spleen: a quantitative X-ray microanalytical study. Acta Histochem. Cytochem. 24, 201-208, 1991.

Olea, M. T., Nagata, T.: X-ray microanalysis of cerium in mouse spleen cells demonstrating acid phosphatase activity using high voltage electron microscopy, Cell. Mol. Biol. 37, 155-163, 1991.

Olea, M. T., Nagata, T. : Simultaneous localization of ^{3}H-thymidine incorporation and acid phosphatase activity in mouse spleen: EM radioautography and cytochemistry. Cell. Mol. Biol. 38, 115-122, 1992a.

Olea, M. T., Nagata, T.: A radioautographic study on RNA synthesis in aging mouse spleen after ^{3}H-uridine labeling in vitro. Cell. Mol. Biol. 38, 399-405, 1992b.

Oliveira, S. F., Nagata, T., Abrahamsohn, P. A., Zorn, T. M. T.: Electron microscopic radioautographic study on the incorporation of ^{3}H-proline by mouse decidual cells. Cell. Mol. Biol. 37, 315-323, 1991.

Oliveira, S. F., Abrahamsohn, P. A., Nagata, T., Zorn, T. M. T.: Incorporation of ^{3}H-amino acids by endometrial stromal cells during decidualization in the mouse. A radioautographical study. Cell. Mol. Biol. 41, 107-116, 1995.

Pearse, A. G. E.: Histochemistry, Theoretical and Applied. 4th Ed. Vol. 1. 439 pp., 1980, Vol. 2. 1055 pp., 1985, Vol. 3. Ed. with P. Stoward, 728 pp. Churchill Livingstone, Edinburgh, London and New York, 1991.

Sakai, Y., Ikado, S., Nagata, T.: Electron microscopic radioautography of satellite cells in regenerating muscles. J. Clin. Electr. Microsc. 10, 508-509, 1977.

Sato, A.: Quantitative electron microscopic studies on the kinetics of secretory granules in G-cells. Cell Tissue Res. 187, 45-59, 1978.

Sato, A., Iida, F., Furihara, R., Nagata, T.: Electron microscopic raioautography of rat stomach G-cells by means of ^{3}H-amino acids. J. Clin. Electron Microsc. 10, 358-359, 1977.

Shimizu, T., Usuda, N., Yamanda, T., Sugenoya, A., Iida, F.: Proliferative activity of human thyroid tumors evaluated by proliferating cell nuclear antigen/cyclin immnohistochemical studies. Cancer 71, 2807-2812, 1993.

Sun, L.: Age related changes of RNA synthesis in the lungs of aging mice by light and electron microscopic radioautography. Cell. Mol. Biol. 41, 1061-1072, 1995.

Sun, L., Gao, F., Duan, H., Nagata, T.: Light microscopic radioautography of DNA synthesis in pulmonary cells in aging mice. In, Radioautography in Medicine, Nagata, T. Ed., pp. 201-205, Shinshu University Press, Matsumoto, 1994.

Sun, L., Gao, F., Nagata, T.: Study on the DNA synthesis of pulmonary cells in aging mice by light microscopic radioautography. Cell. Mol. Biol. 41, 851-859, 1995a.

Sun, L., Gao, F., Jin, C., Duan, H., Nagata, T.: An electron microscopic radioautographic study on the DNA synthesis of pulmonary tissue cells in aging mice. Med. Electron. Microsc. 28, 129-131, 1995b.

Sun, L., Gao, F., Jin, C., Nagata, T.: DNA synthesis in the tracheae of aging mice by means of light and electron microscopic radioautography. Acta Histochem. Cytochem. 30, 211-220, 1997a.

Sun, L., Gao, F., Nagata, T.: A Light Microscopic radioautographic study on protein synthesis in pulmonary cells of aging mice. Acta Histochem. Cytochem. 30, 463-470, 1997b.

Suzuki, K., Imada, T., Gao, F., Ma, H., Nagata, T.: Radioautographic study of benidipine hydrochloride: localization in the mesenteric artery of spontaneously hypertensive rat. Drug Res. 44, 129-133, 1994.

Terauchi, A., Mori, T., Kanda, H., Tsukada, M., Nagata, T.: Radioautographic study of 3H-taurine uptake in mouse skeletal muscle cells. J. Clin. Electron Microsc. 21, 627-628, 1988.

Terauchi, A., Nagata, T.: Observation on incorporation of ^{3}H-taurine in mouse skeletal muscle cells by light and electron microscopic radioautography. Cell. Mol. Biol. 39, 397-404, 1993.

Terauchi, A., Nagata, T.: In corporation of ^{3}H-taurine into the blood capillary cells of mouse skeletal muscle. Radioautography in Medicine, Nagata, T. ed., Shinshu University Press, Matsumoto, 1994.

Toriyama, K.: Study on the aging changes of DNA and protein synthesis of bipolar and photo-receptor cells of mouse retina by light and electron microscopic radioautography. Cell. Mol. Biol. 41, 593-601, 1995.

Tsukahara, S., Yoshida, K., Nagata, T.: A radioautographic study on the incorporation of ^{14}C-bupranolol (beta-blocking agent) into the rabbit eye. Histochemistry 68, 237-244, 1980.

Usuda, N., Nagata, T.: Electron microscopic radioautography of acyl-CoA mRNA by in situ hybridization. J. Clin. Electron Microsc. 25, 332-333, 1992.

Usuda, N., Nagata, T.: The immunohistochemical and in situ hybridization studies on hepatic peroxisomes. Acta Histochem. Cytochem. 28, 169-172, 1995.

Usuda, N., Hanai, T., Morita, T., Nagata, T.: Radioautographic demonstration of peroxisomal acyl-CoA oxidase mRNA by in situ hybridization. In, Recent advances in cellular and molecular biology, Vol. 6. Molecular biology of nucleus, peroxisomes, organelles and cell movement. Wegmann, R. J., Wegmann, M., Eds, pp.181-184, Peeters Press, Leuven, 1992.

Uwa, H., Nagata, T.: Cell population kinetics of the scleroblast during ethisterone-induced anal-fin process formation in adult females of the Medaka. Dev. Growth Different. 9, 693-694, 1976.

Watanabe, I., Makiyama, M. C. K., Nagata, T.: Electron microscopic radioautographic observation of the submandibular salivary gland of aging mouse. Acta Microscopica 6. 130-131, 1997.

Yamabayashi, S., Gunarso, W., Tsukahara, S., Nagata, T.: Incorporation of ^{3}H-befunolol (beta blocking agent) into melanin granules of ocular tissues in the pigmented rabbits. I. Light microscopic radioautography. Histochemistry 73, 371-375, 1981.

Yamada, A. T.: Timely and topologically defined protein synthesis in the periimplanting mouse endometrium revealed by light and electron microscopic radioautography. Cell. Mol. Biol. 39, 1-12, 1993.

Yamada, A., Nagata, T.: Ribonucleic acid and protein synthesis in the uterus of pregnant mouse during activation of implantation window. Med. Electron Microsc. 27, 363-365, 1992a.

Yamada, A., Nagata, T.: Light and electron microscopic raioautography of DNA synthesis in the endometria of pregnant-ovariectomized mice during activation of implantation window. Cell. Mol. Biol. 38, 763-774, 1992b.

Yamada, A., Nagata, T.: Light and electron microscopic raioautography of RNA synthesis of peri-implanting pregnant mouse during activation of receptivity for blastocyst implantation. Cell. Mol. Biol. 38, 211-233, 1993.

Yoshinaga, K.: Uterine receptivity for blastcyst implantation. Ann. N. Y. Acad. Sci. USA, 541, 424-431, 1988.

Yoshizawa, S., Nagata, A., Honma, T., Oda, M., Murata, F., Nagata, T.: Study of ethionine pancreatitis by means of electron microscopic radioautography. J. Clin. Electron Microsc. 7, 349-350, 1974.

Yoshizawa, S., Nagata, A., Honma, T., Oda, M., Murata, F., Nagata, T.: Radioautographic study of protein synthesis in pancreatic exocrine cells of alcoholic rats. J. Clin. Electron. Microsc. 10, 372-373, 1977.

10

Caveolar Vesicles in Cellular Senescence

Keith Wheaton
Department of Biology, York University, Toronto
Canada

1. Introduction

The replicative limit of human fibroblasts has long provided a model to assess the molecular mechanisms which underlie cellular aging. In culture, fibroblasts which reach the end of their proliferative lifespan acquire profound molecular changes that limit their response to growth factors, and cause permanent exit from the cell cycle. Part of the senescence programme is due to a well established link between telomere attrition which occurs with each population doubling and the subsequent activation of the p53 tumour suppressor. Critical shortening of telomeres is thought to cause a form of DNA damage, that leads to the activation of caretaker proteins ATM, ATR or DNA-PK which activate p53, leading to the initiation of senescence through p53 effector genes. In addition, p53 mediates senescence by many other stimuli including oxidative stress, DNA damaging agents and oncogenic activation.

Caveolins are the main scaffolding proteins driving the formation of caveolae (50-100 nm wide cave like invaginations at the plasma membrane) from lipid rafts and allows the organization of many signalling cascades. This compartmentalization concentrates receptors, proteins with lipid anchors, and the lipids from which second messengers are derived. In this capacity caveolin has also been shown to bind and inactivate many key components of mitogenic pathways through the caveolin scaffolding domain (CSD) and thus is often considered a tumour suppressor. About a decade ago, the original investigations into the relationship between caveolae and senescence showed elevated levels of the caveolin proteins during replicative senescence. Both the ectopic expression or endogenous upregulation of caveolin was shown to lead to p53 mediated senescent arrest. However, this upregulation has not been proven to lead to an increase of caveolae at the cell surface, but does lead to increased number of internalized structures. The mislocalized caveolar vesicles likely influence the regulation of mitogenic signals that normally are integrated through caveolae. In addition the caveolin protein has been shown to regulate many signalling cascades that have an impact on senescence, and on the activation of p53.

1.1 Fibroblast model of senescence

Primary human diploid fibroblasts (HDFs) have traditionally served as an experimental model to investigate cellular and molecular aspects of aging. Normal fibroblasts derived from human donors are neither immortal or transformed and are limited in the number of times they are capable of dividing *in vitro* and *in vivo*. This finite replicative lifespan terminates with the acquisition of a phenotype having distinct morphological and biochemical characteristics termed replicative senescence (Hayflick, 1965a; Hayflick and

Moorhead, 1961). This is a cell fate distinct from apoptosis or differentiation since the cells remain viable but are refractory to mitogenic signals. Senescing fibroblasts progressively acquire a flattened morphology with an increased cellular volume, an irregular shape, and increased accumulation of debris (Hayflick, 1965a; Hayflick and Moorhead, 1961).

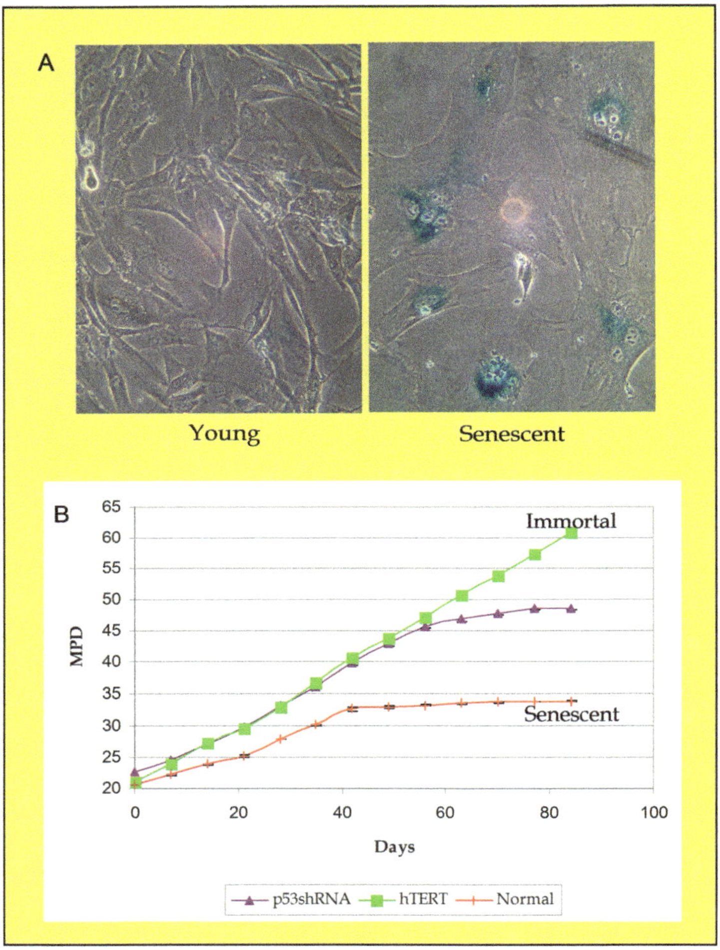

Fig. 1. Cellular Senescence
A) Young and senescent primary fibroblasts prepared with an acidic β-galactosidase assay, staining senescent cells blue. Note the enlarged and flattened morphology of senescent cells compared to young fibroblasts. **B)** A growth curve of AG08470 primary fibroblasts virally transfected with empty vector (Normal), human telomerase (hTERT) or p53 short hairpin RNA (p53 shRNA). This strain of fibroblasts undergo replicative senescence at a mean population doubling (MPD) of 34 (Normal). The knockdown of p53 protein levels using p53 shRNA extends the proliferative lifespan by 13 MPDs. The ectopic expression of hTERT immortalizes the fibroblast strain.

Additional phenotypic changes include increased actin stress fibers and an increase in cells containing irregular and multiple nuclei (Wheaton et al., 1996). These morphological changes were further shown to be dependent on the innate property of the number of mean population doublings of a particular cell strain, and were not influenced by proximity to lower passage cells, splitting ratio, viral contamination, media composition, or periods of storage in liquid nitrogen (Hayflick, 1965b; Hayflick and Moorhead, 1961). Senescent cells can be distinguished from low passage fibroblasts by the innate property to stain blue (Fig. 1A) when processed with an acidic β-galactosidase (β-gal) assay (Dimri, 1995).

Although HDFs in culture are separated from their normal cellular environment, several observations support the use of HDFs grown *in vitro* as a valid model of biological aging. These include: 1) Fibroblasts isolated from young individuals are reproducibly able to undergo more mean population doublings (MPDs) then cells from old donors of the same species (Martin et al., 1970). 2) Fibroblasts from longer lived species undergo more population doublings than short lived species, indicating a relationship between the maximum lifespan of a species and the proliferative capacity of their fibroblasts grown in culture (Goldstein and Singal, 1974; Rohme, 1981). 3) Fibroblasts from individuals with premature aging syndromes such as progeria and Werner's syndrome undergo fewer population doublings in culture than normal cells (Brown, 1990). 4) Normal human fibroblasts in tissues stain with β-gal, indicating senescence *in vivo* (Dimri et al., 1995). These and other observations regarding biochemical differences that accompany senescent HDFs suggest that the HDF model of cellular aging reflects the normal biological processes that occur during *in vivo* aging. In addition to this evidence it has become increasingly apparent that senescence is a natural barrier to oncogenic transformation in many different tumour types (Collado and Serrano, 2010).

2. The post-receptor block

As a result of the refractory nature of senescent fibroblasts to mitogenic stimuli, many of the original studies exploring the molecular details of replicative senescence focused on various growth factor signalling pathways. The growth factors which have been reported to give optimal proliferative response in human fibroblasts are the epidermal growth factor (EGF), insulin like growth factor 1 (IGF-1) and dexamethasone (Phillips and Cristofalo, 1988). Although these tyrosine kinase receptors (RTKs) have many potential down stream mediators, the classical example is a cascade through the small GTPase Ras that leads to the mitogen-activated protein kinase (MAPK) cascade. Early studies investigating the loss of mitogenic response of senescent human diploid fibroblasts focused upon the growth factor receptor:ligand interactions (Gerhard et al., 1991; Phillips et al., 1983; Sell et al., 1993). In the cases of EGF, dexamethasone and IGF-1, it was found that the number of receptors remained the same per unit surface area in senescent cells and that the ligand affinity remained unchanged (Sell et al., 1993). Similarly, the binding kinetics of the glucocorticoid receptor and the insulin receptor have been shown to remain unaltered in senescent cells (Chua et al., 1986). The degradation process of EGF receptor has been shown to be largely unchanged with fibroblast age (Phillips et al., 1984). Studies have also looked at tyrosine autophosphorylation of EGF and PDGF and found no significant differences (Chua et al., 1986; Gerhard et al., 1991). Thus, the general consensus is that decrease in mitogen response is not due to alterations of receptor function or processing (Rattan and Derventzi, 1991).

However, it has been established that the downstream responses of growth factors, such as the induction of the immediate early genes *c-fos* and *egr-1*, are blunted in senescent cells (Meyyappan et al., 1999; Riabowol et al., 1992; Seshadri and Campisi, 1990). Thus, the senescence-specific growth block is thought not to be due to alterations in growth factor-receptor engagement or processing, but to a postreceptor block which leads to reduced immediate early gene responsiveness. Many studies have attempted to identify the points where signalling cascades are modified in senescent cells to account for this block.

2.1 EGFR-Ras-MAPK cascade in senescence

The signal transduction pathway through the epidermal growth factor receptor (EGFR) (and other RTKs) begins with ligand engagement, which initiates dimerization and autophosphorylation within the cytoplasmic domains of the receptors (Fantl et al., 1993). This, in turn, is followed by Src homology 2 domain (SH2) containing molecules interacting with the receptor and recruiting other factors including adaptors. The classic adaptor molecule is Grb2, which binds to the growth factor receptor directly or indirectly through Shc and is associated constitutively with the son of sevenless (SOS) protein. The SOS guanine nucleotide exchange factor catalyses the transformation of inactive GDP-ras to active GTP-ras (Boguski and McCormick, 1993). Activated Ras recruits Raf1 to the membrane allowing its activation (Moodie and Wolfman, 1994). Appropriately localized and activated Raf1 then initiates a kinase cascade through the MAPK cascade. Raf 1 phosphorylates MEK (MAP kinase kinase) which activates Erk1 and Erk2 (MAPK) and leads to MAPK nuclear localization. In the nucleus MAPK phosphorylates transcription factors such as Elk-1 & serum response factor (SRF), stimulating cell proliferation (reviewed in (Wheaton et al., 1996)).

Many studies have attempted to elucidate the role of signalling cascades in senescence, and investigations into the signalling pathways downstream of the EGFR are the most common. It is unclear whether these pathways are compromised in senescence, or whether they need to be intact to promote senescence. Considering that EGFR has been localized to caveolae it is of interest to evaluate studies that have seen a differential regulation in senescent cells. First, it has been found that the EGFR is cleaved to a 100 kD fragment in non-ionic detergent isolates from senescent but not from young primary fibroblasts. This activity appears to be confined to a specialized region of the plasma membrane in senescent cells and was independent of receptor turnover (Carlin et al., 1994). However, it is unclear whether this proteolytic processing is an artefact of the isolation procedure or is physiologically relevant. Second, several studies have reported an age dependent decline in mitogenic stimulation of rat hepatocytes which may be caused by a reduced association between Shc and the EGFR (Hutter et al., 2000; Palmer et al., 1999). Although total tyrosine phosphorylation levels on EGFR appear equivalent after serum stimulation in young and senescent cells, there is a specific reduction of phosphorylation on tyrosine 1173 (a known phosphotyrosine binding (PTB) domain that interacts with Shc (Okabayashi et al., 1994)) in senescent cells. The 1173 residue is known as an autophosphorylation site (Downward et al., 1985), but can also be targeted by Src kinase *in vitro* (Wright et al., 1996) and also is recognized by the SHP-1 phosphatase (Keilhack et al., 1998). This site was subsequently shown to be dephosphorylated by an upregulation of SHP-1 and other phosphatases in senescent human fibroblasts (Tran et al., 2003). Thus, the EGFR in senescence is regulated by differential interaction of adapter molecule Shc at the plasma membrane, which likely compromises its downstream signalling. Interestingly, different splice forms of Shc may have a role in cellular senescence and longevity (Migliaccio et al., 1999).

Studies have also focused on the downstream effectors of EGFR in senescence, particularly on the MAPK (erk 1/2). MAPK is pivotal in transactivation of various transcription factors that drive the cell through proliferation. Although a general reduction in MAPK (Erk1/2) activation has not been seen in senescence (Kim et al., 2000; Tresini et al., 2001), there is a dramatic reduction in localization of activated Erk to the nucleus of senescent fibroblasts (Bose et al., 2004; Kim et al., 2000; Lim et al., 2000; Tresini et al., 2001). However, it has also been found that a phosphatase MKP-2 is stabilized and acts on Erk in senescent cells (Torres et al., 2003), explaining the reduction in active MAPK. This reduced MAPK activity leads to a lack of Elk-1 activation, which loses its ability to bind to the adjacent ets region of the serum response element in *c-fos* promoter. These studies also concede that the reduction in binding of elk-1 alone cannot explain the entire loss of binding at the *c-fos* promoter, or the repression of its transcription. Interestingly, there is an equal amount of non-phosphorylated MAPK (activated) in the nucleus of young and senescent cells, and although reduced, there is still some localization of phospho-MAPK to the senescent cell nucleus. Therefore, these studies raise the question of whether the reduction in MAPK and Elk-1 represents a significant threshold to suppress *c-fos* transcription and explain the post receptor block. This was confirmed when SRF, the primary transcription factor of c-fos and Egr-1, was shown to be hyper-phosphorylated in senescent cells by PKCδ, and suppress immediate early gene transcription (Atadja et al., 1994; Wheaton and Riabowol, 2004). It was shown that immediate early genes could be restored in senescent cells by blocking PKCδ activity independent of MAPK. However, even though immediate early genes could be restored, this was insufficient to allow senescent cells to re-enter the cell cycle (Wheaton and Riabowol, 2004). Not surprisingly, other factors are clearly involved in senescent arrest besides the refractory mitogenic response.

Premature senescence can be induced by the introduction of oncogenic Ras (Serrano et al., 1997) or Raf (Zhu et al., 1998) or constitutively active MAPK (Lin et al., 1998) in primary fibroblasts. In every case, there is an upregulation of p21 and p16 and an induction of senescence which depends on p53 activity. Also the inducible levels of the *c-fos* transcript were found to be repressed in oncogenic Ras induced senescence (Serrano et al., 1997). This oncogenic Ras/Raf/MAPK pathway has been shown to impinge on the $p38^{MAPK}$ stress activated kinase, which is required for the induction of premature senescence (Debacq-Chainiaux et al., 2010). The p38 activity likely impinges directly on p53, or the Rb/p16 pathway to lead to this arrest. The induction of premature senescence likely represents a natural mechanism to block oncogenesis when there is an inappropriately sustained mitogenic signal. Thus, the perturbations of the cellular system as a whole induced by hyperactive Ras/Raf/MAPK are important to inducing premature senescence and only indirectly produce the post receptor block.

2.2 Protein kinase C activity during senescence

Conventional PKCs (α, β, & γ isoforms) and novel PKCs (δ, ε, η, & θ isoforms) are activated by the lipid second messenger 1,2 diacylglycerol (DAG) which is generated downstream of many G-protein-coupled receptors and tyrosine-kinase receptors (Dempsey et al., 2000; Newton, 2001). PKC family members are activated in response to a large number of extracellular signals, and regulate a large number of effectors including receptors, kinases, cytoskeletal components and transcription factors. The outcomes of these signals include changes in proliferation, gene expression, differentiation, permeability, migration, hypertrophy, apoptosis and exocytosis. Each outcome depends on the cellular context, the originating signal and the particular isoform activated (Dempsey et al., 2000). Altered PKC activity has been observed in senescent fibroblasts where a change in serum-induced translocation of PKC from the cytosol to the plasma membrane was reported (De Tata et al., 1993). In contrast, exogenously added phorbol 12-myristate 13-acetate (PMA, a DAG analog) stimulated this translocation identically in young and old cells, implying that the difference in response to serum was due to changes in the production of DAG (De Tata et al., 1993). A similar differential response to serum vs. PMA was also reported for the induction of the *c-fos* gene (Riabowol et al., 1992). Although old fibroblasts have a higher basal level of DAG (Vannini et al., 1994; Venable et al., 1994), the response due to serum induction was reported to be much lower in senescent cells (Chang and Huang, 1994). It should be noted that the specific isoforms were not noted in these studies, but are likely PKC α or β. Attenuated PKC response could contribute to the senescent phenotype in several ways. For example, decreased phosphorylation of substrates that are downstream of PKC could affect the expression of many growth-regulatory genes such as *c-fos* or other MAPK substrates. Alternatively, the higher basal amount of DAG in the resting state of senescent cells compared to young cells may also allow constitutive activation of some kinases. One such kinase, PKCα was found to be required in oxidative stress induced senescence in human fibroblasts. PKCα was shown to induce the senescent phenotype by acting on erk1/2 and Sp1, which converge on and up regulate transcription from the p21 promoter (Kim and Lim, 2009). Interestingly the knockdown of PKCα restored proliferative capacity of senescent cells, implying that the increased basal levels in DAG in senescence constitutively activate PKCα.

The PKCδ isoform has a multifunctional role in various processes including growth inhibition, differentiation, apoptosis and tumour suppression (Basu and Pal, 2010). Many groups have reported that PKC δ activation leads to growth inhibition, including CHO cells, smooth muscle, NIH 3T3 fibroblasts, human glioma cells, and capillary endothelial cells (reviewed in (Gschwendt, 1999)). Studies using ectopic PKCδ implicate p27 as part of the cellular arrest phenotype (Ashton et al., 1999). PKCδ also assists in the sustained up-regulation of p21 message and protein levels in a p53-independent manner (Zezula et al., 1997). However, PKCδ is also capable of phosphorylating and activating p53 in response to genotoxic stress (Johnson et al., 2002; Yoshida et al., 2006). The activity of PKCδ is substantially higher in senescent cells, and has been shown to hyper-phosphorylate SRF, which prevents it from binding to DNA and transactivating the immediate early genes *egr-1* and *c-fos* (Wheaton and Riabowol, 2004). Similarly, increased PKCδ activity during senescence was shown to indirectly inactivate WARTS, a kinase required to exit cytokinesis, and arrest cells after nuclear fission but not before division (Takahashi et al., 2006). A subsequent report also established that constitutively active PKCδ can induce senescent phenotype when introduced to young fibroblasts (Katakura et al., 2009). Interestingly, the activation of PKCδ by PMA is known to be localized to caveolae in cardiomyocytes (Rybin et al., 2008). PKCδ has also been found in the caveolar membrane microdomains of human fibroblasts (Wheaton, 2002). The high activity of PKCδ in senescence may be due to the loss of Src activity in caveolae, which normally inactivates PKCδ through degradation (Blake et al., 1999).

2.3 Phospholipase D activity during senescence

Phospholipase D (PLD) activity is required for an integrated mitogenic response from a variety of receptors including RTKs (reviewed in (Exton, 1998)). PLD catalyses the hydrolysis of phosphatidylcholine (PC) to yield phosphatidic acid (PA) and choline. PA is an important second messenger in many biological processes (Exton, 1997) such as in a membrane localization of Raf-1 (Ghosh et al., 1996). PA can be modified further to the mitogenic lyso-PA by phospholipase A2 or to DAG by PA phosphohydrolase (McDermott et al., 2004). Activation of PLD is complex, involving phosphatidylinositol 4,5-bisphosphate (PIP_2), PKCα/β, Rho family GTPases and ARF GTPase (Exton, 1999). PLD activity has been correlated with mitogenic activity (reviewed in (Foster and Xu, 2003)) and has been shown to be reduced in senescent cells (Venable et al., 1994), possibly due to the reduction of PKC activity or inhibition by ceramide (Venable and Obeid, 1999).

Ceramide acts as a second messenger in a variety of biological processes after being generated from membrane associated sphingomyelin by neutral sphingomyelinase (reviewed in (Foster and Xu, 2003)). The ceramide levels remain stable with fibroblast age, but rise 2-4 fold relative to young cells upon the onset of senescence and are accompanied by an increase in the neutral Mg^{2+} dependent sphingomyelinase activity. The addition of a synthetic, cell permeable analog of ceramide to young cells led to growth arrest and cellular senescence (Venable et al., 1995). These cells showed cellular arrest, lack of Rb phosphorylation, no AP-1 production (Venable et al., 1995) and stained positive with the β-gal test for senescence (Mouton and Venable, 2000). Ceramide acts to inhibit the activation of PLD, by possibly three mechanisms, inhibiting activation by PKC, translocation to the membrane and gene expression, but none of these has been clearly defined (Exton, 1999). Exogenous ceramide inhibits PLD activity in primary fibroblasts (Venable et al., 1994) and this inhibition was shown to depend on PKC interaction but not on phosphorylation or translocation in cell free systems (Venable et al., 1996). However, ceramide has been shown to disrupt the liquid order of lipid rafts, which leads to PLD inhibition (Gidwani et al., 2003). To this end, recent work has identified that the deficiency in PLD activity is specific to the senescent membrane fractions, and not to changes in the cytosol (Webb et al., 2010). The increase in ceramide in senescent membranes likely changes the properties of lipid rafts, which no longer supports an integrated PLD response. Thus, the elevated ceramide levels and PKC regulation are important factors in reduced PLD activity in senescent cell membranes.

2.4 The phosphoinositide 3-kinase pathway during senescence

Due to the role of IGF-1 in the optimal proliferative response of primary fibroblasts (Phillips and Cristofalo, 1988), a common downstream effector of this pathway, phosphoinositide 3-kinase (PI3K) was assessed for its role during senescence. PI3K consists of a p85 regulatory subunit and a p110 catalytic subunit which selectively phosphorylates the 3-OH position of phophoinositides. The enzyme is implicated in both mitogenic and survival signals which are propagated through down stream effectors such as PKB/Akt, IRS-1 and p70 S6K (reviewed in (Vanhaesebroeck et al., 1997). Application of a specific PI3K inhibitor (LY294002) to young primary fibroblasts resulted in cell arrest and the senescent phenotype. Conversely, in the same study, inactivation of the Ras/Raf/MAPK pathway by PD98059 also arrested cell growth, but no induction of senescence was noted (Tresini et al., 1998). Another group using the same approach showed that p27 was highly up-regulated, but that the presumed mediators of senescence, p21, p16, and p53 were all down-regulated. They additionally showed that p27 was the only mediator of the response, since p27-/- mouse embryonic fibroblasts did not become senescent with application of the drug (Collado et al., 2000). Overall, the role of PI3K seems to be in maintaining continued growth, and its inhibition leads to senescence through a non-traditional mechanism of p27 up-regulation.

One of the main effectors of the PI3K pathway is the mammalian target of rapamycin (mTOR), a kinase that integrates signals from multiple mitogenic pathways. The mTOR protein is activated by PI3K through the pathway of PKB/Akt leading to TSC1/TSC2 inactivation, which normally acts as a guainine exchange protein for the small GTPase Rheb. Rheb-GDP holds mTOR in an inactive state, until it is activated through PI3K-Akt-TSC1/2, allowing GTP exchange (Li et al. 2004). The increased mTOR activity has a global effect on translation and cell growth, which is required for rapid cell proliferation (Guertin and Sabatini, 2007). However, it has been shown that serum mitogens are required for the full development of the senescent phenotype (Satyanarayana et al., 2004). The mitogen based induction of senescence has been shown to involve the activation of mTOR, because its inactivation by rapamycin prevents senescence, even in the presence of DNA damage (Leontieva and Blagosklonny, 2010). Recently, treatment with rapamycin has even been shown to extend the proliferative lifespan of human fibroblasts (Cao et al., 2011). These effects were shown to depend on increased autophagic degradation of proteins that enforce senescence, but may also involve a disconnection between mitogen induced translational control and promotion of the cell cycle. Thus, the roles of PI3K and mTOR in senescence would seem to oppose each other, even though they are in the same pathway. These divergent results are difficult to reconcile, but possibly can be explained by other downstream PI3K targets that promote proliferation over senescence. This is supported by the finding that transcriptional targets of the PI3K/PKB are not properly regulated in replicative senescence (Lorenzini et al., 2002). Conversely, in stress or oncogenic induced senescence the PI3K pathway has been shown to have a positive role in the onset of senescence (Binet et al., 2009; Matuoka et al., 2003), consistent with a role for mTOR.

3. P53 and telomere attrition

Replicative senescence is widely accepted to be triggered by critically short telomeres. Stress-induced senescence (also called premature senescence) occurs when cells are exposed to sub-lethal cellular stress such as DNA damage (Di Leonardo et al., 1994; Resnick-Silverman et al., 1998), oxidative stress (Chen and Ames, 1994), or oncogenic stress (Serrano et al., 1997). Both replicative and stress-induced senescence result in the accumulation of cells that are incapable of further divisions and arrest primarily in the G1 phase of the cell cycle. Thus, senescence is thought to represent an intrinsic barrier to oncogenesis by limiting proliferation.

The p53 tumour suppressor protein is a transcription factor that under normal cellular conditions is unstable and inactive. In response to many forms of cellular stress, p53 becomes activated by post-translational modification and is able to transactivate target genes that regulate diverse cellular processes including, cell cycle progression, senescence, DNA repair, metabolism and cell survival (Appella and Anderson, 2001; Vousden and Lane, 2007). Active p53 protein promotes cell cycle arrest through transcriptional activation of many genes including: *p21WAF1* (el-Deiry et al., 1993), *GADD45α* (Kastan et al., 1992), *BTG2* (Rouault et al., 1996), *REPRIMO* (Ohki et al., 2000), *14-3-3σ* (Hermeking et al., 1997), and Prl-3 (Basak et al., 2008). Different subsets of p53-responsive genes regulate DNA repair, metabolism, survival and apoptosis. The p53 protein has been implicated as one of the key mediators of both the onset (Hara et al., 1991; Shay et al., 1991) and maintenance (Beausejour et al., 2003; Gire and Wynford-Thomas, 1998) of cellular senescence. The transcriptional targets of p53 shown to initiate replicative senescence include *p21WAF1* (Brown et al., 1997; el-Deiry et al., 1993; Noda et al., 1994), BTG2 (Wheaton et al., 2010), GADD45α (Jackson and Pereira-Smith, 2006) and PAI (Kortlever et al., 2006). When p53 is knocked down by shRNA (or other method) in fibroblasts it extends the proliferative lifespan by 10-20 population doublings, by avoiding p53 induced senescent program (Fig. 1B).

Various chemical agents and gamma irradiation can cause DNA double stranded breaks, which also lead to premature senescence. The biological process that occurs at a doubled stranded break involves hundreds of H2AX histones becoming phosphorylated (γH2AX) as the signal propagates from the break site. This DNA damage can visualized in the nucleus of a cell as foci by immunofluorescence using a γH2AX phospho-specific antibody (Rogakou et al., 1999; Rogakou et al., 1998). The kinase Ataxia telangiectasia mutated (ATM) acts as a sensor of DNA damage, becomes activated at the γH2AX foci and phosphorylates p53 (Bakkenist and Kastan, 2003). This phosphorylation stabilizes and activates p53, and allows it to transactivate its effector genes.

Telomeres consist of a tandem repeat of 6 nucleotides (TTAGGG) that cap the ends of linear double stranded DNA in mammals. Human telomeres consist of 5-10 kilobase pairs at the end of every chromosome arm and have an overhang of single-stranded telomeric repeat of several hundred bases. If ends of chromosomes were not protected by telomeres the cell would detect the end of the DNA as a double stranded break and initiate a DNA damage response. Telomeres normally protect the double stranded chromosome ends by the formation of telomere-loops (t-loops) in which the telomeric single-stranded overhang hybridizes within the double-stranded telomeric region of a chromosome. T-loops are maintained and stabilized by a complex of proteins referred to as shelterin (de Lange, 1994). Telomeres are know to shorten with each round of DNA replication because of the "end-replication problem"; the inability of DNA polymerases to completely replicate the 3′ end of linear DNA molecules (Harley et al., 1990). Telomeres will shorten to a critical length after a finite number of DNA replication cycles, and lose the ability to form the sheltern/t-loop cap at the ends of chromosomes. The DNA damage signal arising from eroded telomeres will activate ATM, leading in turn to p53 activation (Herbig et al., 2004). Thus, t-loop disruption as a consequence of telomere shortening exposes telomeric DNA that leads to p53 activation and the initiation of replicative senescence (Herbig et al., 2004; Karlseder et al., 2002; Li et al., 2003; Li et al., 2004; Stewart et al., 2003). This model is supported by studies that ectopically expressed telomerase (hTERT) allowing fibroblasts to bypass senescence (Figure 1B) and become immortalized by preventing telomere shortening (Bodnar et al., 1998; Vaziri and

Benchimol, 1998). Hence, critically short telomeres provide a physiological signal for activation of the p53-dependent replicative senescence program in human cells.

4. Caveolae

Caveolae are flask shaped invaginations which form at the plasma membrane from lateral assemblies of cholesterol and sphingolipids known as lipid rafts (reviewed in (Harder and Simons, 1997)). Lipid rafts are composed of sphingomyelin, glycosphingolipids and cholesterol which form microdomain rafts surrounded by the phosphatidylcholine rich plasma membrane. These rafts sequester a specific subset of transmembrane associated proteins (Simons and Ikonen, 1997). Thus, lipid rafts are organelle like environments which have been postulated to have specific functions in various cell types (Simons and Toomre, 2000). Lipid rafts are also found to form higher order structures, caveolae (little caves), when associated with the scaffolding transmembrane proteins (Figure 2A & B). Caveolae are distinct from other vesicle-like invaginations, such as clathrin coated pits, due to their small size and lipid raft composition (Kurzchalia and Parton, 1999). The catalyst of caveolar formation is the oligomerization of a class of sphingolipid and cholesterol binding proteins known as caveolins (Monier et al., 1996), which stabilize and induce the lipid rafts to coalesce (reviewed in (Harder and Simons, 1997)). Three caveolin genes have been identified encoding caveolin proteins 1, 2, and 3 (Glenney, 1992; Kurzchalia et al., 1992; Scherer et al., 1996; Tang et al., 1996 ; Way and Parton, 1996). Both the N- and C-termini of caveolins are exposed to the cytoplasm forming a hairpin structure in the membrane, and are palmitoylated (Dietzen et al., 1995). Caveolin monomers oligomerize into hetero- and homo-dimers and trimers as they are translated in the cytoplasm (Lisanti et al., 1993), traffic through the trans-Golgi network and associate at the plasma membrane (reviewed in (Rietveld and Simons, 1998)). Due to their unique lipid raft composition caveolae can be biochemically purified based on their insolubility in the detergent Triton X-100 as well as on their low buoyant density on sucrose gradients (reviewed in (Anderson, 1998)). The insolubility of caveolae is likely due to the enrichment in sphingolipids and cholesterol in contrast to the glycerophospholipid rich plasma membrane (Hanada et al., 1995).

Recent advances in the characterization of lipid rafts have identified a new class of proteins which are essential to the biogenesis and regulation of caveolae; cavins 1 to 4. Cavin 1 (or polymerase I and transcript release factor: PTRF) has been shown to be essential for the stability and organization of caveolar structures (Hill et al., 2008; Liu and Pilch, 2008). Cavins 1-4 are part of a heterologous complex which bind to mature caveolin proteins (Figure 2B) that have emerged as caveolae after being processed through the trans-golgi system (Hayer et al., 2010). This interaction is cavin 1 and caveolin dependent and is thought to be due to a cavin phosphatidyl-serine interaction domain in all cavin proteins; a lipid which is enriched in caveolae (Bastiani et al., 2009). Cavin-2 is thought to play a role in the curvature of caveolar structures (Hansen et al., 2009). Cavin-3 is known to be involved in caveolae endocytosis and coordinates internalization on microtubules (McMahon et al., 2009). Cavin-4 is specifically expressed in striated muscle and has been shown to have specific roles related to myogenesis (Tagawa et al., 2008) and contraction (Ogata et al., 2008). Caveolin and cavin expression are interdependent on one another, since the depletion of one leads to the suppression of the other. Thus, cavins have emerged as an essential component of caveolae regulation, stability, and biogenesis.

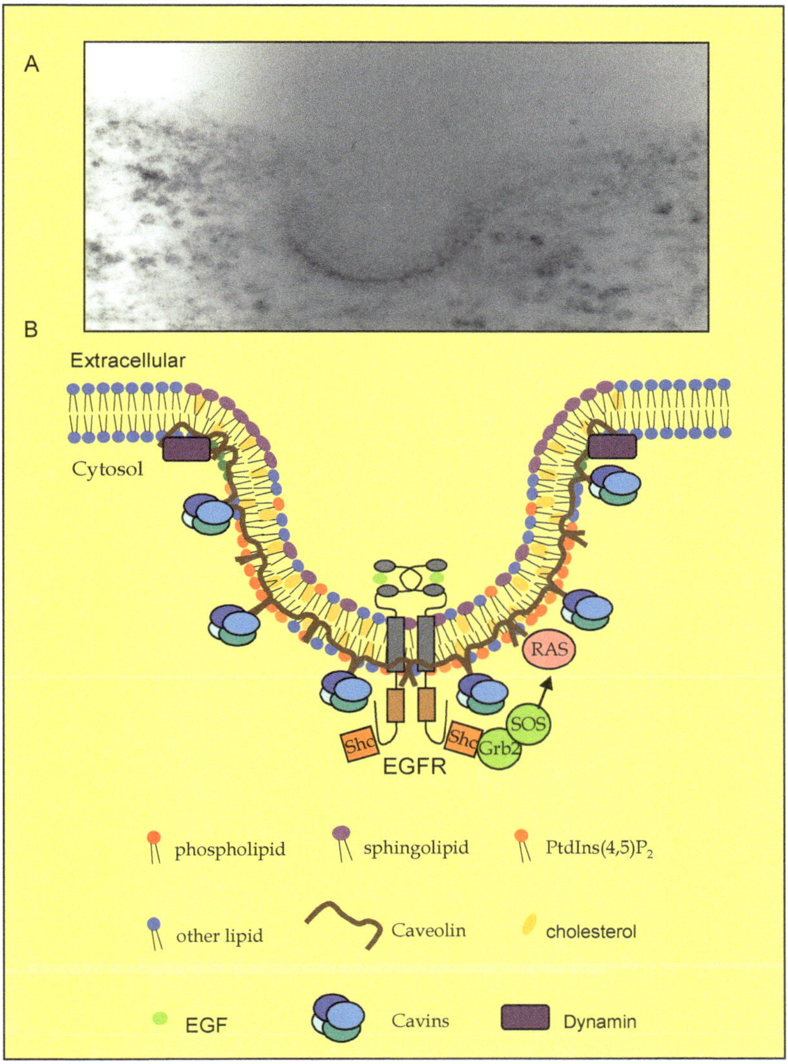

Fig. 2. Caveolae Structure

A) Electron micrograph of a caveolae at the membrane surface of 3T3 L1 fibroblasts.

B) Schematic diagram of the lipid and protein composition of caveolae. Phospholipids and PtdIns(4,5)P2 are concentrated on the inner leaflet, while sphingolipid is found in the outer leaflet of caveolae. Cholesterol is required for the caveolin proteins to bind and create a multimeric scaffold that forms the caveolar structure. Dynamin proteins surround the neck of the caveolae, and play a role in the fission of caveolar vesicles during endocytosis. Many receptor and signalling cascades are compartmentalized in caveolae, including the epidermal growth factor receptor (EGFR), the adaptors Shc, Grb & SOS, and Ras small GTPase. Cavins 1 to 4 bind to caveolin and regulate caveolae biogenesis, shape and trafficking.

Caveolae are a type of endocytic vesicle (much like clathrin coated pits) and can undergo internalization with particular stimuli. The process is dependent on dynamin (a GTPase involved in membrane fission) and cholesterol and leads to the production of enclosed vesicular caveolar structures with in the cytoplasm of the cell (Fig. 3A & B). For example, caveolae can be generated in transformed NIH 3T3 cells when a dominant negative dynamin is introduced, and is blocked in the presence of cholesterol sequestering agent (Li et al., 2001). This implies that even in cells that normally have no detectable caveolae, there is a steady state of caveolar invagination that is stabilized with dominant negative dynamin. Pinched off caveolae can be visualized as self contained vesicles (called cavicles) using electron microscopy (Mundy et al., 2002), and have been shown to fuse with a specialized endosome (that is caveolin positive) called caveosomes (Pelkmans et al., 2002). This pathway has been shown to be utilized by SV-40 virus, cholera toxin and albumin (reviewed in (Lajoie and Nabi, 2007)). Caveolae utilizing this pathway are known as type I (Fig. 3B) and this cavicle trafficking has been shown to depend upon transport along microtubles. These caveolar vesicles form into alternate morphologies in different cell types; in adipocytes, caveolar vesicles cluster into rosettes, and in endothelial cells, the vesicles elongate and fuse to form channels. A second population of caveolae (type 2) are involved in continuous rounds of fusion and invagination (Fig. 3B) but are restricted to remain near the cell membrane by the actin cytoskeleton (Pelkmans and Zerial, 2005). This cycle is highly regulated by various kinases (del Pozo et al., 2005) which can also shift these caveolae into long range type I transport. It is theorized that type 2 caveolae are largely responsible for the potocytosis of vitamins and ions (Anderson et al., 1992). Type 3 caveolae form elongated tubes from the cell surface in certain cell types and are not as well characterized. Several types of caveolae can be seen simultaneously in the same cell lineage (Fig. 3A & B), and often carry out specialized functions that depend on cell type (reviewed in (White and Anderson, 2005)).

In addition to the endocytic transport described above, it has recently been found that the EGFR has a functional role during trafficking through caveolar system. Under normal conditions, activated EGFR is recycled through clathrin coated pit mediated endocytosis and degraded or recycled through endosomes or lysomes respectively (reviewed in (Madshus and Stang, 2009)). However, in circumstances of sustained EGF signalling levels or the presence of oxidative stress or gamma irradiation, the caveolar system transports the EGFR through an endosomal endoplasmic route to the nucleus (reviewed in (Dittmann et al., 2009)). These functions have been shown to be dependent on active EGFR, which is capable of utilizing transport through the nuclear pore and becomes soluble in the nucleoplasm. This localization is thought to be very important in the resolution of DNA damage. To this end, nuclear EGFR has been shown to phosphorylate and activate DNA dependent protein kinase (DNA-PK), an important mediator of non-homologous end joining (Dittmann et al., 2005). In addition, nuclear EGFR has been shown to regulate transcription of cyclin D1, iNOS and B-myb, and regulate PCNA by phosphorylation, all of which correlate with entry into the cell cycle (Dittmann et al., 2009). Many tumours have been reported to have a nuclear population of EGFR, and its presence is linked to continued proliferation. Thus, this cellular transport system is the target of chemotherapy agents that either inactivate EGF signalling by inhibition of src kinase or antibody therapies that promote the uptake of EGFR through caveolae and thus block its activation (Dittmann et al., 2009).

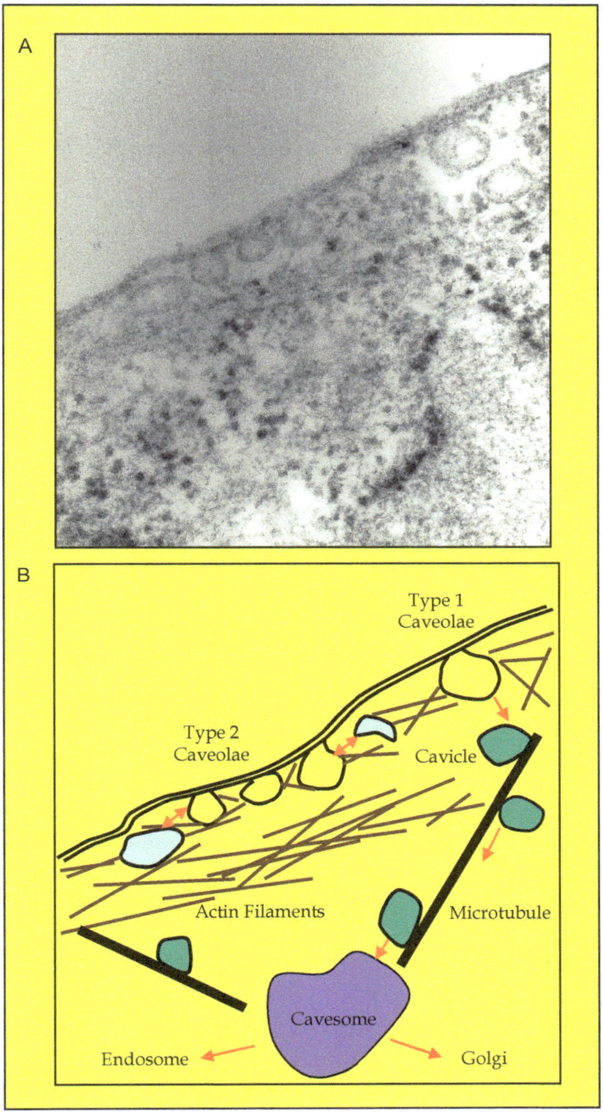

Fig. 3. Caveolar Endocytosis
A) Electron micrograph of caveolar structures at the membrane surface of differentiated 3T3 L1 fibroblasts. **B)** A schematic diagram showing caveolar endocytosis. Type I caveolae invaginate in response to various stimulae and form caveolar vesicles (cavicles) that are transported on microtubules. Cavicles can fuse together in cavesomes, where various contents may be recycled, or are transported to endosomes or the Golgi. Type II caveolae invaginate, fuse, and transport back to the cell surface. They are confined by the actin cytoskeleton and are important for potocytosis.

4.1 Cell signalling through caveolae

There is significant evidence that many well-studied receptor signalling systems are localized and operate through caveolae. Caveolin protein has been shown to directly interact with many of these signalling molecules through a 20 amino acid domain known as the caveolin scaffolding domain (CSD: (Couet et al., 1997)). Interaction with this domain has been shown to hold the signalling molecules in an inactive conformation (Okamoto et al., 1998). The pepitide sequence is rich in aromatic amino acids and has been shown to bind to adenylyl cyclase (AC), heterotrimeric Gα and Gβγ, PI3K, endothelial nitric oxide synthase (eNOS), protein kinase A (PKA), PKC, ERK1/2 and Src family kinases. In each case, the CSD is associated with the inactivation of the particular pathway (reviewed in (Patel et al., 2008)). However, since the CSD also binds lipids, there is the possibility that there are two populations of caveolin, one integral with rafts and another associated with the membrane that is not part of caveolae (Parton and Simons, 2007). Alternatively, the CSD has been conjectured to release the signalling proteins it binds to upon conformational change with activation (Okamoto et al., 1998). Although the exact role of the CSD is controversial, it is clear that it is involved in regulating and possibly sequestering signalling molecules to the caveolae. The inhibitory signalling associated with caveolin proteins has been challenged by many studies which show a requirement for Caveolin 1 in activating signal cascades (Kurzchalia and Parton, 1999, White, 2005 #707). This theory is based on the caveolin driven generation of caveolae, which acts to compartmentalize many signalling cascades. Thus, the caveolar structures and caveolin protein act as a center of cross talk and integration of mitogenic responses in normal cycling cells.

Many proteins have been found to localize to caveolae either through direct interaction with caveolins or sequestered through the properties of the lipid microdomains (Zajchowski and Robbins, 2002). Localization of proteins to caveolae is also enhanced by palmitoylation. A significant number of these are membrane bound receptors and kinase cascades with their adaptor proteins. Lipid moieties in caveolae are often utilized as second messengers (figure 2B) or for docking to plasma membrane by signalling proteins. This allows caveolae to regulate the activation of signalling cascades by concentration or proximity of substrates and second messengers (Cohen et al., 2004). Many receptor signalling systems are localized to and operate through caveolae and include; Src family kinases, nitric oxide synthase, EGFR, PDGFR, PLCγ, PLD, PKCα & β, Ras, trimeric G-proteins, MEK and Erk2, among others. Some of these proteins can also be found to be concentrated in lipid rafts (reviewed in (de Laurentiis et al., 2007)).

The most thoroughly characterized RTK residing in caveolae is EGFR, which is also crucial to mitogenic reposnse in human fibroblasts. Whether the localization of EGFR in caveolae has a negative or positive role in EGFR signalling is controversial. There is strong evidence for both, and it may be difficult to reconcile the observations definitively. The first evidence of the MAPK pathway in caveolae found H-Ras, Grb2 and SOS in Rat-1 cells (Mineo et al., 1996). The EGFR was present in the caveolin fraction in unstimulated Rat-1 cells, but declined rapidly after stimulation by EGF. The same migration was found in human fibroblasts and depended on EGF engagement, active EGFR tyrosine kinase, and phosphorylation of the EGFR receptor (Mineo et al., 1999). These observations implied that the activation of EGFR occurred within the caveolae, and is required to migrate in order to attenuate the EGF response. This was supported by the finding that some common mutations in EGFR that are oncogenic are also incapable of migrating from caveolae when stimulated.

In addition to the MAPK pathway, EGFR has been shown to activate other downstream effectors such as phospholipase Cγ (PLCγ) through PI3K (Jang et al., 2001) and PLD1 (Han et al., 2002)) in caveolae. PLCγ hydrolyses phosphatidylinositol- 4,5-bisphosphate (PIP_2) to produce diacylglycerol and inositol 1,3,5,-trisphosphate (IP_3), which activates PKC and mobilizes intracellular calcium respectively. PLD catalyses the hydrolysis of phosphatidylcholine to yield phosphatidic acid. These second messengers lead to the activation of Akt and mTOR which promote cellular proliferation (reviewed in (de Laurentiis et al., 2007)).

5. Caveolae and cavicles in senescence

The first report examining caveolin protein in replicative senescence (Park et al., 2000) found that the protein was highly elevated as human fibroblasts reached high population doublings. This upregulation was correlated to an increased number of caveolar structures seen in the EM of senescent cells, which were thought to be caveolae. Additionally, the group found EGFR signalling was attenuated in senescence through the binding to caveolin. However, these studies did not address whether caveolin generated bona-fide functional caveolar structures or whether it represented a differential regulation of caveolar pathway. Unfortunately, the EM chosen to examine the senescent cells did not have a clear cross section of the outer membrane, and it was unclear whether the structures being scored were caveolae or cavicles. The paradigm then (as now) was that increased caveolin drives the formation of caveolae and the regulation of cavicle generation had not yet been fully elucidated. This lead to the interpretation that the number of caveolae increase in senescent cells. Additionally, the low numbers of caveolae counted in low passage fibroblasts was in direct contrast to what had been previously seen in human fibroblasts by EM (Rothberg et al., 1992) or by freeze fracture EM (Fujimoto et al., 2000). Since this initial report, other groups have visualized a similar elevation of caveolar structures, which appear to be vesicles, and not classical caveolae in senescent cells (Bai et al., 2011; Volonte and Galbiati, 2011). Figure 3A shows structures consistent with caveolae in differentiated 3T3 L1 fibroblasts. Note the flask like invaginations within the outer membrane and the opening to the extracellular space through a smaller pore like neck. Thus, classically defined caveolae do not increase during senescence, and potentially are lost, while cavicles consistent with increased endocytosis have been seen in all senescent cells examined so far (Figure 3A & B). Although our own work examining caveolae (Wheaton and Riabowol, 2004) originally appeared to radically differ from other published reports, it does support the idea of increased caveolar vesicles in senescent fibroblasts. A comparison of young and senescent cells indicated that senescent cells contained a higher total amount of caveolins 1 and 2, but had significantly less of both proteins in the caveolar fraction obtained by sucrose density centrifugation. Additionally, caveolar fractions from senescent cells completely lacked the tyrosine kinase activity associated with functional caveolae and the EGF response. Furthermore, old cells had little caveolar protein exposed to the outer plasma membrane as estimated by an *in vivo* biotinoylation assay and there was no caveolin 1 detected on the cell surface using immunofluorescence (Fig. 4A & B) and confocal microscopy. Together, these data suggest that a fundamental loss of signal integration at the plasma membrane of senescent cells is due to the loss of signalling competent caveolae. However, they do not rule out the possibility that increased caveolar vesicles are present within the cytoplasm, as has been conjectured above. In fact, the poor response of EGFR and tyrosine phosphorylation of its targets in caveolae could be due the separation of these structures from the extracellular space. Therefore, the increase in caveolar structures in senescence may represent a misregulation of the normal functional caveolae during the aging process. This implies that some part of the endocytic process is blocked, leading to accumulation and stabilization of cavicles during senescence. This is supported by the finding that endocytosis is reduced in senescent cells and is correlated with an upregulation of caveolin 1 (Park et al., 2002).

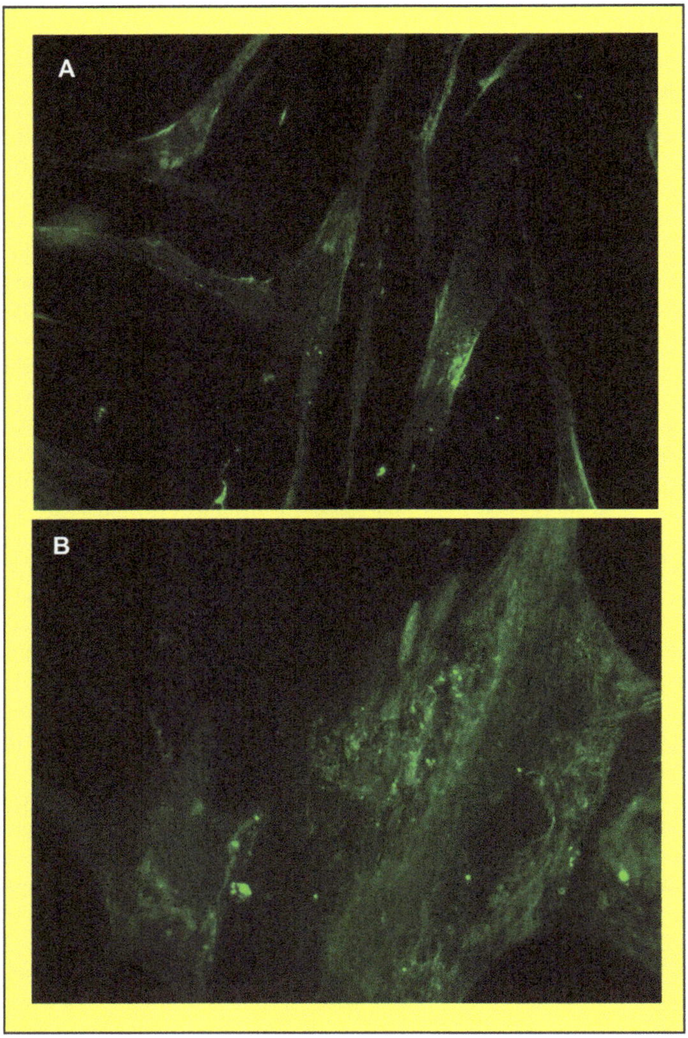

Fig. 4. Caveolin-1 Localization in Young and Senescent Fibroblasts
Young (A) and senescent (B) Hs68 fibroblasts were immuno-stained with anti-caveolin-1 and Alexa 488 flourescent secondary antibodies. Caveolin-1 is concentrated at the cell periphery in young cells (A) and is largely internalized in senescent cells (B).

These data suggest a model in which senescent cells are unable to organize and localize the components of functional caveolae to the plasma membrane. The pinched off caveolar vesicles observed in senescent cells may represent a unique misregulation of caveolae that could explain both the functional differences and the increase in caveolar structures seen previously (Wheaton, 2011). Thus, although increased expression of cavin-1 and caveolin proteins drive the biogenesis of caveolar structures, they unbalance the normal equilibrium between cavicles and cavaolae. This unbalance could promote invagination, but stall cavicle migration at some point in their migration along microtubles before they fuse with caveosomes or endosomes. Consistent with this idea, caveolin 1 staining is found at the cell periphery in young cells, but not at the membrane of senescent cells. However, there is very intense staining of caveolin within the interior of senescent cells (Fig. 4A & B). Using confocal microscopy with a deconvolution algorithm found intense caveolin-1 staining in senescent cells that was associated with fibre like structures (Fig. 5B). Confocal analysis of young cell caveolin -1 found it concentrated in regions of the cell periphery (Fig. 5A). This pattern of fibres is consistent with the caveolae associating with microtubules, and becoming stalled after internalization. The increased number of cavicles could represent a natural response to stress, in which cell arrest is maintained by sequestering key signalling receptors away from their ligands in the extracellular space. This would have a similar effect to EGFR antibody based chemotherapies which drive internalization of the EGFR through caveolae, but prevent its localization to the nucleus and activation (Dittmann et al., 2009). Considering internalization of EGFR through caveolae is a response to gentoxic agents, it is possible that DNA damage downstream of telomere attrition may do the same. Thus, the cell may respond to stress by down regulating mitogen activated cascades through internalization of caveolae. Recent studies in which caveolae are found to be redistributed by internalization during mitosis support this idea (Boucrot et al., 2011) . The caveolae in this case internalize through microtubules and could be a mechanism to block mitogenic signals after the commitment to divide.

5.1 Physical changes in senescent membranes and lipids

It is possible that some of the changes seen in caveolar structures can be explained by the unique properties and composition of the plasma membrane of old cells (Rutter et al., 1996; Schroeder et al., 1984). Rafts are comprised of a high concentration of sphingolipids and cholesterol, which have strong cohesive forces that counteract the entropic force inherent in a fluid mosaic membrane (reviewed in (Harder and Simons, 1997)). Examining senescent cells using proton magnetic resonance has shown that the ratio of cholesterol/ phosphatidylcholine increases as human fibroblasts age in culture, indicating an increased amount of mobile cholesterol (Rutter et al., 1996). Another early report sought to probe the lipid composition of these fibroblasts by using fluorescent probes. These labelled lipids can be used to determine the limiting anisotropy (fluidity) of membranes and showed that the microsomal fraction increased in fluidity with donor age (Schroeder et al., 1984). Lipid composition changes during fibroblast aging lead to higher lipid fluidity and may reflect the inability of significant raft domains to form with age, since rafts by nature are islands of less fluid lipid. This is in part is due to the levels of a species of phosphatidylcholine comprised of stearic acid and arachidonic acid becoming elevated in senescent fibroblasts (Naru et al., 2008). This dilutes the components of lipid rafts in the membrane of senescent cells.

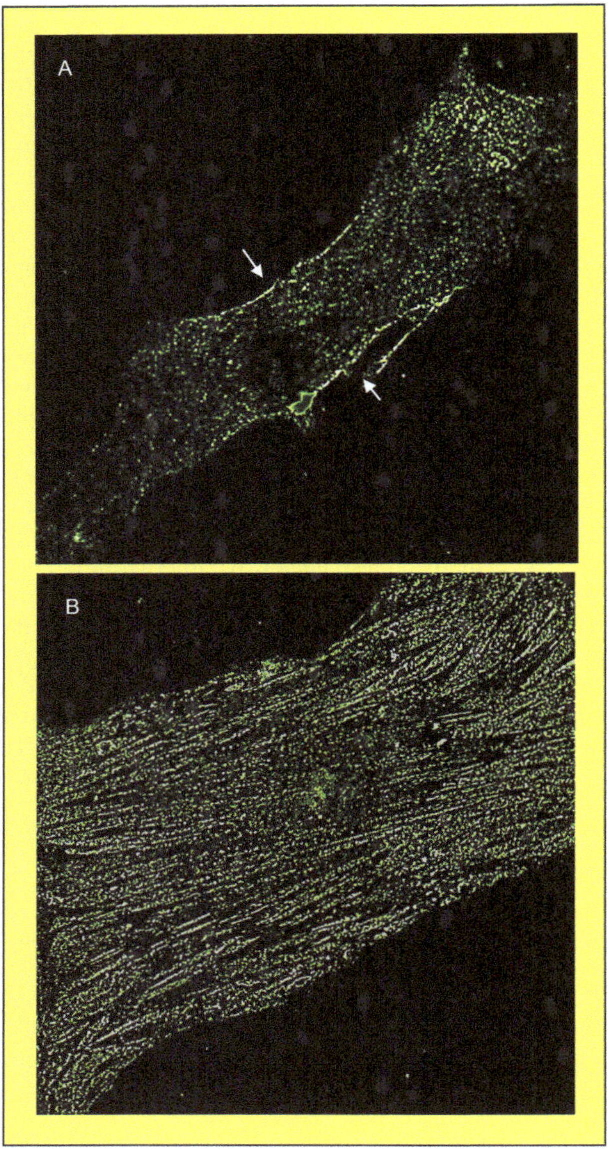

Fig. 5. Confocal Microscopy of Caveolin-1 Staining in Young and Senescent Fibroblasts. Young (A) and senescent (B) Hs68 fibroblasts were immuno-stained with anit-caveolin-1 and Alexa 488 fluorescent secondary antibodies. The images show a single confocal layer that was filtered through a de-convolution protocol to show the localization in sharper detail. The young cells have a high concentration of caveolin at the cell periphery, indicated by arrows. In senescent cells, caveolin is concentrated through out the cell on filaments consistent with microtubules.

In addition, senescent cell membranes are characterized by the specific loss of cholesterol from the microsomal fraction (Nakamura et al., 2003), even if the overall cholesterol may rise with age (Park et al., 2000). As previously mentioned, cholerterol is essential for lipid raft and caveolae formation (Fig. 2B). The altered cholesterol in the membrane also likely affects various signal transduction pathways (Fielding and Fielding, 2004). Alteration of the raft composition or dynamics may interfere with raft coalescence or its ability to sequester caveolin proteins. The loss of caveolae at the surface of the plasma membrane in old cells supports this model, since it would indicate that the proteins are unable to associate with rafts due to altered dynamics or composition, and subsequently they are not transported to the cell surface.

5.2 Upregulation of caveolin proteins and its effect on caveolar signalling

Key signalling pathways that operate through caveolar structures which are inactivated or misregulated could lead to aspects of the senescent phenotype. Since many growth receptors have been found to be localized within, and to signal through caveolar structures, it is possible that one or more of these may modulate the mitogenic response in primary fibroblasts. Many signalling pathways shown to be altered in senescence, including MAPK, PI3K, PLD and PKCδ (reviewed in (Caino et al., 2009)) are localized to caveolae. Since a significant post receptor block of mitogenic signalling is associated with the development of senescence, the existence of microdomains where receptors and signalling cascades are believed to be linked offers a promising functional connection to these observations. Indeed, the absence of a membrane population of caveolae may prevent mitogenic signal propagation. Thus, the post receptor block could be the result of receptors being either uncoupled from internalized caveolae or sequestered from access to ligands in the extracellular space. These kinase cascades would subsequently be unable to maintain integrated downstream responses to ligand and contribute to senescent arrest. Alternatively, the increases in caveolin protein itself may lead directly to the attenuation of many kinase cascades by virtue of the tumour suppressive function of CSD binding to key components of the cascade. Thus, a mechanism linking the post receptor block to caveolae is either through misregulated caveolar internalization or the inhibitory CSD of caveolin, both of which are characterized by increased caveolin-1 protein expression.

The initial observation linking caveolae to repressed signal propagation in senescence noted that EGF signalling was attenuated (Park et al., 2000). These first studies observed an increase in caveolin-1 & 2 proteins, and therefore the CSD domain (Park et al., 2000; Wheaton et al., 2001). Consistent with this observation, ectopic caveolin 1 expression in fibroblasts was shown to induce premature senescence and thus lead directly to cellular arrest (Volonte et al., 2002). This was further supported by the suppression of caveolin-1 using small inferring RNA or antisense oligos in senescent fibroblasts. Knockdown of caveolin in fibroblasts restores the response to EGF and the cells are capable of entering the cell cycle (Cho et al., 2003). It is thought that this response works primarily through the interaction of the caveolin CSD with erk1/2, leading to attenuation of kinase activity (Engelman et al., 1998). Supporting this, the erk1/2 kinase has been previously reported to be misregulated during senescence (see section 2.1).

Oxidative stress has been shown to cause irreversible growth arrest in NIH 3T3 cells and is dependent on the upregulation of caveolin-1 (Volonte et al., 2002). Such stress also leads to tyrosine phosphorylation of caveolin-1 which depends on p38 and Src (Volonte et al., 2001) and drives the internalization of caveolae. The same stress (H_2O_2) was shown to cause premature senescence in human fibroblasts, with a relocalization of caveolin to the cytoplasm and nucleus that required p38 activity (Chretien et al., 2008). Thus stress induced senescence appears to upregulate caveolin and redistribute the protein (and likely caveolar structures). This is similar to what is seen in replicative senescence where caveolin protein is found in the cytoplasm (Wheaton et al., 2001) or when cavicles accumulate (Bai et al., 2011; Park et al., 2000; Volonte and Galbiati, 2011). Thus, p38 activated by stress may drive internalization of caveolae and causes receptor tyrosine receptor signalling to be attenuated.

Senescent cells have a flattened and enlarged morphology that is characterized by an increase in focal adhesions (Cho et al., 2004) and which lead to increased amount of actin stress fibres (Wheaton et al., 1996). Caveolin has also been reported to localize to focal adhesions through integrin and likely plays a role in mitogen signalling during cell adhesion. This interaction appears to be upregulated during senescence since caveolin-1 and paxillin (a focal adhesion marker) co-localize strongly in senescent cells (Cho et al., 2004). Furthermore, when caveolin-1 is knocked down in senescent cells, there is an inhibition of focal adhesion kinase, depolarization of anchored actin and reversion of the cells to a younger morphology (Cho et al., 2004). The origin of these morphological alterations is difficult to verify since caveolin repression affects other aspects of mitogenic signalling. However, it has been shown that breaking cell adhesion contacts influences the internalization of lipid rafts within transformed cells (del Pozo et al., 2004; del Pozo et al., 2005). In addition, the clustering of integrins is known to activate Fyn in a caveolin-1 dependent manner to allow anchorage dependent growth (Wary et al., 1998). Consistent with these studies it has been shown that localization of caveolin-1 to cholesterol enriched microdomains decreases in senescent fibroblasts when they are liberated from the substratum by scraping or trypsin (Inomata et al., 2006). This implies that internalization of lipid rafts occurs in senescent cells when there detached and reflects caveolae misregulation in senescent cells. The absence of caveolae when senescent cells are establishing new focal contacts could represent another way in which mitogenic signals are blocked.

In summary, inappropriate caveolin regulation in senescent cells may contribute to the senescence associated post receptor block in three ways. First, the CSD may bind and neutralize various signalling molecules, and prevent signal propagation. Second, caveolin drives internalization of caveolae which prevents receptors access to ligands. Third, caveolin may no longer be sequestered to lipid rafts preventing the development of caveolae to allow signal integration.

5.3 P53, DNA damage and caveolae

Although the most common interpretation in the literature is that telomere attrition is the origin of DNA damage in replicative senescence, a considerable amount of damage foci (γH2AX) are not localized to telomeres in senescent cells (Sedelnikova et al., 2004). The γH2AX-telomere foci are also dependent on whether these cells are cultured in normoxic (2% O_2) conditions (Herbig et al., 2004). Thus, the possibility exists that other forms of stress cause DNA damage in parallel with telomere erosion, such as elevated reactive oxygen species (ROS) or replication fork collapse. These forms of stress may result from the misregulation of caveolae or the upregulation of caveolin. The upregulation of caveolin leading to a increase in caveolar vesicles causes both the attenuation of the EGFR signalling (de Laurentiis et al., 2007) and an increase in the activity of the p53 tumour suppressor through elevated ROS (Volonte and Galbiati, 2009a). Both of these mechanisms may lead to DNA damage that is not localized to telomeres. This is also consistent with studies in senescent cells which demonstrate higher p53 activity and p21 expression (see section 3).

Overexpression of caveolin 1 protein has been shown to cause G1 arrest, which is dependent on p21 expression induced by increased p53 activity. Mouse embryonic fibroblasts expressing caveolin-1 have a reduced proliferative lifespan and a senescent morphology (Volonte et al., 2002). The aberrant caveolin-1 levels lead to G0/G1 cell cycle arrest, activation of p53 and upregulation of p21 (Galbiati et al., 2001). This shows that caveolin-1 expression mediates premature senescence through a p53/p21 dependent pathway. This pathway is further enhanced by caveolin 1 mediated inactivation of MDM2 and PP2A-C which act as negative regulators of p53 and ATM, respectively (Bartholomew et al., 2009; Volonte et al., 2009). The MDM2 protein is an ubiquitin ligase which targets p53 for degradation and keeps p53 at basal levels in unstressed conditions. Upon activation by stress p53 is phosphorylated at residues which prevent MDM2 association and in turn stabilizes p53. It has been shown during oxidative stress induced senescence that MDM2 is neutralized by caveolin and stabilizes p53 (Volonte et al., 2009). It is thought that the activation of p53 is usually achieved by the kinase ATM, which is sensor of DNA damage (see section 3). Auto-phosphorylation of ATM occurs when it binds to regions of DNA damage and dissociates as active monomers (Bakkenist and Kastan, 2003). ATM is turned off after repair is completed by the phosphatase PP2A-C. When modeling pulmonary emphysema in murine fibroblasts, it was found that oxidative stress caused the sequestration and neutralization of PP2A-C into caveolar enriched microdomains (Bartholomew et al., 2009). Thus, there appears to be a role for CSD in caveolin for the activation of p53 at many levels. To further demonstrate a role for the CSD in caveolin, it was shown that introduction of the peptide of this domain was able to cause premature senescence by itself (Volonte et al., 2009).

The family of cavin proteins have recently been investigated for their potential role during senescence. Cavin-1 (PTRF) and Cavin-3 (SRBC) both appear to be elevated through protein stabilization during replicative senescence, but not quiescence (Bai et al., 2011; Cong et al., 2006) Additionally, it has been shown that cavin-1 is upregulated during oxidative stress induced premature senescence (Volonte and Galbiati, 2011). The ectopic expression of cavin-1 leads to activation of p53, a decrease in MAPK activity, and induced premature senescence. Interestingly, the cavin-1 expression also induced DNA foci that were visualized by γH2AX and did not colocalize with telomeres. The knockdown of cavin-1 extended proliferative lifespan, and reversed the γH2AX in high passage cells (Bai et al., 2011). Similarly, the knockdown of PTRF prevented oxidative stress from inducing senescence and was shown to prevent the MDM2:caveolin 1 interaction leading to p53 activation (Volonte and Galbiati, 2011). Cavin-1 expression is known to have a direct affect on the levels of caveolin (Hansen and Nichols, 2010) and therefore the upregulation of cavin-1 in these systems likely explains the upregulation of caveolin observed during senescence. Although these two studies focused on PTRF, much of the data can be explained by the elevation of caveolin and caveolar vesicles. What is most significant is the generation of DNA damage seen downstream of caveolar vesicles (Wheaton, 2011).

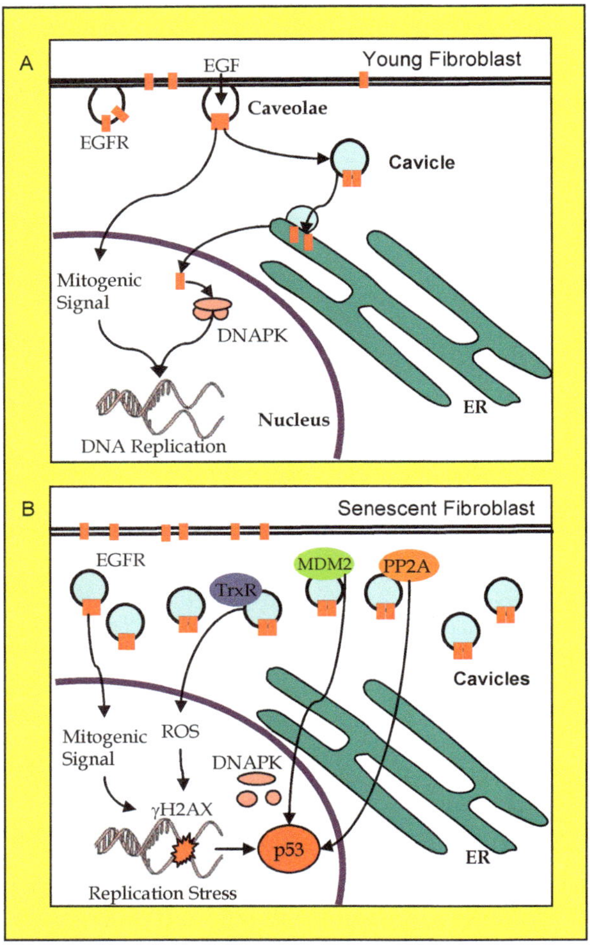

Fig. 6 Model for Cavicle Induced Senescence.
A) In young fibroblasts mitogenic signalling is achieved through a caveolae localized EGFR cascade that activates transcription of growth promoting genes. Signalling competent caveolae are capable of internalizing and traffic to the endoplasmic reticulum (ER) in response to cellular stress. EGFR is transported via cavicles to the ER followed by localization to the nucleus where it can activate DNA protein kinase (DNA-PK). DNA-PK is required to repair DNA damage that is generated by stress or during replication.
B) In senescence fibroblasts mitogenic signalling occurs through EGFR, but is unable to initiate DNA replication. The majority of caveolae are internalized as cavicles, and the increased caveolin protein levels inactivate many signalling molecules or negative regulators of the p53 tumour suppressor (MDM2 and PP2A-C). In theory the EGFR is no longer able to localize to the nucleus and therefore DNA damage generated by reactive oxygen species (ROS), replication stress, and cellular aging cannot be repaired. The activation of p53 through DNA damage causes cellular senescence.

The upregulation of caveolin was shown to block thioredoxin reductase I activity, and thus raise the ROS levels within fibroblasts (Volonte and Galbiati, 2009b). Elevated ROS production is well known to damage DNA, activate p53 and lead to senescence (Chen et al., 1995). Thus, DNA damage could be caused by the presence of ROS being produced in cells that over express caveolin 1 or reach replicative senescence. Furthermore, the negative regulation by caveolin 1 of many key regulatory proteins involved in the p53 mediated DNA damage response could be ensuring that damage signals that lead to a senescent outcome are reinforced.

The transfer of EGFR to the nucleus is well known to influence the resolution of γH2AX damage foci. As previously described (section 4), EGFR is internalized in response to gamma radiation or oxidative stress through a caveolar mediated endocytotic pathway. This pathway leads to EGFR localization to the nucleus and phosphorylates targets such as DNA-PK which facilitate the completion of DNA repair (Dittmann et al., 2005). This function depends on EGFR kinase activity and is blocked by a class of radio-sensitizing agents that work by antagonizing EGFR such as Cetuximab or Gefitinib. Such drugs enhance DNA damage and lead to apoptosis of malignant cells because of unrepaired DNA damage. The EGF signalling pathway is also upstream of the survival kinases MAPK and Akt which assist in the resolution of DNA damage (Golding et al., 2009). Additionally, EGFR activity was found to modulate non-homologous end joining after gamma radiation (Kriegs et al., 2010). Taken together, EGFR signalling plays a major role in controlling cell cycle arrest in response to cellular stress and DNA damage. In support of the role of caveolae in this model, EGFR was found to transport to the nucleus in a caveolin and Src dependent mechanism after oxidative stress (Khan et al., 2006). Similarly, gamma radiation causes Src induced association between caveolin-1 and EGFR leading to internalization. This leads to nuclear localization and to control of DNA-PK activity (Dittmann et al., 2008). Caveolin-1 expression has been demonstrated to be up regulated by ionizing radiation and is required for both homologous recombination and non-homologous end joining (Zhu et al., 2010). In this case, caveolin-1 is pivotal in forming the caveolar vesicles that allow transport of the EGFR, and nuclear DNA-PK activation. Lastly, Gefitinib (an EGFR kinase inhibitor) can generate premature senescence in non-small lung cancer cells (Hotta et al., 2007). Thus, the current evidence strongly suggests mitogenic signals and EGFR internalization to the nucleus are required to resolve DNA damage. Furthermore, blocked EGFR activity leads to apoptosis or senescence.

Although primary fibroblasts are genetically stable, they do undergo transient DNA damage foci as a result of mitogenic stimulation (Ichijima et al., 2005; McManus and Hendzel, 2005). These γH2AX foci occur as a natural part of the synthesis of DNA or division, and are resolved by the time the cell returns to G0 of the cell cycle. Thus, it is possible that a signal from the EGFR could assist cycling cells to resolve this DNA damage. Therefore, EGFR would promote both the initiation of the cell cycle and its continued signalling would be required to resolve DNA damage arising from progress through the cell cycle. However, in senescent cells, the up regulation of caveolin antagonizes EGFR signalling (see section 5.2). Thus, the attenuation of the EGFR pathway during senescence could perpetuate the normally transient DNA damage foci in fibroblasts. The blocking of EGF signalling in this case would prevent the resolution of damage induced by replication stress during normal growth. This could explain the DNA damage seen downstream of increased caveolar vesicles during aging as a form of replication stress (Fig. 6A & B).

6. Conclusion

The inhibition or reduction of components of mitogenic signaling cascades in senescent cell caveolar fractions is consistent with the loss of caveolae playing a causal role in the blunted growth response seen during cellular senescence. Collectively, these observations suggest that localization and integration of signalling cascades in caveolae are disrupted in senescent cells. The mitogenic pathways leading to MAPK, PLD, PKC isoforms and PI3K are all localized to caveolae and have been shown to have deficiencies in senescent cells. In the case of EGFR, the proper coordination with shc is disrupted, the CSD of caveolin blocks kinase activity, and inappropriate retention or internalization in caveolae occurs. It is unclear whether this difference influences the downstream localization of activated erk1/2 to the senescent nucleus, however several groups have noted a decreased erk stimulation by EGF with age. Evidence suggests that the composition of lipid rafts likely change in senescent cells and that this may influence how PI3K, PLD and PKC are regulated. Cholesterol levels are decreased in lipid raft fractions, PC levels increase in the plasma membrane, and ceramide levels increase with fibroblast age. These observations suggest an increase in fluidity of the membrane, which changes the properties of lipid rafts. The increases in ceramide disrupt PLD, and the increase in DAG constitutively activates PKCα in caveolae. The decrease in src kinase activity in caveolae may prevent the inactivation of the PKCδ and induce senescence. The loss of PI3K signal integration through caveolae may be a part of the induction of senescence, explaining why inhibitors of the pathway induce premature senescence. These examples indicate that signal transduction pathways rely upon caveolae for signal integration and propagation that they become disorganized in senescent cells. Collectively, these signalling changes may explain their reduced or absent response to mitogens in senescent cells.

There has been a shift in the understanding of how the DNA damage occurring in senescence is generated. Remarkably, the increase of caveolar vesicles observed in the senescent state can itself lead to the generation of DNA damage foci in parallel with the well known DNA damage localized to eroded telomeres. The exact mechanism by which this is achieved is still unknown, but likely involves the strong inhibitory activities of the scaffolding protein caveolin 1. The proteins theoridoxin, MDM2 and PP2A-C are all inhibited or sequestered by increases in caveolin protein during senescence. Inhibition of theoridoxin leads to an increase in ROS, which leads to DNA damage and p53 activation. The activation of p53 in turn is augmented by inhibition of the negative regulators of p53, MDM2 and PP2A-C by caveolin. The DNA damage that leads to p53 mediated senescence may also be generated through failure to resolve γH2AX foci that occur during normal replication. The increase of caveolin and cavin-1 proteins shift the balance of caveolar structures to internalized cavicles. These cavicles are no longer capable of integrating the EGFR cascade, which is required to transfer EGFR to the nucleus and resolve replication dependent DNA damage through proteins such as DNA-PK. The failure to repair DNA leads to replication stress, the activation of p53 and thus cellular senescence (Fig. 6A & B). Thus, the misregulation of caveolar vesicles interferes with the essential role of these structures in repairing DNA damage.

7. Acknowledgements

I would like to thank Dr. S. Robbins and Dr. Kari Sampsel (University of Calgary) for use of the electron microscopy images. I would also like to thank Densie Campuzano for editing this manuscript.

8. References

Anderson, R.G. (1998). The caveolae membrane system. Annu Rev Biochem *67*, 199-225.

Anderson, R.G., Kamen, B.A., Rothberg, K.G., and Lacey, S.W. (1992). Potocytosis: sequestration and transport of small molecules by caveolae. Science *255*, 410-411.

Appella, E., and Anderson, C.W. (2001). Post-translational modifications and activation of p53 by genotoxic stresses. Eur J Biochem *268*, 2764-2772.

Ashton, A.W., Watanabe, G., Albanese, C., Harrington, E.O., Ware, J.A., and Pestell, R.G. (1999). Protein kinase Cdelta inhibition of S-phase transition in capillary endothelial cells involves the cyclin-dependent kinase inhibitor p27(Kip1). J Biol Chem *274*, 20805-20811.

Atadja, P.W., Stringer, K.F., and Riabowol, K.T. (1994). Loss of serum response element-binding activity and hyperphosphorylation of serum response factor during cellular aging. Mol Cell Biol *14*, 4991-4999.

Bai, L., Deng, X., Li, J., Wang, M., Li, Q., An, W., A, D., and Cong, Y.S. (2011). Regulation of cellular senescence by the essential caveolar component PTRF/Cavin-1. Cell Res *21*, 1088-1101.

Bakkenist, C.J., and Kastan, M.B. (2003). DNA damage activates ATM through intermolecular autophosphorylation and dimer dissociation. Nature *421*, 499-506.

Bartholomew, J.N., Volonte, D., and Galbiati, F. (2009). Caveolin-1 regulates the antagonistic pleiotropic properties of cellular senescence through a novel Mdm2/p53-mediated pathway. Cancer Res *69*, 2878-2886.

Basak, S., Jacobs, S.B., Krieg, A.J., Pathak, N., Zeng, Q., Kaldis, P., Giaccia, A.J., and Attardi, L.D. (2008). The metastasis-associated gene Prl-3 is a p53 target involved in cell-cycle regulation. Mol Cell *30*, 303-314.

Bastiani, M., Liu, L., Hill, M.M., Jedrychowski, M.P., Nixon, S.J., Lo, H.P., Abankwa, D., Luetterforst, R., Fernandez-Rojo, M., Breen, M.R., *et al.* (2009). MURC/Cavin-4 and cavin family members form tissue-specific caveolar complexes. J Cell Biol *185*, 1259-1273.

Basu, A., and Pal, D. (2010). Two faces of protein kinase Cdelta: the contrasting roles of PKCdelta in cell survival and cell death. ScientificWorldJournal *10*, 2272-2284.

Beausejour, C.M., Krtolica, A., Galimi, F., Narita, M., Lowe, S.W., Yaswen, P., and Campisi, J. (2003). Reversal of human cellular senescence: roles of the p53 and p16 pathways. Embo J *22*, 4212-4222.

Binet, R., Ythier, D., Robles, A.I., Collado, M., Larrieu, D., Fonti, C., Brambilla, E., Brambilla, C., Serrano, M., Harris, C.C., *et al.* (2009). WNT16B is a new marker of cellular senescence that regulates p53 activity and the phosphoinositide 3-kinase/AKT pathway. Cancer Res *69*, 9183-9191.

Blake, R.A., Garcia-Paramio, P., Parker, P.J., and Courtneidge, S.A. (1999). Src promotes PKCdelta degradation. Cell Growth Differ *10*, 231-241.

Bodnar, A.G., Ouellette, M., Frolkis, M., Holt, S.E., Chiu, C.P., Morin, G.B., Harley, C.B., Shay, J.W., Lichtsteiner, S., and Wright, W.E. (1998). Extension of life-span by introduction of telomerase into normal human cells [see comments]. Science *279*, 349-352.

Boguski, M.S., and McCormick, F. (1993). Proteins regulating Ras and its relatives. Nature *366*, 643-654.

Bose, C., Bhuvaneswaran, C., and Udupa, K.B. (2004). Altered mitogen-activated protein kinase signal transduction in human skin fibroblasts during in vitro aging:

differential expression of low- density lipoprotein receptor. J Gerontol A Biol Sci Med Sci *59*, 126-135.

Boucrot, E., Howes, M.T., Kirchhausen, T., and Parton, R.G. (2011). Redistribution of caveolae during mitosis. J Cell Sci *124*, 1965-1972.

Brown, J.P., Wei, W., and Sedivy, J.M. (1997). Bypass of senescence after disruption of p21CIP1/WAF1 gene in normal diploid human fibroblasts. Science *277*, 831-834.

Brown, W.T. (1990). Genetic diseases of premature aging as models of senescence. Annu Rev Gerontol Geriatr *10*, 23-42.

Caino, M.C., Meshki, J., and Kazanietz, M.G. (2009). Hallmarks for senescence in carcinogenesis: novel signaling players. Apoptosis *14*, 392-408.

Cao, K., Graziotto, J.J., Blair, C.D., Mazzulli, J.R., Erdos, M.R., Krainc, D., and Collins, F.S. (2011). Rapamycin reverses cellular phenotypes and enhances mutant protein clearance in Hutchinson-Gilford progeria syndrome cells. Sci Transl Med *3*, 89ra58.

Carlin, C., Phillips, P.D., Brooks-Frederich, K., Knowles, B.B., and Cristofalo, V.J. (1994). Cleavage of the epidermal growth factor receptor by a membrane-bound leupeptin-sensitive protease active in nonionic detergent lysates of senescent but not young human diploid fibroblasts. J Cell Physiol *160*, 427-434.

Chang, Z.F., and Huang, D.Y. (1994). Decline of protein kinase C activation in response to growth stimulation during senescence of IMR-90 human diploid fibroblasts. Biochem Biophys Res Commun *200*, 16-27.

Chen, Q., and Ames, B.N. (1994). Senescence-like growth arrest induced by hydrogen peroxide in human diploid fibroblast F65 cells. Proc Natl Acad Sci U S A *91*, 4130-4134.

Chen, Q., Fischer, A., Reagan, J.D., Yan, L.J., and Ames, B.N. (1995). Oxidative DNA damage and senescence of human diploid fibroblast cells. Proc Natl Acad Sci U S A *92*, 4337-4341.

Cho, K.A., Ryu, S.J., Oh, Y.S., Park, J.H., Lee, J.W., Kim, H.P., Kim, K.T., Jang, I.S., and Park, S.C. (2004). Morphological adjustment of senescent cells by modulating caveolin-1 status. J Biol Chem *279*, 42270-42278.

Cho, K.A., Ryu, S.J., Park, J.S., Jang, I.S., Ahn, J.S., Kim, K.T., and Park, S.C. (2003). Senescent phenotype can be reversed by reduction of caveolin status. J Biol Chem *278*, 27789-27795.

Chretien, A., Piront, N., Delaive, E., Demazy, C., Ninane, N., and Toussaint, O. (2008). Increased abundance of cytoplasmic and nuclear caveolin 1 in human diploid fibroblasts in H(2)O(2)- induced premature senescence and interplay with p38alpha(MAPK). FEBS Lett *582*, 1685- 1692.

Chua, C.C., Geiman, D.E., and Ladda, R.L. (1986). Receptor for epidermal growth factor retains normal structure and function in aging cells. Mech Ageing Dev *34*, 35-55.

Cohen, A.W., Hnasko, R., Schubert, W., and Lisanti, M.P. (2004). Role of caveolae and caveolins in health and disease. Physiol Rev *84*, 1341-1379.

Collado, M., Medema, R.H., Garcia-Cao, I., Dubuisson, M.L., Barradas, M., Glassford, J., Rivas, C., Burgering, B.M., Serrano, M., and Lam, E.W. (2000). Inhibition of the phosphoinositide 3- kinase pathway induces a senescence-like arrest mediated by p27Kip1. J Biol Chem *275*, 21960-21968.

Collado, M., and Serrano, M. (2010). Senescence in tumours: evidence from mice and humans. Nat Rev Cancer *10*, 51-57.

Cong, Y.S., Fan, E., and Wang, E. (2006). Simultaneous proteomic profiling of four different growth states of human fibroblasts, using amine-reactive isobaric tagging reagents and tandem mass spectrometry. Mech Ageing Dev *127*, 332-343.

Couet, J., Li, S., Okamoto, T., Ikezu, T., and Lisanti, M.P. (1997). Identification of peptide and protein ligands for the caveolin-scaffolding domain. Implications for the interaction of caveolin with caveolae-associated proteins. J Biol Chem *272*, 6525-6533.

de Lange, T. (1994). Activation of telomerase in a human tumor. Proc Natl Acad Sci U S A *91*, 2882-2885.

de Laurentiis, A., Donovan, L., and Arcaro, A. (2007). Lipid rafts and caveolae in signaling by growth factor receptors. Open Biochem J *1*, 12-32.

De Tata, V., Ptasznik, A., and Cristofalo, V.J. (1993). Effect of the tumor promoter phorbol 12-myristate 13-acetate (PMA) on proliferation of young and senescent WI-38 human diploid fibroblasts. Exp Cell Res *205*, 261-269.

Debacq-Chainiaux, F., Boilan, E., Dedessus Le Moutier, J., Weemaels, G., and Toussaint, O. (2010). p38(MAPK) in the senescence of human and murine fibroblasts. Adv Exp Med Biol *694*, 126-137.

del Pozo, M.A., Alderson, N.B., Kiosses, W.B., Chiang, H.H., Anderson, R.G., and Schwartz, M.A. (2004). Integrins regulate Rac targeting by internalization of membrane domains. Science *303*, 839-842.

del Pozo, M.A., Balasubramanian, N., Alderson, N.B., Kiosses, W.B., Grande-Garcia, A., Anderson, R.G., and Schwartz, M.A. (2005). Phospho-caveolin-1 mediates integrin-regulated membrane domain internalization. Nat Cell Biol *7*, 901-908.

Dempsey, E.C., Newton, A.C., Mochly-Rosen, D., Fields, A.P., Reyland, M.E., Insel, P.A., and Messing, R.O. (2000). Protein kinase C isozymes and the regulation of diverse cell responses. Am J Physiol Lung Cell Mol Physiol *279*, L429-438.

Di Leonardo, A., Linke, S.P., Clarkin, K., and Wahl, G.M. (1994). DNA damage triggers a prolonged p53-dependent G1 arrest and long-term induction of Cip1 in normal human fibroblasts. Genes Dev *8*, 2540-2551.

Dietzen, D.J., Hastings, W.R., and Lublin, D.M. (1995). Caveolin is palmitoylated on multiple cysteine residues. Palmitoylation is not necessary for localization of caveolin to caveolae. J Biol Chem *270*, 6838-6842.

Dimri, G.P., Lee, X., Basile, G., Acosta, M., Scott, G., Roskelley, C., Medrano, E.E., Linskens, M., Rubelj, I., Pereira-Smith, O., *et al.* (1995). A biomarker that identifies senescent human cells in culture and in aging skin in vivo. Proc Natl Acad Sci U S A *92*, 9363-9367.

Dittmann, K., Mayer, C., Fehrenbacher, B., Schaller, M., Raju, U., Milas, L., Chen, D.J., Kehlbach, R., and Rodemann, H.P. (2005). Radiation-induced epidermal growth factor receptor nuclear import is linked to activation of DNA-dependent protein kinase. J Biol Chem *280*, 31182-31189.

Dittmann, K., Mayer, C., Kehlbach, R., and Rodemann, H.P. (2008). Radiation-induced caveolin-1 associated EGFR internalization is linked with nuclear EGFR transport and activation of DNA-PK. Mol Cancer *7*, 69.

Dittmann, K., Mayer, C., and Rodemann, H.P. (2009). Nuclear EGFR as novel therapeutic target: insights into nuclear translocation and function. Strahlenther Onkol *186*, 1-6.

Downward, J., Waterfield, M.D., and Parker, P.J. (1985). Autophosphorylation and protein kinase C phosphorylation of the epidermal growth factor receptor. Effect on tyrosine kinase activity and ligand binding affinity. J Biol Chem *260*, 14538-14546.

el-Deiry, W.S., Tokino, T., Velculescu, V.E., Levy, D.B., Parsons, R., Trent, J.M., Lin, D., Mercer, W.E., Kinzler, K.W., and Vogelstein, B. (1993). WAF1, a potential mediator of p53 tumor suppression. Cell *75*, 817-825.

Engelman, J.A., Chu, C., Lin, A., Jo, H., Ikezu, T., Okamoto, T., Kohtz, D.S., and Lisanti, M.P. (1998). Caveolin-mediated regulation of signaling along the p42/44 MAP kinase cascade in vivo. A role for the caveolin-scaffolding domain. FEBS Lett *428*, 205-211.

Exton, J.H. (1997). New developments in phospholipase D. J Biol Chem *272*, 15579-15582.

Exton, J.H. (1998). Phospholipase D. Biochim Biophys Acta *1436*, 105-115.

Exton, J.H. (1999). Regulation of phospholipase D. Biochim Biophys Acta *1439*, 121-133.

Fantl, W.J., Johnson, D.E., and Williams, L.T. (1993). Signalling by receptor tyrosine kinases. Annu Rev Biochem *62*, 453-481.

Fielding, C.J., and Fielding, P.E. (2004). Membrane cholesterol and the regulation of signal transduction. Biochem Soc Trans *32*, 65-69.

Foster, D.A., and Xu, L. (2003). Phospholipase D in cell proliferation and cancer. Mol Cancer Res *1*, 789- 800.

Fujimoto, T., Kogo, H., Nomura, R., and Une, T. (2000). Isoforms of caveolin-1 and caveolar structure. J Cell Sci *113 Pt 19*, 3509-3517.

Galbiati, F., Volonte, D., Liu, J., Capozza, F., Frank, P.G., Zhu, L., Pestell, R.G., and Lisanti, M.P. (2001). Caveolin-1 expression negatively regulates cell cycle progression by inducing G(0)/G(1) arrest via a p53/p21(WAF1/Cip1)-dependent mechanism. Mol Biol Cell *12*, 2229-2244.

Gerhard, G.S., Phillips, P.D., and Cristofalo, V.J. (1991). EGF- and PDGF-stimulated phosphorylation in young and senescent WI-38 cells. Exp Cell Res *193*, 87-92.

Ghosh, S., Strum, J.C., Sciorra, V.A., Daniel, L., and Bell, R.M. (1996). Raf-1 kinase possesses distinct binding domains for phosphatidylserine and phosphatidic acid. Phosphatidic acid regulates the translocation of Raf-1 in 12-O-tetradecanoylphorbol-13-acetate-stimulated Madin-Darby canine kidney cells. J Biol Chem *271*, 8472-8480.

Gidwani, A., Brown, H.A., Holowka, D., and Baird, B. (2003). Disruption of lipid order by short-chain ceramides correlates with inhibition of phospholipase D and downstream signaling by FcepsilonRI. J Cell Sci *116*, 3177-3187.

Gire, V., and Wynford-Thomas, D. (1998). Reinitiation of DNA synthesis and cell division in senescent human fibroblasts by microinjection of anti-p53 antibodies. Mol Cell Biol *18*, 1611-1621.

Glenney, J.R., Jr. (1992). The sequence of human caveolin reveals identity with VIP21, a component of transport vesicles. FEBS Lett *314*, 45-48.

Golding, S.E., Morgan, R.N., Adams, B.R., Hawkins, A.J., Povirk, L.F., and Valerie, K. (2009). Pro- survival AKT and ERK signaling from EGFR and mutant EGFRvIII enhances DNA double- strand break repair in human glioma cells. Cancer Biol Ther *8*, 730-738.

Goldstein, S., and Singal, D.P. (1974). Senescence of cultured human fibroblasts: mitotic versus metabolic time. Exp Cell Res *88*, 359-364.

Gschwendt, M. (1999). Protein kinase C delta. Eur J Biochem *259*, 555-564.

Guertin, D.A., and Sabatini, D.M. (2007). Defining the role of mTOR in cancer. Cancer Cell *12*, 9-22.

Han, J.M., Kim, Y., Lee, J.S., Lee, C.S., Lee, B.D., Ohba, M., Kuroki, T., Suh, P.G., and Ryu, S.H. (2002). Localization of phospholipase D1 to caveolin-enriched membrane via

palmitoylation: implications for epidermal growth factor signaling. Mol Biol Cell *13*, 3976-3988.
Hanada, K., Nishijima, M., Akamatsu, Y., and Pagano, R.E. (1995). Both sphingolipids and cholesterol participate in the detergent insolubility of alkaline phosphatase, a glycosylphosphatidylinositol-anchored protein, in mammalian membranes. J Biol Chem *270*, 6254-6260.
Hansen, C.G., Bright, N.A., Howard, G., and Nichols, B.J. (2009). SDPR induces membrane curvature and functions in the formation of caveolae. Nat Cell Biol *11*, 807-814.
Hansen, C.G., and Nichols, B.J. (2010). Exploring the caves: cavins, caveolins and caveolae. Trends Cell Biol *20*, 177-186.
Hara, E., Tsurui, H., Shinozaki, A., Nakada, S., and Oda, K. (1991). Cooperative effect of antisense-Rb and antisense-p53 oligomers on the extension of life span in human diploid fibroblasts, TIG-1. Biochem Biophys Res Commun *179*, 528-534.
Harder, T., and Simons, K. (1997). Caveolae, DIGs, and the dynamics of sphingolipid-cholesterol microdomains. Curr Opin Cell Biol *9*, 534-542.
Harley, C.B., Futcher, A.B., and Greider, C.W. (1990). Telomeres shorten during ageing of human fibroblasts. Nature *345*, 458-460.
Hayer, A., Stoeber, M., Bissig, C., and Helenius, A. (2010). Biogenesis of caveolae: stepwise assembly of large caveolin and cavin complexes. Traffic *11*, 361-382.
Hayflick, L. (1965a). The Limited in Vitro Lifetime of Human Diploid Cell Strains. Exp Cell Res *37*, 614- 636.
Hayflick, L. (1965b). The limited in vitro lifetime of human diploid cell strians. Exp Cell Res *37*, 614-636.
Hayflick, L., and Moorhead, P.S. (1961). The serial cultivation of human diploid cel strains. Exp Cell Res *25*, 585-621.
Herbig, U., Jobling, W.A., Chen, B.P., Chen, D.J., and Sedivy, J.M. (2004). Telomere shortening triggers senescence of human cells through a pathway involving ATM, p53, and p21(CIP1), but not p16(INK4a). Mol Cell *14*, 501-513.
Hermeking, H., Lengauer, C., Polyak, K., He, T.C., Zhang, L., Thiagalingam, S., Kinzler, K.W., and Vogelstein, B. (1997). 14-3-3 sigma is a p53-regulated inhibitor of G2/M progression. Mol Cell *1*, 3-11.
Hill, M.M., Bastiani, M., Luetterforst, R., Kirkham, M., Kirkham, A., Nixon, S.J., Walser, P., Abankwa, D., Oorschot, V.M., Martin, S., *et al.* (2008). PTRF-Cavin, a conserved cytoplasmic protein required for caveola formation and function. Cell *132*, 113-124.
Hotta, K., Tabata, M., Kiura, K., Kozuki, T., Hisamoto, A., Katayama, H., Takigawa, N., Fujimoto, N., Fujiwara, K., Ueoka, H., *et al.* (2007). Gefitinib induces premature senescence in non-small cell lung cancer cells with or without EGFR gene mutation. Oncol Rep *17*, 313-317.
Hutter, D., Yo, Y., Chen, W., Liu, P., Holbrook, N.J., Roth, G.S., and Liu, Y. (2000). Age-related decline in Ras/ERK mitogen-activated protein kinase cascade is linked to a reduced association between Shc and EGF receptor. J Gerontol A Biol Sci Med Sci *55*, B125-134.
Ichijima, Y., Sakasai, R., Okita, N., Asahina, K., Mizutani, S., and Teraoka, H. (2005). Phosphorylation of histone H2AX at M phase in human cells without DNA damage response. Biochem Biophys Res Commun *336*, 807-812.
Inomata, M., Shimada, Y., Hayashi, M., Kondo, H., and Ohno-Iwashita, Y. (2006). Detachment- associated changes in lipid rafts of senescent human fibroblasts. Biochem Biophys Res Commun *343*, 489-495.

Jackson, J.G., and Pereira-Smith, O.M. (2006). p53 is preferentially recruited to the promoters of growth arrest genes p21 and GADD45 during replicative senescence of normal human fibroblasts. Cancer Res *66*, 8356-8360.

Jang, I.H., Kim, J.H., Lee, B.D., Bae, S.S., Park, M.H., Suh, P.G., and Ryu, S.H. (2001). Localization of phospholipase C-gamma1 signaling in caveolae: importance in EGF-induced phosphoinositide hydrolysis but not in tyrosine phosphorylation. FEBS Lett *491*, 4-8.

Johnson, C.L., Lu, D., Huang, J., and Basu, A. (2002). Regulation of p53 stabilization by DNA damage and protein kinase C. Mol Cancer Ther *1*, 861-867.

Karlseder, J., Smogorzewska, A., and de Lange, T. (2002). Senescence induced by altered telomere state, not telomere loss. Science *295*, 2446-2449.

Kastan, M.B., Zhan, Q., el-Deiry, W.S., Carrier, F., Jacks, T., Walsh, W.V., Plunkett, B.S., Vogelstein, B., and Fornace, A.J., Jr. (1992). A mammalian cell cycle checkpoint pathway utilizing p53 and GADD45 is defective in ataxia-telangiectasia. Cell *71*, 587-597.

Katakura, Y., Udono, M., Katsuki, K., Nishide, H., Tabira, Y., Ikei, T., Yamashita, M., Fujiki, T., and Shirahata, S. (2009). Protein kinase C delta plays a key role in cellular senescence programs of human normal diploid cells. J Biochem *146*, 87-93.

Keilhack, H., Tenev, T., Nyakatura, E., Godovac-Zimmermann, J., Nielsen, L., Seedorf, K., and Bohmer, F.D. (1998). Phosphotyrosine 1173 mediates binding of the protein-tyrosine phosphatase SHP- 1 to the epidermal growth factor receptor and attenuation of receptor signaling. J Biol Chem *273*, 24839-24846.

Khan, E.M., Heidinger, J.M., Levy, M., Lisanti, M.P., Ravid, T., and Goldkorn, T. (2006). Epidermal growth factor receptor exposed to oxidative stress undergoes Src- and caveolin-1-dependent perinuclear trafficking. J Biol Chem *281*, 14486-14493.

Kim, H.S., and Lim, I.K. (2009). Phosphorylated extracellular signal-regulated protein kinases 1 and 2 phosphorylate Sp1 on serine 59 and regulate cellular senescence via transcription of p21Sdi1/Cip1/Waf1. J Biol Chem *284*, 15475-15486.

Kim, K., Nose, K., and Shibanuma, M. (2000). Significance of nuclear relocalization of ERK1/2 in reactivation of c-fos transcription and DNA synthesis in senescent fibroblasts. J Biol Chem *275*, 20685-20692.

Kortlever, R.M., Higgins, P.J., and Bernards, R. (2006). Plasminogen activator inhibitor-1 is a critical downstream target of p53 in the induction of replicative senescence. Nat Cell Biol *8*, 877-884.

Kriegs, M., Kasten-Pisula, U., Rieckmann, T., Holst, K., Saker, J., Dahm-Daphi, J., and Dikomey, E. (2010). The epidermal growth factor receptor modulates DNA double-strand break repair by regulating non-homologous end-joining. DNA Repair (Amst) *9*, 889-897.

Kurzchalia, T.V., Dupree, P., Parton, R.G., Kellner, R., Virta, H., Lehnert, M., and Simons, K. (1992). VIP21, a 21-kD membrane protein is an integral component of trans-Golgi-network-derived transport vesicles. J Cell Biol *118*, 1003-1014.

Kurzchalia, T.V., and Parton, R.G. (1999). Membrane microdomains and caveolae. Curr Opin Cell Biol *11*, 424-431.

Lajoie, P., and Nabi, I.R. (2007). Regulation of raft-dependent endocytosis. J Cell Mol Med *11*, 644-653.

Leontieva, O.V., and Blagosklonny, M.V. (2010). DNA damaging agents and p53 do not cause senescence in quiescent cells, while consecutive re-activation of mTOR is associated with conversion to senescence. Aging (Albany NY) *2*, 924-935.

Li, G.Z., Eller, M.S., Firoozabadi, R., and Gilchrest, B.A. (2003). Evidence that exposure of the telomere 3' overhang sequence induces senescence. Proc Natl Acad Sci U S A *100*, 527-531.

Li, G.Z., Eller, M.S., Hanna, K., and Gilchrest, B.A. (2004). Signaling pathway requirements for induction of senescence by telomere homolog oligonucleotides. Exp Cell Res *301*, 189-200.

Li, W.P., Liu, P., Pilcher, B.K., and Anderson, R.G. (2001). Cell-specific targeting of caveolin-1 to caveolae, secretory vesicles, cytoplasm or mitochondria. J Cell Sci *114*, 1397-1408.

Lim, I.K., Won Hong, K., Kwak, I.H., Yoon, G., and Park, S.C. (2000). Cytoplasmic retention of p-Erk1/2 and nuclear accumulation of actin proteins during cellular senescence in human diploid fibroblasts. Mech Ageing Dev *119*, 113-130.

Lin, A.W., Barradas, M., Stone, J.C., van Aelst, L., Serrano, M., and Lowe, S.W. (1998). Premature senescence involving p53 and p16 is activated in response to constitutive MEK/MAPK mitogenic signaling. Genes Dev *12*, 3008-3019.

Lisanti, M.P., Tang, Z.L., and Sargiacomo, M. (1993). Caveolin forms a hetero-oligomeric protein complex that interacts with an apical GPI-linked protein: implications for the biogenesis of caveolae. J Cell Biol *123*, 595-604.

Liu, L., and Pilch, P.F. (2008). A critical role of cavin (polymerase I and transcript release factor) in caveolae formation and organization. J Biol Chem *283*, 4314-4322.

Lorenzini, A., Tresini, M., Mawal-Dewan, M., Frisoni, L., Zhang, H., Allen, R.G., Sell, C., and Cristofalo, V.J. (2002). Role of the Raf/MEK/ERK and the PI3K/Akt(PKB) pathways in fibroblast senescence. Exp Gerontol *37*, 1149-1156.

Madshus, I.H., and Stang, E. (2009). Internalization and intracellular sorting of the EGF receptor: a model for understanding the mechanisms of receptor trafficking. J Cell Sci *122*, 3433-3439.

Martin, G.M., Sprague, C.A., and Epstein, C.J. (1970). Replicative life-span of cultivated human cells. Effects of donor's age, tissue, and genotype. Lab Invest *23*, 86-92.

Matuoka, K., Chen, K.Y., and Takenawa, T. (2003). A positive role of phosphatidylinositol 3-kinase in aging phenotype expression in cultured human diploid fibroblasts. Arch Gerontol Geriatr *36*, 203-219.

McDermott, M., Wakelam, M.J., and Morris, A.J. (2004). Phospholipase D. Biochem Cell Biol *82*, 225-253.

McMahon, K.A., Zajicek, H., Li, W.P., Peyton, M.J., Minna, J.D., Hernandez, V.J., Luby-Phelps, K., and Anderson, R.G. (2009). SRBC/cavin-3 is a caveolin adapter protein that regulates caveolae function. Embo J *28*, 1001-1015.

McManus, K.J., and Hendzel, M.J. (2005). ATM-dependent DNA damage-independent mitotic phosphorylation of H2AX in normally growing mammalian cells. Mol Biol Cell *16*, 5013-5025.

Meyyappan, M., Wheaton, K., and Riabowol, K.T. (1999). Decreased expression and activity of the immediate-early growth response (Egr-1) gene product during cellular senescence. J Cell Physiol *179*, 29-39.

Migliaccio, E., Giorgio, M., Mele, S., Pelicci, G., Reboldi, P., Pandolfi, P.P., Lanfrancone, L., and Pelicci, P.G. (1999). The p66shc adaptor protein controls oxidative stress response and life span in mammals. Nature *402*, 309-313.

Mineo, C., Gill, G.N., and Anderson, R.G. (1999). Regulated migration of epidermal growth factor receptor from caveolae. J Biol Chem *274*, 30636-30643.

Mineo, C., James, G.L., Smart, E.J., and Anderson, R.G. (1996). Localization of epidermal growth factor- stimulated Ras/Raf-1 interaction to caveolae membrane. J Biol Chem *271*, 11930-11935.

Monier, S., Dietzen, D.J., Hastings, W.R., Lublin, D.M., and Kurzchalia, T.V. (1996). Oligomerization of VIP21-caveolin in vitro is stabilized by long chain fatty acylation or cholesterol. FEBS Lett *388*, 143-149.

Moodie, S.A., and Wolfman, A. (1994). The 3Rs of life: Ras, Raf and growth regulation. Trends Genet *10*, 44-48.

Mouton, R.E., and Venable, M.E. (2000). Ceramide induces expression of the senescence histochemical marker, beta-galactosidase, in human fibroblasts. Mech Ageing Dev *113*, 169-181.

Mundy, D.I., Machleidt, T., Ying, Y.S., Anderson, R.G., and Bloom, G.S. (2002). Dual control of caveolar membrane traffic by microtubules and the actin cytoskeleton. J Cell Sci *115*, 4327-4339.

Nakamura, M., Kondo, H., Shimada, Y., Waheed, A.A., and Ohno-Iwashita, Y. (2003). Cellular aging- dependent decrease in cholesterol in membrane microdomains of human diploid fibroblasts. Exp Cell Res *290*, 381-390.

Naru, E., Takanezawa, Y., Kobayashi, M., Misaki, Y., Kaji, K., and Arakane, K. (2008). Increased levels of a particular phosphatidylcholine species in senescent human dermal fibroblasts in vitro. Hum Cell *21*, 70-78.

Newton, A.C. (2001). Protein kinase C: structural and spatial regulation by phosphorylation, cofactors, and macromolecular interactions. Chem Rev *101*, 2353-2364.

Noda, A., Ning, Y., Venable, S.F., Pereira-Smith, O.M., and Smith, J.R. (1994). Cloning of senescent cell- derived inhibitors of DNA synthesis using an expression screen. Exp Cell Res *211*, 90-98.

Ogata, T., Ueyama, T., Isodono, K., Tagawa, M., Takehara, N., Kawashima, T., Harada, K., Takahashi, T., Shioi, T., Matsubara, H., *et al.* (2008). MURC, a muscle-restricted coiled-coil protein that modulates the Rho/ROCK pathway, induces cardiac dysfunction and conduction disturbance. Mol Cell Biol *28*, 3424-3436.

Ohki, R., Nemoto, J., Murasawa, H., Oda, E., Inazawa, J., Tanaka, N., and Taniguchi, T. (2000). Reprimo, a new candidate mediator of the p53-mediated cell cycle arrest at the G2 phase. J Biol Chem *275*, 22627-22630.

Okabayashi, Y., Kido, Y., Okutani, T., Sugimoto, Y., Sakaguchi, K., and Kasuga, M. (1994). Tyrosines 1148 and 1173 of activated human epidermal growth factor receptors are binding sites of Shc in intact cells. J Biol Chem *269*, 18674-18678.

Okamoto, T., Schlegel, A., Scherer, P.E., and Lisanti, M.P. (1998). Caveolins, a family of scaffolding proteins for organizing "preassembled signaling complexes" at the plasma membrane. J Biol Chem *273*, 5419-5422.

Palmer, H.J., Tuzon, C.T., and Paulson, K.E. (1999). Age-dependent decline in mitogenic stimulation of hepatocytes. Reduced association between Shc and the epidermal growth factor receptor is coupled to decreased activation of Raf and extracellular signal-regulated kinases. J Biol Chem *274*, 11424-11430.

Park, S.C., Park, J.S., Park, W.Y., Cho, K.A., Ahn, J.S., and Jang, I.S. (2002). Down-regulation of receptor- mediated endocytosis is responsible for senescence-associated hyporesponsiveness. Ann N Y Acad Sci *959*, 45-49.

Park, W.Y., Park, J.S., Cho, K.A., Kim, D.I., Ko, Y.G., Seo, J.S., and Park, S.C. (2000). Up-regulation of Caveolin Attenuates Epidermal Growth Factor Signaling in Senescent Cells. J Biol Chem *275*, 20847-20852.

Parton, R.G., and Simons, K. (2007). The multiple faces of caveolae. Nat Rev Mol Cell Biol *8*, 185-194.

Patel, H.H., Murray, F., and Insel, P.A. (2008). Caveolae as organizers of pharmacologically relevant signal transduction molecules. Annu Rev Pharmacol Toxicol *48*, 359-391.

Pelkmans, L., Puntener, D., and Helenius, A. (2002). Local actin polymerization and dynamin recruitment in SV40-induced internalization of caveolae. Science *296*, 535-539.

Pelkmans, L., and Zerial, M. (2005). Kinase-regulated quantal assemblies and kiss-and-run recycling of caveolae. Nature *436*, 128-133.

Phillips, P.D., and Cristofalo, V.J. (1988). Classification system based on the functional equivalency of mitogens that regulate WI-38 cell proliferation. Exp Cell Res *175*, 396-403.

Phillips, P.D., Kaji, K., and Cristofalo, V.J. (1984). Progressive loss of the proliferative response of senescing WI-38 cells to platelet-derived growth factor, epidermal growth factor, insulin, transferrin, and dexamethasone. J Gerontol *39*, 11-17.

Phillips, P.D., Kuhnle, E., and Cristofalo, V.J. (1983). [125I]EGF binding ability is stable throughout the replicative life-span of WI-38 cells. J Cell Physiol *114*, 311-316.

Rattan, S.I., and Derventzi, A. (1991). Altered cellular responsiveness during ageing. Bioessays *13*, 601- 606.

Resnick-Silverman, L., St Clair, S., Maurer, M., Zhao, K., and Manfredi, J.J. (1998). Identification of a novel class of genomic DNA-binding sites suggests mechanism for selectivity in target gene activation by the tumor suppressor protein p53. Genes Dev *12*, 2102-2107.

Riabowol, K., Schiff, J., and Gilman, M.Z. (1992). Transcription factor AP-1 activity is required for initiation of DNA synthesis and is lost during cellular aging. Proc Natl Acad Sci U S A *89*, 157-161.

Rietveld, A., and Simons, K. (1998). The differential miscibility of lipids as the basis for the formation of functional membrane rafts. Biochim Biophys Acta *1376*, 467-479.

Rogakou, E.P., Boon, C., Redon, C., and Bonner, W.M. (1999). Megabase chromatin domains involved in DNA double-strand breaks in vivo. J Cell Biol *146*, 905-916.

Rogakou, E.P., Pilch, D.R., Orr, A.H., Ivanova, V.S., and Bonner, W.M. (1998). DNA double-stranded breaks induce histone H2AX phosphorylation on serine 139. J Biol Chem *273*, 5858-5868.

Rohme, D. (1981). Ageing and the fusion sensitivity potential of human cells in culture: relation to tissue origin, donor age, and in vitro culture level and condition. Mech Ageing Dev *16*, 241-253.

Rothberg, K.G., Heuser, J.E., Donzell, W.C., Ying, Y.S., Glenney, J.R., and Anderson, R.G. (1992). Caveolin, a protein component of caveolae membrane coats. Cell *68*, 673-682.

Rouault, J.P., Falette, N., Guehenneux, F., Guillot, C., Rimokh, R., Wang, Q., Berthet, C., Moyret-Lalle, C., Savatier, P., Pain, B., *et al.* (1996). Identification of BTG2, an antiproliferative p53- dependent component the DNA damage cellular response pathway. Nat Genet *14*, 482-486.

Rutter, A., Mackinnon, W.B., Huschtscha, L.I., and Mountford, C.E. (1996). A proton magnetic resonance spectroscopy study of aging and transformed human fibroblasts. Exp Gerontol *31*, 669-686.

Rybin, V.O., Guo, J., Gertsberg, Z., Feinmark, S.J., and Steinberg, S.F. (2008). Phorbol 12-myristate 13- acetate-dependent protein kinase C delta-Tyr311 phosphorylation in cardiomyocyte caveolae. J Biol Chem *283*, 17777-17788.

Satyanarayana, A., Greenberg, R.A., Schaetzlein, S., Buer, J., Masutomi, K., Hahn, W.C., Zimmermann, S., Martens, U., Manns, M.P., and Rudolph, K.L. (2004). Mitogen stimulation cooperates with telomere shortening to activate DNA damage responses and senescence signaling. Mol Cell Biol *24*, 5459-5474.

Scherer, P.E., Okamoto, T., Chun, M., Nishimoto, I., Lodish, H.F., and Lisanti, M.P. (1996). Identification, sequence, and expression of caveolin-2 defines a caveolin gene family. Proc Natl Acad Sci U S A *93*, 131-135.

Schroeder, F., Goetz, I., and Roberts, E. (1984). Age-related alterations in cultured human fibroblast membrane structure and function. Mech Ageing Dev *25*, 365-389.

Sedelnikova, O.A., Horikawa, I., Zimonjic, D.B., Popescu, N.C., Bonner, W.M., and Barrett, J.C. (2004). Senescing human cells and ageing mice accumulate DNA lesions with unrepairable double- strand breaks. Nat Cell Biol *6*, 168-170.

Sell, C., Ptasznik, A., Chang, C.D., Swantek, J., Cristofalo, V.J., and Baserga, R. (1993). IGF-1 receptor levels and the proliferation of young and senescent human fibroblasts. Biochem Biophys Res Commun *194*, 259-265.

Serrano, M., Lin, A.W., McCurrach, M.E., Beach, D., and Lowe, S.W. (1997). Oncogenic ras provokes premature cell senescence associated with accumulation of p53 and p16INK4a. Cell *88*, 593- 602.

Seshadri, T., and Campisi, J. (1990). Repression of c-fos transcription and an altered genetic program in senescent human fibroblasts. Science *247*, 205-209.

Shay, J.W., Pereira-Smith, O.M., and Wright, W.E. (1991). A role for both RB and p53 in the regulation of human cellular senescence. Exp Cell Res *196*, 33-39.

Simons, K., and Ikonen, E. (1997). Functional rafts in cell membranes. Nature *387*, 569-572.

Simons, K., and Toomre, D. (2000). Lipid rafts and signal transduction. Nat Rev Mol Cell Biol *1*, 31-39.

Stewart, S.A., Ben-Porath, I., Carey, V.J., O'Connor, B.F., Hahn, W.C., and Weinberg, R.A. (2003). Erosion of the telomeric single-strand overhang at replicative senescence. Nat Genet *33*, 492- 496.

Tagawa, M., Ueyama, T., Ogata, T., Takehara, N., Nakajima, N., Isodono, K., Asada, S., Takahashi, T., Matsubara, H., and Oh, H. (2008). MURC, a muscle-restricted coiled-coil protein, is involved in the regulation of skeletal myogenesis. Am J Physiol Cell Physiol *295*, C490-498.

Takahashi, A., Ohtani, N., Yamakoshi, K., Iida, S., Tahara, H., Nakayama, K., Nakayama, K.I., Ide, T., Saya, H., and Hara, E. (2006). Mitogenic signalling and the p16INK4a-Rb pathway cooperate to enforce irreversible cellular senescence. Nat Cell Biol *8*, 1291-1297.

Tang, Z., Scherer, P.E., Okamoto, T., Song, K., Chu, C., Kohtz, D.S., Nishimoto, I., Lodish, H.F., and Lisanti, M.P. (1996). Molecular cloning of caveolin-3, a novel member of the caveolin gene family expressed predominantly in muscle. J Biol Chem *271*, 2255-2261.

Torres, C., Francis, M.K., Lorenzini, A., Tresini, M., and Cristofalo, V.J. (2003). Metabolic stabilization of MAP kinase phosphatase-2 in senescence of human fibroblasts. Exp Cell Res *290*, 195-206.

Tran, K.T., Rusu, S.D., Satish, L., and Wells, A. (2003). Aging-related attenuation of EGF receptor signaling is mediated in part by increased protein tyrosine phosphatase activity. Exp Cell Res *289*, 359-367.

Tresini, M., Lorenzini, A., Frisoni, L., Allen, R.G., and Cristofalo, V.J. (2001). Lack of Elk-1 phosphorylation and dysregulation of the extracellular regulated kinase signaling pathway in senescent human fibroblast. Exp Cell Res *269*, 287-300.

Tresini, M., Mawal-Dewan, M., Cristofalo, V.J., and Sell, C. (1998). A phosphatidylinositol 3-kinase inhibitor induces a senescent-like growth arrest in human diploid fibroblasts [see comments]. Cancer Res *58*, 1-4.

Vanhaesebroeck, B., Leevers, S.J., Panayotou, G., and Waterfield, M.D. (1997). Phosphoinositide 3- kinases: a conserved family of signal transducers. Trends Biochem Sci *22*, 267-272.

Vannini, F., Meacci, E., Vasta, V., Farnararo, M., and Bruni, P. (1994). Involvement of protein kinase C and arachidonate signaling pathways in the alteration of proliferative response of senescent IMR-90 human fibroblasts. Mech Ageing Dev *76*, 101-111.

Vaziri, H., and Benchimol, S. (1998). Reconstitution of telomerase activity in normal human cells leads to elongation of telomeres and extended replicative life span. Curr Biol *8*, 279-282.

Venable, M.E., Bielawska, A., and Obeid, L.M. (1996). Ceramide inhibits phospholipase D in a cell-free system. J Biol Chem *271*, 24800-24805.

Venable, M.E., Blobe, G.C., and Obeid, L.M. (1994). Identification of a defect in the phospholipase D/diacylglycerol pathway in cellular senescence. J Biol Chem *269*, 26040-26044.

Venable, M.E., Lee, J.Y., Smyth, M.J., Bielawska, A., and Obeid, L.M. (1995). Role of ceramide in cellular senescence. J Biol Chem *270*, 30701-30708.

Venable, M.E., and Obeid, L.M. (1999). Phospholipase D in cellular senescence. Biochim Biophys Acta *1439*, 291-298.

Volonte, D., and Galbiati, F. (2009a). Caveolin-1, cellular senescence and pulmonary emphysema. Aging (Albany NY) *1*, 831-835.

Volonte, D., and Galbiati, F. (2009b). Inhibition of thioredoxin reductase 1 by caveolin 1 promotes stress- induced premature senescence. EMBO Rep *10*, 1334-1340.

Volonte, D., and Galbiati, F. (2011). Polymerase I and Transcript Release Factor (PTRF)/Cavin-1 Is a Novel Regulator of Stress-induced Premature Senescence. J Biol Chem *286*, 28657-28661.

Volonte, D., Galbiati, F., Pestell, R.G., and Lisanti, M.P. (2001). Cellular stress induces the tyrosine phosphorylation of caveolin-1 (Tyr(14)) via activation of p38 mitogen-activated protein kinase and c-Src kinase. Evidence for caveolae, the actin cytoskeleton, and focal adhesions as mechanical sensors of osmotic stress. J Biol Chem *276*, 8094-8103.

Volonte, D., Kahkonen, B., Shapiro, S., Di, Y., and Galbiati, F. (2009). Caveolin-1 expression is required for the development of pulmonary emphysema through activation of the ATM-p53-p21 pathway. J Biol Chem *284*, 5462-5466.

Volonte, D., Zhang, K., Lisanti, M.P., and Galbiati, F. (2002). Expression of caveolin-1 induces premature cellular senescence in primary cultures of murine fibroblasts. Mol Biol Cell *13*, 2502-2517.

Vousden, K.H., and Lane, D.P. (2007). p53 in health and disease. Nat Rev Mol Cell Biol *8*, 275-283.

Wary, K.K., Mariotti, A., Zurzolo, C., and Giancotti, F.G. (1998). A requirement for caveolin-1 and associated kinase Fyn in integrin signaling and anchorage-dependent cell growth. Cell *94*, 625-634.

Way, M., and Parton, R.G. (1996). M-caveolin, a muscle-specific caveolin-related protein [corrected and republished with original paging, article originally printed in FEBS Lett 1995 Nov 27;376(1- 2):108-12]. FEBS Lett *378*, 108-112.

Webb, L.M., Arnholt, A.T., and Venable, M.E. (2010). Phospholipase D modulation by ceramide in senescence. Mol Cell Biochem *337*, 153-158.

Wheaton, K. (2002). Regulation of Serum Response Factor during Cellular Senescence. In Biochemistry & Molecular Biology (Calgary, University of Calgary), pp. 208.

Wheaton, K. (2011). Caveolar vesicles generate DNA damage and perpetuate cellular aging. Cell Res *21*, 993-994.

Wheaton, K., Atadja, P., and Riabowol, K. (1996). Regulation of transcription factor activity during cellular aging. Biochem Cell Biol *74*, 523-534.

Wheaton, K., Muir, J., Ma, W., and Benchimol, S. (2010). BTG2 antagonizes Pin1 in response to mitogens and telomere disruption during replicative senescence. Aging Cell *9*, 747-760.

Wheaton, K., and Riabowol, K. (2004). Protein kinase C delta blocks immediate-early gene expression in senescent cells by inactivating serum response factor. Mol Cell Biol *24*, 7298-7311.

Wheaton, K., Sampsel, K., Boisvert, F.M., Davy, A., Robbins, S., and Riabowol, K. (2001). Loss of functional caveolae during senescence of human fibroblasts. J Cell Physiol *187*, 226-235.

White, M.A., and Anderson, R.G. (2005). Signaling networks in living cells. Annu Rev Pharmacol Toxicol *45*, 587-603.

Wright, J.D., Reuter, C.W., and Weber, M.J. (1996). Identification of sites on epidermal growth factor receptors which are phosphorylated by pp60src in vitro. Biochim Biophys Acta *1312*, 85-93.

Yoshida, K., Liu, H., and Miki, Y. (2006). Protein kinase C delta regulates Ser46 phosphorylation of p53 tumor suppressor in the apoptotic response to DNA damage. J Biol Chem *281*, 5734-5740.

Zajchowski, L.D., and Robbins, S.M. (2002). Lipid rafts and little caves. Compartmentalized signalling in membrane microdomains. Eur J Biochem *269*, 737-752.

Zezula, J., Sexl, V., Hutter, C., Karel, A., Schutz, W., and Freissmuth, M. (1997). The cyclin-dependent kinase inhibitor p21cip1 mediates the growth inhibitory effect of phorbol esters in human venous endothelial cells. J Biol Chem *272*, 29967-29974.

Zhu, H., Yue, J., Pan, Z., Wu, H., Cheng, Y., Lu, H., Ren, X., Yao, M., Shen, Z., and Yang, J.M. (2010). Involvement of Caveolin-1 in repair of DNA damage through both homologous recombination and non-homologous end joining. PLoS One *5*, e12055.

Zhu, J., Woods, D., McMahon, M., and Bishop, J.M. (1998). Senescence of human fibroblasts induced by oncogenic Raf. Genes Dev *12*, 2997-3007.

11

Cellular Senescence and Its Relation with Telomere

Diego Julio Arenas-Aranda[1],
Elena Hernández-Caballero[2] and Fabio Salamanca-Gómez[1]
[1]*Unit of Medical Research in Human Genetics, Medical National Center 21st Century, Mexican Institute of Social Insurance,*
[2]*Section of Posgrado's Studies and Investigation, High School of Medicine, Technical National Institute. Mexico City, Mexico*

1. Introduction

For years it was thought that cells under culture conditions were immortal; however, from the publication of the works of Leonard Hayflick, this concept changed. Hayflick demonstrated that somatic cells under culture conditions had a limited capacity to proliferate; they stop dividing and enter into permanent arrest in the cell cycle, known as senescence. A particular characteristic of this state is that the cell maintains its viability and metabolic activity, and despite the presence of nutrients and mitogens, it does not divide (Ouellette et al., 2000). There are reports of cells that have been maintained in this state during several decades (Michaloglou et al., 2005). Regarding the origin of this process in mammalian cells, two different and apparently contradictory hypotheses have been proposed: senescence as a mechanism that suppresses tumor development, and senescence as the loss of the cells' regenerative capacity *in vivo*. In terms of the first hypothesis, senescence possesses a beneficial effect for the organism because it would avoid the development of cancer; otherwise, the second hypothesis would exert a harmful effect on organisms in that it would favor aging. Commentary will appear later on that both hypotheses joined together in the antagonist pleiotropic hypothesis (Williams, 1957; Campisi & Adda di Fagagna, 2007; Campisi, 2011).

2. Hayflick experiments

It was 50 years ago that Hayflick first reported on the condition of mortality that cells maintain (Hayflick & Moorhead, 1961), which the researcher himself cited in 1998 (Hayflick, 1998). To the present day, few have dared to enter into the study of Biogerontology. The generalized belief was that cells kept under culture conditions could replicate themselves indefinitely; if this were not possible, it would be due to lack of knowledge on the appropriate conditions for maintaining cells under culture (Hayflick, 2003). Carrel in 1921 stated that it was possible to keep chicken-heart fibroblasts indefinitely, maintaining these with embryonic tissue extract. However, as Hayflick noted, this experiment involved a great technical error because the extract with which the culture was nourished throughout 34 years was supplied with fresh cells throughout the entire time that the experiment lasted (Hayflick, 1998).

Hayflick identified three phases in his cellular proliferation experiments: phase I or that of primary culture, in which cells initiate their proliferation; phase II, that of rapid and continuous proliferation, and phase III, in which proliferation velocity diminishes and is finally detained (Hayflick & Moorhead, 1961). Hayflick concludes that cells possess some type of counting mechanism because they stop dividing after a certain number of duplications, between 40 and 60. This counting mechanism is conserved even after the cells are frozen and cultured anew (Hayflick & Moorhead, 1961; Hayflick, 1965).

Some years later, thanks to the work of McClintock on chromosomal ends (McClintock, 1941), Olovnikov formulated his theory concerning the problem of replication and the solution to this. Suggesting that the inability of polymerase to replicate chromosomal ends totally was what could lead to cellular senescence, he proposed that the ends could function as a buffer, avoiding the loss of important sequences, but that in turn, this buffer function could also could be lost with successive replications (Olovnikov, 1996). Later, with the discovery of the telomeric repeats sequence of and telomerase by Blackburn, the study of telomeres and their participation in the senescence process began.

Currently, the existence is accepted of a limit of normal somatic cell replication, denominated the Hayflick limit. However, once the importance is established of telomerase as the enzyme that synthesizes telomeres (Greider & Blackburn, 1985), it was discovered that this enzyme is found to be active in immortalized cell lines (Morin, 1989), tumor cells (Kim et al., 1994), stem cells (Chiu et al., 1996), and embryonic and germinal cell lines (Mantell & Greider, 1994; Wright et al., 1996).

3. The antagonistic pleiotropy hypothesis

Antagonistic pleiotropy is a concept pertaining to Evolutionary Biology that proposes that some genes can have an impact on the physical state of the organism differentially throughout its lifetime (Williams, 1957; Tuminello & Han, 2011). It is suggested that senescence evolved as an example of antagonistic pleiotropy; thus, its characteristics are beneficial in a reproductively active organism; later in this organism's lifespan, these characteristics become deteriorating. That is, senescence is the result of the random accumulation, whether passive or active, of harmful mutations (Kirkwood, 1977), reducing vigor and longevity, after the individual has passed reproductive age (Walker, 2011), although there is no definitive evidence that supports the negative effect of senescence in old persons.

Diverse processes have been found that are considered examples of antagonistic pleiotropy. Inflammation is a vital process to fight against infections and for cicatrization at all ages, but is also a process that can have negative effects if it becomes chronic in old age (Hornsby, 2010). On analyzing the influence of TP53 polymorphisms on cancer with respect to age, Cherdyntseva et al. (2010) found a relationship between the presence of polymorphisms in both genes and the increase of the risk of lung cancer in young, but not in older, smokers, while the combination of wild-type alleles increased the risk of lung cancer in individuals aged >60 years. The researchers concluded that p53 protects the organism against cancer at the beginning of life, but that it promotes the aging phenotype in older persons, including the appearance of cancer at the end of life. Another recent study suggests that allele ε4 carriers enjoy a beneficial cognitive effect in youth, and that later at an advanced age present cognitive diminution, which could increase the risk of presenting Alzheimer disease, although it could be that this allele is not pleiotropically antagonistic, but rather that it interacts with other risk factors for Alzheimer disease (Tuminello & Han, 2011). Feltes et al. (2011) suggest that aging and age-associated diseases could be the result of a program of development that is activated from the embryo stage, that persists throughout life, and that is regulated by the interaction of protein networks that connect environmental with molecular signals. The protein networks of the immune system, the epigenetic network, and aerobic metabolism are subject to great selection pressure during embryogenesis. However, this pressure becomes more relaxed in the adult, which allows the initiation of aging-associated diseases.

4. Somatic cells and senescence

After gametes are fused during the fertilization process, that primordial cell denominated the zygote begins a long journey in the formation of the individual. However, this journey begins with an accelerated expansion in the number of cells, which later decelerates. Once the organism has been formed, it will utilize cellular replication during its entire lifetime to grow, regenerate its tissues, cure wounds, or during the immune response.

Somatic cells possess a limited number of possible cell divisions, after which these become refractory to mitogeneic stimuli and enter into replicative senescence. In 2001, Sin et al. cite that aging at the cellular level is the result of cell function alterations, such as the response to DNA structural changes that is reflected in the expression of genes. A senescent cell remains arrested at cell cycle stage G1, and although it does not divide again, it remains metabolically active for a long time. The accumulation of mutations and damage to the DNA, together with inefficient repair mechanisms, become critical with each cell division and cause genetic heterogeneity in aging cells (De, 2011). There are somatic cells that possess the capacity to renew themselves, such as epithelial and blood cells, while there are cells that once differentiated, do not divide again, such as neurons. The equilibrium in death maintains homeostasis in the individual; thus, excessive death can lead to tissue degeneration and the inability to die can lead to hyperplasia and finally, cancer. The accumulated errors in a senescent cell's DNA can alter this balance and cause diverse diseases (Hotchkiss et al., 2009).

5. Senescent phenotype

To date, there is no senescence marker that is totally specific for this stage because not all senescent cells express the same markers. These cells can exhibit diverse changes that in their entirety can aid us in defining the senescent phenotype (Rodier & Campisi, 2011).

Change in cell volume is one of the most evident characteristics of a senescent cell, because cells may be observed that range from 1,000 μm^2 in an early passage of human fetal fibroblasts up to 9,000 μm^2 in terminal passages of the culture. Size increase has been correlated with changes in cytoskeletal organization (Wang & Gundersen, 1984), which leads to modification in cell shape. For example, fibroblasts lose their typical tapered form to acquire a flat appearance (Greenberg et al., 1977), apparently due to changes in the expression of diverse cytoskeletal proteins. Nishio et al. (2001) found that senescent fibroblasts contain three times the amount of the cytoskeletal protein vimentin as embryonic fibroblasts. Vimentin presents as dense filament bundles that are parallel to the longest cell-body axis in senescent cells, while in young cells vimentin formation is observed as a network of short and thin filaments. The authors also demonstrated that young fibroblasts acquire a senescent phenotype once they are transfected with a vector that over expresses the vimentin gene, while actin levels diminish in senescent fibroblasts (Nishio & Inoue, 2005). This diminution causes rigidity in old donor cells and increases in the kinesis of cell reorganization (Zahn et al., 2011). On the other hand, cellular adhesion of aging fibroblasts increases. It was found recently that the senescence of vascular epithelial cells induces an increase in cell adhesion proteins, which in turn increases the adhesion of monocytes to endothelial cells through their binding with Intracellular adhesion molecule 1 (ICAM1), contributing to the appearance of atherosclerosis (Yanaka et al., 2011).

Another characteristic of cellular senescence is change in diverse organelles; a very common occurrence comprises the increase in lysosome number and size. In lysosomes, granules of lipofuscin, the so-called aging pigment, accumulate (Brunk & Terman, 2002; Gutteridge, 1984). This material cannot be degraded by the cell's proteolytic machinery, is highly toxic, and inhibits the degradation of oxidized proteins (Bader et al., 2007; Höhn et al., 2011). Another lysosome-linked senescence biomarker is β-galactosidase, which is derived from the β-D-galactosidase and whose activity increases in senescent cells. In non-senescent cells, lysosomes possess a pH4-optimal function, while when the cell ages, the lysosomal compartment expands and β-galactosidase increases; thus, it is possible to detect a suboptimal pH of 6, a change known as senescence-associated β-galactosidase activity (Dimri et al., 1995; Lee et al., 2006). Senescence is also implicated in the deterioration of mitochondrial function and in the appearance of mutations in mitochondrial DNA, due to the lack of a repair system (Percy et al., 2008), and it is considered that aberrant production of Reactive oxygen species (ROS) can increase the mitochondrial mass (Hwang et al., 2009) and on the other hand accelerate telomere shortening and contribute to cellular aging (Liu et al., 2002) due to damage to the DNA. With respect to the nucleus, the increase in chromatin condensation is the most evident nuclear change. Regions are formed that are known as Senescence-associated heterochromatin foci (SAHF); these DNA regions are associated with heterochromatin proteins such as HP1 and H3K9m (Narita et al., 2003). SAHF are also evident in cells that become senescent because of oncogenic stress with H-Ras (Kosar et al., 2011).

Changes in senescent cells are also reflected in their functions; for example, fibroblasts under culture conditions adopt a matrix-degrading phenotype, while adrenal cortex epithelial cells produce an altered steroid-hormone profile (Campisi, 2000). Due to the increase in the secretion of pro-inflammatory proteins such as interleukins and chemokines, it is said that senescent cells are found in a pro-inflammatory state (Freund et al., 2010), which in an aging organism can favor tissue deterioration.

5.1 Changes in gene expression, growth arrest, and apoptosis resistance

Cell cycle arrest-associated replicative senescence is related with telomere shortening and, as previously noted, is a response for suppressing tumor formation and that can have aging of the organism as a side effect (Harley et al., 1990; Campisi, 2011). The cell population gradually stops dividing. (Thomas et al., 1997). Cell growth inhibition can be the result of cellular quiescence, whether due to lack of growth or nutrient factors or to other in- and extrinsic factors. This leads to the cell's exhibiting a low metabolic rate, low protein synthesis and cell functions, and the absence of growth (Blagosklonny, 2011). The senescent cell is defined by the permanent lack of replicative potential despite receiving a mitogenic stimulus (Rodier & Campisi, 2011). The cells remain arrested in G1, although on some occasions can be stopped in G2 (Harley et al., 1990). The central signaling pathways for senescence are represented by the p16-pRb Retinoblastoma (Rb) protein and the p53 tumor suppressor (Lowe & Sherr, 2003). The p14-p53-p21 pathway is partially telomere-dependent, while the p16-pRb pathway is independent of the presence of dysfunctional telomeres (Campisi & d'Adda di Fagagna, 2007). The product of *p53* gene accumulates in response to cellular stress, which activates a specific gene target program to restrict the growth of abnormal or damaged cells; the result can be apoptosis, transitory cell cycle arrest, or permanent arrest (Beausejour et al., 2003). Thus, the product of *p53 gene* possesses anti-cancerous as well as pro-aging effects depending upon the context of the individual's age (Campisi, 2005). Among p53 target genes are found the Cyclin-dependent kinase (CDK) inhibitor p21, the pro-apoptotic genes *BAX* and *APAF1*, and the E3 ubiquitin ligase, MDM2 (Vousden & Lu, 2002). *p53* expression is controlled by p19 (Arf). On the other hand, Rb expression is controlled by p16Ink4a, whose protein levels increase in senescent cells. p16Ink4a maintains pRb in a hypophosphorylated state, which inhibits cell proliferation and induces growth arrest through the pRB effect on E2F; this is necessary to activate the genes implicated in cell cycle progression (Campisi & d'Adda di Fagagna, 2007). Both signaling pathways interact and are reciprocally regulated. However, each can interrupt the cell cycle independently.

As Blagosklonny (2011) cites, cell arrest is only one part of the senescence equation, because senescent cells also become resistant to apoptosis. There are mechanisms of defense that augment apoptosis resistance, increasing anti-apoptotic signaling and avoiding the death of damaged cells. The increase of apoptosis resistance is a cell safety mechanism, because if cells have an acute stress due to some damage, they possess the capacity of recovering their homeostasis. However, in aging, when stress becomes persistent apoptosis resistance can cause the survival of undesired cells (Hampel et al., 2004; Salminen et al., 2011). It has been observed that the equilibrium between apoptotic and pro-apoptotic proteins changes with age. Bcl-2 and Bcl-xL protein levels are higher in aging than in young fibroblasts, while pro-apoptotic Bax levels are higher in young cells (Rochette & Brash, 2008). Apoptosis markers such as FasL and cytochrome C decrease in serum and, on the other hand, levels of soluble Fas (an apoptosis inhibitor) increase (Kavathia et al., 2009). Salminen et al. (2011) suggest that apoptosis resistance can affect the host's defenses in age-related fashion, a situation that meets promoted by the chronic inflammation that senescent cells develop.

6. Senescence markers

There are some molecular senescence markers that are characteristic of damage to DNA, including the nuclear foci of phosphorylated histones H2AX and DNA damage response factors such as 53BP1, MDC1, and NBS1 (d'Adda di Fagagna, 2008), which explains why culture shock can trigger senescence without the participation of telomeres. p16 is used for identification of senescent cells due to that it is overexpressed in the majority of these cells (Krishnamurthy et al., 2004). Although it has been demonstrated that in *Caenorhabditis elegans* alterations in microRNA´s (miRNAs) expression are associated with aging, the role that these could play in mammals remains unknown. In *C. elegans*, lin-4 over expression leads to an extension in its life span, while loss of lin-4 reduces the life span (Boehm & Slack, 2005). Lin-4 acts on the lin-14 messenger, affecting not only the life span, but also the insulin signaling pathway (Hung et al., 2010).

7. Causes of cellular senescence

In stem cells, it has been observed that premature differentiation and senescence are alternatives to DNA damage repair that can exert a beneficial effect on restricting the accumulation of defective stem cells. However, sensitivity to DNA damage and p53-related apoptosis induction differ widely among stem cells (Blanpain et al., 2011). The stem cells of melanocytes, for example, undergo premature differentiation, which reduces the stem cell pool and causes graying of the hair (Inomata et al., 2009). On the other hand, loss of TP63 (a member of the p53 family) in dermal precursors leads to skin ulceration and cicatrization defects due to genomic instability and the induction of senescence (Su et al., 2009).

Recently, a chromatin remodeling factor was found to be implicated in replicative senescence, the Jun 2 dimerization protein (JDP2), which binds to histones and inhibits the binding of Polycomb repressor complexes (PRC1 and PRC2) to p16 gene promotor (Huang et al., 2011).

7.1 Telomere-induced senescense

The loss of proliferative potential can be a genetically programmed process. The telomere-associated aging theory suggests that progressive telomere sequence loss triggers chronic p53 activation, which consequently leads the cell to halt its proliferation (Lee et al., 1998). Telomere shortening is a stochastic event; thus, telomere length varies greatly among individuals (Halaschek-Wiener et al., 2008).

Dysfunctional telomeres trigger the response to DNA damage, which includes activation of ATM, 53BP1, Mdc1, Chk2, and H2AX, in addition to over expression of cell cycle inhibitors p21Cip1/Waf1 and p16INK4a and under expression of different cell cycle proteins (Herbig et al., 2004).

7.2 Senescence induced by non-telomeric chromatin alterations

Non-genotoxic stress can cause perturbations in chromatin, that is, epigenetic changes that can alter the genetic schema of the cell. It has been demonstrated that global genome methylation diminishes with age. However, it has also been observed that the promoters of certain genes can be hypermethylated, thus silenced. Histones themselves undergo modifications during aging, whether because of methylation or hypoacetylation; all of these chromatin regions in senescent cells are observed as transcriptionally inactive domains

(SAHF) (Funayama, 2007). One characteristic of the senescence program is observed in cultured fibroblasts in chromatin reorganization through H3 methylation in the Lys9 residue and by protein recruitment in the heterochromatin. Some agents that interact with DNA, such as doxorubicin, cisplatin, taxol, vincristine, cytarabine, and etoposide, can produce the senescent phenotype in tumor cells (Chang et al., 1999).

It has been suggested that the condensation of genes implicated in proliferation through SAHF formation can contribute directly to senescence-associated silencing. However, it was recently suggested that SAHF are the result of persistent damage to the DNA and that it is condensation of the genes that promote proliferation, to a greater extent than large-scale SAHF, which detains senescence-associated proliferation (Rai & Adams, 2011).

7.3 Senescence induced by stress and other factors

Cellular senescence can additionally be induced prior to telomere shortening, which is also known as premature senescence or stress-induced premature senescence. One example are fibroblasts under culture, which are exposed to stress by abnormal concentrations of nutrients and growth factors, in addition to the absence of neighbor cells and extracellular matrix, which can lead to the senescent phenotype (Sherr & DePinho, 2000). It also results from exposure to mutagens, such as ionizing radiation, ROS, chemotherapeutic agents, or bacterial toxins (Campisi & d'Adda di Fagagna, 2007); thus, the molecular mechanisms of induction are nearly identical to replicative senescence, that is, it is mediated by ATM-p53-p21. Another type is oncogene-induced senescence (Vavrova & Rezacova, 2011). However, it has been observed that moderate stress leads to telomere shortening and that the main cause of shortening is due to the presence of damaged bases, which interferes with the replication fork in telomeres, increasing the extension of the non-replicated ends (von Zglinicki, 2002; Duan et al., 2005). Non-genotoxic stress induces senescence by means of a telomere-independent mechanism, which involves p16-pRB pathway activation by $p16_{INK4a}$ over regulation (Ben-Porath & Weinberg, 2005).

Matos et al. (2011) recently analyzed the role of copper in inducing senescence in WI-38 fibroblasts and found that on exposing these to subcytotoxic copper sulfate concentrations, the fibroblasts exhibited the appearance of the senescent phenotype and an increase in senescence-associated genes such as *p21*, *apoJ*, fibronectin, *TGF β1*, *IGFBP3*, and Hemo oxigenase-1 (*HO-1*). These results are interesting because of the participation that copper can have in the establishment and progression of diseases such as Alzheimer and that of Wilson. On the other hand, it is possible that oxidative stress produces telomere shortening, therefore senescence, as confirmed by Brandl et al. (2011), because on exposing articular chondrocytes to oxidative stress with a sublethal dose of H_2O_2, the authors observed accelerated telomere exhaustion with over regulation of p21 expression and sub expression of SIRT1 and XRCC5, once the cells had acquired the senescent phenotype. Cellular senescence also can be induced by stress in the cultured cells, such as continuous mitogenic stimulation (Serrano & Blasco, 2001). Sustained exposure of melanocytes to an aberrant mitotic stimulus causes senescence after an initial proliferative burst, such as that observed by Michaloglou et al. (2005) suggesting that oncogene-induced senescence represents a factor of protection against cancer.

8. The mammalian telomere

Telomeres are restricted to chromosomal ends and present in eukaryotes as diverse protozoans, fungi, flagellates, plants, and animals. The greater part of telomere DNA is double-stranded; however, the terminal 3′ end is single-stranded. Each telomere is composed of a great region of short repeats rich in G. The sequence comprising a telomere varies in length and complexity depending on the organism (Greider, 1996). In the case of humans and other mammals, the sequence is TTAGGG, while organisms such as yeasts possess irregular repeat sequences in which a T is followed by one, two, or three Gs (TG_{1-3}), while other organisms lack A in their repeats, as occurs in the ciliate Tetrahymena, which presents the TTGGGG sequence, and the Paramecium, which is distinguished by the TTGGGG and TTTGGG alteration.

Telomere DNA consists of two regions: one is double-stranded, and the other is single-stranded at its terminal end. The G-rich chain is that which projects further than the C-rich chain in the 3′ direction. This salient is essential for telomere formation, due to that linear chromosomes need to protect their ends from degradation. The 3′-OH salient invades double-stranded telomere repeats, forming a loop-like structure called the T-loop (Telomere loop), in such a way that the salient remains hidden in the double chain (Griffith et al., 1999). The T-loop avoids that the ends are considered as DNA breaks and preserves genome integrity. The exact structure of the T-loop's base is unknown, but there is a short, double-stranded DNA segment that forms a D-loop (Displaced loop) of TTAGGG repeats, which is displaced by the invasion of the 3′ salient (De Lange, 2002). The D-loop region can include Holliday-type binding (the intermediate state in homolog recombination) or a quadruple G fold (Neidle & Parkinson, 2003). The 3′ salient, which in humans is between 35 and 600 nucleotides long, is the result of the impossibility of replicating the last fragment of Okasaki and of post-replicative processing events (Stewart et al., 2003).

9. The shelterin

Maintenance of telomeric structure and regulation of its functions are supplied by diverse proteins that stabilize it and that permit the cell to distinguish between a natural chromosomal end and a DNA break. Shelterin or Telosome is a six-protein complex whose function is to form and maintain the T-loop. TRF1 and TRF2 are the main shelterin proteins. TRF1 modules telomere length, while TRF2 stabilizes T-loop structure (Xin et al., 2008). Other proteins have been described that, in addition to associating with the telomere, possess other cellular functions, such as XRCC5, which participates in double-stranded DNA damage repair (Thacker & Zdzienicka, 2004). SIRT1 is a negative regulator of p53 and that which avoids growth arrest, senescence, and apoptosis (Guarente, 1999). Double-stranded telomere DNA is wrapped in protein complexes that specifically bind to double-stranded proteins and that participate in the regulation of their length, while the 3′ salient is wrapped in one or more single-stranded binding proteins that protect it (McEachern et al., 2000). The TRF protein family has a similar architecture, defined by two characteristic sequences: both have a DNA-binding motif in their helix-turn-helix carboxyl-terminus (highly related with the Myb domain of cMyb), and both possess a centrally localized sequence motif known as the TRF homolog domain (TRFH), unique for this protein family, which allows it to form homodimers (Fairall et al., 2001). However, the TRFH domain does not form heterodimers, which leads to the presence of two protein complexes on the telomeres: one formed through TRF1, and the other by means of a paralog, TRF2 (Karlseder, 2003). TRF1 forms homodimers in order to bind stably to the DNA thanks to its Myb domain and, by means of T-loop formation, its binding with Rap1 at telomere repeats induces superficial double-strand folding, which indicates that it participates in loop formation (Bianchi et al., 1999). In cis, TRF1 acts in as a negative telomere length regulator. Its over expression produces telomere shortening and a dominant negative allele produces inappropriate lengthening in such a way that the amount of protein affects telomere size (Smogorzewska & de Lange, 2002). TRF1 can control telomerase access through its interaction with proteins TIN2, PTOP/PIP1, and POT1 and regulates their activity on interacting with PINX1 (Zhou et al., 2001). It also binds to TANK1 and 2 (Smith et al., 1998). Elimination of TRF1 produces telomere lengthening, but the extension stabilizes eventually due to that now the TIN2/TINT1 complex associates with TRF2, blocking access to telomerase (Houghtaling et al., 2004).

On the other hand, the complex formed by TRF2 is particularly important for protecting single-strand of the degradation and DNA repair processes (van Steensel et al., 1998; Smogorzewska & De Lange, 2004). TRF2 couples in the binding between double- and single-stranded repeats to facilitate T-loop formation, thus protecting its ends. In this manner, it is responsible for linear telomeric folding for T-loop formation (Griffith et al., 1999) and is found in >100 copies per chromosome (de Lange, 2002); additionally, given that it is the stabilizer of this structure, a lesser amount of TRF2 leads to T-loop opening, an event that can lead to senescence (Karlseder, 2003). Additionally, loss of TRF2 activates the Ataxia-Telangiectasia protein (ATM) kinase pathway, because while this is present it impedes autophosphorylation. ATM activation leads to p53 over regulation and G1/S arrest by means of p21 (Karlseder et al., 1999). When this is displaced from the telomere employing a dominant negative allele, the cell loses its ability to recognize the difference between a natural DNA end and a broken end. On the other hand, its over expression accelerates telomere shortening, which can be the result of the increase in the activity of a nuclease, whose activity is mediated by TRF2 (van Steelsen et al., 1998); Karlseder et al. (2002) suggest that an increase in TRF2 can protect critically short telomeres, delaying induction of cellular senescence even when the telomeres have been reduced. It can protect the single chain indirectly on recruiting Pot1.

It can bind to proteins such as Rap1 and to others involved in DNA damage repair responses, such as the MRE11/RAD50/NBS1 complex, Ku86, and ERCC1/XPF. Among its activities in blocking DNA repair is found that of avoiding that the T-loop insertion site is treated as a Holliday structure. On the other hand, inhibiting the binding of Non-homologous end-joining (NHEJ) and homologous recombination in telomeres and probably in non-telomeric breaks allows determination of which repair pathway the cell should use (Wright & Shay, 2005). On the other hand, TRF2 facilitates the degradation of telomeric DNA on interacting with the WRN exonuclease, whose loss-of-function is implicated in premature cellular senescence, increasing the frequency of cancer and genomic instability (Machwe et al., 2004).

There are diverse proteins that bind indirectly with telomeres; TIN2 has emerged as an important component of the telomere complex. It interacts with the telomere through the

TRFH domain of TRF1, negatively regulating telomere length. A truncated form of TIN2 produces abnormal telomerase-independent telomere lengthening; therefore, it is a TRF1-function mediator on potentiating the pairing of telomere repeats in a TRF1-dependent manner; in addition, it can lead to the telomerase-inhibiting telomere (Kanoh & Ishikawa, 2003). In addition to this telomere-size regulator function, another mutant form generates DNA damage response and senescence (Kim et al., 2004). Through this protein, TRF1 and TRF2 can interact, while TIN2 can stabilize TRF2-complex binding to the telomere on acting as a liaison between this and TRF1. On the other hand, thanks to its third domain, it binds to PIP1/PTOP/TINT1, which in turn serves to recruit POT1 (Liu et al., 2004; Ye & de Lang, 2004). On controlling the Poly (ADP-ribose) polymerase (PARP) activity of tankirase 1, TIN2 protects TRF1 on its removal from the telomere (Ye & de Lang, 2004).

Tankirase 1 interacts with the TRF1. It is a PARP telomere that adds poly ADP-ribose to TRF1, diminishing its affinity for telomeric DNA (Smith & de Lange, 2000). More than diminishing TRF1 affinity for telomere repeats, what TANK1 does on separating TRF1 from the DNA is to expose its myb domain, which is recognized and marked by ubiquitin, leading to degradation of the protein by the proteosome. If this were performed otherwise, TRF1 would bind again to the telomere and its separation would permit the telomerase to gain access to and extend it (Chang et al., 2003). TIN2 is what stabilizes TANK1 binding to the TRF1 complex, and although TANK1 is the lesser abundant of these two proteins, it is necessary for controlled dismantling of the telomere complex during S phase (Ye & de Lange, 2004).

TANK 2, recently identified as a Golgi-associated protein, shares 80% identity in the amino-acid sequence with TANK1, in addition to similar distribution; however, when this is over expressed, it induces rapid cell death with necrotic characteristics (Kaminker et al., 2001). In addition and similar to TANK1, it is a PARP modifier of TRF1 and possibly possesses little effect on telomerase activity (d'Adda di Fagagna et al., 2004).

There are a variety of other proteins that bind to telomeres, such as Ku, Rap1, PIP1/PTOP/TINT1, WRN, PINX1 (Stellwagen et al., 2003; Espejel et al., 2002; Kanoh & Ishikawa, 2003; Köning & Rhodes, 1997; Lei et al., 2000; Ye & de Lange, 2004; Crabbe et al., 2004), and ATM. On the other hand, repair-machinery proteins such as the Mre11/Rad50/ /Nsb1 complex (the MRN complex), which participates in double-stranded DNA repair, can play a role in telomere maintenance, although it does not directly bind with it but rather interacts through TRF2 and possibly participates in T-loop formation (Saldanha et al., 2003). Among these single-stranded binding proteins is found POT1; this binds to a telomeric salient with exceptionally high specificity. It adopts an oligosaccharide-oligonucleotide (OB) joining fold with two forks that overhang to form a clamp for binding to DNA (Lei et al., 2003). When the PTO1 binding domain is mutated, there is no telomeric end fusion, but rather an increase in telomere extension by the telomerase; thus, one of its functions is to block the access of this enzyme to the DNA. TRF1 interacts with POT1; thus, it is the length control terminal transductor for TRF1. The more TRF1, more POT1, which leads to an increase in telomerase inhibition (Mattern et al., 2004). POT1 is necessary to maintain structure in telomere salients, protecting the cell against apoptosis, avoiding chromosomal instability and senescence, and interacting with TRF2 at the T-loop formation point, with which it cooperates for maintaining telomere integrity (Yan et al., 2005).

Recently, it was discovered that Rap1 is an important factor for avoiding telomere recombination and fragility (Martínez & Blasco, 2011).

10. The telomerase

Telomerase is a Ribonucleoprotein (RNP) composed of two units: the catalytic subunit, TERT, and a RNA template, TERC. TERT is a member of the family of reverse transcriptases related with non-LTR retrotransposons and group II introns. Its reverse transcriptase domain is found at the middle of the carboxyl-terminus and it supplies the active site for catalysis (Cech, 2004). TERC, also called TR, is highly expressed in all tissues, it not being important whether they possess telomerase activity or not, and it contains a short sequence that acts as a template from which DNA repeats are copied (Cong et al., 2002). Telomerase carries its own template and is restricted to copying solely a small segment of its RNA. Thus, implicit in telomerase polymerization activity is its ability to specify the template region and its limits, and also a mechanism for maintaining itself as a stable RNP while carrying out synthesis, because it allows the template to move through the active site during the synthesis process of a repeat, to later translocate itself and initiate the synthesis of another repeat. (O'Reilly et al., 1999; Cech, 2004).

When the telomere catalytic subunit in mouse mutates, the first generation that lacks telomerase activity is phenotypically normal, with long telomeres. After four to six generations, its telomeres become very short, and the mice suffer from infertility, proliferation defects, and the risk of apoptosis in organs that undergo constant turnover, which diminishes their life span, while a *TERC* gene mutation causes Dyskeratosis congenita (DKC), a disease in which, among other characteristics, the telomeres are abnormally short (Blasco, 2005).

There are accessory factors that aid the telomerase in acting on the telomere, such as dyskerin and TP1. Dyskerin is important for ribosomal processing because it binds to many small nucleolar RNA, and it is implicit in TERC and even TERT processing or stability because its small nucleolar RNP domain reached maturation in the nucleolus and later binds to the dyskerin (Cech, 2004). TP1 can form one or more structures that mediate interactions with other telomerase- or telomere-binding proteins such as TRF. The TP1 pattern of expression is not restricted to tissues and cell lines that express telomerase activity; therefore, it is not an essential subunit (Harrington et al., 1997).

Chai et al. (2006) found that it is possible that telomerase preferentially extends the leader chain of 20-30 nucleotides per replication round in order to produce a salient similar to that of the delayed chain, because the leader-chain salient is smaller, which otherwise would affect the conformation of telomere structure.

Telomerase-to-telomere access regulation is carried out by telomere-associated proteins, for example, TRF1 and TRF2; on forming the T-loop, these inhibit telomerase binding, while POT1 binding to the salient does not permit coupling of the enzyme so that this would extend it.

Mutations in telomerase components produce premature dysfunctions in adult stem cells and reduce longevity (Mitchell et al., 1999).

11. Telomere position effect (TPE)

At present, it is accepted that telomere clipping can affect gene expression in subtelomeric regions, which can lead to modification of the Biology of the cell prior to initiation of replicative senescence (Baur et al., 2001).

The Position variegation effect (PVE) refers to inactivation of a gene, which occurs when it is removed from its normal context by means of a rearrangement or by transgene insertion. The best known example of this is the result of the expression of the euchromatic white gene in Drosophila, which is responsible for the red color of the eye. Provoked by a rearrangement in chromosome X that causes its relocation near the heterochromatin region, it eliminates function in some cells, which produces mottled pigmentation. The explanation for silencing of the white gene in some cells and not in others is that condensed and inactive conformation of pericentric heterochromatin is dispersed on the rearrangement break and randomly inactivates nearby genes (Henikoff, 1990). This chromosomal position effect affects up to a distance of approximately 1Mb and reflects a genetic inactivation gradient that is inversely correlated with distance (Wakimoto, 1998). The PVE phenomenon suggests that heterochromatin forms a transcriptionally repressor environment within which the presence of heterochromatically active resident genes is somewhat paradoxical, because these genes exhibit reciprocal heterochromatic PVE, that is, a heterochromatic gene will undergo variation if it is translocated to an euchromatic ambit. From this arises the suggestion that these genes have developed transcriptional dependence in factors that normally silence the expression of other genes (Schulze et al., 2005). Among proteins known to possess an important function in gene silencing in Drosophila is found HP1, which is a protein associated with pericentric heterochromatin. It has unequal distribution through the genome and is principally associated with the fourth chromosome. Specifically, it exerts an impact on structural organization and does not only cover the DNA or serves to direct all repetitive DNA sequences toward the repressive structure (Cryderman et al., 1999).

In the case of *Saccharomyces cerevisiae*, the heterochromatin is not cytologically visible; however, it presents position effects in three places: the telomeres; the rDNA locus, and the silent "mating-type" loci (HML and HMR). When a gene is found in one of these sites, its transcription is repressed. As in PVE, silencing depends on gene localization and not on its sequence (Chen & Widom, 2004). Several proteins are involved in the repressive chromatin, including H3 and H4 histones, their acetylases and associated deacetylases, the molecular regulators Sir1-Sir4, and the Origin recognition complex (ORC) (Pryde & Louis, 1999). The four Sir proteins are necessary for transcriptional silencing in HML and HMR, while only Sir2p, Sir3p, and Sir4p are required for telomere silencing and only Sir2p is required for rDNA silencing (Chen & Widom, 2004). TPE was described for the first time in 1990 through generation of a terminal deletion that caused the *URA3* gene (the gene necessary for uracil synthesis) to be localized 6 kb from a telomere. This new gene position provoked its transcriptional repression, which was lost when this was localized 20 kb from the telomere (Gottschling et al., 1990).

It was in 1992 that Wright and Shay set forth the possibility that TPE exists in humans. However, the first experimental proposal to identify TPE in humans, carried out by Bayne et al. (1994), did not yield positive results because changes were not found in the hygromycin-resistant gene on generating deletions in the long arm of chromosome X. Again, this time in

1996, Sprung et al. (1996) sought to study the telomere-length effect on HSV-tk promotor expression, utilizing the KB319 cell line (SV40-transformed fibroblasts); the authors integrated a plasmid with the neo-gene at the end of chromosome 13 and found no effect on neo-gene expression when telomere length ranged from 25-0.5 kb; this leads them to suggest that chromatin structural differences conferred by telomere length do not affect the expression of nearby genes. A 130-kb microdeletion in the end of chromosome 22q in cells from a child with mental retardation; the broken end had been repaired by telomere addition, and consequently a unique DNA region that was normally localized at a distance of >100 kb from the telomere was now adjacent to it. This was the model employed by Ofir et al. in 1999 to demonstrate that telomeric sequences influence the activation of nearby replication origins, delaying the synchronization of replication at mid-S phase. The latter suggests that if human telomere repeats silence genes adjacent to repaired ends, then it is also conceivable that nearby genes may be epigenetically inactivated. On the other hand, Kilburn et al. (2001) found that the presence of a telomere sequence in an *APRT* gene intron in hamster ovary cells only had a modest effect on its expression. Finally, it was Baur et al. in 2001 who, employing a luciferase reporter, supplied convincing evidence of transcriptional silencing near telomeres in humans. The authors found that on placing the reporter adjacent to telomere repeats, there was 10-times lower expression than when they placed it at non-telomeric sites. However, the authors also found that TPE in humans required a histone deacetylase, because on treating the telomeric clones with trichostatin A (a histone deacetylase inhibitor), luciferase expression was restored. Koering et al. (2002) obtained similar results utilizing the *EGFP* reporter gene in C33-A cells (undifferentiated cervical carcinoma), reverting the repressor effect on employing trichostatin A, and they suggested that the position effect depends on the organization of telomeric chromatin, due to that they encountered the release of heterochromatin proteins HP1α and HP1β. Pedram et al. (2006) developed embryonic stem cell clones of mouse with unique-copy gene markers and found that telomeric transgenes were not silenced in cells obtained from 3-day-old embryos as a result of their demethylation during early development, which led the authors to suggest that TPE also plays a role in embryo development. On the other hand, Wright and Shay (1992) also propose that progressive changes in presenescent cells can be the result of the reorganization of telomere chromatin and of the corresponding silencing or desilencing of subtelomere genes. This hypothesis has been extended to include the possibility that silenced proteins can be released from telomeres when the latter are shortened, in order to affect the expression of genes at internal non-telomere sites (Wood and Sinclair, 2002). This idea was based on studies with *S. cerevisiae*, in which it was demonstrated that gene markers inserted <4 kb from telomere repeats are frequently repressed and replicated at the end of the S phase (Dubrana et al., 2001). Although the loss of TPE is not the senescence trigger, it can be responsible for progressive changes in gene expression as a function of replicative age (Wood and Sinclair, 2002).

12. Role of telomere length in subtelomeric gene expression and its possible relation to cellular senescence

To date, there is only one report to our knowledge that studies *in vivo* TPE in human subtelomere genes. The results obtained suggest that the expression of these genes can be influenced by alterations in local heterochromatin structure so that this obstructs access to transcriptional factors (Ning et al., 2003). Due to the fact that few evidence exist about of how age-related telomere length affects the expression of specific human subtelomeric genes, we analyzed the relationship between telomere length and gene expression levels in fibroblasts derived from human donors at ages ranging from 0-70 years. We studied three groups of genes localized 100-150 kb, 200-250 kb, and >300kb away from telomeres. We found that chromatin modifier-encoding genes *Eu-HMTase1*, *ZMYND11*, and *RASA3* were over expressed in adults and implicated in chromatin restructuring (Hernández-Caballero et al., 2009). These genes are interesting because can participate in cellular senescence through the p53-p21Cip1 pathway and can also participate in chromatin restructuring, interacting with remodeling factors including ATP-dependent helicases, histone deacetylases, and histone methyltransferases (Velasco et al., 2006; Zhang et al., 2007). On the other hand, *Eu-HMTase1* can regulate H3-K9 mono- and dimethylation in euchromatin (Tachibana et al., 2005).

13. Conclusions

Undoubtedly the Antagonistic pleiotropy helped to understand the seemingly contradictory functions of the cellular senescence, nevertheless still it is not clear how does the senescence response balance tumour suppression, tissue regeneration and ageing phenotypes, for which it will be needed of a major number of studies.

Our results suggest that the expression of the subtelomeric genes modifies with the age, probably as result of decrease of the telomere length. How the changes on telomere length affected the expression of subtelomeric genes? Recently Martínez et al (2010) demonstrated that the RAP1 protein associated with the telomere, also it associates to the subtelomeric genes. Probably this protein might be involved in TPE.

The Senescence is a complex phenomenon where different factors are involved, the changes in expression of subtelomeric genes, as result of the age, is another variable that will help to the understanding cellular senescence.

14. Acknowledgment

The authors are grateful to Dra. Martha Ruíz and Maggie Brunner for the support in the preparation of the manuscript. This work was supported by grant Salud-2009-C01-115296, from Consejo Nacional de Ciencia y Tecnología (CONACyT, Mexico).

15. References

Bader, N., Jung, T. & Grune, T. (2007). The proteasome and its role in nuclear protein maintenance. *Exp Gerontol*, Vol.42, No.9, pp. 864-870, ISSN 1873-6815

Baur, JA., Zou, Y., Shay, JW. & Wright, WE. (2001). Telomere position effect in human cells. *Science*, Vol.292, No.5524, pp. 2075-2077, ISSN 1095-9203

Bayne, RA., Broccoli, D., Taggart, MH., Thomson, EJ., Farr, CJ. & Cooke, HJ. (1994). Sandwiching of a gene within 12 kb of a functional telomere and alpha satellite does not result in silencing. *Hum Mol Genet*, Vol.3, No.4, pp. 539-546, ISSN 0964-6906

Beausejour, CM., Krtolica, A., Galimi, F., Narita, M., Lowe, SW., Yaswen, P. & Campisi, J. (2003). Reversal of human cellular senescence: roles of the p53 and p16 pathways. *EMBO J*, Vol. 22, No.16, pp. 4212–4222, ISSN 1460-2075

Ben-Porath, I & Weinberg, RA. (2005). The signals and pathways activating cellular senescence. *Int J Biochem Cell Biol*, Vol.37, No.5, (Epub2004 Dec 30), pp. 961-976, ISSN 1878-5875

Bianchi, A., Stansel, RM., Fairall, L., Griffith, JD., Rhodes, D. & de Lange, T. (1999). TRF1 binds a bipartite telomeric site with extreme spatial flexibility. *EMBO J*, Vol.18, No.20, pp. 5735-5744, ISSN 1460-2075

Blagosklonny, M. (2011). Cell cycle arrest is not senescence. *Aging*, Vol.3, No.2, pp. 94-101, ISSN 1945-4589

Blanpain, C., Mohrin, M., Sotiropoulou, PA. & Passegue, E. (2011). DNA-damage response in tissue-specific and cancer stem cells. *Cell Stem Cell*, Vol.8, No.1, pp. 16–29, ISSN 1875-9777

Blasco, MA. (2005). Telomeres and human disease: ageing, cancer and beyond. *Nat Rev Genet*, Vol. 6, No.8, pp. 611–622, ISSN 1471-0064

Boehm, M. & Salck, F. (2005). A developmental timing microRNA and its target regulate life span in C. elegants. *Science*, Vol.310, No.5756, pp. 1954-1957, ISSN 0036-8075

Brandl, A., Hartmann, A., Bechmann, V., Graf, B., Nerlich, M. & Angele, P. (2011). Oxidative Stress Induces Senescence in Chondrocytes. *J Orthop Res*, Vol.29, No.7, (Epub 2011 Jan 31), pp. 1114–1120, ISSN 1554-527X

Brunk, UT. & Terman, A. (2002). Lipofuscin: mechanisms of age-related accumulation and influence on cell function. *Free Radic Biol Med*, Vol.33, No.5, (Sep 1), pp. 611–619, ISSN 1873-4596

Campisi, J. (2000). Cancer, aging and celular senescence. *In vivo*, Vol.14, No.1, (Jan-Feb), pp. 183-188, ISSN 1791-7549

Campisi, J. (2005). Aging, tumor suppression and cancer: high wire-act! *Mech Ageing Dev*, Vol.126, No.1, (Jan), pp. 51–58, ISSN 1872-6216

Campisi, J. & d'Adda di Fagagna, F. (2007). Cellular senescence: when bad things happen to good cells. *Nat Rev Mol Cell Biol*, Vol. 8, No.9, (Sep), pp. 729-740, ISSN 1471-0080

Campisi, J. (2011). Cellular senescence: Putting the paradoxes in perspective. *Curr opin Genet Dev*, Vol.21, No.1, (Feb), pp. 107- 112, ISSN 1879-0380

Cech, TR. (2004). Beginning to understand the end of the chromosome. *Cell*, Vol.116, No.2, (Jan 23), pp. 273-279, ISSN 1097- 4172

Chai, W., Du, Q,, Shay, JW. & Wright, WE. (2006). Human telomeres have different overhang sizes al leading versus lagging strands. *Mol Cell*, Vol.21, No.3, (Feb 3), pp. 427-435, ISSN 1097-4164

Chang, BD., Xuan, Y., Broude, EV., Zhu, H., Schott, B., Fang, J. & Roninson, IB. (1999). Role of p53 and p21waf1/cip1 in senescence-like terminal proliferation arrest induced in human tumor cells by chemotherapeutic drugs. *Oncogene*, Vol.18, No.34, (Aug 28), pp.4808-4818, ISSN 1476-5594

Chang, W., Dynek, JN. & Smith, S. (2003). TRF1 is degraded by ubiquitina-mediated proteolysis after release from telomeres. *Genes Dev*, Vol.17, No.11, (Jun 1), pp. 1328-1333, ISSN 1549-5477

Chen, L. & Widom, J. (2004). Molecular basis of transcriptional silencing in budding yeast. Biochem *Cell Biol*, Vol.82, No.4, (Aug), pp. 413-418, ISSN 1208-6002

Cherdyntseva, NV., Gervas, PA., Litvyakov, NV., Stakcheeva, MN., Ponomaryeva, AA., Dobrodeev, AY., Denisov, EV., Belyavskaya, VA. & Choinzonov, EL. (2010). Age-related function of tumor suppressor gene tp53: contribution to cancer risk and progression. *Exp Oncol*, Vol.32, No.3, (Sep), pp. 205-208, ISSN 1812-9269

Chiu, CP., Dragowska, W., Kim, NW., Vaziri, H., Yui, J., Thomas, TE., Harley, CB. & Lansdorp, PM. (1996). Differential expression of telomerase activity in hematopoietic progenitors from adult human bone marrow. *Stem Cells*, Vol.14, No.2, pp. 239-248, ISSN 0250-6793

Cong, Y., Wright, WE. & Shay, JW. (2002). Human telomerase and its regulation. *Microb Mol Biol Rev*, Vol.66, No.3, pp. 407- 425, ISSN 1098-5557

Crabbe, L., Verdun, RE., Haggblom, CI. & Karlseder, J. (2004). Defective telomere lagging strand synthesis in cells lacking WRN helicase activity. *Science*, Vol.306, No.5703, (Dec 10), pp. 1951-1953, ISSN 1095-9203

Cryderman, DE., Morris, EJ., Biessmann, H., Elgin, S. & Wallrath, LL. (1999). Silencing at Drosophila telomeres: Nuclear organization and chromatin structure play critical roles. *EMBO J*, Vol.18, No.13, (Jul 1), pp. 3724-3735, ISSN 1460- 2075

d'Adda di Fagagna, F., Teo, S. & Jackson, SP. (2004). Functional links between telomeres and proteins of the DNA-damage response. *Genes Dev*, Vol.18, No.15, (Aug 1), pp. 1781-1799, ISSN 1549-5477

d'Adda di Fagagna, F. (2008). Living on a break: cellular senescence as a DNA-damage response. *Nat Rev Cancer*, Vol.8, No.7, pp. 512-522, ISSN 1474-1768

de Lange, T. (2002). Protection of mammalian telomeres. *Oncogene*, Vol.21, No.4, (Jan 21), pp. 532-540, ISSN 1476-5594

De, S. (2011). Somatic mosaicism in healthy human tissues. *Trends Genet* Vol.27, No. 6, pp. 217-223, ISSN 1476-5594

Dimri, GP., Lee, X., Basile, G., Acosta, M., Scott, G., Roskelley, C., Medrano, EE., Linskens, M., Rubelj, I., Pereira-Smith, O., Peacocke, M. & Campisi, J. (1995). A biomarker that identifies senescent human cells in culture and in aging skin in vivo. *Proc Natl Acad Sci USA*, Vol.92, No.20, (Sep 26), pp. 9363–9367, ISSN 1091-6490

Duan, J., Zhang, Z. & Tong, T. (2005). Irreversible cellular senescence induced by prolonged exposure to H2O2 involves DNA-damage and- repair genes and telomere shortening. *Int J Biochem Cell Biol*, Vol.37, No. 7, pp. 1407–1420, ISSN 1878-5875

Dubrana, K., Perrod, S. & Gasser, SM. (2001). Turning telomeres on and off. *Curr Opin Cell Biol*, Vol.13, No.3, pp. 281-289, ISSN 1879-0410

Espejel, S., Franco, S., Rodríguez-Perales, S., Bouffler, SD., Cigudosa, JC. & Blasco, MA. (2002). Mammalian Ku86 mediates chromosomal fusions and apoptosis caused by critically short telomeres. *EMBO J*, Vol.21, pp. 2207-2219, ISSN 1460-2075

Fairall, L., Chapman, L., Moss, H., de Lange, T. & Rhodes, D. (2001). Structure of the TRFH dimerization domain of the human telomeric proteins TRF1 and TRF2. *Mol Cell*, Vol.8, No.2, pp. 351-361, ISSN 1097-4164

Feltes, BC., de Faria Poloni, J. & Bonatto, D. (2011). The developmental aging and origins of health and disease hypotheses explained by different protein networks. *Biogerontology*, Vol.12, No.4, (Epub 2011 Mar 5), pp. 293-308, ISSN 1389- 5729

Freund, A., Orjalo, AV., Desprez, PY. & Campisi, J. (2010). Inflammatory networks during cellular senescence: causes and consequences. *Trends Mol Med*, Vol.16, No.5, (Epub 2010 May 3), pp. 238–246, ISSN 1471-499X

Funayama, R. & Ishikawa, F. (2007). Cellular senescence and chromatin structure. *Chromosoma*, Vol.116, No.5, (Epub 2007 Jun 20), pp. 431–440, ISSN 1432-0886

Gottschling, DE., Aparicio, OM., Billington, BL. & Zakian, VA. (1990). Position effect at S. cerevisiae telomeres: Reversible repression of Pol II transcription. *Cell*, Vol.63, No.4, pp. 751-762, ISSN 1097-4172

Greenberg, SB., Grove, GL. & Cristofalo, VJ. (1977). Cell size in aging monolayer cultures. *In Vitro*, Vol.13, No. 5, pp. 297–300, ISSN 0073-5655

Greider, CW. & Blackburn, EH. (1985). Identification of a specific telomere terminal transferase enzyme with two kinds of primer specificity. *Cell*, Vol.51, No.6, pp. 405–413, ISSN 1097-4172

Greider, CW. (1996). Telomere length regulation. *Annu Rev Biochem*, Vol.65, pp. 337-365, ISSN 1545-4509

Griffith, JD., Comeau, L., Rosenfield, S., Stansel, RM., Bianchi, A., Moss, H. & de Lange, T. (1999). Mammalian telomeres end in a large duplex loop. *Cell*, Vol. 97, No.4, pp. 503-514, ISSN 1097-4172

Guarente, L. (1999). Mutant mice live longer. *Nature*, Vol.402, No.6759, pp. 243-245, ISNN 1476-4687

Guarente, L. (1999). Diverse and dynamic functions of the Sir silencing complex. *Nat Genet*, Vol.23, No.3, pp. 281–285, ISSN 1546-1718

Gutteridge, JM. (1984). Age pigments: role of iron and copper salts in the formation of fluorescent lipid complexes. *Mech Ageing Dev*, Vol.25, No.1-2, pp. 205–214, ISSN 1872-6216

Halaschek-Wiener, J., Vulto, I., Fornika, D., Collins, J., Connors, JM., Le, ND., Lansdorp, PM. & Brooks-Wilson, A. (2008). Reduced telomere length variation in healthy oldest old. *Mech Ageing Dev*, Vol.29, No.11, (Epub 2008 Aug 14), pp. 638–641, ISSN 1872-6216

Hampel, B., Malisan, F., Niederegger, H., Testi, R. & Jansen-Durr, P. (2004). Differential regulation of apoptotic cell death in senescent human cells. *Exp Gerontol*, Vol.39, No.11-12, pp. 1713–1721, ISSN 1873-6815

Harley, CB., Futcher, AB. & Greider, CW. (1990). Telomeres shorten during ageing of human fibroblasts. *Nature*, Vol.345, No.6274, pp. 458–460, ISSN 1476-4687

Harrington, L., McPhhail, T., Mar, V., Zhou, W., Oulton, R., Bass, MB., Arruda, I. & Robinson, M. (1997). A mammalian telomerase-associated protein. *Science*, Vol.275, No.5302, pp. 973-977, ISSN 1095-9203

Hayflick, L. & Moorhead, PS. (1961). The serial cultivation of human diploid cell strains. *Exp Cell Res* 1961; Vol.25, pp. 585- 621, ISSN 1090-2422

Hayflick, L. (1998). A brief history of the mortality and immortality of cultured cells. *Keio J Med*, Vol.47, No.3, pp. 174-182, ISSN 1880-1293

Hayflick, L. (2003). Living for ever and dying in the attempt. *Exp Gerontol*, Vol.38, No.11-12, pp. 1231-1241, ISSN 1873-6815

Henikoff, S. (1990). Position-effect variegation after 60 years. *Trends Genet*, Vol.6, No.12, pp. 422–426, ISSN 0168-9525

Herbig, U., Jobling, WA., Chen, BP., Chen, DJ. & Sedivy, JM. (2004). Telomere shortening triggers senescence of human cells through a pathway involving ATM, p53, and p21(CIP1), but not p16(INK4a). *Mol Cell*, Vol.14, No.4, pp. 501-513, ISSN 1097-4164

Hernández-Caballero, E., Herrera-González, NE., Salamanca-Gómez, F. & Arenas-Aranda, DJ. (2009). Role of telomere length in subtelomeric gene expression and its possible relation to cellular senescence. *BMB Rep*, Vol.42, No.11, pp. 747- 751, ISSN 1976-670X

Höhn, A., Jung, T., Grimm, S., Catalgol, B., Weber, D. & Grune, T. (2011). Lipofuscin inhibits the proteasome by binding to surface motifs. *Free Radic Biol Med*, Vol.50, No.5, (Epub 2010 Dec 16), pp. 585-591, ISSN 1873-4596

Hornsby, PJ. (2011). Cellular aging and cancer. *Crit Rev Oncol Hematol*, Vol.79, No.2, pp. 189-195, ISSN 1040-8428

Hotchkiss, RS., Strasser, A., McDunn, JE. & Swanson, PE. (2009). Cell death. *N Engl J Med*, Vol.361, No.16, pp. 1570-1583, ISSN 1533-4406

Houghtaling, BR., Cuttonaro, L., Chang, W. & Smith, S. (2004). A dynamic molecular link between the telomere length regulator TRF1 and the chromosome end protector TRF2. *Current Biology*, Vol.14, No.18, pp. 1621–1631, ISSN 1879- 0445

Huang, YC., Hasegawa, H., Wang, SW., Ku, CC., Lin, YC., Chiou, SS., Hou, MF., Wu, DC., Tsai, EM., Saito, S., Yamaguchi, N. & Yokoyama, KK. (2011). Jun dimerization protein 2 controls senescence and differentiation via regulating histone modification. *J Biomed Biotechnol*, Epub 2010 Dec 12, ISSN 1110-7251

Hung, YC., Lee, JH., Chen, HM. & Huang, GS. (2010). Effects of static magnetic fields on the development and aging of Caenorhabditis elegans. *J Exp Biol*, Vol.213, No.12, pp. 2079-2085, ISSN 1477-9145

Hwang, ES., Yoon, G. & Kang, HT. (2009). A comparative analysis of the cell biology of senescence and aging. *Cell Mol Life Sci*, Vol.66, No.15,(Epub 2009 May 7), pp. 2503-2524, ISSN 1420-9071

Inomata, K., Aoto, T., Binh, NT., Okamoto, N., Tanimura, S., Wakayama, T., Iseki, S., Hara, E., Masunaga, T., Shimizu, H. & Nishimura, EK. (2009). Genotoxic stress abrogates renewal of melanocyte stem cells by triggering their differentiation. *Cell*, Vol.137, No.6, pp. 1088–1099, ISSN 1097-4172

Kaminker, PG., Kim, S., Taylor, RD., Zebarjadian, Y., Funk, WD., Morin, GB., Yaswen. P. & Campisi, J. (2001). TANK2, a new TRF1-associated Poly(ADP-ribose) polymerase, causes rapid induction of cell death upon overexpression. *J Biol Chem*, Vol.276, No.38, (Epub 2001 Jul 13), pp. 35891–35899, ISSN 1083-351X

Kanoh, J. & Ishikawa, F. (2003). Composition and conservation of the telomeric complex. *Cell Mol Life Sci*, Vol.60, No.11, pp. 2295-2302, ISSN 1420-9071

Karlseder, J., Broccoli, D., Day, D., Hardy, S. & de Lange, T. (1999). p53- and ATM-dependent apoptosis induced by telomeres lacking TRF2. *Science*, Vol.283, No.5406, pp. 1321-1325, ISSN 1095-9203

Karlseder, J., Smogorzweska, A. & de Lange, T. (2002). Senescence induced by altered telomere state, not telomere loss. *Science*, Vol.295, No.5564, pp. 2446-2449, ISSN 1095-9203

Karlseder, J. (2003). Telomere repeat binding factors: Keeping the ends in check. *Cancer Letters*, vol.194, No.2, pp. 189-197, ISSN 1872-7980

Kavathia, N., Jain, A., Walston, J., Beamer, BA., Fedarko, NS. (2009). Serum markers of apoptosis decrease with age and cancer stage. *Aging* Vol.1, No.7, pp. 652–663, ISSN 1945-4589

Kilburn, AE., Shea, MJ., Sargent, G. & Wilson. JH. (2001). Insertion of a telomere repeat sequence into a mammalian gene causes chromosome instability. *Mol Cell Biol*, Vol.21, No.1, pp. 126-135, ISSN 1097-4164

Kim, NW., Piatyszek, MA., Prowse, KR., Harley, CB., West, MD., Ho, PL., Coviello. GM., Wright, WE., Weinrich, SL. & Shay, JW. (1994). Specific association of human telomerase activity with immortal cells and cancer. *Science*, Vol.266, No.5193, pp. 2011–2015, ISSN 1095-9203

Kim, S., Beausejour, C., Davalos, AR., Kaminker, P., Heo, S. & Campisi, J. (2004). TIN2 mediates functions of TRF2 at Human Telomeres. *J Biol Chem*, Vol. 279, No.42, (Epub 2004 Aug 3) pp. 43799–43804, ISSN 1083-351X

Kirkwood, TB. (1977). Evolution of ageing. *Nature*, Vol.170, No.5635, pp. 201–204, ISSN 1476-4687

Koering, CS., Pollice A., Zibella, MP., Bauwens, S. Puisieux, A., Brunori, M., Brun, C., Martins, L., Sabatier, L., Pulitzer, JF. & Gilson, E. (2002). Human telomeric position effect is determined by chromosomal context and telomeric chromatin integrity. *EMBO Rep*, Vol.3, No. 11, pp. 1055-1061, ISSN 1469-221X

König, P. & Rhodes, D. (1997). Recognition of telomeric DNA. *TIBS*, Vol.22, No.2, pp. 43-47, ISSN 0968-0004

Kosar, M., Bartkova, J., Hubackova, S., Hodny, Z., Lukas, J. & Bartek, J. (2011). Senescence-associated heterochromatin foci are dispensable for cellular senescence, occur in a cell type- and insult-dependent manner and follow expression of p16 (ink4a). *Cell Cycle*, Vol.10, No.3, pp. 457-468, ISSN 1551-4005

Krishnamurthy, J., Torrice, C., Ramsey, MR., Kovalev, GI., Al-Regaiey, K., Su, L. & Sharpless, NE. (2004). Ink4a/Arf expression is a biomarker of aging. *J Clin Invest*, Vol.114, No.9, pp. 1299-1307, ISSN 1558-8238

Lee, BY., Han, JA., Im, JS., Morrone, A., Johung, K., Goodwin, EC., Kleijer, WJ., DiMaio, D. & Hwang, ES. (2006). Senescence-associated b-galactosidase is lysosomal b-galactosidase. *Aging Cell*, Vol.5, No.2, pp. 187–195, ISSN 1474-9726

Lee, HW., Blasco, MA., Gottlieb, GJ., Horner, JW., Greider, CW. & DePinho, RA. (1998). Essential role of mouse telomerase in highly proliferative organs. *Nature*, Vol.392, No. 6676, pp. 569–574, ISSN 1476-4687

Lei, M., Podell, ER., Baumann, P. & Cech, TR. (2003). DNA self-recognition in the structure of Pot1 bound to telomeric single-stranded DNA. *Nature*, Vol.426, No.6963, pp. 198-203, ISSN 1476-4687

Liu, D., O'Connor, M., Qin, J. & Songyang, Z. (2004). Telosome. A mammalian telomere associated complex formed by multiple telomeric proteins. *J Biol Chem*, Vol.279, No.49, (Epub 2004 Sep 20), pp. 51338-51342, ISSN 1083-351X

Liu, L., Trimarchi, JR., Smith, PJ. & Keefe, DL. (2002). Mitochondrial dysfunction leads to telomere attrition and genomic instability. *Aging Cell*, Vol.1, No.1, pp. 40–46, ISSN 1097-4172

Lowe, SW. & Sherr, CJ. (2003). Tumor suppression by Ink4a-Arf: progress and puzzles. *Curr Opin Genet Dev*, Vol.13, No.1, pp. 77–83, ISSN 1879-0380

Machwe, A., Xiao, L. & Orren, DK. (2004). TRF2 recruits the Werner syndrome (WRN) exonuclease for processing of telomeric DNA. *Oncogene*, Vol.23, No.1, pp. 149-156, ISSN 1476-5594

Mantell, LL. & Greider, CW. (1994). Telomerase activity in germline and embryonic cells of Xenopus. *EMBO J*, Vol.13, No.13, pp. 3211-3217, ISSN 1460-2075

Martínez, P., Thanasoula, M., Caslos, AR., Gómez-López, G., Tejera, AM., Schoefther, S., Dominguez, O., Pisano, DG., Tarsounas, M. & Blasco, MA. (2010). Mammalian Rap1 controls telomere function and gene expression through binding to telomeric and extratelomeric sites. *Nat Cell Biol*, Vol.12, No.8, pp. 768-780, ISSN 1097-6256

Martínez, P. & Blasco MA. (2011). Telomeric and extra-telomeric roles for telomerase and the telomere-binding proteins. *Nat Rev Cancer*, Vol.11, pp. 161-176, ISSN 1474-175X

Matos, L., Gouveia, A. & Almeida, H. (2011). Copper ability to induce premature senescence in human fibroblasts. *Age (Dordr)*, Jun 22, ISSN 1574-1587

Mattern, KA., Swiggers, JJ., Nigg, AL., Löwenberg, B., Houtsmuller, AB. & Zijlmans, JM. (2004). Dynamics of protein binding to telomeres in living cells: Implications for telomere structure and function. *Mol Cell Biol*, Vol.24, No.12, pp. 5587-5594, ISNN 1098-5549

McClintock, B. (1941). The stability of broken ends of chromosomes in Zea mays. *Genetics*, Vol.26, No.2, pp. 234-282, ISSN 1943-2631

McEachern, MJ., Krauskopf, A. & Blackburn, EH. (2000). Telomeres and their control. *Annu Rev Genet*, Vol. 34, pp. 331-358, ISSN 1545-2948

Michaloglou, C., Vredeveld, LC., Soengas, MS., Denoyelle, C., Kuilman, T., van der Horst, CM., Majoor, DM., Shay, JW., Mooi, WJ. & Peeper, DS. (2005). BRAFE600-associated senescence-like cell cycle arrest of human naevi. *Nature*, Vol.436, No.7051, pp. 720-724, ISSN 1476-4687

Mitchell, JR., Wood, E. & Collins, K. (1999). A telomerase component is defective in the human disease dyskeratosis congenita. *Nature*, Vol.402, No.6761, pp. 551-555, ISSN 1476-4687

Morin, GB. (1989). The human telomere terminal transferase enzyme is a ribonucleoprotein that synthesizes TTAGGG repeats. *Cell*, Vol.59, No.3, pp. 521–529, ISSN 1097-4172

Narita, M., Nuñez, S., Heard, E., Narita, M., Lin, AW., Hearn, SA., Spector, DL., Hannon, GL. & Lowe, SW. (2003). Rb- mediated heterochromatin formation and silencing of E2F target genes during cellular senescence. *Cell*, Vol.113, No.6, pp. 703–716, ISSN 1097-4172

Neidle, S. & Parkinson, GN. (2003). The structure of telomeric DNA. *Curr Opin Struct Biol*, Vol.13, No.3, pp. 275-283, ISSN 1879-033X

Ning, Y., Xu, JF., Li, Y., Chavez, L., Riethman, HC., Lansdorp, PM. & Weng, NP. (2003). Telomere length and the expression of natural telomeric genes in human fibroblasts. *Hum Mol Genet*, Vol.12, No.11, pp. 1329-1336, ISSN 1460-2083

Nishio, K., Inoue, A., Qiao, S., Kondo, H. & Mimura, A. (2001). Senescence and cytoskeleton: overproduction of vimentin induces senescent-like morphology in human fibroblasts. *Histochem Cell Biol*, Vol.116, No.4, pp. 321–327, ISSN 1432-119X

Nishio, K. & Inoue, A. (2005) Senescence-associated alterations of cytoskeleton: extraordinary production of vimentin that anchors cytoplasmic p53 in senescent human fibroblasts. *Histochem Cell Biol*, Vol.123, No.3, pp. 263–273, ISSN 1432-119X

O'Reilly, M., Teichmann, SA. & Rhodes, D. (1999). Telomerases. *Curr Opin Struc Biol*, Vol.9, No.1, pp. 56-65, ISSN 1879-033X

Ofir, R., Wong, AC., McDermind, HE., Skorecki, KL. & Selig, S. (1999). Position effect of human telomeric repeats on replication timing. *Proc Natl Acad Sci USA*, Vol.96, No.20, pp. 11434-11439, ISSN 1091-6490

Olovnikov, AM. (1996). Telomeres, telomerase and aging: Origin of the Theory. *Exp Gerontol*, Vol.31, No.4, pp. 443-448, ISSN 1873-6815

Ouellette, MM., Liao, M., Herbert, BS., Johnson, M., Holt, SE., Liss, HS., Shay, JW. & Wright WE. (2000). Subsenescent telomere lengths in fibroblasts immortalized by limiting amounts of telomerase. *J Biol Chem*, Vol.275, No.14, pp. 10072-10076, ISSN 0021-9258

Pedram, M., Sprung, CN., Gao, Q., Lo, AW., Reynolds, GE. & Murnane JP. (2006). Telomere position effect and silencing of transgenes near telomeres in the mouse. *Mol Cell Biol*, Vol.26, No.5, pp. 1865-1878, ISSN 1098-5549

Percy, CJ., Power, D. & Gobe, GC. (2008). Renal ageing: changes in the cellular mechanism of energy metabolism and oxidant handling. *Nephrology*, Vol.13, No.2, pp. 147–152, ISNN 1440-1797

Pryde, FE. & Louis, EJ. (1999). Limitations of silencing at native yeast telomeres. *EMBO J*, Vol.18, No.9, pp. 2538-2550, ISSN 1460-2075

Rai, TS. & Adams, PD. (2011). Lessons from senescence: Chromatin maintenance in non-proliferating cells. *Biochim Biophys Acta*. Aug 3, ISSN 0006-3002

Riley, T., Sontag, E., Chen, P. & Levine, A. (2008). Transcriptional control of human p53-regulated genes. *Nat Rev Mol Cell Biol*, Vol. 9, No.5, pp. 402–412, ISSN 1097-4164

Rochette, PJ. & Brash, DE. (2008). Progressive apoptosis resistance prior to senescence and control by the anti-apoptotic protein BCL-xL. *Mech Ageing Dev*, Vol.129, No.4, pp. 207–214, ISSN 1872-6216

Rodier, F. & Campisi, J. (2011). Four faces of cellular senescence. *J Cell Biol*, Vol.192, No.4, (Epub 2011 Feb 14), pp. 547-556, ISSN 1540-8140

Saldanha, SN., Andrews, LG. & Tollefsbol, TO. (2003). Assessment of telomere length and factors that contribute to its stability. *Eur J Biochem*, Vol.270, No.3, pp. 389-403, ISSN 1432-1033

Salminen, A., Ojala, J. & Kaarniranta, K. (2011). Apoptosis and aging: increased resistance to apoptosis enhances the aging process. *Cell Mol Life Sci*, Vol.68, No.6, (Epub 2010 Nov 30), pp. 1021-1031, ISSN 1420-9071

Schulze, SR., Sinclair, D., Fitzpatrick, KA. & Honda, BM. (2005). A Genetic and molecular characterization of two Proximal heterochromatic genes on chromosome 3 of Drosophila melanogaster. *Genetics*, Vol.169, No.4, pp. 2165–2177, ISSN 1943-2631

Serrano, M. & Blasco, MA. (2001). Putting the stress on senescence. *Curr Opin Cell Biol*, Vol.13, No.6, pp. 748–753, ISSN 1879-0410

Sherr, CJ. & DePinho, RA. (2000). Cellular senescence: Mitotic clock or culture shock? *Cell*, Vol.102, No.4, pp. 407–410, ISSN 1097-4172

Sin, D., Kucia, M. & Ratajcsak, MZ. (2011). Nuclear and chromatin reorganization during cell senescence and aging- A mini-review. *Gerontology*, Vol.57, No.1, (Epub 2010 Feb 4), pp. 76-84, ISSN 1423-0003

Smith, S., Giriat, I., Schmitt, A. & de Lange, T. (1998). Tankyrase, a poly (ADP-ribose) polymerase at human telomeres. *Science*, Vol.282, No.5393, pp. 1484-1487, ISSN 1095-9203

Smith, S. & de Lange, T. (2000). Tankyrase promotes telomere elongation in human cells. *Curr Biol*, Vol.10, No.20, pp. 1299-1302, ISSN 1879-0445

Smogorzewska, A. & de Lange, T. (2002). Different telomere damage signaling pathways in human and mouse cells. *EMBO J*, Vol.21, No.16, pp. 4338-4348, ISSN 1460-2075

Smogorzewska, A. & de Lange, T. (2004). Regulation of telomerase by telomeric proteins. *Annu Rev Biochem*, Vol.73, pp. 177-208, ISSN 0066-4154

Sprung, CN., Sabatier, L. & Murnane, JP. (1996). Effect of telomere length on telomeric gene expression. Nucleic Acids Res, Vol.24, No.21, pp. 4336-4340, ISSN 1362-4962

Stellwagen, AE., Haimberger, ZW., Veatch, JR. & Gottschling, DE. (2003). Ku interacts with telomerase RNA to promote telomere addition at native and broken chromosome ends. *Genes Dev*, Vol.17, No.19, (Epub 2003 Sep 15), pp. 2384-2395, ISSN 1549-5477

Stewart, SA., Ben-Porath, I., Carey, VJ., O'Connor, BF., Hahn, WC. & Weinberg, RA. (2003). Erosion of the telomeric single-strand overhang at replicative senescence. *Nat Genet*, Vol.33, No.4, (Epub 2003 Mar 24), pp. 492-496, ISSN 1546-1718

Su, X., Paris, M., Gi, YJ., Tsai, KY., Cho, MS., Lin, YL., Biernaskie, JA., Sinha, S., Prives, C., Pevny, LH., Miller. FD. & Flores ER. (2009). TAp63 prevents premature aging by promoting adult stem cell maintenance. *Cell Stem Cell*, Vol.5, No.1, pp. 64–75, ISSN 1875-9777

Tachibana, M., Ueda, J., Fukuda, M., Takeda, N., Ohta, T., Iwanari, H., Sakihama, T., Kodama, T., Hamakubo, T. & Shinkai, Y. (2005). Histone methyltransferases G9a and GLP form heteromeric complexes and are both crucial for methylation of euchromatin at H3-K9. *Genes & Dev*. Vol.19, No.7, pp. 815-826, ISSN 1549-5477

Thacker, J. & Zdzienicka, MZ. (2004). The XRCC genes: expanding roles in DNA double-strand break repair. *DNA Repair*, Vol.3, No. 8-9, pp. 1081-1090, ISNN 1568-7856

Thomas, E., al-Baker, E., Dropcova, S., Denyer, S., Ostad, N., Lloyd, A., Kill, IR. & Faragher, RG. (1997). Different kinetics of senescence in human fibroblasts and peritoneal mesothelial cells. *Exp Cell Res*, Vol.236, No.1, pp. 355-358, ISSN 1090-2422

Tuminello ER & Han SD. (2011). The apolipoprotein e antagonistic pleiotropy hypothesis: review and recommendations. *Int J Alzheimers Dis*, Vol.24, pp. 1-12, ISSN 2090-0252

Van Steensel, B., Smogorzewska, A. & de Lange, T. (1998). TRF2 protects human telomeres from end-to-end fusions. *Cell*, Vol.92, No.3, pp. 401-413, ISSN 1097-4172

Vavrova, J. & Rezacova, M. (2011). The importance of senescence in ionizing radiation-induced tumour suppression. *Folia Biologica*, Vol.57, No.2, pp. 41-46, ISSN 0015-5500

Velasco, G., Grkovic, S. & Ansieau, S. (2006). New insights into BS69 functions. *J Biol Chem*, Vol.281, No.24, (Epub 2006 Mar 24), pp. 16546-16550, ISSN 1083-351X

von Zglinicki, T. (2002). Oxidative stress shortens telomeres. *Trends Biochem Sci*, Vol.27, No.7, pp.339–344, ISSN 0968-0004

Vousden, KH. & Lu, L. (2002). Live or let die: the cell's response to p53. *Nat Rev Cancer*, Vol.2, No.8, pp. 594-604, ISSN 1474-1768

Wakimoto, BT. (1998). Beyond the nucleosome: Epigenetic aspects of position-effect variegation in Drosophila. *Cell*, Vol.93, No.3, pp. 321-324, ISSN 1097-4172

Walker, RF. (2011). Developmental theory of aging revisited: focus on causal and mechanistic links between development and senescence. *Rejuvenation Res*, Vol.14, No.4, (Epub 2011 Jul 18), pp. 429-36, ISSN 1557-8577

Wang, E. & Gundersen, D. (1984). Increased organization of cytoskeleton accompanying the aging of human fibroblasts in vitro. *Exp Cell Res*, Vol.154, No.1, pp. 191–202, ISSN 1090-2422

Williams, G. (1957). Pleiotropy, natural selection, and the evolution of senescence. *Evolution*, Vol.11, pp. 398-411, ISSN 1558-5646

Wood, JG. & Sinclair, DA. (2002). TPE or not TPE? It's no longer a question. *TRENDS Pharmacol Sci*, Vol.23, No.1, pp. 1-4, ISSN 1873-3735

Wright, WE. & Shay, JW. (1992). Telomere positional effects and the regulation of cellular senescence. *Trends Genet*, Vol.8, No.6, pp. 193-197, ISSN 0168-9525

Wright, WE., Piatyszek, MA., Rainey, WE., Byrd, W. & Shay, JW. (1996). Telomerase activity in human germline and embryonic tissues and cells. *Dev Genet*, Vol.18, No.2, pp. 173-179, ISSN 0192-253X

Wright, WE. & Shay, JW. (2005). Telomere-binding factors and general DNA repair. *Nat Genet*, Vol.37, No.2, pp. 116-118, ISSN 1546-1718

Xin, H., Liu, D. & Songyang, Z. (2008). The telosome/shelterin complex and its functions. *Genome Biol*, Vol.9, No.9, (Epub 2008 Sep 18), pp. 232-236, ISSN 1465-6914

Yan, Q., Zheng, Y. & Harris. CC. (2005). POT1 and TRF2 cooperate to maintain telomeric integrity. *Mol Cell Biol*, Vol.25, No.3, pp. 1070-1080, ISSN 1098-5549

Yanaka, M., Honma, T., Sato, K., Shinohara, N., Ito, J., Tanaka, Y., Tsuduki, T. & Ikeda, I. (2011). Increased monocytic adhesion by senescence in human umbilical vein endothelial cells. *Biosci Biotechnol Biochem*, Vol.75, No.6, (Epub 2011 Jun 13), pp. 1098-1103, ISSN 1347-6947

Ye, JZ. & de Lange, T. (2004). TIN2 is a tankyrase 1 PARP modulator in the TRF1 telomere length control complex. *Nat Genet*, Vol.36, No.6, pp. 618-623, ISSN 1546-1718

Zahn, JT., Louban, I., Jungbauer, S., Bissinger, M., Kaufmann, D., Kemkemer, R. & Spatz, JP. (2011). Age-dependent changes in microscale stiffness and mechanoresponses of cells. *Small*, Vol.7, No.10, (Epub 2011 May 2), pp. 1480-1487, ISSN 1613-6829

Zhang, W., Chan, HM., Gao, Y., Poon, R. & Wu, Z. (2007). BS69 is involved in cellular senescence through the p53-p21Cip1 pathway. *EMBO Rep*, Vol.8, No.10, (Epub 2007 Aug 24), pp.952-958, ISSN 1469-3178

Zhou, XZ. & Lu, KP. (2001). The PIN2/TRF1 interacting protein PinX1 is a potent telomerase inhibitor. *Cell*, Vol.107, No.3, pp. 347-359, ISSN 1097-4172

12

Quantification of Elastin, Collagen and Advanced Glycation End Products as Functions of Age and Hypertension

Milena Atanasova[1], Aneliya Dimitrova[2], Boryana Ruseva[3], Angelina Stoyanova[4], Miglena Georgieva[5] and Emiliana Konova[5,6]
[1]Department of Biology, Medical University of Pleven, Pleven,
[2]Department of Pathophysiology, Medical University of Pleven, Pleven,
[3]Department of Physiology, Medical University of Pleven, Pleven,
[4]Department of Chemistry, Medical University of Pleven, Pleven,
[5]Medical Center "Clinical Institute for Reproductive Medicine", Pleven,
[6]Center for Reproductive Health; Medical University of Pleven, Pleven,
Bulgaria

1. Introduction

Elastin and collagen are the major extracellular matrix (ECM) proteins that make up the framework of the elastic arteries structure. These two fibrous proteins are the main structural components of arterial walls, and they provide the strength and resilience needed by the aorta to accommodate the pressure and volume variations during each heartbeat (Coquand et al., 2011). Elastin is a polymer of linear polypeptide chains and constituting more than 50% of the dry weight of the proximal parts of the aorta. It is the main provider of tissue elasticity, while collagen acts to stiffen the wall and to limit its extensibility (Samila & Carter, 1981). During aging the aorta diameter and stiffness increases because of degenerative changes in elastin, resulting in a transfer of stress to less extensible collagenous components of the aorta (Coquand et al., 2011).

Elastin production takes place during late gestation and before the end of childhood, but expression in very low rate persists in adulthood. The turnover rate for collagen and elastin is low in healthy arteries, but vascular pathology upsets the regulatory pathways that maintain this balance. In response to hypertension, the overexpression of both proinflammatory and proteinase-inhibitory molecules dramatically increases arterial ECM synthesis (Arribas et al., 2006; Jacob et al., 2001; Jacob, 2003). Elastin and elastic fibers are progressively degraded by enzymatic processes involving an age-related imbalance between anti-proteases and proteases. In particular, elastin degrading enzymes, i.e. elastases, include several matrix metalloproteases (MMP), such as MMP-2 and MMP-9 (Jacob, 2003; McNulty et al., 2005). Imbalance in matrix metalloproteinase/tissue inhibitors of metalloproteinases may contribute to alteration in collagen turnover and extracellular matrix remodelling. However, the ECM proteins synthesized in response to hypertension have a three-dimensional architecture that is functionally less optimal than those deposited during fetal development and may play an important role in determining the modulus of pathological elastin tissue (Arribas et al., 2006; Jacob et al., 2001; Jacob, 2003). The increase of aortic stiffness is not only because of enzymatic degradation of elastin, but also is due to other mechanisms mainly including age-dependent increase in the collagen content and arterial wall thickening, non-enzymatic glycation of proteins (elastin and collagens), leading to the formation of deleterious advanced glycation end-products (AGEs) and related molecular cross-links which modify the tissue mechanical properties (Gibbons & Dzau, 1994; Corman et al., 1998; Lakatta, 2003; O'Routke, 2007).

Advanced glycation is a major pathway for the posttranslational modifications of tissue proteins and begins with non-enzymatic addition of sugars to the primary amino groups of proteins. These early glucose-derived Schiff bases and Amadori products undergo a series of inter- and intramolecular rearrangement, dehydration, and oxidation-reduction reactions and produce the late products termed advanced glycation end products (AGEs). Excessive accumulation of AGE on tissue proteins has been implicated in the pathogenesis of many of the sequels of diabetes and normal aging. Protein-linked AGEs act to crosslink connective tissue proteins and to chemically inactive nitric oxide activity and thus are associated with endothelial dysfunction. They also act as a recognition signals for AGE receptor systems that are present on diverse cell types (Yang et al.,1991; Vlassara, 1994). Accumulation of AGEs in vascular walls increases intimal medial thickening, collagen is impaired, more cross-linked and elastin fibrils are broken. Large arteries are stiffen and mechanical stress increases, leading to hypertension and all its deleterious consequences (Dart & Kingwell, 2001; Zieman & Kass, 2004)

Excessive accumulation of AGEs on tissue proteins changes their structure and respectively functions, reduces their susceptibility to degradation and none of the last place immunogenicity. Specifically, the interaction of AGEs with vessel wall components increases vascular permeability, the expression of procoagulant activity and the generation of ROS (Yan et al., 1994). Glycated proteins form common immunological epitopes which raise formation of population of anti-AGE autoantibodies (AGEAb). These antibodies recognize and react with AGE-epitopes regardless of the proteins they have been formed on. Assessment of the levels of these antibodies shows that they are present at low titres even in sera of healthy subjects, perhaps as a part of homeostatic mechanism which clears glycated structures via *in situ* destruction or via opsonization of the gflycated proteins and products of their degradation (Baydanoff et al., 1996). However, in the conditions of increased non-enzymatic glycation the homeostatic control is inefficient and the generation of these antibodies increases (Baydanoff et al., 1996).

The aim of our study was to investigate the effects of age and hypertension on quantity and quality of elastin and collagen in the aortic wall of the rats. In order to reach this aim we used the spontaneously hypertensive rats (SHR), that are appropriate genetical model for studying essential hypertension compared to normotensive Wistar-Kyoto rats (WKR) at 2-, 4- and 8-months of age.

2. Materials and methods

2.1 Animals

Female and male Spontaneously Hypertensive Rats (SHR, n=30) and Wistar Kyoto Rats (WKY, n=33) were used. Animals were born and raised under conventional conditions in the animal facility of the Medical University of Pleven and were allowed free access to tap water and a standard laboratory chow. Animals were housed and kept under a normal 12 h light/dark cycle at 22 °± 2°C. They were divided into 6 groups: 2-month-old SHR (2mSHR, n=7, 4 female and 3 male); 4-month-old SHR (4m SHR, n=9, 5/4); 8-month-old SHR (8mSHR, n=14, 9/5); 2-month-old WKY (2mWKY, n=12, 5/7); 4-month-old SHR (4mWKY, n=9, 5/4); 8-month-old WKY (8mWKY, n=12, 5/7). Our experimental design was approved by the Animal Care and Use of Laboratory Animals group of the Ethical Committee of the University, based on the principles described in the Guide for the Care and Use of Laboratory Animals (Bayne, 1996).

2.2 Blood collection

At the end of the 2nd, 4th and 8th month, following overnight fasting, the abdominal cavity of rats was opened under pentobarbitone sodium anesthesia (26 mg/kg body weight, i.p.). Blood was collected from the bifurcation of the aorta and put at 37 °C to clot. After separation 0.02% NaN_3 was added to each serum sample and stored at - 20 °C prior to use.

2.3 Quantification of elastin in the thoracic aorta

The quantity of elastin was measured after dissection of the descending thoracic aorta and cleaning of blood and surrounding adipose tissue. Length and width (both sides) of the vessels were recorded by using a grid in the eyepiece, after opening of the vessel in length.

Elastin was then quantified by using a protocol deriving from a previously described method (Wolinski, 1972) with small modifications. Briefly, after delipidation in acetone/diethyl ether (1:1, vol/vol) and drying, the dry weight was recorded by using a Kern ALS 120-4 balance (precision: 0.01 mg). Cell proteins were extracted by gentle agitation in 0.3% SDS for 24 h and then 3 times in 5 M guanidinium chloride with preservative 0.02% sodium azide for 2 h. After washing 3 times in distilled water the extracellular proteins, other than elastin remaining in the aortic segments, were solubilized by three 15-minute extractions in 1 ml of 0.1 M NaOH in a boiling water bath. Elastin was quantified by determining the dry weight of the residue as percent dry weight of the aorta or mg/cm.

2.4 Quantification of collagen in the thoracic aorta

The content of the aortic collagen was assayed by determination the hydroxyproline presented in NaOH solution. The solution was evaporated to dryness and hydrolyzed in 6N HCl under vacuum conditions for 24 hours at 110 °C. A colorimetric assay according Woessner (1961) was applied for determination of the hydroxyproline. Assuming that collagen contains 12.77% hydroxyproline by weight its quantity was presented as mg collagen per cm aorta (Keeley et al., 1984).

2.5 Direct determination of advanced glycation end-products (AGEs) formed *in vivo*

Soluble α-elastin was obtained from the descending thoracic aorta by the method of Partridge (1955). Insoluble elastin was hydrolyzed 5 times in 0.25 M oxalic acid in boiling water bath, and then dialyzed against phosphate buffered saline (PBS). Protein content of each sample was determined by measurement of the absorption of UV light at 280 nm wavelength and calculated according to a standard curve, constructed on the basis of different dilutions of rat α-elastin (EPC, St. Louis, USA).

Maillard reaction-related fluorescence (FC), representative of AGEs formed *in vivo*, was measured as an index of advanced glycation in 360/450 nm excitation/emission (Baydanoff et al., 1994; Monnier et al., 1984) with Corning-EEL fluorimeter. Quinine sulfate 1 μM in 0.1N H_2SO_4 was used as a standard. The levels of AGEs were expressed as arbitrary fluorescence units (AU) per mg protein.

2.6 Measurement of circulating AGEAb

2.6.1 Glycation of KLH

Keyhole Limpets Hemocyanin (KLH) (Sigma) - 20mg/ml was glycated *in vitro* with 3.33 M glucose in 0.4 M phosphate buffer, pH 7.5 with preservative 0.04% NaN_3, at 37°C, for 12 weeks. The formation of advanced glycated end product of KLH (AGE-KLH) was determined via measuring the fluorescence at 360/440 nm excitation/emission. After dialyzing against phosphate buffer, the obtained AGE-KLH was used as antigen in home-made ELISA.

2.6.2 ELISA for determination of circulating anti-AGE antibodies

AGEAb were assessed by home-made ELISA, as described in Dimitrova et al. (2009). The assay was performed as follows: Microtiter 96 well plates (Greiner Microlon) were coated with 100 μl of AGE-KLH per well in carbonate buffer (pH 9.6) with concentration 5 μg/ml. The plates were incubated 2 hours at 37 °C and then overnight at 4 °C to complete binding. Then the plates were washed three times with PBS with 0.05 % Tween 20 (PBS-Tween) before their blocking with 0.1 % BSA and incubation for 1 h at 37 °C. The next step was addition of 100 μl of tested rat serum, diluted 1:10 with PBS-Tween. Bound antibodies were reacted with anti-mouse IgG peroxidase conjugate, diluted 1:3200 in 1% human serum albumin for 1 h at 37 °C (BulBio - Sofia, Bulgaria). Ortho-phenyene diamine was used as colorimetric substrate. The reaction was terminated by 50 μl 8N H_2SO_4 and the absorbance (A) was read at 492 nm on automatic micro-ELISA plate reader. Serum samples were analyzed in triplicate and the average calculated.

2.7 Statistics

Comparisons of SHR and control rats were assessed using one- or two-way ANOVA followed when necessary by Fisher's least significant difference test (LSD) for paired value comparisons. The results are presented as mean values ±SEM, and p values ≤0.05 were considered as statistically significant. The correlations between investigated parameters were tested by the method of Pearson. The statistical package used was SPSS v. 15.

3. Results

3.1 Elastin content in the thoracic aorta

Figure 1 shows elastin content in the thoracic aortas of both rat strains aged 2, 4 and 8 months. When the elastin quantity was presented as percentage of the dry weight of the aorta (fig. 1A) SHR at different ages did not differ significantly, whereas in the WKR groups factor "age" was found to have a significant effect (p<0.05) - elastin quantity decreased with age. Elastin content was significantly larger (p<0.032) in 8-month-old hypertensive animals compared to age matched normotensive group. When the elastin quantity was presented as milligrams per centimeter of aorta (fig. 1B) the absolute amount of elastin in SHR at different ages did not differ from age matched WKR.

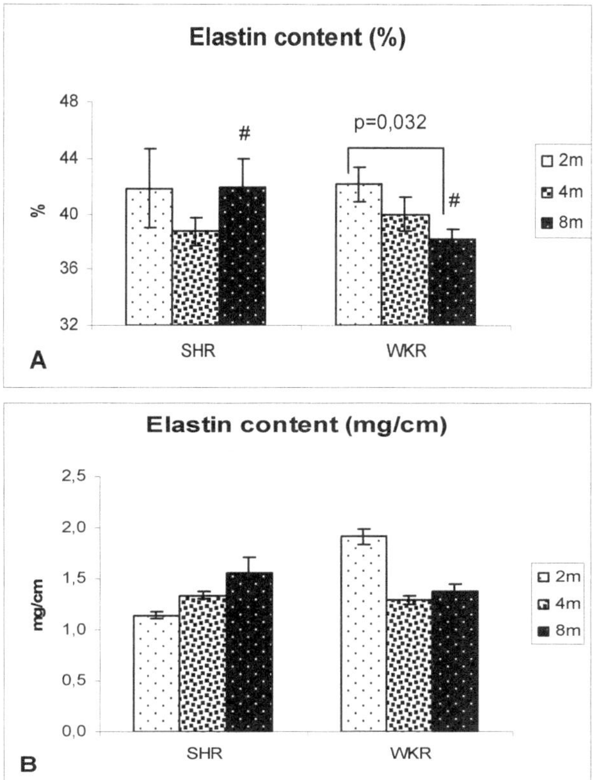

Fig. 1. Quantity of elastin in the thoracic aorta of SHR and WKR at three different ages, expressed as (A) percent dry weight of the aorta and (B) mg elastin per cm aorta.
Significant difference between 8-months-old SHR and WKR (p = 0.032).

3.2 Collagen content in the thoracic aorta

The collagen quantity showed a trend towards increase with age, but without reaching the statistically significant threshold (fig. 2). SHR at 2 and 4 months of age had significantly larger collagen content then age matched WKR. The groups of 8-month-old animals did not differ significantly. Hypertension was found to have significant effect (p=0.019) on the collagen content of the thoracic aorta of the investigated rats, but age was not.

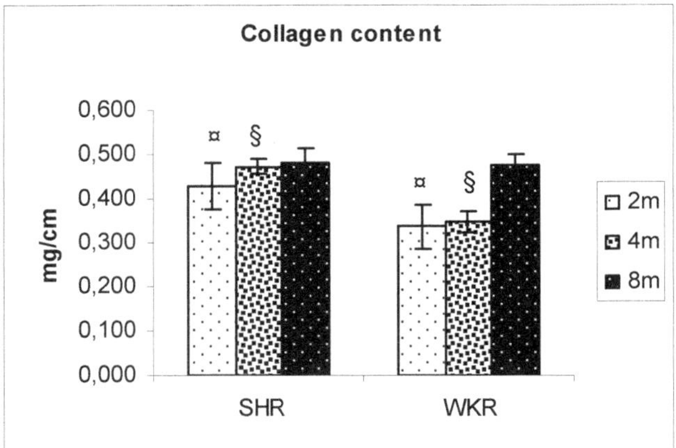

Fig. 2. Quantity of collagen in the thoracic aorta of SHR and WKR at three different ages, expressed as mg collagen per cm aorta. ¤ - Significant difference between 2-months-old SHR and WKR ($p = 0.056$); § - Significant difference between 4-months-old SHR and WKR ($p = 0.007$).

3.3 Direct determination of AGEs formed *in vivo*

Results obtained by direct determination of AGEs formed *in vivo* are presented in AU per mg elastin (fig. 3). The direct measurement of Maillard reaction-related fluorescence of obtained aortic α-elastin was found to increase with age in both strains and to be significantly higher ($p=0.036$) in elastin obtained by 8-month-old SHR samples compared to WKR at the same age.

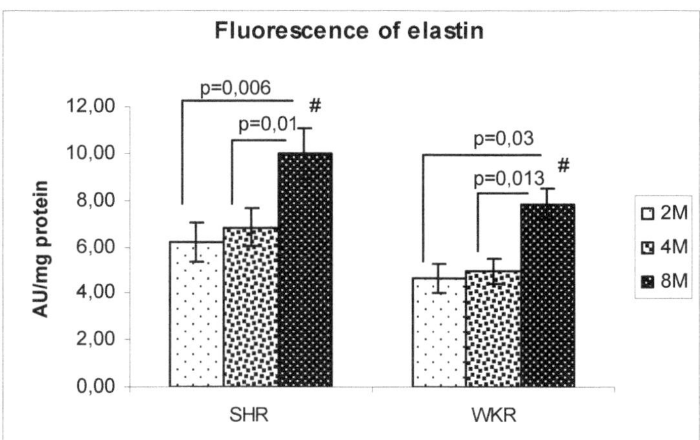

Fig. 3. Direct determination of AGEs formed *in vivo*. Maillard reaction-related fluorescence, measured for the soluble α-elastin obtained from the three aged groups of SHR and WKR. #- Significant difference between 8-months-old SHR and WKR ($p = 0.036$). AU – arbitrary units

ANOVA analysis showed that fluorescence of the purified elastin was influenced significantly by the age and hypertension (p=0.001 and p=0.0026).

3.4 Circulating anti-AGE antibodies

AGEAb were presented in all the sera tested. The serum levels of these antibodies (fig.4) in SHR substantially increased with age. Their mean values in WKR showed a trend towards increase with age, but without being significant. The youngest animals group from the two strains did not differ in contrast with the other two age groups in which antibody levels were significantly higher in hypertensive animals, compared to normotensive ones. The correlation (Pearson coefficient) between the serum levels of anti-AGEAb and the age and hypertension was highly significant (r = 0.553, p = 0.001 for age and r = - 0.440, p = 0.009 for hypertension).

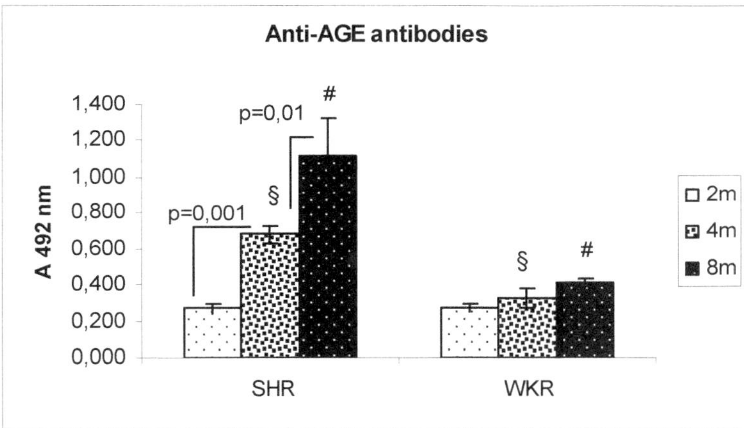

Fig. 4. Serum levels of AGEAb, measured by indirect home-made ELISA. § - Significant difference between 4-months-old SHR and WKR (p = 0.012). #- Significant difference between 8-months-old SHR and WKR (p = 0.001).

4. Discussion

Elastin and collagen fibrils are the structures that endow the vascular wall with elasticity and strength. But similarly to most of the connective tissue proteins they are proved to be pliable of cross-linking and thus their physiological turnover is changed and they become resistant to hydrolytic degradation (Zieman 2003). Altered by non-enzymatic glycation they accumulate in the vessel matrix in unorganized and non-functional pattern (Bailey 2001). This structural disorganization affects mainly the media and leads to mechanical alteration of the aortic wall especially in the condition of hypertension (Laurent, 1995; Bezie et al.,1998). Marque et al., (1999) explained age-linked aortic stiffening in SHR with more rapid increase with age of the aortic collagen content in and accelerated accumulation of advanced glycation end products on elastin and collagen fibers.

In the present study, we compared the elastin and collagen content of thoracic aortas from Wistar-Kyoto rats (WKR) and spontaneously hypertensive rats (SHR) at three different ages. In addition, glycation of purified aortic elastin and antibodies against AGEs were determined. The investigated parameters were compared as functions of age and hypertension.

Elastin production was investigated at the protein level, through the measurement of aortic elastin content. The relative elastin content (expressed as a percentage of dry weight) was significantly lower in the 8-month-old WKR, compared to the 2-month-old WKR and 8-month-old SHR. However, no significant differences were observed when the elastin content in the aorta was expressed as weight per aorta length (mg/cm). This can be explained with variations in body size and, thus, aortic dry weight among individual rats from both sexes, the absolute amount of elastin varied widely (Sauvage et al., 1999). The expression of elastin content relative to dry weight eliminates considerations of these large individual variations in aortic dry weight. It appeared that age had no general significant effect on the joint distribution of elastin content ($p>0.05$), but there is a slight decrease with age in aortic elastin content (-9%) in the oldest WKR (Fig. 1).

In the literature there is not clear age-related trend of decrease or increase of both elastin and collagen quantity and during hypertension development. Tsoporis et al., (1998) followed the biosynthesis of collagen and elastin during the development of spontaneous hypertension in SHR. They found that both collagen and elastin synthesis (as revealed by specific hydroxyproline activity) exceeded WKR control levels in the prehypertensive period (at the age of 4 weeks), decreased in the development of hypertension (to the age of 14-16 weeks), and increased again in the period of the established hypertensive state (beyond the age of 16 weeks). Several studies showed that this second increase in connective tissue proteins may be prevented, depending on the choice of antihypertensive therapy (Tsoporis et al., 1998; Han et al., 2009).

From the two main collagen types present, synthesis of collagen type III exceeded that of type I in the prehypertensive period (at the age of 4 weeks) and this relation was reversed during the period of established hypertension. They suggested that the vascular connective tissue metabolism in SHR differs from that in strain-matched controls, and the reverse rate of collagen type III to collagen type I synthesis during hypertension development may be considered an adaptive response to the increasing pressure load which may alter the mechanical properties of the vessel wall (Deil et al., 1987).

Han et al (2009) investigated medial and adventitial layers of the thoracic aorta 4- and 8-month-old SHR and WKR. They discovered that compared with WKR, SHR exhibited greater collagen and elastin content in the media, but decreased collagen and elastin content in the adventitial layer. Both medial and adventitial collagen and elastin content increased significantly with age in both strains and was greater in 8-month-old rats compared to 4-month-old rats (Han 2009).

It has been shown that the elastin network plays a major role in the maintenance of aortic elastic properties in adult SHR, not through variations of its total amount but through increases of the extent of its anchorage to the smooth muscle cells (Laurent, 1995, Bezieet al., 1998). Our data also supported the finding that quantity of elastin is not generally changed in the conditions of hypertension that is why we tested the Maillard reaction-related fluorescence to check the index of advanced glycation.

Hypertension is characterized by insulin resistance, and a number of studies have suggested that it plays a major role in its etiology (DeLano & Schmidt-Schonbein, 2007).

Other study of DeLano & Schmid-Schönbein (2008) states emphatically that in hypertensive rats, proteases cleave extracellular portions of several protein receptors, such as the insulin receptor, so that insulin can no longer bind and facilitate normal metabolism of glucose. In insulin resistance, alterations in glucose and lipid turnover lead to the production of excess AGEs (Vasdev et al.,2007; Potenza et al., 2005). AGEs are complex group of compounds, and the structure of a lot of them is identified - N-carboxymethil-lysine (CML), pentosidine, imidasolones, pirraline, etc. Although the chemistry of AGEs is not fully discovered, increased levels of circulating and tissue AGEs have been demonstrated in both animal and human studies in pathologies and aging (Brüel & Oxlund, 1996; Zieman & Kass, 2004). Increase of non-enzymatic glycation with aging is due to the life-long exposure to glucose even in normoglycemia. In pathology or aging conditions, when non-enzymatic glycation of proteins is increased, the capacity of normal homeostasis seems to be inefficient. This way AGEs are accumulated and contribute to the development of long-term aging process especially in long-lived proteins (Konova et al., 2004). Our results from the fluorescence studies of obtained aortic α-elastin showed increase with age especially in the hypertensive animals. The highest fluorescence was measured in the group of 8-month-old SHR which was significantly higher ($p=0.036$) in comparison to WKR at the same age. These results suggest an increase of non-enzymatic glycation with aging, even in normoglycemic animals and are in agreement with our previous investigations of age-dependent glycation of human aortic elastin (Konova et al., 2004). Moreover, we demonstrated the accelerating role of hypertension for the increase of formation of irreversible late advanced glycation end products in long-lived connective tissue proteins.

We did not find significant differences in AGEAb levels between SHR and WKR rats at 2 months of age. At about two months of age in arterial walls of SHR occur alterations that are consequences of developed hypertension. Observed vascular wall hypertrophy also plays an important role in hypertension by reducing tissue flexibility and elasticity (Mecham & Davis, 1994). During the early phases of hypertension development, wall distensibility might even be augmented in the SHR rats but long-term hypertension decreases arterial distensibility (Zanchi et al., 1997). Formed AGEs affect the biochemical and physical properties of proteins and ECM, including the charge, hydrophobicity, turnover and elasticity of collagen and elastin (Baydanov et al, 1994), and the cell adhesion, permeability and pro-inflammatory properties of the ECM (Baines, 2001). These structural and functional changes alter the antigenicity and immunogenicity of vascular elements. This provokes the immune system to react by producing different types of antibodies including AGEAb, directed to their own but altered structures. In our study AGEAb were presented in all the investigated rat sera but their levels were in positive correlation to the increasing age only in hypertensive animals. We observed significantly higher levels of AGE Abs in 8-month-old SHR compared to other age groups and age matched WKR. Our previous studies in patients with occupational vegetative polyneuropathy of upper limbs and in male SHR demonstrated similar results (Dimitrova et al., 2009, 2010).

5. Conclusion

In general, there were not clearly drawn age-related differences in the elastin and collagen quantity of the thoracic aortas of the investigated rats. But there were significant differences in the late products of glycation of the purified elastin and serum levels of antibodies against them. The quantity of obtained elastin did not change but the quality did. It appeared that the investigated factors had a generally significant effect and we observed a substantial increase of AGEs and AGEAb with age especially in the condition of hypertension.

Although AGEs in proteins are probably correlative, rather than causative, with respect to aging, they accumulate to high levels in tissues in age-related chronic diseases, such as atherosclerosis, diabetes, arthritis and neurodegenerative diseases. Inhibition of AGE formation in these diseases may limit oxidative and inflammatory damage in tissues, retarding the progression of pathophysiology and improve the quality of life during aging. It is necessary to investigate genetic, pharmacologic, and nutritional factors, which could decrease the levels of AGEs, to break cross-links in the glycated structures of ECM and to restore the turnover balance in order to improve the functionality of the vessel walls.

6. References

Arribas, S. Hinek, A. & Gonzalez, M. (2006). Elastic Fibres and Vascular Structure in Hypertension. *Pharmacol Ther* No.111, pp. 771-791

Atanasova, M. Konova, E. Georgieva, M. Dimitrova, A. Coquand-Gandit, M. Faury, G, & Baydanoff, S. (2010). Age-related Changes of Anti-elastin Antibodies in Senescence-Accelerated Mice. *Gerontology*, No. 56, pp. 310-318 DOI: 10.1007/BF00731464

Bailey, A. (2001). Molecular Mechanisms of Aging in Connective Tissues. *Mech Ageing Dev*, No. 122, pp. 735-757

Baydanoff, S. Konova, E. & Ivanova, N. (1996). Determination of Anti-AGE antibodies in Human Serum. *Glycoconj J*, No.13, pp. 335-339

Baydanoff, S. Konova, E. Dosheva, I. & Dorovski, P. (1994). Non-enzymatic glycation of elastin. *Glycosylation Dis.* No. 1, pp. 53-58

Bayne, K. (1996). Revised guide for the care and use of laboratory animals available. *Physiologist*, Vol. 39, No. 199, pp., 208-11, ISSN

Baynes, J. (2001). The Role of AGEs in Aging: *Causation or Correlation Experimental Gerontology* Vol. 36, Issue 9, pp. 1527-1537

Bezie, Y Lacolley, P. Laurent, S. & Gabella, G. (1998). Connection of Smooth Muscle Cells to Elastic Lamellae in Aorta of Spontaneously Hypertensive Rats. *Hypertension*, No.32, pp. 166 -169

Cattell, M. Anderson, J. & Hasleton, P. (1996). Age-related Changes in Amounts and Concentrations of Collagen and Elastin in Normotensive Human Thoracic Aorta. *Clin Chim Acta*, No.245, pp.73-84

Coquand-Gandit, M. Jakob, MP. Romero, B. Georgieva, M. Andrieu, JP. Bouilot, S. Garsia-Honduvilla, N. Starcher, B. Atanasova, M. Bujan, J. & Faury, G. (2011). Large Vessels of SAM-P8 and SAM-R1 Mice Exhibit Structural and Functional Differencses Related to implementation of Cardiovascular Therapies. *In Press*

Corman B, Duriez, M. Poitevin, P. Heudes, D. Bruneval, P. Tedgui, A. & Levy, B. (1998). Aminoguanidine Prevents Age-related Arterial Stiffening and Cardiac Hypertrophy. *Proc Natl Acad Sci USA*, No.95, pp. 1301-1306

Dart, A. & Kingwell, B. (2001). Pulse Pressure - a Review of Mechanisms and Clinical Relevance. *J Am Coll Cardiol* No. 37, pp. 975-984

Deil, Z. Jelínek, J. Macek, K. Chaldakov, G. & Vankov, V. (1987). Collagen and Elastin Synthesis in the Aorta of Spontaneously Hypertensive Rats. *Blood Vessels*, Vol. 24, No.6, pp. 313-320

DeLano, F. & Schmid-Schonbein, G. (2007). Proteinase Activity In The Spontaneously Hypertensive Rat: Free Radical Production, Receptor Cleavage, and Insulin Resistance. *FASEB J* , No.21,pp. 422

DeLano, F. & Schmid-Schönbein, G. (2008). Proteinase Activity and Receptor Cleavage. Mechanism for Insulin Resistance in the Spontaneously Hypertensive Rat. *Hypertension*, No.52, pp. 415-423

Dimitrova, A. Stoilova, I. Atanasova, M. Georgieva, M. Russeva, A. Strashimirov, D. & Tzachev, K. (2009). Anti-AGE Antibodies and Serum Concentration of Zinc, Copper and Selenium in Patients with Occupational Vegetative Polyneuropathy of Upper Limbs. *J Biomed Clin Res*. No.2, pp. 31-35

Dimitrova, A. Atanasova, M. Georgieva, M. Ruseva, A. Strashimirov, D. & Baydanoff, S. (2010). Anti-AGE Antibodies and Markers of Oxidative Stress in Spontaneously Hipertensive Rats at Different Ages. *J Biomed Clin Rec*. Vol.3, No.1, pp. 33-39

Gibbons, G. & Dzau V. (1994).The Emerging Concept of Vascular Remodeling. *N Engl J Med*; 330:1431-1438

Han, W. Wu, L. Zhou, H. Zhang, J. Che, Z. Wu, Y. Liu, J. Zhu, D. & Gao, P. (2009). Changes in the Composition of the Thoracic Aortic Wall in Spontaneously Hypertensive Rats Treated with Losartan or Spironolactone. *Clin Exp Pharmacol Physiol*. May;36(5-6), pp. 583-538

Jacob, MP. Badier-Commander, C. Fontaine, V. Benazzoug, Y. Feldman, L. & Michel, JB. (2001). Extracellular Matrix Remodeling in the Vascular Wall. *Pathol Biol (Paris)*, No.49, pp. 326–332

Jacob, MP. (2003). Extracellular Matrix Remodeling and Matrix Metalloproteinases in the Vascular Wall During Aging and in Pathological Conditions. *BiomedPharmacother*, No.57, pp. 195–202

Jacob, MP. (2006). Extracellular Matrix and Vascular Ageing. *Med Sci* (Paris); No. 22, pp. 273-278

Keeley, F. Morin, J. &Vesely, S. (1984). Characterization of Collagen from Normal Human Sclera. *Exp Eye Res*, No.39, pp. 533-542

Konova, E. Baydanoff, S. Atanasova, M. and Velkova, A. (2004). Age-related changes in the glycation of human aortric elastin. *Exp Gerontol* , Vol. 39, No.2, pp. 249-254

Lakatta, E. (2003). Arterial and Cardiac Aging: Major Shareholders in Cardiovascular Disease Enterprises. Part III: Cellular and Molecular Clues to Heart and Arterial Aging. *Circulation*, No.107, pp. 490-497.

Laurent, S. (1995). Arterial Wall Hypertrophy and Stiffness in Essential Hypertensive patients. *Hypertension*, No.26, pp. 355–362

Marque, V. Kieffer, P. Atkinson, J. Lartaud-Idjouadiene, I. (1999). Elastic Properties and Composition of the Aortic Wall in Old Spontaneously Hypertensive Rats. *Hypertension*. No.34, pp. 415-422

McNulty, M. Spiers, P. McGovern, E., Feely, J. (2005). Aging is Associated with Increased Matrix Metalloproteinase-2 Activity in the Human Aorta. *Am J Hypertens*, No.18, pp. 504-509

Mecham, R. & Davis, E. (1994). Elastic Fiber Structure and Assembly. In: *Extracellular Matrix Assembly and Structure.* Yurchenko, P. Birk, D. Mecham, R. Ed. Academic Press, pp. 281-310

Monnier, V. Kohn, R. & Cerami, A. (1984). Accelerated Age-Related Browning of Human Collagen in Diabetes Mellitus. *Proc Natl Acad Sci,* USA, No. 81, pp. 583-587

O'Rourke, M. (2007). Arterial Aging: Pathophysiological Principles. *Vasc Med,* No.12, pp. 329-341

Partridge, S. Davis, H. & Adair, G. (1955). The chemistry of connective tissues soluble proteins derived from partial hydrolysis of elastin. *Biochem J,* Vol.61, pp. 11-15

Potenza, M. Marasciulo, F. Chieppa, D. Brigiani, G. Formoso, G. Quon, M. Montagnani, M. (2005). Insulin Resistance in Spontaneously Hypertensive Rats (SHR) Is Associated With Endothelial Dysfunction Characterized by Imbalance Between NO and ET-1. *Am J Physiol Heart Circ Physiol* , No.289, pp. H813-H822

Samila, Z. & Carter, S. (1981). The Effect of Age on the Unfolding of Elastin Lamellae and Collagen Fibers with Stretch in Human Carotid Arteries. *Can J Physiol Pharmacol,* No.59, pp. 1050-1057

Sauvage, M. Osborne-Pellegrin, M. Deslandes-Le Flohic, F. & Jacob, MP. (1999). Influence of Elastin Gene Polymorphism on the Elastin content of the Aorta. *Arterioscler Thromb Vasc Biol.* No.19 pp. 2308-2315

Tsoporis, J. Keeley, F. Lee, W. & Leenen, H. (1998). Arterial Vasodilatation and Vascular Connective Tissue Changes in Spontaneously Hypertensive Rats. *Journal of cardiovascular pharmacology,* Vol.31, No.6, pp. 960-62

Vasdev, S. Gill, V. Singal, P. (2007). Role of Advanced Glycation End Products in Hypertension and Atherosclerosis: Therapeutic Implications. *Cell Biochem Biophys,* Vol.49, No.1, pp. 148-63

Vlassara, H. (1994). AGE-receptors and in vivo biological effects of AGEs. In: Labuza T., Reineccius, G. Monnier, V. O'Brien, J. Baynes, J. (eds). Maillard Reactions in Chemistry, Food and Health. *The Royal Society of Chemistry,* UK. Vol.151, pp. 254-261

Woessner, J. (1961).The Determination of hydroxyproline in Tissue and Protein Samples Containing Small Proportions of This Amino Acid. *Arch Biochem Biophys.* No. 93, pp. 440-447

Wolinsky, H. (1972). Effects of Androgen Treatment on the Male Rat Aorta. *J Clin Invest,* No.51, pp. 2552-2555

Yan, S. Schmidt, A. Anderson, G. et al. (1994). Enhanced Cellular Oxidant Stress by the Interaction of Advanced Glycation End Products with Their Receptors. *J Biol Chem.* No. 269, pp. 9889-9897

Yang, Z. Makita, Z. Horii, Y. et al. (1991). Two Novel Rat Liver Membrane Proteins that Bind Glycosylation End Products: Relationship to Macrophage Receptor for Glucose-Modified Proteins. *J Exp Med,* No.174, pp. 515-524

Zanchi, A. Brunner, H. & Hayoz, D. (1997). Age-related Changes of the Mechanical Properties of the Carotid Artery in Spontaneously Hypertensive Rats. *J Hypertens,* No.15, pp. 1415–1422

Zieman, S. & Kass, D. (2003). Advanced Glycation end Products Cross-Linking: Pathophysiologic Role and Therapeutic Target in Cardiovascular Disease. *Congestive Heart Failure* , No. 10, pp. 144-151, ISSN 1751-7133

13

Alternative Splicing in Endothelial Senescence: Role of the TGF-β Co-Receptor Endoglin

Francisco J. Blanco and Carmelo Bernabéu
Centro de Investigaciones Biológicas, CSIC,
Centro de Investigación Biomédica en Red de Enfermedades Raras,
Spain

1. Introduction

The vascular endothelium is the thin monolayer of specialized cells that line the blood vessels of the cardiovascular system. This endothelium is more than a simple protective barrier since it possesses anticoagulatory properties, mediates the metabolites exchange and regulates the vascular tone and homeostasis maintenance. These functions are finely tuned by endothelial cells that, in the absence of any stimuli, remain in a quiescent stage (Conway & Carmeliet, 2004). In fact, endothelial cells occasionally divide in a normal vessel, displaying a very low turnover rate except for localized areas (Foteinos et al., 2008). Thus, the endothelium is quite sensitive to a variety of signals including shear stress and circulating factors that lead to endothelial activation. As a result of their own physiology along the lifespan, endothelial cells progressively accumulate reactive oxygen species and pro-oxidant metabolites due to an increased oxidative stress, damages in DNA and advanced cellular replication involving shortening of telomeres. Altogether, these alterations lead endothelial cells to reach senescence (Brandes et al., 2005; Foreman & Tang, 2003), which has been proposed to be at the cellular basis of most of the vascular pathologies associated with ageing, such as atherosclerosis or hypertension (Minamino & Komuro, 2008; Rodríguez-Mañas et al., 2009).

The major aspect of endothelial physiology implies the growth or formation of new blood vessels from pre-existing ones, process named angiogenesis which is mainly induced by metabolic requests (Fraisl et al., 2009). Angiogenesis plays a key role from the first steps during the embryonic development to the adult stage, and is involved in numerous physiological processes such as wound repair or the growth of the tissues (Carmeliet & Jain, 2011). However, angiogenesis and vascular remodelling decline with age and several lines of evidence indicate that ageing and endothelial dysfunction progress in parallel (Brandes et al., 2005; Ferrari et al., 2003; Minamino et al., 2004). In this sense, numerous efforts are addressed to elucidate the molecular mechanisms that underlie vascular ageing.

2. TGF-β in angiogenesis - Role of endoglin

The angiogenesis process consists of two separate but balanced phases, activation and resolution, that are finely arranged by a suite of cytokines, among which the transforming growth factor (TGF)-β plays a dual role (Pardali et al., 2010). TGF-β is the prototypic member of a large family of multifunctional and evolutionarily conserved cytokines, including also activins and bone morphogenetic proteins (BMPs). Upon proteolytic activation, TGF-β circulates as a 25 kDa homodimer that elicits its cellular functions by binding to a membrane complex of type II (TβRII) and type I (TβRI or ALKs) receptors with cytoplasmic serine-threonine kinase activity (Kang et al., 2009). Endothelial cells express two different TβRIs, named ALK5 and ALK1, with distinct affinity for the ligand and different signalling pathways mediated mainly by Smad proteins (Smad2/3 and Smad1/5/8, respectively) (Massague et al., 2005). Moreover, endothelial cells also express endoglin, or CD105, an auxiliary TGF-β receptor that modulates the balance between ALK1 and ALK5 signalling. Endoglin is mainly expressed as a homodimeric protein of 180 kDa and is associated to the activation phase of angiogenesis, acting as a modulator between both phases. In this context, endoglin interacts with ALK1 and promotes the TGF-β/ALK1 signalling pathway (Blanco et al., 2005; Lebrin et al., 2004).

The TGF-β/endoglin pairing has been studied in different contexts such as differentiation (Tang et al., 2011), cancer (Bernabeu et al., 2009; Perez-Gomez et al., 2010) and other pathologies including liver fibrosis (Meurer et al., 2011) or preeclampsia (Venkatesha et al., 2006). However, endoglin plays a major role in angiogenesis as well as in vascular remodelling and homeostasis (Lopez-Novoa & Bernabeu, 2010; ten Dijke et al., 2008). Heterozygous mutations in the endoglin gene (*ENG*) are responsible for the vascular dysplasia named hereditary haemorrhagic telangiectasia (HHT) type 1 (McAllister et al., 1994; Shovlin, 2010), a rare genetic disease with autosomal dominant inheritance. These mutations lead to the development of abnormal vascular structures that are the basis of the characteristic HHT symptoms, including frequent and recurrent nosebleeds, telangiectases in the nasal and gastrointestinal tracts and large arteriovenous malformations in different organs such as lung, liver or brain (Mahmoud et al., 2010; Shovlin, 2010). Nonetheless, the HHT symptoms are not present at birth and normally appear during adolescence, getting worse with age. This is in line with the functional role of endoglin in angiogenesis and with previous observation that angiogenesis becomes impaired with ageing (Rivard et al., 1999).

2.1 Two alternatively spliced endoglin isoforms

Most of published studies about endoglin are referred to L-endoglin (long endoglin) that is the predominantly expressed isoform. However, the expression of a short variant (S-endoglin) was described first in humans (Bellon et al., 1993) and later in mouse (Perez-Gomez et al., 2005). In humans, both isoforms share the identical large extracellular region and the transmembrane domain, so that the only difference resides in their cytoplasmic tails (Figure 1A). In the case of L-endoglin, this region is composed by 47 amino acids with a high frequency of serine and threonine residues susceptible to be phosphorylated. Also, the sequence serine-methionine-alanine, SMA, in the C-terminal end is a docking site for proteins with a PDZ domain and is involved in the cytoskeleton organization (Koleva et al., 2006). By contrast, the sequence of the S-endoglin cytoplasmic tail is 14 amino acids long and contains only one serine and threonine residues; also the last 7 residues are specific for this isoform (Figure 1B). These data suggest that L-endoglin and S-endoglin may elicit different functional effects on the endothelial cell.

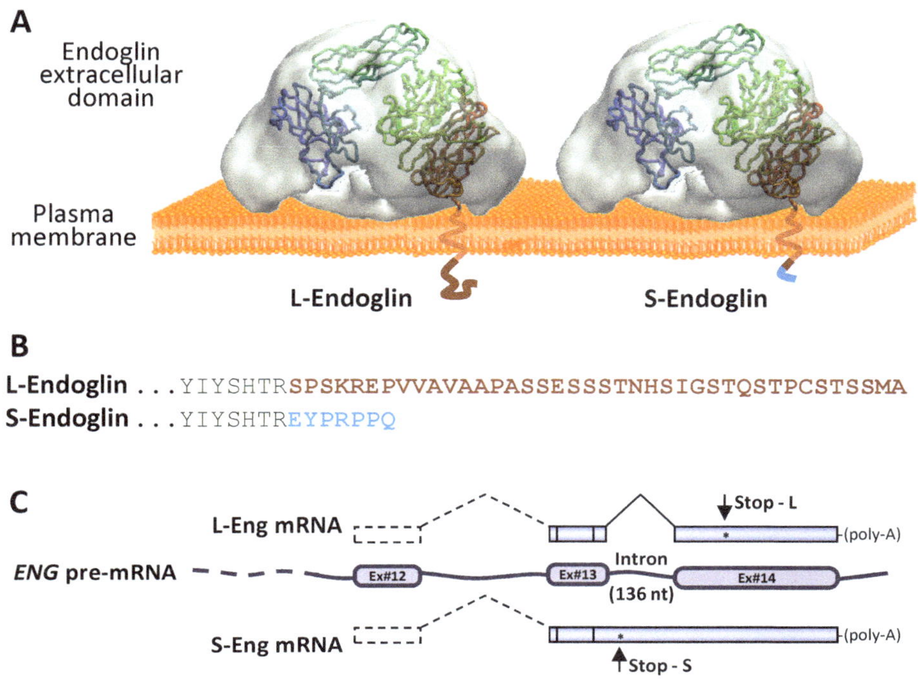

Fig. 1. The two endoglin isoforms. (A) The electron microscopy density map (grey) of the endoglin extracellular region shows the overall structure. The backbone of a theoretical atomic model of the endoglin monomer is fitted inside (adapted from Llorca et al., 2007). This structure is common to both endoglin variants. The transmembrane domain (red) and cytoplasmic tails (brown, L; blue, S) are schematized. (B) The amino acid sequence of the cytoplasmic domain is detailed for both isoforms. (C) The endoglin pre-mRNA is represented in the middle of the mature transcripts that originate each isoform. The retention of the final intron by an alternative splicing process leads to S-endoglin expression.

S-endoglin arises as the result of an alternative splicing mechanism by which the last intron, between exons #13 and #14, is retained in the mature mRNA (Figure 1C). Consequently, an early stop codon appears in the open reading frame and truncates the mature protein in the cytoplasmic region. Although this mechanism of intron retention normally involves a rapid degradation by the nonsense-mediated decay machinery (Lareau et al., 2004; Nott et al., 2003), under certain conditions it may also lead to a biologically active isoform (Sakabe & de Souza, 2007); and this is the case of endoglin. Thus, when endothelial cells become senescent during the ageing process, they show an up-regulation of S-endoglin (Blanco et al., 2008). At this senescent stage, both endoglin isoforms are co-expressed likely forming heterodimers, as it occurs in mice (Perez-Gomez et al., 2005), and some of the cellular responses to TGF-β are oppositely regulated by each isoform. Indeed, the S-endoglin increase has an antiangiogenic role in the blood vessels and contributes to vascular pathology (Blanco et al., 2008; Perez-Gomez et al., 2005; Velasco et al., 2008).

3. Endothelial senescence and TGF-β

It is well known that ageing *per se* is the major risk factor for the development of cardiovascular diseases. Thus, senescence has been widely and mainly analyzed in *in vitro* studies but there are also evidences that this process takes place *in vivo* (Erusalimsky & Kurz, 2005; Minamino & Komuro, 2007). The first evidence of cellular senescence in primary cultures *in vitro* is the deceleration in the proliferation, that is, an increase in the doubling time of the cell population. In parallel, cells experience morphological changes along theses passages that involve the augment of the cellular size and shape. However, these observations are usually complemented with a useful tool based on the abnormal behaviour associated with senescent cells of the lysosomal hydrolase β-galactosidase. Thus, the senescence-associated β-galactosidase (SA β-gal) activity at pH 6 is widely accepted as an easily detectable senescence histochemical marker (Dimri et al., 1995).

Endothelial senescence is a cellular process that is clearly linked to both ageing and the development of vascular pathologies as well (Brandes et al., 2005; Erusalimsky, 2009; Minamino & Komuro, 2007). Basically, senescence constitutes a stress and damage response phenomenon that involves a permanent growth arrest (Campisi & d'Adda di Fagagna, 2007). Consequently, senescent cells undergo diverse changes in gene and protein expression that lead to an impairment of cellular functions (Foreman & Tang, 2003; Young & Narita, 2009). Thus, these changes usually affect to the endothelial phenotype favouring a pro-inflammatory, pro-atherosclerotic, or a prothrombotic state (Erusalimsky, 2009).

Here, TGF-β plays an important role owing to its ability to prompt senescence in a variety of cell types (Cipriano et al., 2011; Kordon et al., 1995; Tremain et al., 2000; van der Kraan et al., 2011; Wu et al., 2009). In the vascular context, it has been reported, e. g., elevated levels of TGF-β in the aging varicose veins that likely favour the fibrous process and the consequent venous insufficiency (Pascual et al., 2007). In this sense, the profibrotic effect of TGF-β is mediated by the stimulation via Smad3 signalling of the plasminogen activator inhibitor (PAI)-1 expression, a key regulator of the synthesis and deposition of the extracellular matrix in the tissue homeostasis (Ghosh & Vaughan, 2011). Thus, the increase of TGF-β up-regulates PAI-1 expression, which contributes to the accumulation of collagen and other extracellular matrix components. This PAI-1 increase is also in line with the decrease of the antithrombogenic properties of a senescent endothelium due to the inhibition of the urokinase- and tissue-type plasminogen activator (uPA and tPA, respectively)/plasmin axis (Comi et al., 1995; Schneiderman et al., 1992).

3.1 Replicative senescence

Senescence was initially considered to reflect the finite capacity for division that normal diploid cells exhibit when propagated in culture. This statement is based on the successive rounds of cell division that imply the progressively shortening and eventual dysfunction of telomeres, the physical ends of chromosomes, in a phenomenon known as Hayflick's limit (Hayflick, 2003; Shay & Wright, 2007). Thus, the down-regulation of telomerase, the enzyme responsible for maintaining the telomeres length, is clue for the senescence program. Besides, because telomerase is re-activated in the majority of neoplastic processes, it is postulated that inhibiting telomerase activity should result in senescence induction by telomere shortening which can cause the death of cancer cells (Folini et al., 2011). Interestingly, the senescence inducer TGF-β down-regulates the telomerase activity. Thus, upon TGF-β treatment, Smad3 is able to interact with the transcription factor c-myc, so repressing the promoter of the hTERT gene, encoding the catalytic subunit of telomerase (Figure 2). Thus, the c-myc activity is blocked in the Smad3 complexes which negatively affects to the cell cycle (Li & Liu, 2007; Li et al., 2006). In addition, this repression of the hTERT promoter mediated by TGF-β can be alternatively reinforced by the activation of the TGF-β activated kinase (TAK)-1 pathway that abrogates the transcriptional activity of Sp1 on the hTERT promoter (Fujiki et al., 2007).

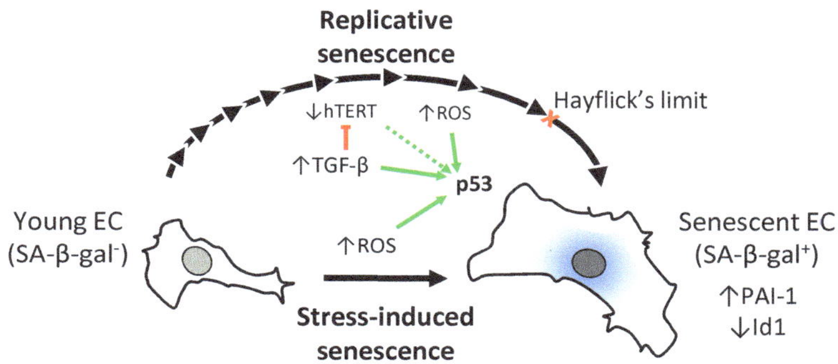

Fig. 2. The endothelial senescence. Endothelial cells extensively cultured *in vitro* enlarge their size and shape, showing a positive blue staining for the SA-β-gal activity. Endothelial senescence is reached by, at least, two different routes, including replicative or oxidative stress-induced. Both pathways involve the activation of p53 and are characterized by an increase in PAI-1 expression and the repression of Id1.

The characteristic and irreversible growth arrest observed in senescent cells occurs in the transition from phase G1 to phase S of the cell cycle and is dependent on the retinoblastoma family proteins, playing the tumour suppressor p53 a key role which senses the telomeric DNA damage (Wesierska-Gadek et al., 2005). In this transition, the abolition of p53 expression interferes with the senescence process that would be related to the low levels of PAI-1, one of the p53 target genes (Kortlever et al., 2008). Conversely, it is well known that the p53 overexpression or activation is able to arrest the cell cycle and launch the senescence program, suggesting that this process could be useful in cancer therapy (Chen & Goligorsky, 2006; Ewald et al., 2010; Rosso et al., 2006; Sugrue et al., 1997). Furthermore, it was demonstrated that the prolonged treatment with interferon (IFN)-γ induces cellular senescence in endothelial cells, involving cell cycle arrest and an up-regulation of p53 and p21 proteins cells (Kim et al., 2009).

Another TGF-β target protein that is associated with endothelial senescence is the helix-loop-helix (HLH) transcription factor Id1, or inhibitor of DNA binding 1. Id1 lacks a basic DNA-binding domain, but is able to form heterodimers with other HLH proteins, thereby inhibiting DNA binding, a process that is essential for cellular proliferation (Benezra et al., 1990). In epithelial cells, TGF-β induces the formation of a Smad3/ATF3 heteromeric complex that represses the Id1 expression and negatively regulates the cell cycle (Kang et al., 2003). Hence, the decrease in the Id1 expression is considered a biomarker of endothelial senescent cells (Tang et al., 2002).

3.2 Oxidative stress-induced senescence

Endothelial senescence can also be triggered by telomere-independent events that in general involve damages in the DNA. In this sense, the oxidative stress is a major stimulus for the induction of this type of senescence, which is due to the generation of reactive oxygen species (ROS, including oxygen ions and peroxides) in the mitochondria (Collins & Tzima, 2011; Erusalimsky & Skene, 2009). Thus, the cellular metabolism is the central source of ROS, but often they have an extracellular origin such as the one induced by radiation. In any case, ROS can either provoke or accelerate the development of senescence by damaging the DNA (Figure 2), which triggers multiple response mechanisms that usually act through the retinoblastoma protein family pathways, the final effectors of the senescence program (Campisi & d'Adda di Fagagna, 2007; Erusalimsky, 2009).

In cell culture, ROS induce an acute form of senescence termed stress-induced premature senescence, which does not require extensive cell culture but which resembles somehow the replicative one (Toussaint et al., 2000). This type of senescence is relatively easy to analyze in *in vitro* assays because the sole treatment with hydrogen peroxide (H_2O_2) for a short lapse of time is enough to prompt this type of senescence (Chen et al., 1998). By contrast, using antioxidant agents such as the grape stilbenoid resveratrol protect from the oxidative stress-induced premature senescence (Kao et al., 2010). Also, several lines of evidence show that ROS can interact and deplete the nitric oxide (NO) generated by the endothelium in the vasodilator responses, so contributing to the endothelial dysfunction associated to ageing (Grisham et al., 1998; Steiner et al., 2002). This is in line with the availability of NO-donors to inhibit endothelial cell senescence (Hayashi et al., 2006). In fact, comparing elderly with young adults one can find that the NO levels, or its bioavailability, are decreased in the first group but, interestingly, without any difference regarding to the expression levels or activation state of the endothelial nitric oxide synthase (eNOS), the enzyme responsible of the NO generation (Sun et al., 2004; Taddei et al., 2001). In parallel, this decrease in the NO levels attenuates the negative interference that it exerts on the TGF-β signalling pathway (Saura et al., 2005), which contributes to prompt the senescence program.

On the other hand, radiation is an exogenous trigger for ROS. In human skin fibroblasts, repeated exposure to ultraviolet-B light at subcytotoxic level is able to prompt premature senescence. Interestingly, this effect is mediated by the increase in the TGF-β expression and consequently by its downstream signalling pathway (Debacq-Chainiaux et al., 2005). In the vascular context, this source of ROS has been poorly studied beyond the methodological interest to induce premature senescence because endothelial cells enter rapidly in apoptosis due to their high sensitivity to radiation (Paris et al., 2001). In this regard, a recent study has demonstrated that ionizing radiation suppresses angiogenesis in mice and this effect is mediated through the TGF-β/ALK5-dependent inhibition of endothelial cell sprouting (Imaizumi et al., 2010).

4. Induction of S-endoglin and its role in endothelial senescence

The molecular changes involved or associated to the senescent program not only concern to the induction or repression of a specific set of genes. Many of the changes described in the literature report post-translational modifications, e. g., the advanced glycation endproducts

(AGEs) which have been implicated in age-related disease and aging itself; as well as the p53 acetylation in stress-induced senescence (Furukawa et al., 2007). In addition, a growing body of evidence supports the involvement of the post-transcriptional modifications that occur in senescence, i. e., the alternative splicing processes associated with senescence (Harries et al., 2011; Meshorer & Soreq, 2002). Thus, alterations in the splicing pattern have been described for several age-related diseases, such as the Hutchison Gilford progeria syndrome (Eriksson et al., 2003), or the Alzheimer's disease-related tauopathies (Chen et al., 2010).

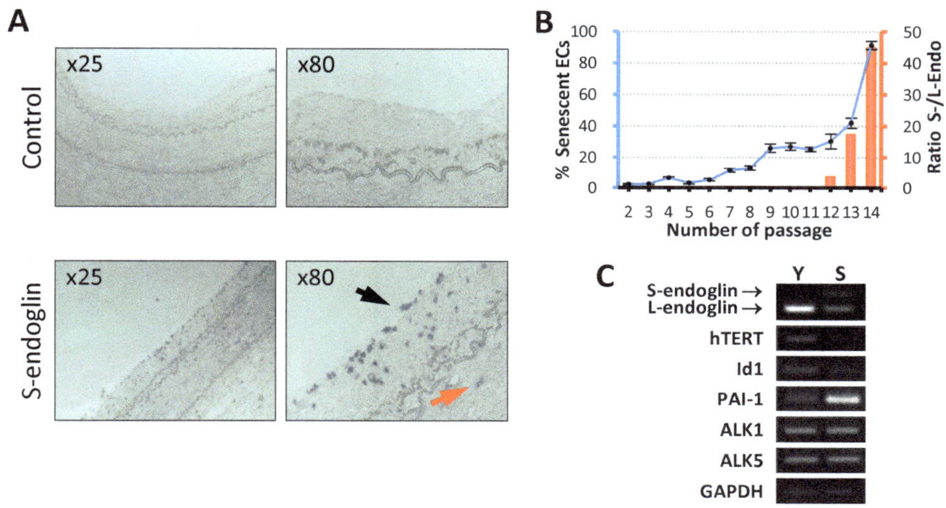

Fig. 3. S-endoglin expression in senescence. (A) The expression of S-endoglin in blood vessels can be revealed by *in situ* hybridization in the endothelium of human coronary artery (black arrow) and in some smooth muscle cells (red arrow). (B) The increase in the percentage of senescent endothelial cells *in vitro* (blue graph) is concomitant with the induction of S-endoglin (red graph). (C) Primary cultures of human umbilical vein endothelial cells (HUVECs) maintained *in vitro* along passages co-express both endoglin isoforms comparing young (Y) versus senescent (S) cells in RT-PCR assays. In parallel, PAI-1 is increased, while Id1 and telomerase (hTERT) are down-regulated in senescent cell. As a control, the expression levels of the TGF-β type I receptors ALK1 and ALK5 are not altered. (Figure adapted from Blanco et al., 2008).

Nonetheless, little is known about the role of splicing in the vascular context during senescence. A recent study demonstrates that TGF-β induces the distal splice-site selection leading to an antiangiogenic variant of the vascular endothelial growth factor (VEGF) (Nowak et al., 2008), and this could be one of the reasons why there is a reduced capability to form tubular-like structure by senescent endothelial *in vitro* (Chang et al., 2005).

As described above, the role of TGF-β in senescence has been clearly established, modulating specific intracellular effectors and leading to the cell growth arrest. In a first step, TGF-β binds to the specific receptor complex at the endothelial cell surface. Then, the signal is transmitted into the cytoplasm by different pathways depending on the type I receptor present in the complex. Thus, ALK5 signals via Smad2 and Smad3, whereas ALK1 mainly activates Smad1 and Smad5. In the TGF-β receptor complex, the presence of the predominantly expressed isoform, L-endoglin, favours the ALK1/Smad1 pathway and is related to the activation phase of the angiogenesis (Blanco et al., 2005; Lebrin et al., 2004). However, a post-transcriptional change during endothelial senescence, such as the retention of the last and small intron in the endoglin mRNA, has important consequences. Thus, the up-regulation of S-endoglin *in vitro* and *in vivo* is clearly associated with the ageing (Figures 3A and 3B). The co-expression of S- and L-endoglin in the senescent endothelial cells is able to tilt the angiogenic balance toward the resolution phase (ALK5/Smad3 pathway) in detriment of the ALK1/Smad1 route (Blanco et al., 2008). Also, S-endoglin induces the up-regulation of the PAI-1 and the repression of Id1, changes clearly associated to the cell cycle arrest in senescence (Figure 3C and 4).

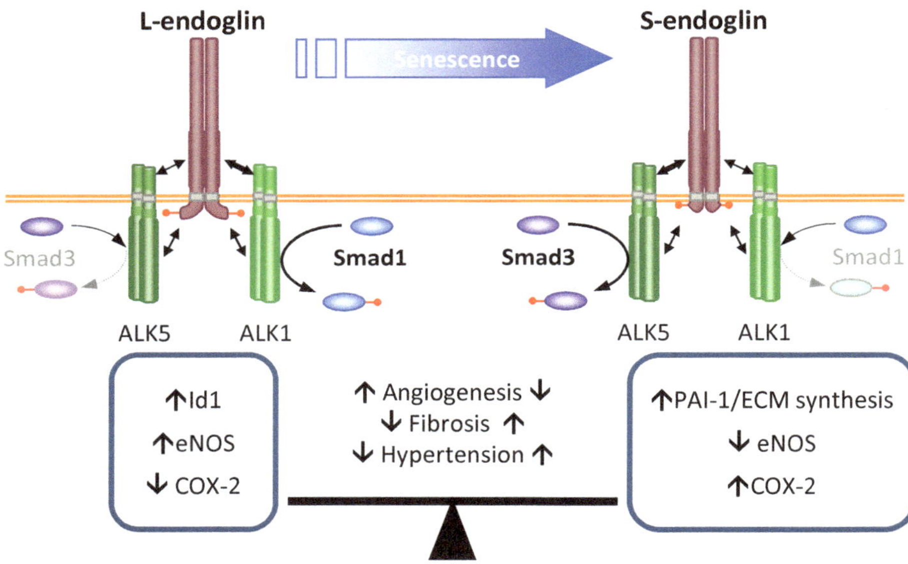

Fig. 4. Functional effects of S-endoglin in endothelial senescence. The S-endoglin up-regulation in aged endothelial cells promotes the ALK5/Smad3 signalling pathway. As a consequent, the vascular physiology is affected decreasing the angiogenesis, increasing the fibrosis and unbalancing the eNOS/COX-2 system which is related to hypertension. (Figure adapted from Blanco et al., 2008)

Furthermore, transgenic mice that overexpress the human S-endoglin isoform (*S-Eng*$^{+}$) experience a significant increase in the mean arterial pressure and a failure in the control on the NO-dependent vascular homeostasis, similarly to what happens in the endoglin deficient mouse model (*Eng*$^{+/-}$) that resembles the HHT disease (Blanco et al., 2008; Santibanez et al., 2007). Supporting this, a common compensatory mechanism takes place in *S-Eng*$^{+}$ and *Eng*$^{+/-}$ mice involving the up-regulation of the cyclooxigenase (COX)-2 enzyme

(Blanco et al., 2008; Jerkic et al., 2006). Taken together, the induction of S-endoglin during endothelial senescence might be at the basis of the development of cardiovascular pathologies associated with ageing, including atherosclerosis and hypertension (Figure 4).

4.1 Regulation of endoglin alternative splicing in senescence

Briefly, the alternative splicing is a molecular process by which organisms notably increase the diversity and functionality of their proteome from a finite number of genes. This process is carried out by the spliceosome, a huge ribonucleoprotein complex that works with amazing fidelity: i) skipping or shuffling exons; ii) selecting alternative splice sites; or iii) retaining introns (Graveley, 2001; Kwan et al., 2007). In humans, there are two distinct spliceosome complexes, named the major (M-Sp) and the minor (m-Sp) spliceosome. The M-Sp is involved in the vast majority of the splicing events and comprises five snRNPs named U1, U2, U4, U5, and U6 and a multitude of non-snRNP splicing factors (Jurica & Moore, 2003; Matlin et al., 2005; Zhou et al., 2002). Likewise, the m-Sp is composed by four unique snRNPs, U11, U12, U4atac, and U6atac, besides the U5 snRNP shared by both spliceosomes (Hall & Padgett, 1996; Tarn & Steitz, 1996). The m-Sp was first associated with the maturation of the so-called non-canonical introns but its role on standard splicing has been recently reported (Sheth et al., 2006; Will & Luhrmann, 2005). Interestingly, the difference between the major spliceosome and the minor spliceosome is their spatial segregation. While the M-Sp is in the nucleus, the m-Sp can be detected in the cytosol (Caceres & Misteli, 2007; Konig et al., 2007). In both cases, the spliceosome assembly is driven by a set of snRNPs that sequentially recognize the 5′ and 3′ splice sites, as well as the branch point element in between them (Burge et al., 1999). These snRNPs constitute the basal machinery of the spliceosome, besides a number of essential proteins that takes part in the spliceosome assembly. Moreover, there are several groups of auxiliary proteins that may regulate the alternative splicing. These splicing factors, or *trans*-elements, recognize binding sites, or *cis*-elements, spatially distributed inside the introns or exons and act as silencers or enhancers (Moore & Silver, 2008; Singh & Valcarcel, 2005; Sperling et al., 2008; Wang et al., 2006). Unfortunately, the alternative splicing during endothelial senescence has been poorly studied so far, but its importance has been suggested by the lifespan extension provoked by the overexpression of the splicing factor SNEV (Voglauer et al., 2006).

One of the best characterized groups of splicing factors is the serine/arginine (SR) protein family, from which the alternative splicing factor/splicing factor 2 (ASF/SF2) is the prototypical member (Graveley, 2000). ASF/SF2 is involved in both constitutive and alternative splicing processes. Although ASF/SF2 is mainly found in the nuclear speckles, it continuously shuttles between the nucleus and the cytoplasm depending on the phosphorylation and/or methylation states, which in turn determines its activity (Sanford et al., 2008; Sanford et al., 2005; Sinha et al., 2010). In this context, it has been recently reported the role of ASF/SF2 in the regulation of the S-endoglin intron retention during endothelial senescence (Blanco & Bernabeu, 2011). In endothelial senescent cells, the subcellular pattern of ASF/SF2 is mainly cytoplasmic, where ASF/SF2 interferes with the minor spliceosome inhibiting the elimination of the last intron of endoglin mRNA. The role of cytoplasmic ASF/SF2 as a senescent inductor is supported by its antiangiogenic properties, because the inhibition of the ASF/SF2 phosphorylation promotes its cytoplasmic localization and this is associated with increased expression levels of the antiangiogenic isoform VEGF165b (Nowak et al., 2010).

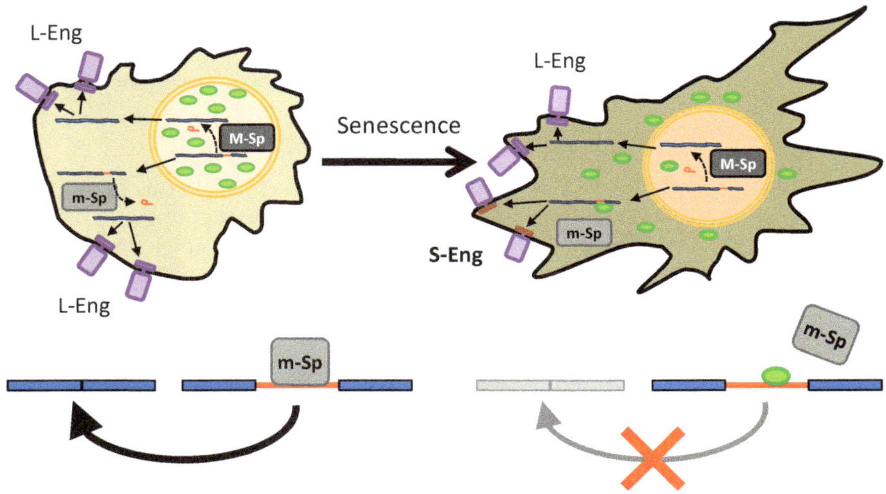

Fig. 5. Regulation of the alternative splicing of endoglin in senescent endothelial cells. In this hypothetical model, the last intron of the *ENG* gene is eliminated in the mature mRNA, so that L-endoglin is the predominantly expressed isoform. In this mRNA processing, both spliceosomes (nuclear M-Sp and cytoplasmic m-Sp) can be involved. However, in senescent endothelial cells, the splicing factor ASF/SF2 (green) is translocated to the cytoplasm, stabilizing the S-endoglin mRNA and interfering with the m-Sp activity. Consequently, ASF/SF2 promotes the intron retention, thus up-regulating the levels of S-endoglin mRNA (adapted from Blanco & Bernabeu, 2011).

5. Conclusions

Vascular physiology progressively declines with age due to multiple factors including an increase in oxidative stress, DNA damage, and advanced cellular replication involving telomere attrition. All these events converge in the key molecule p53, which acts typically arresting the cell cycle and triggering the endothelial senescence. At this stage, the expression of many specific genes is modulated, regarding not only to their expression levels but also the post-translational modifications and alternative processing of their premature mRNA molecules, which give rise to interesting protein variants. Nowadays, it can be postulated that this phenomenon is at the cellular basis of several age-associated cardiovascular pathologies, such as hypertension or atherosclerosis.

TGF-β is able to induce endothelial senescence via a cell surface receptor complex that includes the type I (ALK1 and ALK5) and the type II signalling receptors as well as endoglin. Endoglin is a TGF-β co-receptor highly expressed as L-(long)-endoglin by endothelial cells which is associated with active angiogenesis foci and vascular remodelling processes. Conversely, an alternative spliced and shorter isoform (S-endoglin) with opposite effects to those of L-endoglin in the context of the TGF-β system has been described. Usually, S-endoglin is almost undetectable in endothelial cells, but is induced during senescence. In this up-regulation, the senescence-induced cytoplasmic localization of the splicing factor ASF/SF2 plays a key role favouring the retention of the intron between exons

#13 and #14. Thus, the up-regulated expression of S-endoglin is considered to be part of the endothelial senescence program. Moreover, *in vitro* and *in vivo* studies suggest that S-endoglin contributes to vascular pathology associated with ageing. In this regard, mutations in the human *ENG* gene are responsible for HHT-1, an autosomic dominant vascular disease whose symptoms increase and become worse with age. Currently, the haploinsufficiency of the predominantly expressed L-endoglin isoform is widely accepted as the pathogenic mechanism of the disease. Because S-endoglin is up-regulated in aged mice as well as during senescence of endothelial cells and S-endoglin counteracts the function of L-endoglin, the increased S-endoglin expression during ageing would increase the functional L-endoglin haploinsufficiency in HHT-1 and could explain why the symptoms become worse with ageing. Therefore, one could predict that the age-dependent penetrance of the HHT-1 is due, at least in part, to the S-endoglin induction mediated by ASF/SF2.

In summary, these data suggest an important role for the TGF-β co-receptor endoglin as a modulator of the vascular pathology associated with endothelial senescence.

6. Acknowledgments

This work was supported by grants of the Spanish Ministry of Science and Innovation to CB (SAF2010-19222) and *Genoma España* (MEICA). FJB is a post-doctoral researcher of the *Centro de Investigación Biomédica en Red* (CIBER) *de Enfermedades Raras*, ISCIII, Spain. CB is a research professor in the *Centro de Investigaciones Biológicas*, CSIC, Spain.

7. References

Bellon, T.; Corbi, A.; Lastres, P.; Cales, C.; Cebrian, M.; Vera, S.; Cheifetz, S.; Massague, J.; Letarte, M. & Bernabeu, C. (1993). Identification and expression of two forms of the human transforming growth factor-beta-binding protein endoglin with distinct cytoplasmic regions. *European Journal of Immunology*, Vol. 23, No. 9, (September 1993), pp. 2340-2345, ISSN 0014-2980.

Benezra, R.; Davis, R.L.; Lockshon, D.; Turner, D.L. & Weintraub, H. (1990). The protein Id: a negative regulator of helix-loop-helix DNA binding proteins. *Cell*, Vol. 61, No. 1, (April 1990), pp. 49-59, ISSN 0092-8674.

Bernabeu, C.; Lopez-Novoa, J.M. & Quintanilla, M. (2009). The emerging role of TGF-beta superfamily coreceptors in cancer. *Biochimica et Biophysica Acta*, Vol. 1792, No. 10, (July 2009), pp. 954-973, ISSN 0006-3002.

Blanco, F.J. & Bernabeu, C. (2011). Alternative splicing factor or splicing factor-2 plays a key role in intron retention of the endoglin gene during endothelial senescence. *Aging Cell*, Vol. 10, No. 5, (June 2011), pp. 896-907, ISSN 1474-9726.

Blanco, F.J.; Grande, M.T.; Langa, C.; Oujo, B.; Velasco, S.; Rodriguez-Barbero, A.; Perez-Gomez, E.; Quintanilla, M.; Lopez-Novoa, J.M. & Bernabeu, C. (2008). S-endoglin expression is induced in senescent endothelial cells and contributes to vascular pathology. *Circulation Research*, Vol. 103, No. 12, (November 2008), pp. 1383-1392, ISSN 1524-4571.

Blanco, F.J.; Santibanez, J.F.; Guerrero-Esteo, M.; Langa, C.; Vary, C.P. & Bernabeu, C. (2005). Interaction and functional interplay between endoglin and ALK-1, two components of the endothelial transforming growth factor-beta receptor complex. *Journal of Cellular Physiology*, Vol. 204, No. 2, (February 2005), pp. 574-584, ISSN 0021-9541.

Brandes, R.P.; Fleming, I. & Busse, R. (2005). Endothelial aging. *Cardiovascular Research*, Vol. 66, No. 2, (May 2005), pp. 286-294, ISSN 0008-6363.

Burge, C.; Tuschl, T. & Sharp, P. (1999). Splicing of precursors to mRNAs by the spliceosoomes, In: *The RNA World*. R. Gesteland, T. Cech, and J. Atkins (eds.), pp. 525-560. Cold Spring Harbor Laboratory Press, ISBN 0-87969-589-7, Plainview, NY, United States of America.

Caceres, J.F. & Misteli, T. (2007). Division of labor: minor splicing in the cytoplasm. *Cell*, Vol. 131, No. 4, (November 2007), pp. 645-647, ISSN 0092-8674.

Campisi, J. & d'Adda di Fagagna, F. (2007). Cellular senescence: when bad things happen to good cells. *Nature Reviews Molecular Cell Biology*, Vol. 8, No. 9, (August 2007), pp. 729-740, ISSN 1471-0080.

Carmeliet, P. & Jain, R.K. (2011). Molecular mechanisms and clinical applications of angiogenesis. *Nature*, Vol. 473, No. 7347, (May 2011), pp. 298-307, ISSN 1476-4687.

Cipriano, R.; Kan, C.E.; Graham, J.; Danielpour, D.; Stampfer, M. & Jackson, M.W. (2011). TGF-beta signaling engages an ATM-CHK2-p53-independent RAS-induced senescence and prevents malignant transformation in human mammary epithelial cells. *Proceedings of the National Academy of Sciences of the United States of America*, Vol. 108, No. 21, (May 2011), pp. 8668-8673, ISSN 1091-6490.

Collins, C. & Tzima, E. (2011). Hemodynamic forces in endothelial dysfunction and vascular aging. *Experimental Gerontology*, Vol. 46, No. 2-3, (October 2010), pp. 185-188, ISSN 1873-6815.

Comi, P.; Chiaramonte, R. & Maier, J.A. (1995). Senescence-dependent regulation of type 1 plasminogen activator inhibitor in human vascular endothelial cells. *Experimental Cell Research*, Vol. 219, No. 1, (July 1995), pp. 304-308, ISSN 0014-4827.

Conway, E.M. & Carmeliet, P. (2004). The diversity of endothelial cells: a challenge for therapeutic angiogenesis. *Genome Biology*, Vol. 5, No. 2, (January 2004), pp. 207, ISSN 1465-6914.

Chang, M.W.; Grillari, J.; Mayrhofer, C.; Fortschegger, K.; Allmaier, G.; Marzban, G.; Katinger, H. & Voglauer, R. (2005). Comparison of early passage, senescent and hTERT immortalized endothelial cells. *Experimental Cell Research*, Vol. 309, No. 1, (June 2005), pp. 121-136, ISSN 0014-4827.

Chen, J. & Goligorsky, M.S. (2006). Premature senescence of endothelial cells: Methusaleh's dilemma. *American Journal of Physiology. Heart and Circulatory Physiology*, Vol. 290, No. 5, (April 2006), pp. H1729-1739, ISSN 0363-6135.

Chen, Q.M.; Bartholomew, J.C.; Campisi, J.; Acosta, M.; Reagan, J.D. & Ames, B.N. (1998). Molecular analysis of H2O2-induced senescent-like growth arrest in normal human fibroblasts: p53 and Rb control G1 arrest but not cell replication. *Biochemical Journal*, Vol. 332 (Pt 1), (May 1998), pp. 43-50, ISSN 0264-6021.

Chen, S.; Townsend, K.; Goldberg, T.E.; Davies, P. & Conejero-Goldberg, C. (2010). MAPT isoforms: differential transcriptional profiles related to 3R and 4R splice variants. *Journal of Alzheimer's Disease*, Vol. 22, No. 4, (October 2010), pp. 1313-1329, ISSN 1875-8908.

Debacq-Chainiaux, F.; Borlon, C.; Pascal, T.; Royer, V.; Eliaers, F.; Ninane, N.; Carrard, G.; Friguet, B.; de Longueville, F.; Boffe, S.; Remacle, J. & Toussaint, O. (2005).

Repeated exposure of human skin fibroblasts to UVB at subcytotoxic level triggers premature senescence through the TGF-beta1 signaling pathway. *Journal of Cell Science*, Vol. 118, No. 4, (January 2005), pp. 743-758, ISSN 0021-9533.

Dimri, G.P.; Lee, X.; Basile, G.; Acosta, M.; Scott, G.; Roskelley, C.; Medrano, E.E.; Linskens, M.; Rubelj, I.; Pereira-Smith, O. & et al. (1995). A biomarker that identifies senescent human cells in culture and in aging skin in vivo. *Proceedings of the National Academy of Sciences of the United States of America*, Vol. 92, No. 20, (September 1995), pp. 9363-9367, ISSN 0027-8424.

Eriksson, M.; Brown, W.T.; Gordon, L.B.; Glynn, M.W.; Singer, J.; Scott, L.; Erdos, M.R.; Robbins, C.M.; Moses, T.Y.; Berglund, P.; Dutra, A.; Pak, E.; Durkin, S.; Csoka, A.B.; Boehnke, M.; Glover, T.W. & Collins, F.S. (2003). Recurrent de novo point mutations in lamin A cause Hutchinson-Gilford progeria syndrome. *Nature*, Vol. 423, No. 6937, (April 2003), pp. 293-298, ISSN 0028-0836.

Erusalimsky, J.D. (2009). Vascular endothelial senescence: from mechanisms to pathophysiology. *Journal of Applied Physiology*, Vol. 106, No. 1, (November 2008), pp. 326-332, ISSN 8750-7587.

Erusalimsky, J.D. & Kurz, D.J. (2005). Cellular senescence in vivo: its relevance in ageing and cardiovascular disease. *Experimental Gerontology*, Vol. 40, No. 8-9, (June 2005), pp. 634-642, ISSN 0531-5565.

Erusalimsky, J.D. & Skene, C. (2009). Mechanisms of endothelial senescence. *Experimental Physiology*, Vol. 94, No. 3, (October 2008), pp. 299-304, ISSN 1469-445X.

Ewald, J.A.; Desotelle, J.A.; Wilding, G. & Jarrard, D.F. (2010). Therapy-induced senescence in cancer. *Journal of the National Cancer Institute*, Vol. 102, No. 20, (September 2010), pp. 1536-1546, ISSN 1460-2105.

Ferrari, A.U.; Radaelli, A. & Centola, M. (2003). Invited review: aging and the cardiovascular system. *Journal of Applied Physiology*, Vol. 95, No. 6, (December 2003), pp. 2591-2597, ISSN 8750-7587.

Folini, M.; Venturini, L.; Cimino-Reale, G. & Zaffaroni, N. (2011). Telomeres as targets for anticancer therapies. *Expert Opinion on Therapeutic Targets*, Vol. 15, No. 5, (February 2011), pp. 579-593, ISSN 1744-7631.

Foreman, K.E. & Tang, J. (2003). Molecular mechanisms of replicative senescence in endothelial cells. *Experimental Gerontology*, Vol. 38, No. 11-12, (November 2003), pp. 1251-1257, ISSN 0531-5565.

Foteinos, G.; Hu, Y.; Xiao, Q.; Metzler, B. & Xu, Q. (2008). Rapid endothelial turnover in atherosclerosis-prone areas coincides with stem cell repair in apolipoprotein E-deficient mice. *Circulation*, Vol. 117, No. 14, (April 2008), pp. 1856-1863, ISSN 1524-4539.

Fraisl, P.; Mazzone, M.; Schmidt, T. & Carmeliet, P. (2009). Regulation of angiogenesis by oxygen and metabolism. *Developmental Cell*, Vol. 16, No. 2, (February 2009), pp. 167-179, ISSN 1878-1551.

Fujiki, T.; Miura, T.; Maura, M.; Shiraishi, H.; Nishimura, S.; Imada, Y.; Uehara, N.; Tashiro, K.; Shirahata, S. & Katakura, Y. (2007). TAK1 represses transcription of the human telomerase reverse transcriptase gene. *Oncogene*, Vol. 26, No. 36, (February 2007), pp. 5258-5266, ISSN 0950-9232.

Furukawa, A.; Tada-Oikawa, S.; Kawanishi, S. & Oikawa, S. (2007). H2O2 accelerates cellular senescence by accumulation of acetylated p53 via decrease in the function of SIRT1 by NAD+ depletion. *Cellular Physiology and Biochemistry*, Vol. 20, No. 1-4, (June 2007), pp. 45-54, ISSN 1015-8987.

Ghosh, A.K. & Vaughan, D.E. (2011). PAI-1 in Tissue Fibrosis. *Journal of Cellular Physiology*, (April 2011), DOI: 10.1002/jcp.22783, ISSN 1097-4652.

Graveley, B.R. (2000). Sorting out the complexity of SR protein functions. *RNA*, Vol. 6, No. 9, (September 2000), pp. 1197-1211, ISSN 1355-8382.

Graveley, B.R. (2001). Alternative splicing: increasing diversity in the proteomic world. *Trends in Genetics*, Vol. 17, No. 2, (February 2001), pp. 100-107, ISSN 0168-9525.

Grisham, M.B.; Granger, D.N. & Lefer, D.J. (1998). Modulation of leukocyte-endothelial interactions by reactive metabolites of oxygen and nitrogen: relevance to ischemic heart disease. *Free Radical Biology & Medicine*, Vol. 25, No. 4-5, (September 1998), pp. 404-433, ISSN 0891-5849.

Hall, S.L. & Padgett, R.A. (1996). Requirement of U12 snRNA for in vivo splicing of a minor class of eukaryotic nuclear pre-mRNA introns. *Science*, Vol. 271, No. 5256, (March 1996), pp. 1716-1718, ISSN 0036-8075.

Harries, L.W.; Hernandez, D.; Henley, W.; Wood, A.R.; Holly, A.C.; Bradley-Smith, R.M.; Yaghootkar, H.; Dutta, A.; Murray, A.; Frayling, T.M.; Guralnik, J.M.; Bandinelli, S.; Singleton, A.; Ferrucci, L. & Melzer, D. (2011). Human aging is characterized by focused changes in gene expression and deregulation of alternative splicing. *Aging Cell*, Vol. 10, No. 5, (June 2011), pp. 868-878, ISSN 1474-9726.

Hayashi, T.; Matsui-Hirai, H.; Miyazaki-Akita, A.; Fukatsu, A.; Funami, J.; Ding, Q.F.; Kamalanathan, S.; Hattori, Y.; Ignarro, L.J. & Iguchi, A. (2006). Endothelial cellular senescence is inhibited by nitric oxide: implications in atherosclerosis associated with menopause and diabetes. *Proceedings of the National Academy of Sciences of the United States of America*, Vol. 103, No. 45, (November 2006), pp. 17018-17023, ISSN 0027-8424.

Hayflick, L. (2003). Living forever and dying in the attempt. *Experimental Gerontology*, Vol. 38, No. 11-12, (December 2003), pp. 1231-1241, ISSN 0531-5565.

Imaizumi, N.; Monnier, Y.; Hegi, M.; Mirimanoff, R.O. & Ruegg, C. (2010). Radiotherapy suppresses angiogenesis in mice through TGF-betaRI/ALK5-dependent inhibition of endothelial cell sprouting. *PLoS One*, Vol. 5, No. 6, (June 2010), pp. e11084, ISSN 1932-6203.

Jerkic, M.; Rivas-Elena, J.V.; Santibanez, J.F.; Prieto, M.; Rodriguez-Barbero, A.; Perez-Barriocanal, F.; Pericacho, M.; Arevalo, M.; Vary, C.P.; Letarte, M.; Bernabeu, C. & Lopez-Novoa, J.M. (2006). Endoglin regulates cyclooxygenase-2 expression and activity. *Circulation Research*, Vol. 99, No. 3, (July 2006), pp. 248-256, ISSN 1524-4571.

Jurica, M.S. & Moore, M.J. (2003). Pre-mRNA splicing: awash in a sea of proteins. *Molecular Cell*, Vol. 12, No. 1, (July 2003), pp. 5-14, ISSN 1097-2765.

Kang, J.S.; Liu, C. & Derynck, R. (2009). New regulatory mechanisms of TGF-beta receptor function. *Trends in Cell Biology*, Vol. 19, No. 8, (August 2009), pp. 385-394, ISSN 1879-3088.

Kang, Y.; Chen, C.R. & Massague, J. (2003). A self-enabling TGFbeta response coupled to stress signaling: Smad engages stress response factor ATF3 for Id1 repression in epithelial cells. *Molecular Cell*, Vol. 11, No. 4, (April 2003), pp. 915-926, ISSN 1097-2765.

Kao, C.L.; Chen, L.K.; Chang, Y.L.; Yung, M.C.; Hsu, C.C.; Chen, Y.C.; Lo, W.L.; Chen, S.J.; Ku, H.H. & Hwang, S.J. (2010). Resveratrol protects human endothelium from H(2)O(2)-induced oxidative stress and senescence via SirT1 activation. *Journal of Atherosclerosis and Thrombosis*, Vol. 17, No. 9, (July 2010), pp. 970-979, ISSN 1880-3873.

Kim, K.S.; Kang, K.W.; Seu, Y.B.; Baek, S.H. & Kim, J.R. (2009). Interferon-gamma induces cellular senescence through p53-dependent DNA damage signaling in human endothelial cells. *Mechanisms of Ageing and Development*, Vol. 130, No. 3, (December 2008), pp. 179-188, ISSN 0047-6374.

Koleva, R.I.; Conley, B.A.; Romero, D.; Riley, K.S.; Marto, J.A.; Lux, A. & Vary, C.P. (2006). Endoglin structure and function: Determinants of endoglin phosphorylation by transforming growth factor-beta receptors. *Journal of Biological Chemistry*, Vol. 281, No. 35, (June 2006), pp. 25110-25123, ISSN 0021-9258.

Konig, H.; Matter, N.; Bader, R.; Thiele, W. & Muller, F. (2007). Splicing segregation: the minor spliceosome acts outside the nucleus and controls cell proliferation. *Cell*, Vol. 131, No. 4, (November 2007), pp. 718-729, ISSN 0092-8674.

Kordon, E.C.; McKnight, R.A.; Jhappan, C.; Hennighausen, L.; Merlino, G. & Smith, G.H. (1995). Ectopic TGF beta 1 expression in the secretory mammary epithelium induces early senescence of the epithelial stem cell population. *Developmental Biology*, Vol. 168, No. 1, (March 1995), pp. 47-61, ISSN 0012-1606.

Kortlever, R.M.; Nijwening, J.H. & Bernards, R. (2008). Transforming growth factor-beta requires its target plasminogen activator inhibitor-1 for cytostatic activity. *Journal of Biological Chemistry*, Vol. 283, No. 36, (July 2008), pp. 24308-24313, ISSN 0021-9258.

Kwan, T.; Benovoy, D.; Dias, C.; Gurd, S.; Serre, D.; Zuzan, H.; Clark, T.A.; Schweitzer, A.; Staples, M.K.; Wang, H.; Blume, J.E.; Hudson, T.J.; Sladek, R. & Majewski, J. (2007). Heritability of alternative splicing in the human genome. *Genome Research*, Vol. 17, No. 8, (August 2007), pp. 1210-1218, ISSN 1088-9051.

Lareau, L.F.; Green, R.E.; Bhatnagar, R.S. & Brenner, S.E. (2004). The evolving roles of alternative splicing. *Current Opinion in Structural Biology*, Vol. 14, No. 3, (June 2004), pp. 273-282, ISSN 0959-440X.

Lebrin, F.; Goumans, M.J.; Jonker, L.; Carvalho, R.L.; Valdimarsdottir, G.; Thorikay, M.; Mummery, C.; Arthur, H.M. & ten Dijke, P. (2004). Endoglin promotes endothelial cell proliferation and TGF-beta/ALK1 signal transduction. *The EMBO Journal*, Vol. 23, No. 20, (September 2004), pp. 4018-4028, ISSN 0261-4189.

Li, H. & Liu, J.P. (2007). Mechanisms of action of TGF-beta in cancer: evidence for Smad3 as a repressor of the hTERT gene. *Annals of the New York Academy of Sciences*, Vol. 1114, (October 2007), pp. 56-68, ISSN 0077-8923.

Li, H.; Xu, D.; Li, J.; Berndt, M.C. & Liu, J.P. (2006). Transforming growth factor beta suppresses human telomerase reverse transcriptase (hTERT) by Smad3 interactions with c-Myc and the hTERT gene. *Journal of Biological Chemistry*, Vol. 281, No. 35, (June 2006), pp. 25588-25600, ISSN 0021-9258.

Lopez-Novoa, J.M. & Bernabeu, C. (2010). The physiological role of endoglin in the cardiovascular system. *American Journal of Physiology. Heart and Circulatory Physiology*, Vol. 299, No. 4, (July 2010), pp. H959-974, ISSN 1522-1539.

Llorca, O.; Trujillo, A.; Blanco, F.J. & Bernabeu, C. (2007). Structural model of human endoglin, a transmembrane receptor responsible for hereditary hemorrhagic telangiectasia. *Journal of Molecular Biology*, Vol. 365, No. 3, (November 2006), pp. 694-705, ISSN 0022-2836.

Mahmoud, M.; Allinson, K.R.; Zhai, Z.; Oakenfull, R.; Ghandi, P.; Adams, R.H.; Fruttiger, M. & Arthur, H.M. (2010). Pathogenesis of arteriovenous malformations in the absence of endoglin. *Circulation Research*, Vol. 106, No. 8, (March 2010), pp. 1425-1433, ISSN 1524-4571.

Massague, J.; Seoane, J. & Wotton, D. (2005). Smad transcription factors. *Genes & Development*, Vol. 19, No. 23, (December 2005), pp. 2783-2810, ISSN 0890-9369.

Matlin, A.J.; Clark, F. & Smith, C.W. (2005). Understanding alternative splicing: towards a cellular code. *Nature Reviews Molecular Cell Biology*, Vol. 6, No. 5, (June 2005), pp. 386-398, ISSN 1471-0072.

McAllister, K.A.; Grogg, K.M.; Johnson, D.W.; Gallione, C.J.; Baldwin, M.A.; Jackson, C.E.; Helmbold, E.A.; Markel, D.S.; McKinnon, W.C.; Murrell, J. & et al. (1994). Endoglin, a TGF-beta binding protein of endothelial cells, is the gene for hereditary haemorrhagic telangiectasia type 1. *Nature Genetics*, Vol. 8, No. 4, (December 1994), pp. 345-351, ISSN 1061-4036.

Meshorer, E. & Soreq, H. (2002). Pre-mRNA splicing modulations in senescence. *Aging Cell*, Vol. 1, No. 1, (October 2002), pp. 10-16, ISSN 1474-9718.

Meurer, S.K.; Tihaa, L.; Borkham-Kamphorst, E. & Weiskirchen, R. (2011). Expression and functional analysis of endoglin in isolated liver cells and its involvement in fibrogenic Smad signalling. *Cellular Signalling*, Vol. 23, No. 4, (December 2010), pp. 683-699, ISSN 1873-3913.

Minamino, T. & Komuro, I. (2007). Vascular cell senescence: contribution to atherosclerosis. *Circulation Research*, Vol. 100, No. 1, (January 2007), pp. 15-26, ISSN 1524-4571.

Minamino, T. & Komuro, I. (2008). Role of telomeres in vascular senescence. *Frontiers in Bioscience*, Vol. 13, (January 2008), pp. 2971-2979, ISSN 1093-4715.

Minamino, T.; Miyauchi, H.; Yoshida, T.; Tateno, K.; Kunieda, T. & Komuro, I. (2004). Vascular cell senescence and vascular aging. *Journal of Molecular and Cellular Cardiology*, Vol. 36, No. 2, (February 2004), pp. 175-183, ISSN 0022-2828.

Moore, M.J. & Silver, P.A. (2008). Global analysis of mRNA splicing. *RNA*, Vol. 14, No. 2, (December 2007), pp. 197-203, ISSN 1469-9001.

Nott, A.; Meislin, S.H. & Moore, M.J. (2003). A quantitative analysis of intron effects on mammalian gene expression. *RNA*, Vol. 9, No. 5, (April 2003), pp. 607-617, ISSN 1355-8382.

Nowak, D.G.; Amin, E.M.; Rennel, E.S.; Hoareau-Aveilla, C.; Gammons, M.; Damodoran, G.; Hagiwara, M.; Harper, S.J.; Woolard, J.; Ladomery, M.R. & Bates, D.O. (2010). Regulation of vascular endothelial growth factor (VEGF) splicing from pro-angiogenic to anti-angiogenic isoforms: a novel therapeutic strategy for angiogenesis. *Journal of Biological Chemistry*, Vol. 285, No. 8, (November 2009), pp. 5532-5540, ISSN 1083-351X.

Nowak, D.G.; Woolard, J.; Amin, E.M.; Konopatskaya, O.; Saleem, M.A.; Churchill, A.J.; Ladomery, M.R.; Harper, S.J. & Bates, D.O. (2008). Expression of pro- and anti-angiogenic isoforms of VEGF is differentially regulated by splicing and growth factors. *Journal of Cell Science*, Vol. 121, No. Pt 20, (October 2008), pp. 3487-3495, ISSN 0021-9533.

Pardali, E.; Goumans, M.J. & ten Dijke, P. (2010). Signaling by members of the TGF-beta family in vascular morphogenesis and disease. *Trends in Cell Biology*, Vol. 20, No. 9, (September 2010), pp. 556-567, ISSN 1879-3088.

Paris, F.; Fuks, Z.; Kang, A.; Capodieci, P.; Juan, G.; Ehleiter, D.; Haimovitz-Friedman, A.; Cordon-Cardo, C. & Kolesnick, R. (2001). Endothelial apoptosis as the primary lesion initiating intestinal radiation damage in mice. *Science*, Vol. 293, No. 5528, (July 2001), pp. 293-297, ISSN 0036-8075.

Pascual, G.; Mendieta, C.; Garcia-Honduvilla, N.; Corrales, C.; Bellon, J.M. & Bujan, J. (2007). TGF-beta1 upregulation in the aging varicose vein. *Journal of Vascular Research*, Vol. 44, No. 3, (March 2007), pp. 192-201, ISSN 1018-1172.

Perez-Gomez, E.; Del Castillo, G.; Santibanez, J.F.; Lopez-Novoa, J.M.; Bernabeu, C. & Quintanilla, M. (2010). The role of the TGF-beta coreceptor endoglin in cancer. *The Scientific World Journal*, Vol. 10, (December 2010), pp. 2367-2384, ISSN 1537-744X.

Perez-Gomez, E.; Eleno, N.; Lopez-Novoa, J.M.; Ramirez, J.R.; Velasco, B.; Letarte, M.; Bernabeu, C. & Quintanilla, M. (2005). Characterization of murine S-endoglin isoform and its effects on tumor development. *Oncogene*, Vol. 24, No. 27, (April 2005), pp. 4450-4461, ISSN 0950-9232.

Rivard, A.; Fabre, J.E.; Silver, M.; Chen, D.; Murohara, T.; Kearney, M.; Magner, M.; Asahara, T. & Isner, J.M. (1999). Age-dependent impairment of angiogenesis. *Circulation*, Vol. 99, No. 1, (January 1999), pp. 111-120, ISSN 0009-7322.

Rodríguez-Mañas, L.; El-Assar, M.; Vallejo, S.; López-Dóriga, P.; Solís, J.; Petidier, R.; Montes, M.; Nevado, J.; Castro, M.; Gómez-Guerrero, C.; Peiró, C. & Sánchez-Ferrer, C.F. (2009). Endothelial dysfunction in aged humans is related with oxidative stress and vascular inflammation. *Aging Cell*, Vol. 8, No. 3, (June 2009), pp. 226-238, ISSN 1474-9726.

Rosso, A.; Balsamo, A.; Gambino, R.; Dentelli, P.; Falcioni, R.; Cassader, M.; Pegoraro, L.; Pagano, G. & Brizzi, M.F. (2006). p53 Mediates the accelerated onset of senescence of endothelial progenitor cells in diabetes. *Journal of Biological Chemistry*, Vol. 281, No. 7, (December 2005), pp. 4339-4347, ISSN 0021-9258.

Sakabe, N.J. & de Souza, S.J. (2007). Sequence features responsible for intron retention in human. *BMC Genomics*, Vol. 8, (February 2007), pp. 59, ISSN 1471-2164.

Sanford, J.R.; Coutinho, P.; Hackett, J.A.; Wang, X.; Ranahan, W. & Caceres, J.F. (2008). Identification of nuclear and cytoplasmic mRNA targets for the shuttling protein SF2/ASF. *PLoS One*, Vol. 3, No. 10, (October 2008), pp. e3369, ISSN 1932-6203.

Sanford, J.R.; Ellis, J.D.; Cazalla, D. & Caceres, J.F. (2005). Reversible phosphorylation differentially affects nuclear and cytoplasmic functions of splicing factor 2/alternative splicing factor. *Proceedings of the National Academy of Sciences of the United States of America*, Vol. 102, No. 42, (October 2005), pp. 15042-15047, ISSN 0027-8424.

Santibanez, J.F.; Letamendia, A.; Perez-Barriocanal, F.; Silvestri, C.; Saura, M.; Vary, C.P.; Lopez-Novoa, J.M.; Attisano, L. & Bernabeu, C. (2007). Endoglin increases eNOS expression by modulating Smad2 protein levels and Smad2-dependent TGF-beta signaling. *Journal of Cellular Physiology*, Vol. 210, No. 2, (October 2006), pp. 456-468, ISSN 0021-9541.

Saura, M.; Zaragoza, C.; Herranz, B.; Griera, M.; Diez-Marques, L.; Rodriguez-Puyol, D. & Rodriguez-Puyol, M. (2005). Nitric oxide regulates transforming growth factor-beta signaling in endothelial cells. *Circulation Research*, Vol. 97, No. 11, (October 2005), pp. 1115-1123, ISSN 1524-4571.

Schneiderman, J.; Sawdey, M.S.; Keeton, M.R.; Bordin, G.M.; Bernstein, E.F.; Dilley, R.B. & Loskutoff, D.J. (1992). Increased type 1 plasminogen activator inhibitor gene expression in atherosclerotic human arteries. *Proceedings of the National Academy of Sciences of the United States of America*, Vol. 89, No. 15, (August 1992), pp. 6998-7002, ISSN 0027-8424.

Shay, J.W. & Wright, W.E. (2007). Hallmarks of telomeres in ageing research. *Journal of Pathology*, Vol. 211, No. 2, (January 2007), pp. 114-123, ISSN 0022-3417.

Sheth, N.; Roca, X.; Hastings, M.L.; Roeder, T.; Krainer, A.R. & Sachidanandam, R. (2006). Comprehensive splice-site analysis using comparative genomics. *Nucleic Acids Research*, Vol. 34, No. 14, (August 2006), pp. 3955-3967, ISSN 1362-4962.

Shovlin, C.L. (2010). Hereditary haemorrhagic telangiectasia: pathophysiology, diagnosis and treatment. *Blood Reviews*, Vol. 24, No. 6, (September 2010), pp. 203-219, ISSN 1532-1681.

Singh, R. & Valcarcel, J. (2005). Building specificity with nonspecific RNA-binding proteins. *Nature Structure Molecular Biology*, Vol. 12, No. 8, (August 2005), pp. 645-653, ISSN 1545-9993.

Sinha, R.; Allemand, E.; Zhang, Z.; Karni, R.; Myers, M.P. & Krainer, A.R. (2010). Arginine methylation controls the subcellular localization and functions of the oncoprotein splicing factor SF2/ASF. *Molecular Cell Biology*, Vol. 30, No. 11, (March 2010), pp. 2762-2774, ISSN 1098-5549.

Sperling, J.; Azubel, M. & Sperling, R. (2008). Structure and function of the Pre-mRNA splicing machine. *Structure*, Vol. 16, No. 11, (November 2008), pp. 1605-1615, ISSN 0969-2126.

Steiner, D.R.; Gonzalez, N.C. & Wood, J.G. (2002). Interaction between reactive oxygen species and nitric oxide in the microvascular response to systemic hypoxia. *Journal of Applied Physiology*, Vol. 93, No. 4, (September 2002), pp. 1411-1418, ISSN 8750-7587.

Sugrue, M.M.; Shin, D.Y.; Lee, S.W. & Aaronson, S.A. (1997). Wild-type p53 triggers a rapid senescence program in human tumor cells lacking functional p53. *Proceedings of the National Academy of Sciences of the United States of America*, Vol. 94, No. 18, (September 1997), pp. 9648-9653, ISSN 0027-8424.

Sun, D.; Huang, A.; Yan, E.H.; Wu, Z.; Yan, C.; Kaminski, P.M.; Oury, T.D.; Wolin, M.S. & Kaley, G. (2004). Reduced release of nitric oxide to shear stress in mesenteric arteries of aged rats. *American Journal of Physiology. Heart and Circulatory Physiology*, Vol. 286, No. 6, (January 2004), pp. H2249-2256, ISSN 0363-6135.

Taddei, S.; Virdis, A.; Ghiadoni, L.; Salvetti, G.; Bernini, G.; Magagna, A. & Salvetti, A. (2001). Age-related reduction of NO availability and oxidative stress in humans. *Hypertension*, Vol. 38, No. 2, (August 2001), pp. 274-279, ISSN 1524-4563.

Tang, J.; Gordon, G.M.; Nickoloff, B.J. & Foreman, K.E. (2002). The helix-loop-helix protein id-1 delays onset of replicative senescence in human endothelial cells. *Laboratory Investigation*, Vol. 82, No. 8, (August 2002), pp. 1073-1079, ISSN 0023-6837.

Tang, Y.; Yang, X.; Friesel, R.E.; Vary, C.P. & Liaw, L. (2011). Mechanisms of TGF-beta-Induced Differentiation in Human Vascular Smooth Muscle Cells. *Journal of Vascular Research*, Vol. 48, No. 6, (August 2011), pp. 485-494, ISSN 1423-0135.

Tarn, W.Y. & Steitz, J.A. (1996). Highly diverged U4 and U6 small nuclear RNAs required for splicing rare AT-AC introns. *Science*, Vol. 273, No. 5283, (September 1996), pp. 1824-1832, ISSN 0036-8075.

ten Dijke, P.; Goumans, M.J. & Pardali, E. (2008). Endoglin in angiogenesis and vascular diseases. *Angiogenesis*, Vol. 11, No. 1, (February 2008), pp. 79-89, ISSN 0969-6970.

Toussaint, O.; Medrano, E.E. & von Zglinicki, T. (2000). Cellular and molecular mechanisms of stress-induced premature senescence (SIPS) of human diploid fibroblasts and melanocytes. *Experimental Gerontology*, Vol. 35, No. 8, (December 2000), pp. 927-945, ISSN 0531-5565.

Tremain, R.; Marko, M.; Kinnimulki, V.; Ueno, H.; Bottinger, E. & Glick, A. (2000). Defects in TGF-beta signaling overcome senescence of mouse keratinocytes expressing v-Ha-ras. *Oncogene*, Vol. 19, No. 13, (April 2000), pp. 1698-1709, ISSN 0950-9232.

van der Kraan, P.M.; Goumans, M.J.; Blaney Davidson, E. & Ten Dijke, P. (2011). Age-dependent alteration of TGF-beta signalling in osteoarthritis. *Cell and Tissue Research*, DOI: 10.1007/s00441-011-1194-6, (June 2011), ISSN 1432-0878.

Velasco, S.; Alvarez-Munoz, P.; Pericacho, M.; Dijke, P.T.; Bernabeu, C.; Lopez-Novoa, J.M. & Rodriguez-Barbero, A. (2008). L- and S-endoglin differentially modulate TGFbeta1 signaling mediated by ALK1 and ALK5 in L6E9 myoblasts. *Journal of Cell Science*, Vol. 121, No. Pt 6, (February 2008), pp. 913-919, ISSN 0021-9533.

Venkatesha, S.; Toporsian, M.; Lam, C.; Hanai, J.; Mammoto, T.; Kim, Y.M.; Bdolah, Y.; Lim, K.H.; Yuan, H.T.; Libermann, T.A.; Stillman, I.E.; Roberts, D.; D'Amore, P.A.; Epstein, F.H.; Sellke, F.W.; Romero, R.; Sukhatme, V.P.; Letarte, M. & Karumanchi, S.A. (2006). Soluble endoglin contributes to the pathogenesis of preeclampsia. *Nature Medicine*, Vol. 12, No. 6, (June 2006), pp. 642-649, ISSN 1078-8956.

Voglauer, R.; Chang, M.W.; Dampier, B.; Wieser, M.; Baumann, K.; Sterovsky, T.; Schreiber, M.; Katinger, H. & Grillari, J. (2006). SNEV overexpression extends the life span of human endothelial cells. *Experimental Cell Research*, Vol. 312, No. 6, (January 2006), pp. 746-759, ISSN 0014-4827.

Wang, Z.; Xiao, X.; Van Nostrand, E. & Burge, C.B. (2006). General and specific functions of exonic splicing silencers in splicing control. *Molecular Cell*, Vol. 23, No. 1, (June 2006), pp. 61-70, ISSN 1097-2765.

Wesierska-Gadek, J.; Wojciechowski, J.; Ranftler, C. & Schmid, G. (2005). Role of p53 tumor suppressor in ageing: regulation of transient cell cycle arrest and terminal senescence. *Journal of Physiology and Pharmacology*, Vol. 56, No. 1, (March 2005), pp. 15-28, ISSN 0867-5910.

Will, C.L. & Luhrmann, R. (2005). Splicing of a rare class of introns by the U12-dependent spliceosome. *Biological Chemistry*, Vol. 386, No. 8, (October 2005), pp. 713-724, ISSN 1431-6730.

Wu, S.; Hultquist, A.; Hydbring, P.; Cetinkaya, C.; Oberg, F. & Larsson, L.G. (2009). TGF-beta enforces senescence in Myc-transformed hematopoietic tumor cells through induction of Mad1 and repression of Myc activity. *Experimental Cell Research*, Vol. 315, No. 18, (September 2009), pp. 3099-3111, ISSN 1090-2422.

Young, A.R. & Narita, M. (2009). SASP reflects senescence. *EMBO Reports*, Vol. 10, No. 3, (February 2009), pp. 228-230, ISSN 1469-3178.

Zhou, Z.; Licklider, L.J.; Gygi, S.P. & Reed, R. (2002). Comprehensive proteomic analysis of the human spliceosome. *Nature*, Vol. 419, No. 6903, (September 2002), pp. 182-185, ISSN 0028-0836.

14

Calcium Regulation in Neuronal Function with Advancing Age: Limits of Homeostasis

John N. Buchholz[1], William J. Pottorf[2], Conwin K. Vanterpool[3], Erik J. Behringer[4] and Sue P. Duckles[5]
[1]*Department of Physiology and Pharmcology, Loma Linda University, School of Medicine, Loma Linda, California*
[2]*Department of Medical Affairs, Aztra Zeneca, Wilmington, Deleware*
[3]*Department of Biological Sciences, Alabama A&M, Normal, Alabama*
[4]*Department of Medical Pharmacology and Physiology, University of Columbia, Columbia, Missouri*
[5]*Department of Pharmacology, College of Medicine, University of California, Irvine, California*
USA

1. Introduction

Neurons are specialized excitable cells that transmit information to other neurons or to end effector cells such as skeletal muscle, heart muscle and blood vessel smooth muscle cells, maintaining posture, locomotion and cardiovascular function. Transmission of information from neurons to effector cells usually involves secretion of neurotransmitters contained in clustered pools of synaptic vesicles. This unique anatomical arrangement allows for rapid response to the stimulus of calcium influx and sustained release of neurotransmitters under low and higher frequency conditions (Peirbone *et al.*, 1995; Siksou *et al.*, 2011).

Calcium is hypothesized to act as a universal second messenger in excitable cells; a large volume of literature spanning more than 30 years supports this idea. Calcium integrates multiple neuronal cell functions including initiation of neurotransmitter release, regulation of gene expression, proliferation, excitability and plasticity, and activation of cell death pathways in apoptosis (Malenka *et al.*, 1989; Choi, 1992; Berridge, 1995, 1998; Clapham, 1995; Ginty, 1997; Wuytack *et al.*, 2002; Cavazzini *et al.*, 2005).

In resting neurons a large electrochemical gradient of approximately 10,000-fold exists between the extracellular environment and the cytosol; this physical arrangement allows for rapid activation of transmitter release via mobilization of release vesicles. In peripheral neurons including sympathetic neurons, calcium signaling begins with rapid opening of voltage-gated L, N and some R type calcium channels allowing calcium to enter the cytosol (Kostyuk, 1989; Kostyuk *et al.*, 1993; Trouslard *et al.*, 1993; Vanterpool *et al.*, 2005). More recently the nomenclature of L, N and R type channels has taken on increased complexity due to definition of differences in their activation voltages, coding genes and hence amino acid sequences of the α_1-subunits that comprise the channel pore (Catterall *et al.*, 2005). For example L and N type channels have been divided into five groups: Ca_v1.1. Ca_v 1.2, Ca_v 1.3 and Ca_v 1.4 (L-type) and Ca_v 2.2 (N-type). R type channels, originally termed "residual" because they are resistant to dihidropyridines, which block L-type channels and omega conotoxin, which blocks N-type channels (Catteral *et al.*, 2005), are now denoted as Ca_v 2.3. Given the importance of precise neuronal signaling and the need for regulatory mechanisms to sustain efficient signaling, a number of additional cellular mechanisms play a critical role in modulating cellular calcium. These complex calcium regulatory mechanisms are illustrated in figure 1(A). Following the activation of calcium influx through voltage-gated calcium channels, much of the calcium increase is immediately dampened by multiple calcium buffering proteins (Dove *et al.*, 2000). Despite this rapid calcium buffering, functional signaling is sustained by rapid release of calcium from smooth endoplasmic reticulum (SER) stores. This process has been termed calcium-induced calcium release (CICR) and is mediated by calcium activation of ryanodine receptor (RyR) channels (Belan *et al.*, 1993; Verkhratsky & Shmigol, 1996; Usachev & Thayer, 1997, 1999a,b; Verkhratsky & Petersen, 1998; Akita & Kuba, 2000; Berridge 2002; Behringer et al., 2009[a]).

A variety of additional calcium buffering systems can be activated by a rise in intracellular calcium ($[Ca^{2+}]$i), depending on both the magnitude and duration of the $[Ca^{2+}]$i transient. Buffering systems include calcium-buffering proteins, energy dependent SER Ca^{2+} ATPases (SERCA), plasmalemmal Ca^{2+}-ATPases (PMCA), mitochondrial Ca^{2+}/H^+ symporters and Na^+/Ca^{2+}-exchangers. All of these calcium buffering systems act to modulate the shape and magnitude of stimulus-evoked $[Ca^{2+}]$i transients and contribute to restoration of $[Ca^{2+}]$i levels (Werth & Thayer, 1994; Buchholz *et al.*, 1996; Werth*et al.*, 1996; Usachev & Thayer, 1999a; Pottorf *et al* ., 2000[a,c], 2002; Wuytack *et al.*, 2002). Furthermore, the SERCA function and the CICR process are interdependent, as SERCA pumps not only buffer $[Ca^{2+}]$i transients but also refill SER calcium stores. This interdependence maintains CICR during repeated neuronal activation (Vanterpool *et al.*, 2005). Disruption of any of these mechanisms with advancing age may result in altered function and health of peripheral neurons. This chapter will focus on the impact of advancing age, from young adult to senescence, on the function of adrenergic nerves and the mechanisms by which alterations in function may be related to age-related changes in modulation of $[Ca^{2+}]$i with ongoing neuronal activity. Data from seminal studies on function of peripheral sensory and central neurons will also be included as important background for understanding the impact of age on adrenergic nerves.

2. Adrenergic neurons arising from the Superior Cervical Ganglion as an important study model

The superior cervical ganglion (SCG) is a peripheral neuroendocrine center, as many tissues receive adrenergic input from the SCG (Cardinali et al., 1981). For example, cardiac output, a function of heart rate and stroke volume, is modulated via the SCG. The SCG relays cardioregulatory signals from the central nervous system (CNS) via the SCG axons terminating in the heart. These nerves release noradrenaline (NA) and increase heart rate and contractile tension through β_1-adrenergic receptors (Wingerd *et al.*, 2004). Furthermore, these nerves are implicated in sudden cardiac death following myocardial infarction (MI), as the density of adrenergic innervation increases after MI resulting in severe arrhythmias (Chen *et al.*, 2001). Adrenergic nerve density within the SCG is maintained by neuronal growth factor (NGF); thus this trophic factor ensures proper function of these nerves in target tissues innervated by the SCG. Indeed, there is an age-related decline in NGF; however, the trophic response of aged peripheral adrenergic nerves to added NGF is maintained (Cowen & Gavazzi, 1998; Isaacson & Crutcher, 1998; Dickason & Isaacson, 2002). Loss of NGF during the aging process may have implications for the function of target organs innervated by the SCG. Nevertheless, peripheral neurons can still respond to trophic influences later in life.

Within the rat cerebrovasculature, sensory and adrenergic innervation of blood vessels is complete within the first 30 days of life (Tsai et al., 1989). In addition, adrenergic innervation of cerebral blood vessels influences vascular development and motor function, and the presence of a functional innervation is critical for maintenance of angiogenesis and the modulation of contractile function. For example, in rabbits aged 3–20 weeks, removal of the SCG results in loss of vascular smooth muscle mass, reduced wall thickness and attenuated contractile responses to NA (Bevan & Tsuru, 1981).

Hemorrhagic stroke is a major issue in human health in terms of morbidity, mortality, and health care costs, accounting for about 10% of cerebrovascular disease. Risk factors include low birth weight, premature birth, maternal hypertension and age over 60 years. (Martyn *et. al.*, 1996; Abbott *et. al.*, 2003; Zhang *et. al.*, 2003; Lawes *et. al.*, 2004; Hermes-Desantis and Clyman, 2006; Barker 2008). Constriction or dilation of cerebral blood vessels in response to changing pressure is termed autoregulation and serves to maintain cerebral blood flow (CBF) constant over a normal systolic pressure range of 60-120 mmHg (Faraci and Heisted, 1998; Zhang *et al.*, 2003). Sympathetic nerves arising from the SCG provide a short-term mechanism to increase the upper limit of autoregulation of cerebral blood flow. When systolic pressures exceed ~140 mmHg, activation of sympathetic nerves is important in activating further constriction of cerebral blood vessels to further protect from forced dilation and rupture (Faraci and Heisted 1998; Furuichi *et al.*, 1999; vanLieshout and Secher, 2008). In humans, acute hypercapnia elevates arterial pressure and CBF, which is attenuated by sympathetic nerve activity (Jordan *et al.*, 2000). In lambs with elevated systolic pressures above 140 mmHg or elevated systolic pressure during REM sleep, sympathetic discharge from the SCG significantly increases (Cassaglia *et. al.*, 2008, 2009). Thus, increased activity of the SCG in response to elevated blood pressure helps to protect cerebral blood vessels from rupture.

Studies on how neural input regulates cerebrovascular tone and blood flow have underscored the importance of the "neurovascular unit" in brain function. The combination of neuronal input to vascular smooth muscle and the inherent contractile properties of smooth muscle provide for the optimum function of cerebral blood vessels (reviewed by Hamel, 2006). Given that adrenergic nerves arising from the SCG innervate numerous organs including cerebral blood vessels and the importance of the homeostatic modulatory function of blood vessels in the brain, our group has chosen to focus our aging studies on this neuronal model.

3. General overview of calcium regulation in aging neurons

The 20th century brought extraordinary gains in public health, medicine and food production. These gains have resulted in increased life span with a progressive decline in the ratio of people less than 20 years old to those aged 65 and older (Gems, 2011). Aging is a natural and progressive process and has been suggested to involve a combination of factors including genetic and environmental influences, altered hormonal levels and an inborn aging process. Exquisite studies have shown that the rate of the aging process in animal models can be altered by genetic manipulations resulting in reduction of insulin growth factor 1 signaling and reduced oxidative stress. (Harman, 1998; Guarente & Kenyon, 2000; Clancy *et al.*, 2001; Tatar *et al.*, 2001; Troen, 2003). More recently androgen replacement has been shown to reverse age-related cognitive declines in male rats (Frye *et al.*, 2010). These data suggest that androgenic hormones play a role in modulating neuronal function, and hormone replacement may attenuate and/or restore changes in cognition that occur with advancing age. Caloric restriction in animal models has also been shown to prolong lifespan and reduce age-related morbidity and organ pathology via reduced oxidative stress that accompanies metabolism (Bodkin *et al.*, 2003; Forster *et al.*, 2003). However, these outstanding studies on caloric restriction, hormone replacement, reduced oxidative stress, and lifespan do not fully explain age-related changes in the function of critical organ and neuronal systems or the vulnerability of particular physiological processes to advancing age.

As discussed above, ionized free calcium is a ubiquitous second messenger in neurons serving as both a charge carrier and chemical intermediate linking physiological stimuli to intracellular effectors (Friel and Chiel, 2008). Subtle age-related declines in mechanisms that modulate stimulation-evoked increases in $[Ca^{2+}]i$ have been hypothesized to contribute to age-related neuronal dysfunction and degeneration; this has became known as the "calcium hypothesis" of neuronal aging (Khachaturian, 1987; Landfield, 1987). Indeed subtle declines in mechanisms that modulate $[Ca^{2+}]i$ levels may be responsible for the elevated resting cytosolic $[Ca^{2+}]i$ that has been seen in several types of peripheral and central neurons in tissue culture, as well as in acutely isolated tissue slices or neurons (Kirischuk & Verkhratsky, 1996; Verkhratsky & Toescu, 1998; Raza *et al.*, 2007). One potential consequence of an age-related decline in regulation of intracellular calcium is that calcium overload increases mitochondrial calcium uptake, in turn activating caspases that mediate neuronal cell apoptosis, reducing neuronal survival with age (Ichas & Mazat, 1998; Thibault *et al* ., 1998; Begley *et al* ., 1999; Toescu, and Verkhratsky, 2007).

As the population ages there is an increased incidence of age-related pathology, leading to a tendency to assume a general age-related deterioration in cellular function, including calcium regulatory processes, leading to increased susceptibility to pathology and cell death (Porter *et al.*, 1997). However, this common assumption does not take into account that human populations in developed countries are living longer and healthier than at any point within history (Gems, 2011). Thus, with regards to the function of cellular mechanisms that regulate $[Ca^{2+}]i$ homeostasis, a change in the function of one system may actuate compensatory calcium regulatory mechanisms maintaining some degree of neuronal function in senescent neurons or during acute insults such as stroke (Murchinson & Griffith, 1998; Verkhratsky & Toescu, 1998; Lee *et al.*, 1999; Griffith *et al.*, 2000; Pottorf *et al.*, 2000[a], 2002; Buchholz *et al.*, 2007). Studying normal aging in the absence of pathology, healthy aging, offers the most promise in trying to understand fundamental aging processes.

4. Aging alters function of adrenergic neurons via multiple mechanisms

There is a correlation between risk of stroke and advancing age, which is associated with an age-related increase in systolic blood pressure (Faraci & Heisted, 1998; Abbott *et al.*, 2003;

Zhang *et al.*, 2003; Lawes *et al.*, 2004). Systolic blood pressure rises with age in both F344-rats and humans; this is correlated with rising plasma catecholamine levels. Thus there may be fundamental age-related changes in the function of peripheral adrenergic nerves, vascular responses to NA and ability to modulate circulating levels of NA (Palmer *et al.*, 1978; Esler *et al.*, 1981[a,b]; Barnes *et al.*, 1982; Yu *et al.*, 1985; Insel, 1993). The concept of altered function of peripheral adrenergic nerves with advancing age is supported by our *in vitro* studies showing that stimulation-evoked fractional NA release increases with age in the rat tail and superior mesenteric arteries (Buchholz & Duckles, 1990; Buchholz *et al.*, 1998). In addition, this age-related increase in NA release occurs over a wide range of stimulation frequencies from 0.5 - 8 Hz (Tsai *et al.*, 1995). An explanation for the observed age-related change in function of vascular adrenergic neurons is complex, involving multiple mechanisms such as changes in density of adrenergic neurons, NA content, re-uptake, function of prejunctional inhibitory α_2-autoreceptors, and calcium regulation (Pottorf *et al.*, 2000[b]). In addition, other mechanisms may also modulate the function of adrenergic nerves, and these complicating factors must also be addressed. For example, our studies and others have shown that the SCG (Dun *et al.*, 1995; Wu *et al.*, 1997) and the cerebral vasculature contain adrenergic and neuronal nitric oxide (nNOS) containing nerves; NO released from nNOS neurons facilitates stimulation-evoked NA release in both blood vessels and the CNS (Montague *et al.*, 1994; Yamamoto *et al.*, 1997; Zhang *et al.*, 1998; Lee *et al.*, 2000; Mbaku *et al.*, 2000, 2003). Facilitated function of adrenergic nerves via nNOS nerves may occur through enhancement of Ca^{2+} influx and/or Ca^{2+}release from internal stores. These mechanisms may also be altered with age and cannot be discounted in future discussions on the impact of age on function of adrenergic nerves.

4.1 Neurotransmitter content and density of adrenergic nerves with advancing age

In peripheral organs and blood vessels NA content serves as an index of adrenergic density. In the rat heart, NA content decreases with age (Martinez *et al.*, 1981; Dawson & Meldrum, 1992), while in blood vessels the story is more opaque. NA content in rat renal, femoral, and saphenous arteries increases with age, while in renal, femoral, and saphenous veins there is no change, and a decline in tail arteries (Handa & Duckles, 1987). Consistent with measurements of NA content in rat arteries, catecholamine histofluorescence, another measure of adrenergic nerve density, increases with age in rat superior mesenteric and renal arteries and portal vein (Mione *et al.*, 1988). In contrast spinal cord blood vessels show no age-related change in adrenergic nerve density (Amenta *et al.*, 1990). In two studies by the same group sympathetic innervation in the internal carotid artery and anterior cerebral arteries was found to decline with age. Interestingly after intracerebral infusion of NGF, the number of sympathetic axons, NA content and numbers of tyrosine hydroxylase-containing nerve fibers increased in these same arteries from aged animals (Isaacson & Crutcher, 1998; Dickason & Isaacson, 2002). These studies would suggest that there is an innate age-related ability to maintain a critical number of functioning sympathetic neurons. The NGF studies cited above open up possibilities for therapeutic interventions to maintain cardiovascular homeostasis with advancing age. However, there is no clear relationship between age-related changes in adrenergic nerve density, increases in circulating plasma NA levels reported by others and increased stimulation-evoked NA release in our earlier studies; age-related changes in NA content appear to be species or vascular bed dependent.

4.2 Transmitter uptake and function of inhibitory prejunctional α_2-adrenoceptors

NA concentration in the synaptic cleft is modulated by NA re-uptake and activation of prejunctional α_2-adrenceptors. The latter mechanism has been shown to attenuate NA release by decreasing stimulation-evoked calcium influx in adrenergic neurons by reducing the open probability of voltage-gated calcium channels (Schofield, 1990; Delcour & Tsien, 1993). These mechanisms control the moment-to-moment biophase concentrations of NA with ongoing vascular adrenergic nerve activity (Illes, 1986; Langer & Arbilla, 1990; Buchholz *et al.*, 1992; Insel, 1993; Esler *et al.*, 1995).

Chemical agents such as cocaine and deoxycorticosterone that block the neuronal and extraneuronal uptake of NA, respectively, are used to estimate the function of these systems. Neuronal and extraneuronal uptake of NA has been shown to be reduced with age in the atria and vas deferens in the pithed rat (Borton & Docherty, 1989; De Avellar *et al.*, 1990). These studies suggest that NA uptake declines with age in peripheral adrenergic nerves. In contrast, direct measurement of ^{3}H-NA uptake supports the idea that the function of NA uptake transporters does not change with age in peripheral adrenergic nerves (Duckles *et al.*, 1985). In other studies in the rat heart (Limas, 1975) and tail artery (Buchholz & Duckles, 1990), the effect of cocaine and deoxycorticosterone on NA uptake increased with advancing age. Reconciling these contrasting studies would appear difficult. However, in the tail artery model, when the measured release of NA in the presence of uptake blockers was corrected for total NA release in the absence of uptake blockers, there was no change with age (Buchholz & Duckles, 1990). Thus, age-related changes in the effectiveness of uptake blockers on NA may merely reflect age-related differences in NA concentrations within the junctional cleft of the model under study. If there is less or more NA released with age in a particular model, then the observed effectiveness of the uptake blockers may only reflect the concentration of NA in the synaptic cleft. Therefore, data in the tail artery model would suggest that NA uptake with advancing age remains as a constant fraction of the amount of NA released.

If the age-related increase in stimulation-evoked fractional NA release in peripheral adrenergic nerves cannot be explained by an age-related change in the function of NA transporters, an explanation must lie elsewhere, such as possible age-related changes in presynaptic inhibition of NA release. There are a number of studies aimed at measuring changes in the feedback function of prejunctional α_2-adrenoceptors. Measurements of NA overflow have shown an age-related decline in the effect of prejunctional α_2 -adrenoceptor antagonists in pithed rats, rat vas deferens, heart, and tail artery (Hyland & Docherty, 1985; Docherty & Hyland, 1986; Daly *et al.*, 1989; Buchholz & Duckles, 1990). On the surface these studies seem to offer support for the idea of a general age-related decline in the function of prejunctional α_2-adrenoceptors. However, the issue appears to be more complicated as the studies cited above used only single concentrations of competitive α_2-antagonists. In the rat tail artery, we measured NA overflow over a full concentration range of the competitive α_2-antagonist, idazoxan. We found that the sensitivity of the prejunctional α_2-adrenoceptors to idazoxan declined with advancing age; however, the maximal response to this drug was not altered with advancing age (Buchholz *et al.*, 1992). As fractional stimulation-evoked NA release increases with age, there would be more NA in the junctional cleft interacting with the α_2-adrenoceptor and increased competition between higher NA levels and an antagonist. Chemical competition between NA and the α_2-adrenoceptor antagonists may possibly reduce the apparent sensitivity of the applied antagonist and account for the decrease in the potency of α_2-adrenoceptor antagonists with age (Pottorf *et al.*, 2000b). In light of the studies cited above, age-related changes in NA release are not adequately explained in terms of age-related alterations in the function of NA uptake mechanisms or α_2-adrenoceptor function. Therefore, we looked at other mechanisms that may account for age-related changes in adrenergic nerve activity, including studies of stimulation-evoked NA release with altered extracellular calcium, calcium influx and altered calcium buffering capacity.

4.3 Role of calcium influx in transmitter release and the effect of altering buffering capacity on transmitter release in aged peripheral neurons

Our studies of the rat tail artery cited above showed that stimulation-evoked NA release increased with advancing age. However, altered function of NA uptake or prejunctional α_2-adrenoceptors appears to reflect increased concentrations of NA in the junctional cleft as opposed to real changes in function of these two systems. Thus, our attention was directed toward the possibility that age-related changes in stimulation-evoked NA release are due to changes in calcium regulation in the nerve endings. Calcium entry through voltage-gated calcium channels provides the secretory potential that initiates NA release in sympathetic neurons. Thus we carried out experiments to examine the effects of increased or lowered extracellular calcium, which alters the chemical driving force of calcium, on stimulation-evoked NA release in tail arteries of young and old rats using short or long stimulation protocols (Buchholz *et al.*, 1994). When extracellular calcium was reduced or elevated, stimulation-evoked fractional NA release declined or increased, respectively in tail arteries from both young and old animals. However, in senescent animals the reduction in extracellular calcium reduced stimulation-evoked NA release to a lesser degree while elevated extracellular calcium increased NA release to a greater degree than in the young animals. In addition, at the highest elevated extracellular calcium concentration, 7.5 mM, the increase in stimulation-evoked NA release was maintained in tail arteries from young animals but declined in tail arteries from senescent animals. The explanation for these results may be attributable a variety of possible functional changes, including altered calcium influx, buffering capacity, or sensitivity of the NA release mechanisms to stimulation-evoked increases in $[Ca^{2+}]i$. Indeed higher extracellular calcium significantly elevated NA release by a greater magnitude in old as compared to young tail arteries suggesting that stimulation-evoked $[Ca^{2+}]i$ transients may be greater and more sustained in adrenergic nerves in senescent animals. The significant decline in stimulation-evoked NA release from adrenergic nerves in tail arteries from old animals at the highest extracellular calcium concentrations suggests that older nerves no longer possess the ability to modulate high $[Ca^{2+}]i$ loads which may then become toxic.

Studies on the impact of age on calcium influx are mixed; calcium influx has been reported to increase with age in central neurons (Landfield & Pitler, 1984; Pitler & Landfield, 1990), decrease in peripheral neurons (Kostyuk *et al.*, 1993), or remain unchanged in central neurons (Murchinson & Griffith, 1996). Thus, we measured the impact of age on stimulation-evoked NA release employing three protocols, two of which involved calcium influx through calcium channels and a third where calcium channels were bypassed using the calcium ionophore, ionomycin (Tsai *et al.*, 1997). When we measured NA release from rat-tail arteries via transmural nerve stimulation or depolarization of nerves with high KCl,

NA release was greater in tail arteries from old as compared to young animals. These data could be explained in part by an age-related increase in calcium influx with age. However, when we bypassed voltage-gated calcium channels using the calcium ionophore, ionomycin, we found that the age-related increase in NA release persisted (Tsai *et al.*, 1997). In these experiments [Ca^{2+}]i is elevated via an ionophore, so that [Ca^{2+}]i buffering systems would be the primary mechanisms modulating the magnitude and duration of [Ca^{2+}]i transients and hence transmitter release. While these data do not rule out the possibility that stimulation-evoked calcium increases with age in the rat tail artery model, these data also suggest that other mechanisms, such as [Ca^{2+}]i buffering systems and/or the sensitivity of NA release to calcium, may also be altered with age.

We continued to test the hypothesis that an alteration in the function of [Ca^{2+}]i buffering systems may contribute to the increase in NA release from rat tail arteries (Tsai *et al.*, 1997). In this set of experiments we measured stimulation-evoked NA release before and after the addition of a membrane permeant, non-endogenous [Ca^{2+}]i chelator, 1,2-bis(o-aminophenoxy)ethane-N,N,N',N'-tetraacetic acid acetomethoxy ester (BAPTA). In the presence of BAPTA, stimulation-evoked NA release was attenuated to a greater extent in old arteries as compared to young, suggesting that age-related changes in NA release are in part due to altered calcium-buffering capacity. Overall, these studies suggest that age-related changes in NA release from peripheral adrenergic neurons appear to be due in part to altered [Ca^{2+}]i handling mechanisms. However, age-related alterations in the sensitivity of NA release mechanisms per se have not been ruled out.

5. Alterations in neuronal calcium buffering and extrusion during aging in peripheral neurons

The magnitude and temporal properties of calcium signals are critical, and sustained high levels of [Ca^{2+}]i can be toxic. Thus, calcium is tightly regulated by channels, pumps, exchangers and protein buffers that control the shape and duration of stimulation-evoked [Ca^{2+}]i transients (Fig. 1A; Friel and Chiel 2008). Age-related changes in the function of any one or a combination of the components of this [Ca^{2+}]i control system may alter the function of neurons and/or contribute to neuronal degeneration (Meldrum & Garthwait, 1990; Peterson, 1992; Verkhratsky *et al.*, 1994; Pottorf *et al.*, 2000b, 2002). Given the complex interplay between calcium influx and numerous buffering systems, declining function of one or more systems may possibly be compensated for by increased function of other mechanisms (Pottorf *et al.*, 2000b). Thus, in aging studies investigators cannot assume that a decline in one or more buffering systems automatically leads to cellular dysfunction, as other systems may be able to compensate for decline in a particular calcium buffering system. In this sense, senescent neurons may still maintain a degree of homeostasis even though the limits of that ability are narrower as compared to younger neurons.

The advent in the last 25 years of cell permeable dyes capable of rapid and reversible binding to calcium as [Ca^{2+}]i levels change has contributed greatly to understanding the dynamics of cellular [Ca^{2+}]i signaling. Quantification of dynamic [Ca^{2+}]i transients is possible because the emission fluorescence intensity of these dyes is altered by reversible binding of calcium (Tsien *et al.*, 1985; Roe *et al.*, 1990; Thayer & Miller, 1990; Neher, 1995; Baylor and Hollingworth, 2000; Friel and Chiel, 2008). Microfluorometry coupled with calcium sensitive dyes has allowed investigators to measure the impact of advancing age on the mechanisms that modulate stimulation-evoked $[Ca^{2+}]i$ transients and overall calcium homeostasis (Kirischuk *et al.*, 1992; Kirischuk & Verkhratsky, 1996; Neher, 1998; Baylor & Hollingworth, 2000; Pottorf *et al.*, 2002).

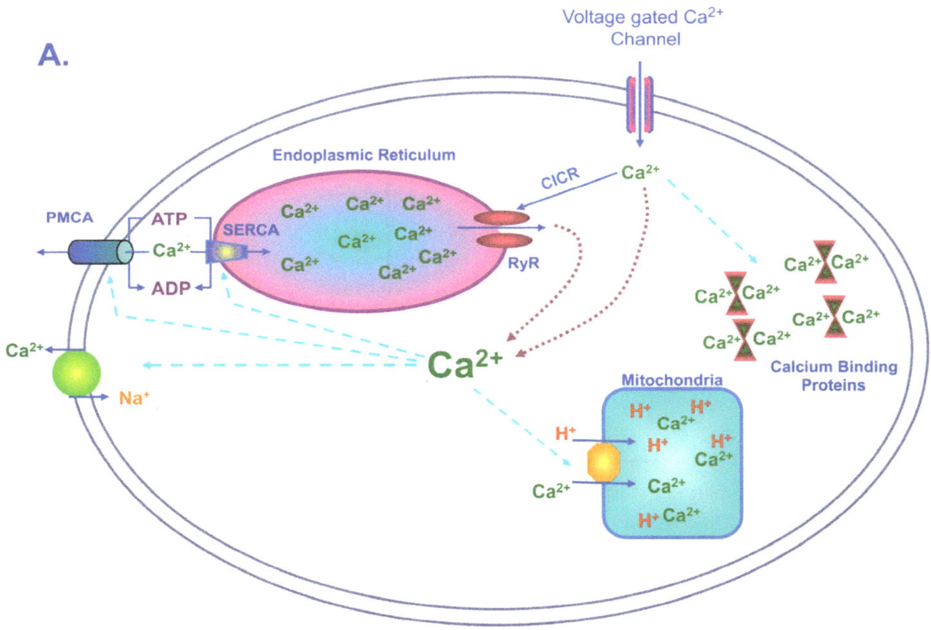

Fig. 1. (A) Representation of mechanisms that modulate stimulation-evoked $[Ca^{2+}]i$ transients in peripheral adrenergic nerves. Depolarization elevates $[Ca^{2+}]i$ via calcium influx through voltage-gated calcium channels. The calcium influx is immediately attenuated via calcium-binding proteins; however, the residual elevated calcium exceeding particular threshold will act on ryanodine receptors (RyR) to evoke "all or none" release of calcium from the endoplasmic reticulum known as calcium-induced calcium release (CICR). The elevation in $[Ca^{2+}]$ is controlled by a dynamic interplay of multiple buffering systems, including: (1) smooth endoplasmic reticulum calcium ATPases (SERCA) sequestration of calcium into the ER that serves to buffer and refill ER calcium stores thus maintaining the ability of the neuron to undergo repetitive CICR; (2) mitochondrial calcium uptake by a H^+/Ca^{2+} uniporter; (3) removal of calcium via plasma membrane calcium ATPase (PMCA) pumps and the Na^+/Ca^{2+} exchanger. Plum dotted lines represent calcium influx and release pathways that elevate $[Ca^{2+}]i$. Light blue dashed lines represent calcium-buffering pathways that control increases in $[Ca^{2+}]i$ and restoration to basal levels.

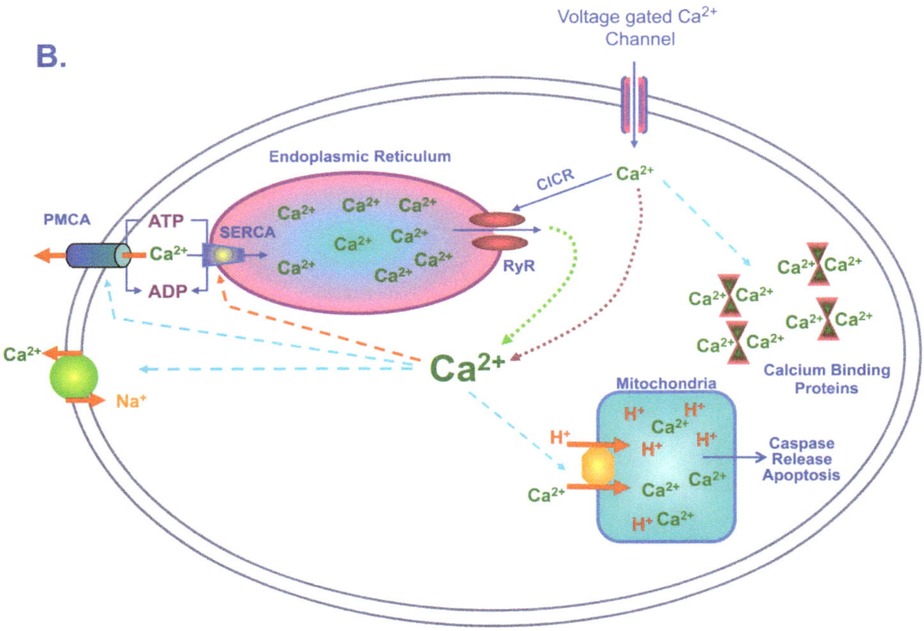

Fig. 1. (B) A model of the current hypothesis that advancing age, in the absence of pathology, results in a subtle decline in the control of [Ca^{2+}]I in peripheral adrenergic neurons. Compensation by other control mechanisms may allow neurons to adapt to an age-related decline in control of [Ca^{2+}]i. This model illustrates the mechanisms that lead to elevated [Ca^{2+}]i in aged peripheral adrenergic neurons. The rise in [Ca^{2+}]i mediated via calcium influx and release from the SER is buffered by SERCA whose function declines with age (broken red line). In response to the decline in SERCA function, mitochondria, PMCA and Na^+/Ca^{2+} exchangers compensate (thick red solid lines) by increasing Ca^{2+} uptake and removal so as to preserve neuronal viability. However, increased mitochondrial calcium uptake may also reduce long-term neuronal survival by activation of caspases leading to apoptosis. Another feature of this model is that the decline in SERCA function in senescent neurons reduces ER Ca^{2+} filling levels and reduces CICR (bright green line).

5.1 Age-related changes in SERCA function in peripheral and central neurons

It has been suggested that an age-related decline in SERCA function contributes to calcium dysregulation in peripheral sympathetic and sensory neurons (Kirischuk *et al.*, 1992; Kirischuk & Verkhratsky, 1996; Pottorf *et al.*, 2002). Such a decline in SERCA function with age may also have additional consequences with regards to calcium signaling in neurons. The SER serves as a source of calcium that can be released by a process called CICR. SERCA buffer [Ca^{2+}]i transients and reload the SER calcium stores with ongoing neuronal activity. Thus an age-related decline in SERCA function may possibly also cause alterations in CICR from the SER (Murchinson & Griffith, 1999; Usachev & Thayer, 1999a; Behringer *et al.*, 2009).

Studies of stimulation-evoked NA release in rat tail artery sympathetic neurons suggested age-related changes in SERCA function (Tsai *et al.*, 1998). When SERCA were blocked with the irreversible or reversible SERCA antagonists, thapsigargin or CPA respectively, stimulation-evoked NA release from sympathetic nerves increased only in tail arteries from young animals with no change in arteries from old animals. These data suggest that in sympathetic neurons from young animals SERCA-mediated calcium buffering modulates depolarization-induced $[Ca^{2+}]i$ transients and hence transmitter release. In contrast, in old animals, SERCA-mediated calcium buffering does not appear to affect transmitter release. To further explore this hypothesis, we measured intracellular calcium in isolated SCG cells and quantified the function of SERCA using reversible or irreversible SERCA blockers, cyclopiazonic acid and thapsigargin, respectively. Each of these agents caused a decline in the rate of recovery of high K^+-evoked $[Ca^{2+}]i$ transients only in cells from young animals with no significant change in old cells (Tsai *et al.*, 1998), supporting our conclusion that SERCA function is lost in sympathetic nerves of old animals.

In a more complex study using SCG cells, we blocked the contribution of PMCA's, mitochondria and the Na^+/Ca^{2+} exchanger and measured $[Ca^{2+}]i$ transients evoked by high K^+. To block PMCA and mitochondrial calcium buffering we utilized low concentrations of vanadate and dinitrophenol (DNP), respectively. To block the Na^+/Ca^{2+} exchanger we replaced Na^+ with the charge carrier tetra-ethyl ammonium. Under these conditions, neurons must rely on SERCA to modulate the rate of recovery of high K^+- evoked transients. Remarkably, with the loss of three calcium buffering systems, SERCA were still able to control the magnitude of $[Ca^{2+}]i$ transients and return elevated $[Ca^{2+}]i$ to resting levels. However, the rate of recovery of high $[Ca^{2+}]i$ transients was significantly slowed in SCG cells from both young and old animals with a significantly greater slowing in SCG cells from old animals (Pottorf *et al.*, 2000c). Thus one important mechanism that can account for a decline in $[Ca^{2+}]i$ homeostasis in aged peripheral neurons is a loss of SERCA function. One possible explanation for a decline in SERCA function is a decline in the genetic expression of SERCA isoforms. Indeed myocardial SERCA-mediated $^{45}Ca^{2+}$ uptake declines with age and there is a decline in genetic expression of SERCA isoforms (Maciel *et al.*, 1990). However, paradoxically, SERCA protein levels remain stable. This study suggests that, even if the genetic expression of SERCA may decline with age, changes in translation of mRNA coding for SERCA or alterations in mRNA stability may be able to compensate for the age-related decline in SERCA gene expression. Alternatively, alterations in mechanisms that modulate on-going SERCA activity, such as phosphorylation, may offer an explanation for the decline in SERCA function (Gafni & Yuh, 1989; Xu & Narayanan, 1998). Indeed, more in-depth studies using a wider range of excitable cell models will be necessary to uncover the mechanisms that account for an age-related alteration in SERCA function.

5.2 Age and mitochondrial calcium buffering

The role of mitochondria in modulation of $[Ca^{2+}]i$ has been an area of intense study. Mitochondria can sequester calcium and act as a buffer using the energy contained in the proton motive force and a Ca^{2+}/H^+ symporter to transport cytosolic calcium into the mitochondria. Early studies suggested that mitochondria only buffer $[Ca^{2+}]i$ transients in excess of 2 μM which may occur only under excessive stimulation or pathological conditions (Nicholls, 1978, 1985). In contrast to this view, however, mitochondrial calcium uptake has been shown to produce a clear "mitochondrial set point" or shoulder in the shape of depolarization evoked $[Ca^{2+}]i$, transients in central and peripheral sensory neurons (Thayer *et al.*, 1994; Murchison *et al.*, 2004). Furthermore, proton motive force disruptors such as dinitrophenol (DNP) and carbonyl cyanide *m*-chlorophenyl hydrazone (CCCP) demonstrate the role of mitochondria in buffering normal $[Ca^{2+}]i$ transients (Thayer & Miller, 1990; Miller, 1991; Werth and Thayer, 1994; Buchholz *et al.*, 1996; Murchinson & Griffith, 1999; Colegrove *et al.*, 2000, Murchison *et al.*, 2004). Thus, it is clear that mitochondria play a role in shaping stimulation-evoked $[Ca^{2+}]i$ transients in central and peripheral sensory and sympathetic neurons under normal $[Ca^{2+}]i$ concentrations in the mid to high nanomolar ranges. Since mitochondria are an integral part of neuronal cellular calcium buffering mechanisms, either a decrease in mitochondrial calcium uptake or an increased reliance on mitochondria to control $[Ca^{2+}]i$ levels could potentially disrupt mitochondrial function leading to calcium overload and apoptosis.

There are studies in central neurons supporting an age-related decline in mitochondrial calcium uptake (Vitorica & Satrustegui, 1985; Villalba *et al.*, 1995; Satrustegui *et al.*, 1996). In a more recent study using basal forebrain neurons the addition of CCCP to dissipate the mitochondrial proton motive force significantly elevated both high K^+ and caffeine-evoked $[Ca^{2+}]i$ transients in neurons from 1-7 month-old rats with little effect in neurons from 24-27 month-old (Murchison *et al.*, 2004). These studies suggest a general decline in mitochondrial calcium buffering capacity accompanying the process of advancing age in CNS neurons.

However, the picture is quite different in peripheral neurons, such as those from the SCG or adrenergic nerve endings in blood vessels, where we have shown that mitochondrial calcium uptake actively participates in buffering high-K^+ evoked $[Ca^{2+}]i$ transients in neurons from old animals. Our studies suggest that, in peripheral neurons, the function of mitochondrial calcium uptake is preserved with advancing age. We first measured stimulation-evoked NA release from adrenergic nerve endings in tail arteries and blocked mitochondrial calcium uptake with DNP. In this case, stimulation-evoked NA release increased in arteries from old animals with no significant effect in young arteries (Tsai *et al.*, 1995). To follow up, SCG neurons were exposed to DNP to block mitochondrial calcium uptake, and $[Ca^{2+}]i$ transients were measured. The peak and rate of rise of high K^+-evoked $[Ca^{2+}]i$ transients were both increased in neurons from old animals with no significant effect in young neurons (Buchholz *et al.*, 1996). In another study we blocked PMCA, SERCA, and Na^+/Ca^{2+} exchangers from participating in calcium buffering in SCG neurons. Under these conditions the neurons must rely primarily on mitochondrial calcium uptake to regulate high K^+-evoked $[Ca^{2+}]i$ transients (Pottorf *et al.*, 2000a). A remarkable aspect is that, despite compromising multiple calcium modulatory systems, mitochondria were still capable of controlling $[Ca^{2+}]i$ transients in SCG neurons, and mitochondrial calcium uptake appeared to be preserved with advancing age. Overall, our studies in the SCG and adrenergic nerve endings suggest that mitochondrial calcium buffering is maintained with advancing age, and there is an increased tendency with age to rely on mitochondria to control $[Ca^{2+}]i$ transients. This increased reliance on mitochondria to modulate elevated $[Ca^{2+}]i$ may possibly place an added stress on aged neurons. Indeed in experiments using brain slices from mice, mitochondrial depolarization served as an index of mitochondrial calcium uptake in response to high K^+-evoked $[Ca^{2+}]i$ transients. Mitochondrial depolarization was observed in brain slices from both young and old mice; however, in old animals the rate of mitochondrial repolarization was significantly slower corresponding to slowed recovery of high K^+-evoked $[Ca^{2+}]i$ transients (Xiong *et al.*, 2002). Thus, increased reliance on mitochondrial calcium uptake with advancing age has a downside in the case of high neuronal activity in CNS neurons. As mitochondrial repolarization is slowed with advancing age, over time the ability of the mitochondria to sustain its buffering of repeated $[Ca^{2+}]i$ transients may be compromised.

All of the studies cited above are consistent in suggesting that mitochondrial calcium uptake is essential for modulation of $[Ca^{2+}]i$ transients with advancing age. However, as more stress is placed on mitochondria to control stimulation-evoked increases in $[Ca^{2+}]i$, the 'polarization state' which maintains the ability of the mitochondria to produce vital cellular energy and modulate $[Ca^{2+}]i$ transients may subtly decline over the lifespan (Toescu and Verkhratsky, 2007). Overall, with advancing age mitochondrial calcium uptake may become more central to controlling neuronal $[Ca^{2+}]i$ transients. Maintenance of mitochondrial function is critical to healthy aging as altered mitochondrial function is thought to play a role in the progression of age related diseases (Miller, 2005). Maintaining mitochondrial function with improved diet and exercise as humans age may reduce the burden of health-related costs that comes with an aging population (Duckles *et al.*, 2006).

5.3 Aging and the potential for PMCA to compensate for age-related declines in SERCA function

The PMCA and Na^+/Ca^{2+} exchange systems are two vital plasma membrane calcium-pumping mechanisms in peripheral and central neurons; each of these plays a role in buffering $[Ca^{2+}]i$ transients via extrusion of calcium from the cytosol. PMCA, like SERCA, requires energy and pumps calcium against the chemical gradient from the cytosol to the extracellular fluid. The Na^+/Ca^{2+} exchanger utilizes energy stored in the Na^+ gradient to pump calcium out of the cellular cytosol (Werth *et al.*, 1996; Blaustein & Lederer, 1999; Pottorf & Thayer, 2002; Pottorf *et al.*, 2006). A noted feature of the PMCA is that its activity can be elevated by a priming stimulus which induces prolonged elevated $[Ca^{2+}]i$ transients. Following prolonged elevation of $[Ca^{2+}]i$, the PMCA-mediated rate of recovery of depolarization-evoked $[Ca^{2+}]i$ is accelerated, and this elevated function can be retained for up to 40 min after the priming stimulus (Pottorf and Thayer, 2002). These studies suggest that some buffering systems are flexible and can elevate their function during times of high neuronal activity to protect the cell from $[Ca^{2+}]i$ overload. There are at least two mechanisms that can account for increased function of PMCA with sustained $[Ca^{2+}]i$ loads: induction of PMCA gene expression and interaction with calmodulin. Blocking SERCA function with SERCA antagonists combined with prolonged $[Ca^{2+}]i$ transients has been shown to induce PMCA gene expression (Kuo et al., 1997). Furthermore, accelerated function of PMCA with prolonged $[Ca^{2+}]i$ transients increased the interaction of calmodulin with PMCA (Pottorf & Thayer, 2002). Thus redundant calcium buffering components, along with functional alteration of some components mediated by molecular modulators, can compensate for a reduction in function of one or more calcium buffering components as illustrated by the case of prolonged $[Ca^{2+}]i$ transients. Further support for this proposition comes from studies showing that, in the failing myocardium, Na^+/Ca^{2+} exchanger levels increase as SERCA levels decline (Hasenfuss *et al.*, 1991).

Our studies of isolated SCG neurons lend direct support to the idea that multiple components of the calcium buffering system can compensate for a loss of SERCA function. As discussed above we found that, with advancing age, SCG neurons become more reliant on mitochondria to control high K^+-evoked $[Ca^{2+}]i$ transients. In a study with isolated SCG neurons, when both SERCA and mitochondrial calcium uptake were blocked with thapsigargin and DNP respectively, PMCA and Na^+/Ca^{2+} exchangers are required to control high K^+-evoked $[Ca^{2+}]i$ transients. In this case SCG neurons from both young and old animals were able to recover from high K^+-evoked $[Ca^{2+}]i$ transients (Tsai *et al.*, 1998). These data suggest that, in peripheral adrenergic neurons, plasma membrane calcium extrusion systems can by themselves control $[Ca^{2+}]i$, and their function is maintained with advancing age. In support of this study, we used a unique ATPase antagonist, vanadate, a compound that, at low concentrations (0.25 μM), selectively blocks PMCA function but does not significantly alter SERCA function. Under these conditions, the rate of recovery of $[Ca^{2+}]i$ transients was diminished to a greater extent in SCG neurons from old as compared to young animals (Pottorf *et al.*, 2000a). This shows again that PMCA function may be elevated to meet the calcium buffering demands that occur as SERCA function declines with advancing age. Although PMCA function has been reported to decline with age in synaptosomes derived from neurons in the CNS (Qin *et al.*, 1998), our studies suggest that during normal aging PMCA function in peripheral adrenergic neurons is maintained with age.

6. Consequences of age-related changes in SERCA function: Aging and SER calcium levels and the role of CICR in neuronal function

Neuronal function depends in part on release of calcium from the SER in response to $[Ca^{2+}]i$ elevation (Usachev & Thayer, 1997, 1999a,b; Behringer *et al.*, 2009). Rapid calcium buffers immediately dampen much of the calcium influx through voltage-gated calcium channels while rapid release of calcium from the SER sustains depolarization-evoked increases in $[Ca^{2+}]i$. This process has been termed CICR, is mediated by RyR channels and has been shown to be relevant in modulating release of neurotransmitters and hormones. For example, when RyR are blocked and hence CICR, excitatory or inhibitory post synaptic currents, indices of presynaptic transmitter release, significantly decline (Emptage *et al.*, 2001; Galante and Marty, 2003).

A unique feature of the CICR process, the "all or none" release of calcium from the SER, has been observed in peripheral sensory neurons (Usachev and Thayer, 1997). Using a combination of patch clamp and microfluorometry, $[Ca^{2+}]i$ was elevated with graded step depolarizations of varied time length. This method produces small graded elevations in $[Ca^{2+}]i$ until a "threshold" of $[Ca^{2+}]i$ is reached, and the full $[Ca^{2+}]i$ transient due to combined calcium influx and CICR is achieved. As CICR is mediated via the opening of RyR channels, agents such as caffeine, which sensitize RyR to calcium, reduce the threshold of $[Ca^{2+}]$ necessary to evoke CICR (Usachev and Thayer, 1997; reviewed by Usachev and Thayer, 1999a,b). In addition, ryanodine, which blocks RyR channel function can ablate CICR so that stimulation-evoked $[Ca^{2+}]i$ transients reflect calcium influx through calcium channels only (Usachev and Thayer, 1997; Behringer *et al.*, 2009).

As CICR is a regenerative process it is sustained during ongoing neuronal activity via refilling of the SER calcium stores through calcium influx and subsequent uptake into the

SER via SERCA pumps (Kostyuk & Verkhratsky, 1994; Verkhratsky *et al.*, 1994; Shmigol *et al.*, 1996). Therefore, it is clear that calcium pumps such as SERCA serve at least two functions, buffering of elevated $[Ca^{2+}]i$ and refilling $[Ca^{2+}]i$ stores to maintain calcium signaling via CICR. In CNS, SCG, and sensory neurons, SER calcium stores can be rapidly refilled by activation of voltage gated calcium channels or following depletion with caffeine they can spontaneously refill within 3–10 min via activation of store operated calcium channels (SOCC) (Friel & Tsien, 1992; Shmigol *et al.*, 1994, 1996; Usachev & Thayer, 1999b; Baba *et al.*, 2003; Vanterpool *et al.*, 2005).

SOCC channels have received much attention recently, and their function and modulation is an ongoing topic in excitable cell models (Cahalan, 2009). SOCC channels on the plasmalemma mediate a process called "store operated calcium entry." This is a complex process involving calcium-sensing proteins within the SER, STIM1 and STIM2, that activate SOCC. The STIM proteins contain calcium-binding sites, the "EF-hand domain", within the SER lumen and act as "sensors" of SER calcium levels. When SER calcium levels are high, STIM proteins form dimers, but when SER calcium levels fall calcium dissociates from the EF-hand domain, and the STIM proteins form oligomers and aggregate in puncta in the SER. STIM oligomers then interact with SOCC dimers to form SOCC tetramers, which conduct calcium (Cahalan, 2009). Calcium influx through SOCC coupled with SERCA-mediated calcium uptake allows for refilling of SER calcium stores.

Our work with adrenergic nerve endings and isolated SCG cells has consistently shown that SERCA-mediated calcium uptake is compromised in senescent animals. As SERCA function declines with age in the SCG, we studied the impact of advancing age on refilling and release of calcium from the SER. Two protocols were used to measure rapid depolarization-evoked refilling of SER calcium stores and spontaneous refilling following caffeine-evoked depletion of SER calcium stores in isolated Fura-2 loaded SCG neurons from rats aged 6-24 months. Following caffeine-evoked depletion of SER calcium stores, rapid high K^+-evoked refilling of these stores markedly declined in SCG cells from 20 and 24 month-old animals but not in 6 and 12 month-old animals. Next we measured spontaneous refilling of SER via the opening of SOCC channels and subsequent SERCA-mediated uptake of calcium into the SER following selective caffeine-evoked depletion of SER calcium stores. To ensure that the rising $[Ca^{2+}]i$ following caffeine-evoked depletion of SER calcium stores was mediated via activated SOCC, we used the SOCC blocker lanthanum which, indeed, abolished the influx of calcium. Spontaneous refilling of SER calcium stores significantly declined in SCG cells from 20 and 24 month-old animals. Overall, these data suggest that a functional consequence of reduced SERCA activity with advancing age is a compromise in the ability of SCG neurons to sustain release of $[Ca^{2+}]i$ during ongoing neuronal activity (Vanterpool *et al.*, 2005).

In a more recent series of experiments we tested the hypothesis that reduced refilling of SER calcium stores with advancing age would reduce the contribution of CICR to electrical field stimulation (EFS)-evoked $[Ca^{2+}]i$ transients in isolated SCG cells from 6, 12 and 24 month-old rats (Behringer et al., 2009). We used small, graded increases of applied EFS current at a constant stimulation train length. With this protocol and measuring the peak and rate of rise of $[Ca^{2+}]i$ we were able to estimate the contribution of CICR to EFS-evoked elevations in $[Ca^{2+}]i$. In the presence of ryanodine to block the contribution of CICR, peak and rate of rise of EFS-evoked $[Ca^{2+}]i$ transients were significantly reduced in SCG cells from both 6 and 12 month-old animals. In contrast, in SCG cells from 24 month-old animals ryanodine had no effect on peak or rate of rise of EFS-evoked $[Ca^{2+}]i$ transients. These data suggest that the contribution of CICR to calcium signaling in senescent SCG neurons is virtually absent, and function of calcium signaling in senescent SCG neurons is maintained only by calcium influx through voltage gated calcium channels (Behringer *et al.*, 2009). There are at least two possible mechanisms accounting for this precipitous loss of CICR in senescent neurons. The first is our demonstration of an age-related decline in SERCA function, which would result in depleted SER calcium levels with age. The second mechanism is that the function of RyR channels may also decline with age. Indeed caffeine is a ryanodine agonist and has been used to sensitize CICR to EFS-evoked $[Ca^{2+}]i$ transients (Usachev and Thayer, 1997). Thus, we tested the hypothesis that CICR may still possibly contribute to EFS-evoked $[Ca^{2+}]i$ transients in senescent neurons in the presence of the RyR agonist caffeine. Indeed, caffeine significantly increased EFS-evoked $[Ca^{2+}]i$ transients with short stimulation train lengths in SCG cells from 6, 12 and 24 month-old animals (Behringer *et al.*, 2009). These data suggest that although CICR from the SER may not contribute to calcium signaling in senescent SCG neurons, possibly due to a combination of reduced SERCA and RyR function, senescent neurons still retain some capacity for CICR in the presence of a RyR agonist.

7. Altered expression of ryanodine receptors (RyR) and modulators of CICR with advancing age

We have shown that the contribution of CICR to EFS-evoked $[Ca^{2+}]i$ transients is abolished with advancing age in SCG neurons; this may be a consequence of reduced SERCA and/or RyR function (Behringer et al., 2009). In this section we will discuss RyR function in more depth so as to illuminate possible age-related changes. As discussed above, release of calcium from the SER is mediated via RyR ion channels. RyR function is controlled both by gene expression and by secondary molecular modulators, which regulate the sensitivity of RyR to $[Ca^{2+}]i$ levels. For example, regulators of RyR play an important role in other excitable cells such as in the heart, and altered regulation of RyR appears to be a root cause of cardiac arrhythmias that occur in the elderly as well as in other pathological states (Marx and Marks, 2002). Thus, RyR are an important clinical target for therapeutic intervention in cardiovascular challenges faced by aging populations.

Modulators of RyR include FK506 binding protein, phosphorylation and intracellular molecules including cyclic adenosine diphosphate ribose (cADPr). FK506 inhibits the channel when bound to RyR, while phosphorylation of RyR reduces the binding of FK506, activating RyR by increasing its open probability (Danila and Hamilton, 2004). In the case of cADPr, in excitable cells including sympathetic neurons, this modulator can mobilize calcium from SER calcium stores and sensitize the RyR to elevated $[Ca^{2+}]i$, thus reducing the threshold of elevated $[Ca^{2+}]i$ necessary to evoke CICR (Galione, 1993; Hua *et al.*, 1994; Ogawa *et al.*, 2000; Hagashida et al., 2001; Marx & Marks, 2002). Furthermore, there are reports that RyR function is influenced by neuronal nitric oxide synthase (nNOS), via nitosylation of RyR by NO itself and modulation of cADPr levels (Galione, 1993; Eu *et al.*, 1999; Hagashida *et al.*, 2001; Balshaw *et al.*, 2002; Meissner, 2002).

Three isoforms of RyR are known; these are RyR1 (skeletal muscle), RyR2 (cardiac muscle and neurons), and RyR3 (neurons and other tissue types). All three have been shown to participate as calcium release channels responsible for CICR (Ogawa et al., 2000). Because RyR are integral components in [Ca^{2+}]i signaling in SCG neurons, we tested the hypothesis that, with advancing age, expression of the predominant RyR isoform(s) in adult rat SCG, along with selective modulators, is altered (Vanterpool *et al.*, 2006). For this study we used F-344 rats aged 6, 12 and 24 months and molecular techniques of reverse transcription polymerase chain reaction and enzyme-linked immunosorbent assays (ELISA). ELISA was used to measure RyR protein levels as these proteins are larger than 500 kDa, limiting the accuracy of Western blots.

Much to our surprise we found that RyR1 mRNA was undetectable in the rat SCG, in direct contrast with other studies demonstrating that RyR1 mRNA is expressed in excitable cells, including neurons (Fill & Copello, 2002). Thus, RyR1 does not appear to play a role in mediating CICR from SER calcium stores in the SCG, regardless of animal age. However, RyR2 and RyR3 mRNA are amply present in rat SCG, appearing to be the major receptor isoforms regulating calcium release from RyR-sensitive stores in all age groups. Furthermore, RyR3 mRNA and protein levels increased from 6 to 12 months and then decreased in senescent (24 month) animals, while RyR2 mRNA and protein levels remained constant with advancing age (Vanterpool *et al.*, 2006). From these data showing the impact of age on mRNA and protein levels of RyR3 it is difficult to make a straightforward conclusion. However, as discussed above, protein levels are not the only important measure of RyR function, as activity of these channels is be regulated by several factors.

Therefore, we extended our studies to investigate possible age-related changes in selected modulators of the RyR, including phosphorylation and nNOS levels which are known to modulate levels of cADPr (Vanterpool *et al.*, 2006). We found that total phosphorylation of RyR channels was not altered with age, suggesting that steady state phosphorylation of RyR and hence regulation by this mechanism does not necessarily change with advancing age. In contrast, however, nNOS protein levels increased from 6 to 12 months and significantly declined from 12 to 24 months. Since nNOS function modulates cADPr levels a reasonable inference is that cADPr levels may also decline with age. In the SCG, as there appear to be only two RyR contributing to calcium release, it is possible that the decline in RyR3 coupled with a decline in nNOS levels resulting in reduced cADPr may account in part for a decline in RyR function with age.

We have shown that caffeine-evoked release of calcium declines with age suggesting that the capacity of RyR to release calcium from the SER declines with advancing age. In addition, the contribution of CICR to EFS-evoked [Ca^{2+}]i transients is abolished in SCG cells from senescent animals (Vanterpool *et al.*, 2005; Behringer et al., 2009). Both agonist-mediated release of calcium from the SER and CICR activated via calcium influx depend on SER calcium filling levels, as well as modulation of RYR. Indeed, we have shown that SERCA function declines in the SCG (Tsai *et al.*, 1998; Pottorf *et al.*, 2000c). In light of these former studies a reasonable inference is that filling levels of the SER may be compromised, altering both agonist-mediated capacity of calcium release and activation of CICR from the SER.

8. Proposed model for age-related changes in [Ca2+]i regulation in peripheral adrenergic neurons

We have cited various reports and reviews suggesting that advancing age leads to [Ca^{2+}]i dysregulation and neuronal loss (Choi, 1992; Kirischuk & Verkhratsky, 1996; Ichas & Mazat,

1998; Thibault *et al.*, 1998; Begley *et al.*, 1999; Toescu and Verkhratsky, 2007). However, this hypothesis may be more applicable to pathological conditions as opposed to normal aging. For example, in our studies of peripheral adrenergic neurons, despite age-related changes in $[Ca^{2+}]i$ buffering which in turn alter $[Ca^{2+}]i$ signaling, the cells remain both viable and functional within reasonable limits. Thus, we pose another hypothesis: In the absence of any visible pathology, advancing age leads to subtle changes in the control of $[Ca^{2+}]i$ that may lead to altered neuronal function, but adjustments of other $[Ca^{2+}]i$ control mechanisms may allow neurons to adapt. This model is summarized in Fig. 1(B), emphasizing that loss of SERCA function may be compensated for by increased mitochondrial calcium uptake and plasma membrane calcium extrusion, so as to preserve some degree of cell viability in the face of advancing age. However, increased reliance on mitochondrial calcium buffering may have consequences for long-term viability as increased mitochondrial calcium may increase the chance of mitochondrial dysfunction and release of the apoptotic signals, caspases. This model also suggests that decline with age in SERCA function reduces SER calcium levels, thus reducing the contribution of CICR to overall calcium signaling in aging peripheral neurons.

9. Summary and conclusions

These data from our laboratory and supporting studies from others suggest that, in peripheral autonomic neurons in the absence of recognizable pathology, advancing age is a subtle and complex process that does necessarily lead to dramatic deterioration. Moreover, it is clear that advancing age does not alter the function of excitable neurons in a uniform manner, as we point out differences in the function of peripheral and central neurons with advancing age. With regards to age-related changes in $[Ca^{2+}]i$ regulation, we presented the idea that cell viability of peripheral neurons is maintained as they compensate for an age related decline in the function of at least one calcium-buffering system, SERCA, by increased function of other calcium-buffering systems, namely, the mitochondria and plasmalemmal calcium extrusion. Increased mitochondrial calcium uptake as a compensatory mechanism may leave the cell more susceptible to apoptosis which contributes to cell death (Ichas & Mazat, 1998; Thibault *et al.*,1998; Begley *et al.*, 1999; Toescu and Verkhratsky, 2007). This review summarizes the major findings in our work on the dynamics of $[Ca^{2+}]i$ regulation and possible consequences for autonomic nerve function with advancing age.

Based on results of our most current studies and reviews (Vanterpool *et al.*, 2005, 2006; Buchholz *et al.*, 2007; Behringer *et al.*, 2009) and our previous work and that of others we propose the following: With advancing age an alteration in $[Ca^{2+}]i$ signaling and function of peripheral adrenergic neurons results from a complex interplay of mechanisms, including increased sensitivity of the NA release mechanism to calcium, decline in SERCA function that alters calcium buffering and refilling of SER calcium stores, reduced RyR3 and decline in nNOS levels, which in turn modulates cADPr levels. This combination ultimately reduces the capacity of the SER to release calcium and abolishes the CICR process. With respect to the loss of CICR, this apparently leaves calcium influx as the primary mechanism to elevate $[Ca^{2+}]i$ in response to nerve activation, allowing some function to continue in senescent neurons. In addition, despite loss in the contribution of CICR to EFS-evoked $[Ca^{2+}]i$ transients, some residual CICR capacity is still retained if RyR are sensitized to elevated $[Ca^{2+}]i$ with an agonist. This latter observation offers some intriguing clinical possibilities to treat an age-related alteration in calcium signaling and hence peripheral neuronal function.

10. Future directions

Given advances in molecular techniques, including the explosion of knowledge of the function of micro RNA as inhibitors of mRNA translation and hence gene function (Saugstad, 2010), future studies are limitless. These may include investigations on the impact of advancing age on the genetic expression and protein levels of multiple buffering systems, including soluble calcium-buffering proteins, SERCA, mitochondrial calcium pumps, PMCA, and the Na^+/Ca^{2+} exchanger. Our work also strongly suggests that calcium influx through store operated calcium channels is compromised with age. Thus, genetic expression and function with advancing age of store operated channels may provide another fruitful avenue of research. Finally, all of our studies suggest that SER calcium levels decline with age, compromising CICR. New calcium indicators are now available to directly study SER calcium levels; this approach may allow more in depth studies of the impact of advancing age on SER calcium levels. These types of future studies and many more may be expected to provide more direct insight on how CICR may be altered with advancing age, potentially leading to novel therapeutic modalities to prevent and/or treat age- and disease-related alterations in neuronal and cardiovascular function.

11. References

Abbott RD, Curb JD, Rodriguez BL, Masaki KH, Popper JS, Ross GW and Petrovitch H (2003) Age-related changes in risk factor effects on the incidence of thromboembolic and hemorrhagic stroke. *J Clin Epidemiol* 56: 479-486.

Akita T and Kuba K. (2000) Functional triads consisting of ryanodine receptors, Ca^{2+} channels, and Ca^{2+}-activated K^+ channels in bullfrog sympathetic neurons: plastic modulation action potential. *J. Gen. Physiol.* 116: 697–720.

Amenta F, Bronzetti E, Ferrante F, Ricci A (1990) The noradrenergic innervation of spinal cord blood vessels in old rats. *Neurobiol. Aging* 11: 47–50.

Balshaw DM, Yamaguchi N, Meissner G (2002) Modulation of intracellular calcium-release channels by calmodulin. *J. Membr. Biol.* 185: 1–8.

Barker DJ (2008) Human growth and cardiovascular disease. *Nestle Nutr Workshop Ser Pediatr Program* 61:21-38.

Barnes RF, Rashkind M, Gumbrecht G, Halter JB (1982) The effects of age on the plasma catecholamine response to mental stress in man. *J. Clin. Endocrinol. Metab.* 54: 64–69.

Baylor SM and Hollingworth S (2000) Measurement and Interpretation of Cytoplasmic $[Ca^{2+}]i$ Signals From Calcium-Indicator Dyes. *News Physiol. Sci.* 15: 19-26.

Begley JG, Duan W, Chan S, Duff K, Mattson MP (1999) Altered calcium homeostasis and mitochondrial dysfunction in cortical synaptic compartments of presinilin-1 mutant mice. *J. Neurochem.* 72: 1030–1039. *Physiol.* 107: 1223-1234.

Behringer EJ, Leite LD, Buchholz NE, Keeney MG, Pearce WJ, Vanterpool CK, Wilson SM and Buchholz JN (2009[a]) Maturation and long-term hypoxia alters Ca^{2+}-induced Ca^{2+} release in sheep cerbrovascular sympathetic neurons. *J. Appl. Pysiol.* 107(4): 1223-1234.

Behringer EJ, Vanterpool CK, Pearce WJ, Wilson SM and Buchholz JN (2009[b]) Advancing age alters the contribution of calcium release from smooth endoplasmic reticulum stores in superior cervical ganglion cells. J. *Gerontol. A Biol. Sci. Med. Sci.* 64A(1): 34-44.

Belan P, Kostyuk P, Snitsarev V, Tepikin A (1993) Calcium clamp in isolated neurons of the snail *Helix pomatia. J. Physiol. (Lond.)* 462: 47–58.

Berridge MJ (1995) Calcium signaling and cell proliferation. *Bioessays* 17: 491-500.

Berridge MJ (1998) Neuronal calcium signaling. *Neuron* 21: 13-26.

Berridge MJ (2002) The endoplasmic reticulum: a multifunctional signaling organelle. *Cell Calcium* 32(5-6): 235-249, 2002.

Bevan RD, Tsuru H (1981) Functional and structural changes in the rabbit ear artery after sympathetic denervation. *Circ. Res.* 48: 478–485.

Blaustein MP, Lederer WJ (1999) Sodium/calcium exchange: its physiological implications. *Physiol. Rev.* 79: 763–854.

Bodkin NL, Alexander TM, Ortmeyer HK, Johnson E, Hansen BC (2003) Mortality and morbidity in laboratory-maintained Rhesus monkeys and effects of long-term dietary restriction. *J. Gerontol. A Biol. Sci. Med. Sci.* 58: 212–219.

Borton M, Docherty JR (1989) The effects of ageing on neuronal uptake of noradrenaline in the rat. *Naunyn Schmiedebergs Arch. Pharmacol.* 340: 139–143.

Buchholz JN, Behringer EJ, Pottorf WJ, Pearce WJ, Vanterpool CK (2007) Age-dependent changes in Ca^{2+} homeostasis in peripheral neurones: implications for changes in function. *Aging Cell* 6: 285-296

Buchholz J, Duckles SP (1990) Effect of age on prejunctional modulation of NE release. *J. Pharmacol. Exp. Ther.* 252, 159–164.

Buchholz J, Nikkah L, Duckles SP (1994) Age-related changes in the sensitivity of sympathetic nerves to altered extracellular calcium in tail arteries of F-344 rats. *Neurobiol. Aging* 15: 197–201.

Buchholz J, Sexton P, Hewitt W (1998) Impact of age on modulation of norepinephrine release from sympathetic nerves in the rat superior mesentery artery. *Life Sci.* 62: 679–686.

Buchholz J, Tsai H, Foucart S, Duckles SP (1996) Advancing age alters intracellular calcium buffering in rat adrenergic nerves. *Neurobiol. Aging* 17: 885–892.

Buchholz J, Tsai H, Friedman D, Duckles SP (1992) Influence of age on control of norepinephrine release from the rat tail artery. *J. Pharmacol. Exp. Ther.* 260: 722–727.

Cahalan MD (2009) STIMulating store-operated Ca^{2+} entry. *Nat. Cell Biol.* 11(6): 669-677.

Cardinali DP, Vacas MI, Gejman PV (1981) The sympathetic superior cervical ganglia as peripheral neuroendocrine centers. *J. Neural Transm.* 52, 1–21.

Cassaglia PA, Griffiths RI, Walker AM (2008) Sympathetic nerve activity in the superior cervical ganglia increases in response to imposed increases in arterial pressure. *Am J Physiol Regul Integr Comp Physiol* 294:R1255-1261.

Cassaglia PA, Griffiths RI, Walker AM (2009) Cerebral sympathetic nerve activity has a major regulatory role in the cerebral circulation in REM sleep. *J Appl Physiol.* 106:1050-1056.

Catterall WA, Perez-Reyes E, Snutch TP, Striessnig J (2005) International union of pharmacology. XLVII. Nomenclature and structure-function relationships of voltage-gated calcium channels. *Pharmacol. Rev.* 54(4): 411-425.

Cavazzini M, Bliss T, Emptage N (2005) Ca^{2+} and synaptic plasticity. *Cell Calcium* 38: 355-367.

Chen PS, Chen LS, Cao JM, Sharifi B, Karagueuzian HS, Fishbein MC (2001) Sympathetic nerve sprouting, electrical remodeling and the mechanisms of sudden cardiac death. *Cardiovasc. Res.* 50, 409–416.

Choi DW (1992) Excitotoxic cell death. *J. Neurobiol.* 23: 1261-1276.

Clancy DJ, Gems D, Harshman LG, Oldham S, Stocker H, Hafen E, Leevers SJ, Partridge L (2001) Extension of life-span by loss of CHICO, a drosphila insulin receptor substrate protein. *Science* 292, 104–106.

Clapham DE (1995) Calcium signaling. *Cell* 80: 259-268.

Colegrove SL, Albrecht MA, Friel DD (2000) Quantitative analysis of mitochondrial Ca^{2+} uptake and release pathways in sympathetic neurons. *J. Gen. Physiol.* 115: 371–388.

Cowen T, Gavazzi I (1998) Plasticity in adult and aging sympathetic neruons. *Prog. Neurobiol.* 54: 249–288.

Daly RN, Goldberg PB, Roberts J (1989) The effect of age on presynaptic α_2-adrenoceptor auto regulation of norepinephrine release. *J. Gerontol.* 44: B59–B66.

Danila CI, Hamilton SL (2004) Phosphorylation of ryanodine receptors. *Biol. Res.* 37: 521–525.

Dawson R Jr, Meldrum MJ (1992) Norepinephrine content in cardiovascular tissues from the aged Fischer 344 rat. *Gerontology* 38: 185–191.

De Avellar MCW, Kobashi YL, Markus RP (1990) Age-related changes in neuronal uptake of noradrenaline. *Naunyn Schmiedebergs Arch. Pharmacol.* 341: 295–300.

Delcour AH, Tsien RW (1993) Altered prevalence of gating modes in neurotransmitter inhibition of N-type calcium channels. *Science* 259: 980–983.

Dickason AK, Isaacson LG (2002) Plasticity of aged perivascular axons following exogenous NGF: analysis of catecholamines. *Neurobiol. Aging* 23: 125–134.

Docherty JR, Hyland L (1986) Alpha-adrenoceptor responsiveness in the aged rat. *Eur. J. Pharmacol.* 126: 75–80.

Dove LS, Nahm SS, Murchison D, Abbott LC and Griffith WH (2000) Altered calcium homeostasis in cerebellar Purkinje cells of leaner mutant mice. *J Neurophysiol* 84: 513–524.

Duckles SP, Carter BJ, Williams CL (1985) Vascular adrenergic neuroeffector function does not decline in aged rats. *Circ. Res.* 56: 109–116.

Duckles SP, Krause DN, Stirone C, Procaccio V (2006) Estrogen and mitochondria: a new paradigm for vascular protection. *Mol. Interv.* 6: 26–35.

Dun NJ, Dun SL, Chiba T, Forsterman U (1995) Nitric oxide synthase immunoreactive vagal afferent fibers in rat superior cervical ganglia. *Neuroscience* 65: 231–239.

Emptage NJ, Reid CA, Fine A (2001) Calcium Stores in Hippocampal Synaptic Boutons Mediate Short-Term Plasticity, Store-Operated Ca^{2+} Entry, and Spontaneous Transmitter Release. *Neuron* 29: 197-208.

Esler M, Jackman G, Bobik A, Leonard P, Kelleher LD, Skews H, Jennings G, Korner P (1981[a]) Norepinephrine kinetics in essential hypertension. Defective neuronal uptake of norepinephrine in some patients. *Hypertension* 3: 149-156.

Esler M, Skews H, Lionard P, Jackman G, Bobik A, Korner P (1981[b]) Age-dependence of noradrenaline kinetics in normal subjects. *Clin. Sci.* (*Lond*) 60: 217–219.

Esler MD, Thompson JM, Kaye DM, Turner AG, Jennings GL, Cox HS, Lambert GW, Seals DR (1995) Effects of ageing on the responsiveness of human cardiac sympathetic nerves to stressors. *Circulation* 91: 351–358.

Eu JP, Xu L, Stamler JS, Meissner G (1999) Regulation of ryanodine receptors by reactive nitrogen species. *Biochem. Pharmacol.* 57: 1079–1084.

Faraci FM and Heistad DD (1998) Regulation of the cerebral circulation: role of endothelium and potassium channels. *Physiol Rev* 78:53-97.

Fill M, Copello JA (2002) Ryanodine receptor calcium release channels. *Physiol. Rev.* 82: 893–922.

Forster MJ, Morris P, Sohal RS (2003) Genotype and age influence the effect of caloric intake on mortality in mice. *FASEB J.* 17: 690–692.A

Friel DD, and Chiel HJ (2008) Calcium dynamics: analyzing the Ca^{2+} regulatory network in intact cells. Trends Neurosci 31(1): 8-19.

Frye CA, Edinger KL, Edwin D. Lephart ED, Walf AA (2010) 3α-androstanediol, but not testosterone, attenuates age-related decrements in cognitive, anxiety, and depressive behavior of male rats. *Front. Aging Neurosci.* 15(2): 1-21.

Furuichi S, Endo S, Haji A, Takeda R, Nisijima M and Takaku A (1999) Related changes in sympathetic activity, cerebral blood flow and intracranial pressure, and effect of an alpha-blocker in experimental subarachnoid haemorrhage. *Acta Neurochir (Wien)* 141: 415-423; discussion 423-414.

Gafni A, Yuh KC (1989) A comparative study of the Ca^{2+}–Mg^{2+} dependent ATPase from skeletal muscles of young, adult and old rats. *Mech. Ageing Dev.* 49: 105–117.

Galante M, Marty A (2003) Presynaptic Ryanodine-Sensitive Calcium Stores Contribute to Evoked Neurotransmitter Release at the Basket Cell-Purkinje Cell Synapse. *J Neurosci.* 23(35): 11229-11234.

Galione A (1993) Cyclic ADP-ribose: a new way to control calcium. *Science* 259, 325–326.

Gems D (2011) Aging: To treat or not to treat. *Amer. Scientist* 99: 178-280.

Ginty DD (1997) Calcium regulation of gene expression: isn't that spatial? *Neuron* 18: 183-186.

Griffith WH, Jasek MC, Bain SH, Murchison D (2000) Modification of ion channels and calcium homeostasis of basal forebrain neurons during aging. *Behav. Brain Res.* 115: 219–233.

Guarente L, Kenyon C (2000) Genetic pathways that regulate ageing in model organisms. *Nature* 408: 255–262.

Hagashida H, Hashii, M Yokoyama S, Hoshi N, Chen XL, Egorova A, Noda M, Zhang JS (2001) Cyclic ADP-ribose as a second messenger revisited from a new aspect of signal transduction from receptors to ADP-ribosyl cyclase. *Pharmacol. & Therap.* 90: 283– 296

Handa RK, Duckles SP (1987) Influence of age on norepinephrine content in arteries and veins of Fischer 344 rats. *Neurobiol. Aging* 8: 511–516.

Hamel E (2006) Perivascular nerves and the regulation of cerebrovascular tone. *J Appl Physiol* 100:1059-1064.

Harman D (1998) Aging and oxidative stress. *J. Int. Fed. Clin. Chem.* 10: 24–27.

Hasenfuss G, Schillinger W, Lehnart SE, Preuss M, Pieske B, Maier LS, Prestle J, Minami K, Heizmann CW, Hunziker W (1991) Intracellular calcium-binding proteins: more sites than insights. *Trends Biochem. Sci.* 16: 98–103.

Hermes-DeSantis ER and Clyman RI (2006) Patent ductus arteriosus: pathophysiology and management. *J Perinatol* 26 Suppl 1: S14-18; discussion S22-23.

Hua SY, Tokimasa T, Takasawa S, Furuya Y, Nohmi M, Okamoto H, Kuba K (1994) Cyclic ADP-ribose modulates Ca^{2+} release channels for activation by physiological Ca^{2+} entry in bullfrog sympathetic neurons. *Neuron* 12: 1073–1079.

Hyland L, Docherty JR (1985) Further examination of the effects of ageing on the adrenoceptor responsiveness of the rat vas deferens. *Eur. J. Pharmacol.* 110: 241–246.

Ichas F, Mazat JP (1998) From signaling to cell death: two conformations for the mitochondrial permeability transition pore. Switchin from low- to high-conductance state. *Biocim. Biophys. Acta* 1366: 33–50.

Illes P (1986) Mechanisms of receptor-mediated modulation of transmitter release in noradrenergic, cholinergic and sensory neurones. *Neuroscience* 17: 909–928.

Insel PA (1993) Adrenergic receptors, G proteins, and cell regulation: implications for aging research. *Exp. Gerontol.* 28: 341–348.

Isaacson LG, Crutcher KA (1998) Uninjured aged sympathetic neurons response to exogenous NGF. *Neurobiol. Aging* 19: 333–339.

Jordan J, Shannon JR, Diedrich A, Black B, Costa F, Robertson D and Biaggioni I (2000) Interaction of carbon dioxide and sympathetic nervous system activity in the regulation of cerebral perfusion in humans. *Hypertension* 36: 383-388.

Khachaturian ZS (1987) Hypothesis on the regulation of cytosol calcium concentration and the aging brain. *Neurobiol. Aging* 8: 345–346.

Kirischuk S, Pronchuk N, Verkhratsky A (1992) Measurements of intracellular calcium in sensory neurons of adult and old rats. *Neurosci.* 50: 957–951.

Kirischuk S, Verkhratsky A (1996) Calcium homeostasis in aged neurones. *Life Sci.* 59: 451–459.

Kostyuk PG (1989) Diversity of calcium ion channels in cellular membranes. *Neuroscience* 28: 253-261.

Kostyuk PG, Pronchuk N, Savchenko A, Verkhratsky A (1993) Calcium currents in aged rat dorsal root ganglion neurones. *J. Phsyiol. (Lond.)* 461: 467-483.

Kuo TH, Liu BF, Yu Y, Wuytack F, Raeymaekers L, Tsang W (1997) Coordinated regulation of the plasma membrane calcium pump and sarco (endo) plasmic reticular calcium pump gene expression by Ca^{2+}. *Cell Calcium* 21: 399–408

Landfield PW (1987) 'Increased calcium-current' hypothesis of brain aging. *Neurobiol. Aging* 8: 346–347.

Landfield PW, Pitler TA (1984) Prolonged Ca^{2+}-dependent after hyperpolarizations in hippocampal neurons of aged rats. *Science* 226: 1089–1092.

Langer SZ, Arbilla S (1990) Presynaptic receptors on peripheral adrenergic neurons. *Ann. N. Y. Acad. Sci.* 604: 7–16.

Lawes CM, Bennett DA, Feigin VL and Rodgers A (2004) Blood pressure and stroke: an overview of published reviews. *Stroke* 35: 776-785.

Lee TJF, Zhang W, Sarwinski S (2000) Pre-synaptic β2-adrenoceptors mediate nicotine-induced NOergic neurogenic dilation in porcine basilar arteries. *Am. J. Physiol. Heart Circ. Physiol.* 279: H808–H816.

Lee JM, Zipfel GJ, Choi DW (1999) The changing landscape of ischemic brain injury mechanisms. *Nature* 399: A7–A14.

Limas CJ (1975) Comparison of the handling of norepinephrine in the myocardium of adult and old rats. *Cardiovasc. Res.* 9: 664–668.

Maciel LM, Polikar R, Rohrer D, Popovich BK, Dillmann WH (1990) Age induced decreases in the messenger RNA coding for the sarcoplasmic reticulum Ca^{2+}-ATPase of the rat heart. *Circ. Res.* 67: 230–234.

Martinez JL, Vasquez BJ, messing RB, Jennsen RA, Liang KC, McCaugh JL (1981) Age related changes in catecholamine content of peripheral organs in male and female F-344 rats. *J. Gerontol.* 36: 280–284.

Martyn CN, Barker DJ and Osmond C (1996) Mothers' pelvic size, fetal growth, and death from stroke and coronary heart disease in men in the UK. *Lancet* 348: 1264-1268.

Marx SO, Marks AR (2002) Regulation of the ryanodine receptor in heart failure. *Basic Res. Cardiol.* 97 (Suppl. 1): I49–I51.

Mbaku EN, Zhang L, Duckles SP, Buchholz JN (2000) Nitric oxide synthase containing nerves facilitate adrenergic transmitter release in sheep middle cerebral arteries. *J. Pharmacol. Exp. Ther.* 293: 397–402.

Mbaku EN, Zhang L, Pearce WJ, Duckles SP, Buchholz J (2003) Chronic hypoxia alters the function of NOS nerves in cerebral arteries of near term fetal and adult sheep. *J. Appl. Physiol.* 94: 724–732.

Meissner G (2002) Regulation of mammalian ryanodine receptors. *Front. Biosci.* 7: d2072–d2080.

Meldrum B, Garthwaite J (1990) Excitatory amino acid neurotoxicity and neurodegenerative disease. *Trends Pharmac. Sci.* 11: 379–387.

Miller RA (2005)The anti-aging sweepstakes: Catalase runs for the ROSes. *Science* 308: 1875–1876.

Miller, RJ (1991) The control of neuronal Ca^{2+} homeostasis. *Prog Neurobiol.* 37: 255-285.

Mione MC, Erdo SL, Kiss B, Ricci A, Amenta F (1988) Age-related changes of noradrenergic innervation of rat splanchnic blood vessels: a histofluorescence and neurochemical study. *J. Auton. Nerv. Syst.* 25: 27–33.

Montague PR, Gancayco CD, Winn MJ, Marchase RB, Friedlander MJ (1994) Role of NO production in NMDA receptor-mediated neurotransmitter release in cerebral cortex. *Science* 263: 973–977.

Murchinson D, Griffith WH (1996) High-voltage-activated calcium currents in basal forebrain neurons during ageing. *J. Neurophysiol.* 76: 158–174.

Murchinson D, Griffith WH (1998) Increased calcium buffering in basal forebrain neurons during aging. *J. Neurophysiol.* 80: 350–364.

Murchinson D, Griffith WH (1999) Age-related alterations in caffeine sensitive calcium stores and mitochondrial buffering in rat basal forebrain. *Cell Calcium* 25: 439–452.

Murchison D, Zawieja DC, Griffith WH (2004) Reduced mitochondrial buffering of voltage-gated calcium influx in aged rat basal forebrain neurons. *Cell Calcium* 36: 62-75.

Neher E (1995) The use of Fura-2 for estimating Ca buffers and Ca fluxes. *Neuropharmacology* 34: 1423–1442.

Nicholls DG (1978) The regulation of extramitochondrial free calcium ion concentration by rat liver mitochondria. *Biochem. J.* 176: 463-474.

Nicholls DG (1985) A role for the mitochondrion in the protection of cells against calcium overload? *Prog. Brain Res.* 63: 97-106.

Ogawa Y, Kurebayashi N, Murayama T (2000) Putative roles of type 3 ryanodine receptor isoforms (RyR3). *Trends Cardiovasc. Med.* 10: 65–70.

Palmer GJ, Ziegler MG, Lake CR (1978) Response to NE and blood pressure to stress increases with age. *J. Gerontol.* 33: 482–
487.

Peterson C (1992) Changes in calcium's role as a messenger during aging in neuronal and non-neuronal cells. *Ann. N. Y. Acad. Sci.* 663: 279–293.

Pieribone, VA, Shupliakov O, Brodin L, Hilfiker-Rothenfluh S, Czernik AJ and Greengard P (1995) Distinct pools of synaptic vesicles in neurotransmitter release. *Lett. to Nature* 375: 493-496.

Pitler TA, Landfield PW (1990) Aging-related prolongation of calcium spike duration in rat hippocampal slice neurons. *Brain Res.* 508: 1–6.

Porter NM, Thibault O, Thibault V, Chen KY, Landfield PW (1997) Calcium channel density and hippocampal cell death with age in long-term culture. *J. Neurosci.* 17: 5629–5639.

Pottorf WJ, Duckles SP, Buchholz JN (2000[a]) Adrenergic nerves compensate for a decline in calcium buffering during ageing. *J. Auton.Pharmacol.* 20: 1–13.

Pottorf WJ, Duckles SP, Buchholz JN (2000[b]) Mechanisms of calcium buffering in adrenergic neurones and effects of aging: testing the limits of homeostasis. *J. Auton. Pharmacol.* 20: 63–75.

Pottorf WJ, Duckles SP, Buchholz JN (2000[c]) SERCA function declines with age in adrenergic nerves from the superior cervical ganglion. *J. Auton. Pharmacol.* 20: 281–290.

Pottorf WJ, Duckles SP, Buchholz JN (2002) Aging and calcium buffering in adrenergic neurons. *Auton. Neurosci.* 96: 2–7.

Pottorf WJ, Johanns TM, Derrington SM, Strehler EE, Enyedi A, Thayer SA (2006) Glutamate-induced protease-mediated loss of plasma membrane Ca^{2+} pump activity in rat hippocampal neurons. *J. Neurochem.* 98: 1646–1656.

Pottorf WJ, Thayer SA (2002) Transient rise in intracellular calcium produces long lasting increase in plasma membrane calcium pump activity in rat sensory neurons. *J. Neurochem.* 83: 1002–1008.

Qin Z, Zaidi A, Gao J, Krainev AG, Michaelis ML, Squier TC, Bigelow DJ (1998) Decrease in Ca-ATPase activity in age synaptosomal membranes is not associated with changes in fatty acyl chain dynamics. *Mech. Aging Dev.* 105, 291–300.

Raza M, Deshpande LS, Blair RE, Carter DS, Robert J. DeLorenzo RJ (2007) Aging is associated with elevated intracellular calcium levels and altered calcium homeostatic mechanisms in hippocampal neurons. *Neurosci. Letters* 418: 77-81.

Roe MW, Lemasters JJ, Herman B (1990) Assessment of Fura-2 for measurements of cytosolic free calcium. *Cell Calcium* 11: 63–73.

Satrustegui J, Villalba M, Pereira R, Bogonez E, Martinez-Serrano A (1996) Cytosolic and mitochondrial calcium in synaptosomes during ageing. *Life Sci.* 59: 429–434.

Saugstad JA (2010) MicroRNAs as effectors of brain function with roles in ischemia and injury, neuroprotection, and neurodegeneration. J. Cereb. Blood Flow Metab. 30: 1564–1576.

Schofield GG (1990) Norepinephrine blocks a calcium current of adult rat sympathetic neurons via an α2-adrenoceptor. *Eur. J. Pharmacol.* 180: 37–47.

Siksou L, Triller A and Marty S. Ultrastructural organization of presynaptic terminals (2011) *Curr. Opinion in Neurobiol.* 21: 1-8.

Tatar M, Kopelman A, Epstein D, Tu MP, Yin CM, Garofalo RS (2001) A mutant drosophila insulin receptor homolog that extends life span and impairs neuroendocrine function. *Science* 292: 107–110.

Thayer SA, Miller RJ (1990) Regulation of the intracellular free calcium concentration in single rat dorsal root ganglion neurones *in vitro*. *J. Physiol.* 425: 85–115.

Thibault O, Porter NM, Chen KC, Blalock EM, Kaminker PG, Clodfelter GV, Brewer LD, Landfield PW (1998) Calcium dysregulation in neuronal aging and Alzheimers disease: history and new directions. *Cell Calcium* 24: 417–433.

Toescu EC, Verkhratsky A (2007) The importance of being subtle: small changes in calcium homeostasis control cognitive decline in normal aging. *Aging Cell* 6: 267-273.

Troen BR (2003) The biology of aging. *Mt. Sinai J. Med.* 70, 3–22.

Trouslard J, Marsh SJ, Brown DA (1993) Calcium entry through nicotinic receptor channels and calcium channels in cultured rat superior cervical ganglion cells. *J. Physiol. (Lond.)* 468: 53-71.

Tsai H, Hewitt CW, Buchholz JN, Duckles SP (1997) Intracellular calcium buffering declines in ageing adrenergic nerves. *Neurobiol. Aging* 18: 229–233.

Tsai H, Pottorf WJ, Buchholz JN, Duckles SP (1998) Adrenergic nerve smooth endoplasmic reticulum calcium buffering declines with age. *Neurobiol. Aging* 19: 89–96.

Tsai SH, Tew JM, Shipley MT (1989) Cerebral arterial innervation II. Development of calcitonin gene-related peptide and norepinephrine in the rat. *J. Comp. Neurol.* 27:, 1–12.

Tsien RY, Rink TJ, Poenie M (1985) Measurement of cytosolic free Ca^{2+} in individual small cells using fluorescent microscopy with dual excitation wavelengths. *Cell Calcium* 6: 145–157.

Usachev YM, Thayer SA (1997) All-or-none Ca^{2+} release from intracellular stores triggered by Ca^{2+} influx through voltage-gated Ca^{2+} channels in rat sensory neurons. *J. Neurosci.* 17: 7404–7414.

Usachev YM and Thayer SA (1999[a]). Controlling the urge to surge: all or none Ca^{2+} release in neurons. *Bioessays* 21: 743–750.

Usachev YM and Thayer SA (1999[b]) Ca^{2+} influx in resting rat sensory neurones that regulates and is regulated by ryanodine-sensitive Ca^{2+}stores. *J. Physiol.* 519 (Part 1), 115–130.

van Lieshout JJ and Secher NH (2008) Point:Counterpoint: Sympathetic activity does/does not influence cerebral blood flow. Point: Sympathetic activity does influence cerebral blood flow. *J Appl Physiol* 105: 1364-1366.

Vanterpool CK, Pearce WJ, Buchholz JN (2005) Advancing age alters rapid and spontaneous refilling of caffeine-sensitive calcium stores in sympathetic superior cervical ganglion cells. *J. Appl. Physiol.* 99: 963-971.

Vanterpool CK, Vanterpool EA, Pearce WJ, Buchholz JN (2006) Advancing age alters the expression of the ryanodine receptor 3 isoform in adult rat superior cervical ganglia. *J. Appl. Physiol.* 101: 392-400.

Verkhratsky AJ and Petersen OH (1998) Neuronal calcium stores. *Cell Calcium* 24: 333–343.

Verkhratsky AJ and Shmigol A (1996) Calcium-induced calcium release in neurones. *Cell Calcium* 19: 1–14.

Verkhratsky A, Shmigol A, Kirischuk S, Pronchuk N, Kostyuk P (1994) Age-dependent changes in calcium currents and calcium homeostasis in mammalian neurons. *Ann. N. Y. Acad. Sci.* 747: 365–381.

Verkhratsky A, Toescu EC (1998) Calcium and neuronal ageing. *TINS* 21: 2–7.

Vitorica J, Satrustegui J (1985) Impairment of glutamate uptake and absence of alterations in the energy-transducing ability of old rat brain mitochondria. *Mech. Ageing Dev.* 29: 255–266.

Villalba M, Pereira R, Martinez-Serrano A, Satrustegui J (1995) Altered cell calcium regulation in synaptosomes and brain cells of the 30-month-old rat prominent effects in hippocampus. *Neurobiol. Aging* 16: 809–816.

Werth JL, Thayer SA (1994) Mitochondria buffer physiological calcium loads in cultured rat dorsal root ganglion neurons. *J. Neurosci.* 14: 348–356.

Werth JL, Usachev YM, Thayer SA (1996) Modulation of Ca^{2+} efflux from rat dorsal root ganglion neurons. *J. Neurosci.* 16: 1008–1015.

Wingerd KL, Goodman NL, Leu ST, Clegg DO (2004) Expression and function of integrin alpha4beta1 and vascular cell adhesion molecule- 1 (VCAM-1) during sympathetic innervation of the heart. *Dev. Dyn.* 231: 359–369.

Wu SY, Dun SL, Forstermann U, Dun NJ (1997) Nitric oxide and excitatory postsynaptic currents in immature rat sympathetic preganglionic neurons *in vitro*. *Neuroscience* 79: 237–245

Wuytack F, Raeymaekers L, Missiaen L (2002) Molecular physiology of the SERCA and SPCA pumps. *Cell Calcium* 32: 279–305.

Xiong J, Verkhratsky A, Toescu E (2002) Changes in mitochondrial status associated with altered Ca^{2+} homeostasis in aged cerebellar granule neurons in brain slices. *J. Neurosci.* 22: 10761–10771.

Xu A, Narayanan N (1998) Effects of aging on sarcoplasmic reticulum Ca^{2+}-cycling proteins and their phosphorylation in rat myocardium. *Am. J. Physiol. Heart Circ. Physiol.* 275: H2087–H2094.

Yamamoto R, Wada A, Asada Y, Toshihiko Y, Tomoaki Y, Niina H, Sumiyoshi A, Kobayashi H, Lee TJF (1997) Nitric oxide-dependent and independent norepinephrine release in rat mesenteric arteries. *Am. J. Physiol. Heart Circ. Physiol.* 272, H207–H210.

Yu BP, Masoro EJ, McMahan A (1985) Nutritional influences on aging of F-344 rats. *J. Gerontol.* 40: 657–670.

Zhang W, Edvinsson L, Lee TJF (1998) Mechanism of nicotine-induced relaxation in the porcine basilar artery. *J. Pharmacol. Exp. Ther.* 284: 790–797.

Zhang B, Fugleholm K, Day LB, Ye S, Weller RO and Day IN (2003) Molecular pathogenesis of subarachnoid haemorrhage. *Int J Biochem Cell Biol* 35: 1341-1360.

15

All Your Eggs in One Basket: Mechanisms of Xenobiotic Induced Female Reproductive Senescence

Alexander P. Sobinoff, Ilana R. Bernstein and Eileen A. McLaughlin
Priority Research Centre in Chemical Biology,
University of Newcastle
Australia

1. Introduction

The irreplaceable mammalian primordial follicle represents the basic unit of female fertility, serving as the primary source of all developing oocytes in the ovary. These primordial follicles remain quiescent, often for decades, until recruited into the growing pool throughout a woman's adult reproductive years. Once recruited, <1% will reach ovulation, with the remainder undergoing an apoptotic process known as atresia (Hirshfield, 1991). Menopause, or ovarian senescence, occurs when the pool of primordial follicles becomes exhausted.

Pre-mature ovarian failure (POF; or early menopause) is an ovarian defect characterised by the premature loss of menstrual cyclicity before the age of 40, well below the median age of natural menopause (51 years). Approximately 1-4% of the female population suffers from this condition, making POF a significant contributor of female infertility (Coulam et al., 1986). There is now a growing body of evidence which suggests that foreign synthetic chemicals, also known as xenobiotics, are capable of causing POF by inducing premature follicular depletion. Indeed, exposure to pesticides, workplace chemicals, chemotherapeutic agents and cigarette smoke have all been associated with primordial follicle reduction resulting in premature ovarian senescence (Hoyer and Devine, 2001; Mattison et al., 1983a, 1983b; Sobinoff et al., 2010, 2011).

In addition to infertility, the loss of ovarian hormones which accompanies POF has been connected with an increased risk of early morbidity and mortality (Shuster et al., 2010). With current statistics indicating an increasing trend in western women opting to delay childbirth, xenobiotic exposure could have long lasting repercussions for both the fertility and long term health of these women. In this review we discuss the susceptible nature of primordial follicles and the consequences of xenobiotic induced POF. We then examine the mechanisms of ovotoxicity for environmental toxicants and xenobiotics known to target immature follicles, and discuss the development of novel methods of wildlife fertility control utilizing these ovotoxicants.

2. The primordial follicle: Precious and vulnerable

Oocyte development and maturation occurs within ovarian follicles. These follicles assemble when primary oocytes (arrested at meiosis prophase I) are enveloped by a single layer of flattened granulosa cells, forming the most immature stage of follicular development, the primordial follicle. The timing of this event is species-specific, but generally occurs in the primitive ovary during foetal development (McNatty et al., 2000). Due to the nature of follicular formation, the number of oocytes established around the time of birth is finite, and represents the total number of germ cells available to the mammalian female throughout her entire life (Edson et al., 2009). It is therefore the size and persistence of this primordial follicle pool which determines the female reproductive lifespan (Fig. 1).

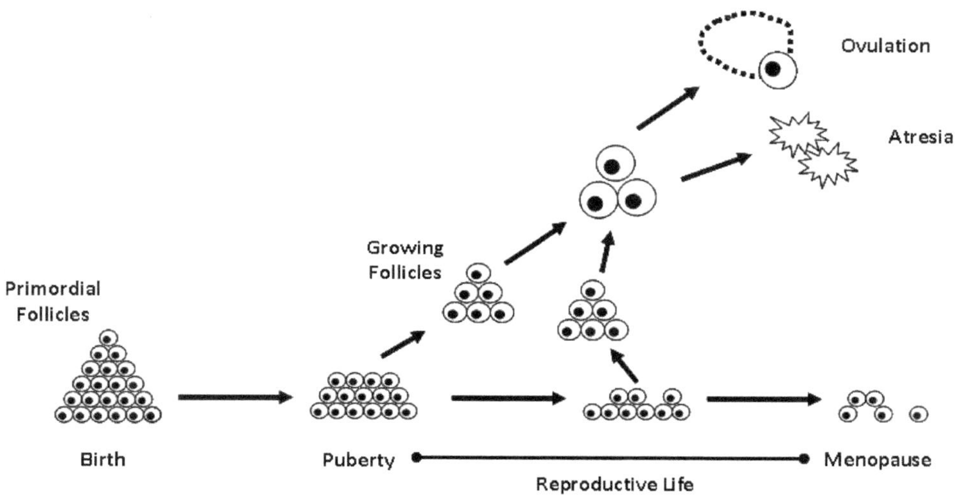

Fig. 1. Simple mechanistic diagram of the human female reproductive lifespan.

The first stage of folliculogenesis involves the recruitment of selected primordial follicles into the growing population. To prolong the length of the female reproductive lifespan, only a few primordial follicles are chosen for recruitment at any one time, with some follicles remaining in a quiescent (non-proliferative) state for months or years (Fig 1). This event occurs in regular waves, and is continuous from birth until ovarian senescence (McGee and Hsueh, 2000). Overall, only a few hundred of all the recruited follicles will complete folliculogenesis and undergo ovulation, with the vast majority being lost to atresia (Hirshfield, 1991). Atresia is thought to be an apoptotic process which selects the healthiest follicles for ovulation, although its mechanism of action is poorly understood. As virtually all follicles are lost during optimal follicular development, it is important that primordial follicles not only survive but also are maintained in a healthy state. Over-stimulation of primordial follicle activation and premature atresia results in the extensive depletion of the primordial follicle pool, resulting in premature ovarian senescence (Reddy et al., 2008).

There is now an increasing volume of studies which link primordial follicle depletion with xenobiotic exposure, suggesting that these irreplaceable follicles are highly sensitive to cytotoxic insult (Hoyer and Devine, 2001; Mattison et al., 1983a, 1983b; Sobinoff et al., 2010, 2011). It is thought that this sensitivity may stem from the primordial follicles quiescent nature. For example, somatic cells which undergo regular rounds of proliferation constantly renew macromolecules and organelles by virtue of mitosis. However, the oocyte and granulosa cells of the primordial follicle are non-proliferative, and do not benefit from mitotic renewal, perhaps making them excessively vulnerable to xenobiotics which cause sub-lethal damage to mitochondria and other structures over time (Tarin, 1996). Similarly the location of the primordial follicle population within a poorly vascularised region of the ovarian cortex also makes them highly susceptible to toxins which damage ovarian blood vessels, with the resulting cortexual fibrosis destroying primordial follicle rich segments of the ovary (Guraya, 1985; Meirow et al., 2007; van Wezel and Rodgers, 1996).

In addition to direct primordial follicle injury, certain xenobiotics which target developing follicles have been shown to cause excessive primordial follicle activation (Keating, 2009; Sobinoff et al., 2010, 2011). This may be due to a homeostatic mechanism of follicular replacement, in which destroyed developing follicles result in primordial follicle activation to replace the developing pool. If the offending xenobiotic is not removed, this could potentially lead to a vicious cycle of primordial follicle depletion.

3. Consequences of xenobiotic induced primordial follicle depletion

The overall impact of xenobiotic induced follicular depletion on female reproduction depends on the type of follicle targeted for destruction, dose, and duration of exposure (Hoyer and Sipes, 1996). For example, xenobiotics which target large developing follicles have an immediately noticeable effect on female fertility. Antral follicles are the primary producers of ovarian estrogen, and therefore play an important role in the FSH-LH negative feedback loop responsible for ovulation. Xenobiotics which selectively target antral follicles consequently have harmful effects on ovarian cyclicity, effectively acting as endocrine disruptors (Jarrell et al., 1991; Mattison and Schulman, 1980). Fortunately, both prolonged and acute exposure to these ovotoxic agents only causes temporary infertility, as these follicles can be replaced by the primordial follicle pool once the harmful xenobiotic is removed from the immediate environment.

Conversely, xenobiotics which target small pre-antral follicles have more permanent effects on female fertility which could potentially go unrecognised for years. Due to the non-renewing nature of primordial follicles, these xenobiotics are particularly damaging to female fertility, causing permanent infertility and premature ovarian senescence. What makes this type of ovotoxicity concerning is that it has a delayed effect on reproduction which is not made apparent until such a time that follicular recruitment cannot be supported (Hooser et al., 1994). Thus this extended period of time between cause and effect means that the detrimental action of xenobiotic contact often goes unnoticed, and consequently steps are not taken to minimise exposure until it is too late. Thus even a systemic low dose of xenobiotics may produce cumulative effects over time, resulting in the same consequences on female fertility as a large single exposure. With current statistics suggesting an increasing trend in developed countries of women opting to delay childbirth until late in their reproductive life (>30 years), accelerated follicle loss resulting from xenobiotic exposure can deprive these women of the chance to start a family in the conventional manner (Martin et al., 2003).

In addition to permanent infertility, the loss of ovarian hormones which accompanies early menopause has been associated with an increased risk for a variety of health problems. For example, estrogen deficiency (a consequence of menopause) is the most common cause of osteoporosis in humans (Cenci et al., 2003). Bone loss results from the absence of estrogen production by maturing ovarian follicles, which leads to a subsequent increase in FSH production due to the negative feedback of estrogen on pituitary gonadotropin secretion. In terms of bone remodelling, increased FSH production stimulates tumor necrosis factor (TNF) secretion, which in turn increases osteoclast formation and bone reabsorption (Cenci et al., 2003). Menopause induced estrogen withdrawal has also been associated with an increase in many traditional cardiovascular risk factors, including body fat redistribution, insulin resistance and high blood pressure, increased plasma triglyceride levels and high-density lipid cholesterol absorption (Bilianou, 2008; Rosano et al., 2007). Increased risk for Alzheimer's disease is also associated with the menopause induced loss of sex steroid hormones as evidenced by various epidemiological and experimental studies, although some clinical findings refute this evidence (Pike et al., 2009).

Over the course of the 20th century, the average life expectancy for women in the developing world has increased by ~40%, resulting in women now living up to a third of their lives in post menopausal years. Unfortunately, this means that women are now spending a larger proportion of their life with increased health risks brought about by the onset of menopause. In addition, increased risk resulting from xenobiotic induced premature menopause means an enhanced chance for problems. It is therefore important to understand the mechanisms behind xenobiotic induced primordial follicle depletion.

4. Mechanisms of xenobiotic induced primordial follicle depletion

4.1 The Aryl Hydrocarbon Receptor

The Aryl Hydrocarbon Receptor (Ahr) is a ligand activated transcription factor implicated in the regulation of a variety of physiological and developmental effects, including xenobiotic metabolism, cell cycle progression, apoptosis and oxidative stress (Denison and Heath-Pagliuso, 1998; Nebert et al., 2000). In its inactivated state, Ahr is found in the cytoplasm bound to a number of molecular chaperones including hsp90, Xap2, and p23 (Carlson and Perdew, 2002; Petrulis and Perdew, 2002). Ligand binding causes conformational changes which expose a nuclear import signal on the Ahr, resulting in its translocation into the nucleus (Pollenz et al., 1994). Once imported the Ahr-ligand receptor complex disassociates with its chaperones and dimerizes with the aryl hydrocarbon receptor nuclear translocator (ARNT) to form an active transcription factor with high affinity to specific DNA sequences known as xenobiotic-response elements (XRE) within the promoter region of a variety of genes, inducing transcription (Fig.2) (Reyes et al., 1992).

The Ahr-ARNT ligand activated transcription factor is known to regulate the toxicity of various xenobiotic compounds such as polycyclic aromatic hydrocarbons, polychlorinated dibenzofurans and polychlorinated biphenyls which are found ubiquitously in the environment and are highly resistant to metabolic breakdown (Nguyen and Bradfield, 2007; Stapleton and Baker, 2003). In an adaptive response to their accumulation in the cell, Ahr induces the expression of a number of xenobiotic metabolising enzymes, including members of the cytochrome P450 A and B families which oxygenate the intruding xenobiotic as part of a three tiered enzymatic detoxification mechanism (Conney, 1982). Unfortunately, this oxygenation often results in the bioactivation of the parent xenobiotic into a more reactive and therefore toxic metabolite (Harrigan et al., 2004; Melendez-Colon et al., 1999). Indeed, many of Ahr's known xenobiotic ligands, such as the polycyclic aromatic hydrocarbons benzo[a]pyrene (BaP), 9:10-dimethyl-1:2-benzanthracene (DMBA), and 3-methyl-cholanthrene (3-MC), cause primordial follicle destruction through Ahr initiated cytochrome P450 induced bioactivation (Borman et al., 2000; Mattison and Thorgeirsson, 1979). For example, BaP is initially metabolised by Ahr regulated cyp1A1 and cyp1B1 enzymes resulting in its biotransformation into 7,8-diol, and 9,10-diol macromolecular-adduct forming metabolites within the ovary. Inhibition of Ahr by α-naphthoflavone nullifies its effects on primordial follicle destruction (Bengtsson et al., 1983; Mattison et al., 1983a).

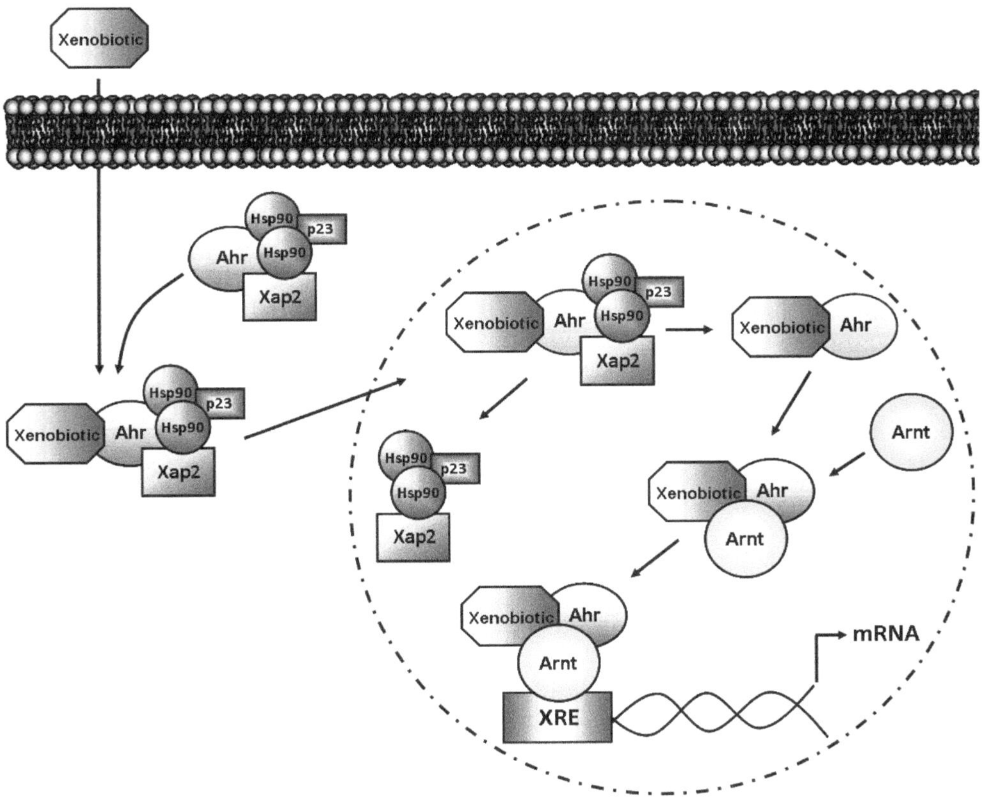

Fig. 2. Molecular mechanism of Arh-Arnt directed gene expression in response to xenobiotic exposure.

In addition to detoxification, the development of Ahr-deficient mice has revealed a physiological role for Ahr in regulating reproduction, growth and development (Benedict et al., 2000, 2003; Nebert et al., 1984; Robles et al., 2000; Schmidt et al., 1996). Benedict et al.

(2000) demonstrated that ovaries from mice deficient for Ahr expression contained significantly more fully formed primordial follicles compared to wild type mice on PND2-3. Robles et al (2000) found similar results, identifying more than a two-fold significant increase in the primordial follicle pool of Ahr deficient PND4 mice compared to wild type mice. These results suggest a developmental role for Ahr in regulating primordial follicle formation and atresia in the mouse. Although the exact details of Ahr role in the regulation of the primordial follicle pool have yet to be determined, given that Ahr xenobiotic ligands cause primordial follicle depletion, we hypothesise that part of these ovotoxic compounds method of ovotoxicity may involve perturbed AhR developmental signalling, inducing premature primordial follicle atresia.

4.2 Bioactivation

Humans come into contact with a variety of xenobiotics over the course of their lifetime, and have evolved a number of physiological mechanisms designed to remove their harmful influence from within the body. Hydrophilic xenobiotics tend to be less toxic, as the body is able to directly excrete them relatively unchanged. However, if the xenobiotic is lipophilic, it will need to be modified by a series of biochemical reactions before it can be eliminated (Pavek and Dvorak, 2008). This series of biochemical reactions is termed biotransformation, and can be divided into two phases. Phase I metabolism involves the introduction or exposure of a reactive polar group on the xenobiotic via oxidation, resulting in a more reactive/water soluble metabolite to facilitate excretion and/or the induction of phase II metabolism. The cytochrome p450 super family of oxidases catalyse the majority of these reactions, although other oxidases, esterases, amidases, and monooxygenases can also be involved (Schroer et al., 2010). Phase II metabolism involves the conjugation of charged species such as glutathione, sulphate, glycine or glucuronic acid to the phase I metabolite to increase its water solubility (Kohalmy and Vrzal, 2011). The addition of these large anionic groups detoxifies reactive electrophiles, resulting in a more polar metabolite which can be actively transported out of the cell. These reactions are carried out by a broad range of transferases, such as glutathione S-transferase, UDP-glucuronosyltransferases, sulfotransferases, N-acetyltransferases, and methyltransferases (Jancovaa et al., 2010).

Unfortunately, phase I metabolism of xenobiotics by the liver and other tissues occasionally results in the production of a more cytotoxic metabolite, a process known as bioactivation (Dekant, 2009). These highly reactive metabolites are electrophilic, and are capable of forming covalent bonds (or adducts) with the nucleophilic centers of cellular macromolecules, such as proteins, DNA, and RNA. Cellular toxicity occurs when these adducts disrupt the normal structure and/or function of these macromolecules, resulting in apoptosis, necrosis or carcinogenesis. The main site of xenobiotic biotransformation within the body is the liver, although the ovary is capable of both phase I and phase II metabolism (Igawa et al., 2009; Rajapaksa et al., 2007a, 2007b; Shimada et al., 2003). Therefore, there is potential for the vulnerable primordial follicle to come into contact with bioactivated ovotoxic metabolites via several routes of exposure. Bioactivated metabolites produced by the liver maybe stable enough to diffuse back into the venous circulatory system, resulting in direct ovarian exposure. Additionally, as the primordial follicle is capable of expressing xenobiotic metabolising enzymes itself, oocytes may be exposed to localised bioactivation. Finally, the xenobiotic may be bioactivated locally into its ovotoxic metabolite by neighbouring somatic ovarian cells and taken up by the primordial oocyte, contributing to localised bioactivation.

A number of studies performed *in vitro* have revealed that the ovary is capable of the localised bioactivation of a number of xenobiotics into ovotoxic intermediates which target primordial follicles for destruction (Rajapaksa et al., 2007a, 2007b). An example of this localised bioactivation is the reported metabolism of the polycyclic aromatic hydrocarbon DMBA (Fig.3). Ovarian exposure to DMBA disrupts folliculogenesis, resulting in the destruction of all follicle populations leading to POF in rodents, although recent evidence suggests an alternate mechanism of ovotoxicity resulting in primordial follicle depletion in the mouse (Mattison and Schulman, 1980; Sobinoff et al., 2011). This toxicity has been attributed to the bioactivation of DMBA into its ultimate DNA-adduct forming intermediate DMBA-3,4-diol-1,2-epoxide (Shiromizu and Mattison, 1985). DMBA is bioactivated by Cyp1B1 to a 3,4-epoxide which is then converted into a 3,4-diol by the microsomal epoxide hydrolase (MeH) phase II enzyme. This intermediate is then further modified by either Cyp1A1 or Cyp1B1 to form the ultimate ovotoxicant DMBA-3,4-diol-1,2-epoxide (Shimada and Fujii Kuriyama, 2004; Shimada et al., 2001). These three enzymes required for DMBA's biotransformation are all expressed and induced by DMBA exposure in the murine ovary (Igawa et al., 2009; Rajapaksa et al., 2007b; Shimada et al., 2003). In further support of localised DMBA bioactivation, inhibition of MeH in cultured rat ovaries inhibited DMBA induced ovotoxicity, while ovarian culture in the presence of DMBA-3,4-diol induced significantly more primordial follicle depletion than DMBA alone (Igawa et al., 2009; Rajapaksa et al., 2007b).

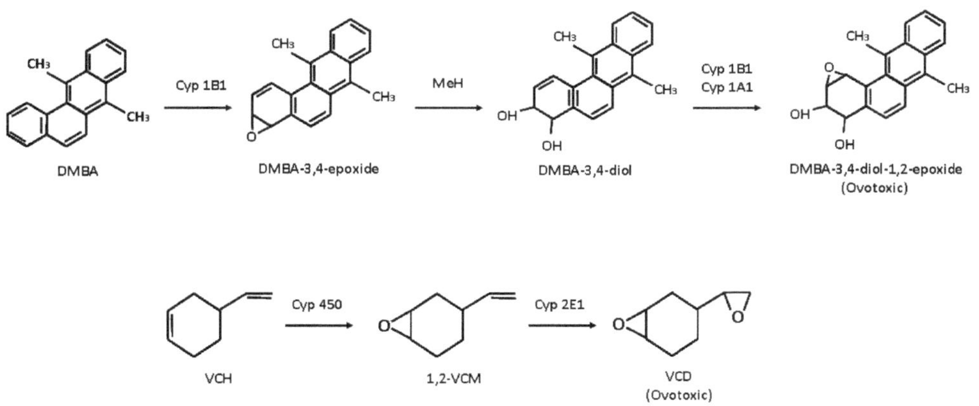

Fig. 3. Metabolism of DMBA and VCD into their ovotoxic metabolites.

Another example of localised bioactivation is the conversion of the industrial chemical 4-Vinylcyclohexene (VCH) into VCH diepoxide (VCD) (Fig. 3). VCH is metabolised by cytochrome P450 phase I enzymes to form VCM-monoepoxide (VCM), which is then converted into VCD. Studies have shown VCD to be the ultimate ovotoxicant, targeting both primordial and primary follicles for depletion (Hu et al., 2001; Smith et al., 1990; Sobinoff et al., 2010). As demonstrated *in vivo* and *in vitro* via knockout studies, VCH/VCM is bioactivated into VCD exclusively by the cyp2e1 isoform in the ovary (Rajapaksa et al.,

2007a). Rajapaksa et al (2007a) cultured neonatal ovaries from both cyp2e1$^{+/+}$ and cyp2e1$^{-/-}$ neonatal mice in VCM and VCD containing media. Both VCH metabolites caused primordial follicle depletion in cyp2e1$^{+/+}$ cultured ovaries. However, unlike VCD, VCM did not produce an ovotoxic affect in cyp2e1$^{-/-}$ cultured ovaries, thus demonstrating its role in VCH induced bioactivation.

4.3 Xenobiotic induced reactive oxygen species generation

Reactive oxygen species (ROS), such as superoxide anion, hydrogen peroxide and the highly toxic hydroxyl free radical, are highly reactive oxygen-containing molecules which are produced naturally as a consequence of oxidative energy metabolism (Valko et al., 2007). These short lived ROS play an important role in regulating signal transduction, selectively oxidizing cysteine residues on proteins resulting in a variety of reversible molecular interactions (Janssen-Heininger et al., 2008). However, in excess these highly unstable molecules may lead to perturbed signal transduction and/or oxidative damage to cellular macromolecules, inducing DNA mutations, lipid peroxidation and premature protein degradation. These molecular lesions coupled with perturbed signal transduction can ultimately result in abnormal cellular function, apoptosis and necrosis (Valko et al., 2006, 2007; Wells et al., 2009).

The ovary is a highly redox sensitive organ, with oocytes themselves being particularly vulnerable to excess ROS exposure due to the low rates of oxidative repair in post-mitotic cells (Cadenas and Davies, 2000; Terman et al., 2006). According to the free radical hypothesis of ageing, non-renewing primordial follicles, which can remain quiescent for many years, gradually produce ROS through electron leakage from the mitochondrial electron transport chain (Tarin, 1996). Over time this excess ROS damages the mitochondrial membranes, leading to more electron leakage and further ROS production. Given the redox sensitive nature of primordial follicles, it is reasonable to assume that the generation of xenobiotic induced ROS formed through detoxification may exacerbate this process, contributing to primordial follicle loss (Bondy and Naderi, 1994; Danielson, 2002; Wells et al., 2009).

Xenobiotic enhanced ROS formation may occur via several mechanisms in the primordial follicle (Fig.4). If the ovotoxic xenobiotic contains a quinone-like structure, it may undergo redox cycling with the corresponding semiquinone radical to produce superoxide anions. Further enzymatic and/or spontaneous dismutation of the superoxide anions produces hydrogen peroxide, which can further react with trace amounts of iron or other transition metals to form hydroxyl free radicals (Bolton et al., 2000). Given the futile cyclical nature of redox cycling, this would allow a relatively small concentration of quinone-like xenobiotics to generate an amplified production of ROS in the ovary (Park et al., 2005). For example, menadione (MEN), a synthetic vitamin K with a quinone-like structure, is a potent toxicant which exerts its cytotoxic affect via quinone cycling (Thor et al., 1982). Recently, we examined the effects of MEN on folliculogenesis in neonatal mouse ovaries *in vitro* (Sobinoff et al., 2010). This study found that MEN caused wide spread oxidative stress and DNA damage resulting in primordial and small developing follicle destruction, as evidenced by the detection of increased levels of the hydroxyl radical–induced mutagenic DNA lesion 8-

hydroxyguanine, and Terminal deoxynucleotidyl transferase dUTP nick end labelling (TUNEL) analysis (Klaunig and Kamendulis, 2004).

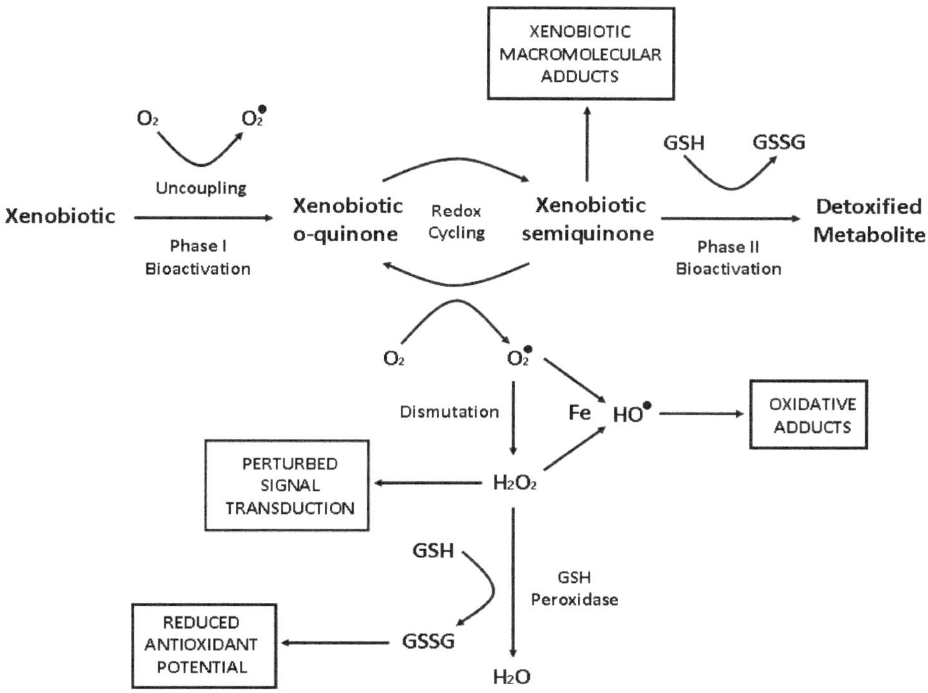

Fig. 4. Biochemical pathways outlining the mechanisms of xenobiotic induced ROS production which may contribute to primordial follicle depletion. Abbreviations: Fe, iron; $O_2^{\bullet}$, superoxide; H_2O_2, hydrogen peroxide; $HO^{\bullet}$, hydroxyl radical.

Another mechanism of xenobiotic induced ROS formation is the phase I bioactivation of the offending xenobiotic into reactive and redox active o-quinone metabolites. As mentioned previously, the PAH BaP is converted into 7,8-diol, and 9,10-diol by Ahr induced cyp1A1 and cyp1B1 enzymes in the ovary (Bengtsson et al., 1983). Studies have also shown that cyp1A1 is also capable of converting BaP into the BaP o-quinones benzo[a]pyrene-3,6-dione and benzo[a]pyrene-6,12-dion (Schwarz et al., 2001). Additionally, cyp 1A1 bioactivated BaP 7,8-diol can be further metabolised via $NAD(P)^+$-dependent oxidation by the phase I dihydrodiol dehydrogenase Akr1c1 enzyme, resulting in the formation of a ketol. This ketol then undergoes tautomerisation to form catechol 7,8-dihydroxybenzo[a]pyrene. Two subsequent 1-electron auto-oxidation events produce a *o*-semiquinone anion, followed by the formation of the o-quinone benzo[a]pyrene-7,8-dione (Trevor et al., 1996). Given the increases observed in cyp 1A1 expression in the ovary in response to BaP exposure, and the relatively high level of dihydrodiol dehydrogenase expression in the ovary compared to the liver, it is reasonable to assume BaP may be exerting part of its ovotoxic affect through o-quinone formation (Hou et al., 1994). Indeed, both benzo[a]pyrene-3,6-dione and benzo[a]pyrene-6,12-dion were detected in rat ovaries after a single dose exposure of BaP in rodents (Ramesh et al., 2010).

The un-natural "uncoupling" of phase I cytochrome P450 enzymes may also contribute to xenobiotic induced ovotoxicity via ROS production. Cytochrome P450 enzymes use H^+ obtained from NADPH to reduce O_2, which leads to the production of hydrogen peroxide and/or superoxide anion radicals as part of phase I oxygenation. Unfortunately, the P450 catalytic cycle can be uncoupled, resulting in the release of the reactive hydrogen peroxide and/or superoxide anion radical from the enzyme substrate complex (Meunier et al., 2004). Although all cytochrome P450 enzymes experience uncoupling, cyp 2E1 experiences a high rate of the phenomenon (Caro and Cederbaum, 2004). Even in the absence of substrate, cyp 2E1 undergoes un-natural "uncoupling" due to its NADPH oxidase activity independent of phase I metabolism (Ekstrom and Ingelman-Sundberg, 1989). As described previously, VCH is exclusively bioactivated by cyp 2E1 to produce the ovotoxic metabolite VCD. It is therefore possible that VCH may partially cause primordial follicle depletion via excess ROS production. Indeed, studies conducted in our laboratory have demonstrated VCD itself, along with the pesticide methoxychlor (MXC) and MEN, is capable of inducing cyp 2E1 expression and oxidative stress in the form of 8-hydroxyguanine adduct formation in primordial follicles (Sobinoff et al., 2010).

Another mechanism by which ovotoxic xenobiotics may cause oxidative stress is through the depletion of glutathione peroxidase (GSH) via detoxification. GSH is the body's most abundant antioxidant, providing protection against all forms of oxidative stress by scavenging ROS by virtue of its reducing thiol group, forming oxidised glutathione disulfide (GSSG) (Kidd, 1997). The glutathione system (GSH/GSSG ratio) acts as a homoeostatic redox buffer that contributes to maintenance of the cellular redox balance, with a reduction in the GSH/GSSG ratio indicating oxidative stress (Schafer and Buettner, 2001). In addition to its function as a ROS scavenger, GSH is also employed in the phase II metabolism of many ovotoxic xenobiotics (Keating et al., 2010; Tsai-Turton et al., 2007; Wu and Berger, 2008). For example, VCD is conjugated to GSH by the glutathione S-transferase Gst isoform pi (Gstp) as part of phase II detoxification in the ovary (Keating et al., 2010).

The mammalian ovary itself is highly redox sensitive, with maturing oocytes containing the highest concentration of GSH compared to any other cell type in the body (Calvin et al., 1986; Clague et al., 1992; Luderer et al., 2001). It is therefore likely that ovarian somatic and germ cell GSH plays an important role in protecting ovarian follicles from damage by ovotoxic xenobiotics. This is especially evident in primordial follicles, where a natural decrease in the GSH/GSSG ratio with advancing reproductive age increases primordial follicle susceptibility to xenobiotic induced destruction (Mattison et al., 1983b). Therefore, we hypothesise that the detoxification of ovotoxic xenobiotics via GSH conjugation reduces the GSH/GSSG ratio in primordial follicles, leaving them vulnerable to oxidative stress and primordial follicle depletion. Indeed, DMBA detoxification involves GSH conjugation, and its ovotoxic ROS production can be reduced through the addition of GSH, curbing its ovotoxicity (Tsai-Turton et al., 2007). There is controversial evidence for this mechanism of ovotoxicity in VCD induced primordial follicle loss. Rodent exposure to VCD was shown to reduce GSH concentrations by 25% and 55% in rat and mouse ovaries 2 hours after VCD administration (Bhattacharya and Keating, 2011). Additionally, rodents given the same dose of VCD over a period of several days caused specific primordial follicle depletion only after 15 days of continual dosing (Springer et al., 1996). The significant decrease in GSH concentrations almost immediately after exposure, coupled with the delayed loss of follicles following chronic exposure suggest that GSH reduction over time due to VCD detoxification could leave the susceptible primordial follicle vulnerable to increasing concentrations of xenobiotic induced ROS, resulting in primordial follicle destruction. It is pertinent that a single dose of a higher concentration of VCD (320 mg/kg) causes significant primordial follicle depletion 6 days after exposure, but is not specific to the primordial follicle pool (Devine et al., 2004). Additionally, VCD *in vitro* culture assays have linked an increase in Gstp expression with the first signs of primordial follicle loss after 6 days of exposure in neonatal rat ovaries (Bhattacharya and Keating, 2011; Keating et al., 2010). As Gstp catalyses VCD-GSH conjugation, the increase in enzymatic expression and therefore activity could have contributed to the observed primordial follicle loss due to a reduction in GSH/GSSG oxidative buffer. Conversely, substituting VCD culture media with antioxidant such as GSH does not prevent primordial follicle depletion, suggesting it is not the ultimate cause of depletion (Devine et al., 2004).

4.4 Xenobiotic induced primordial follicle activation

Traditionally, studies attempting to identify the molecular mechanisms behind xenobiotic induced POF have focused on premature follicular atresia as the main source of primordial follicle depletion. However, there is now a growing body of evidence which suggests that xenobiotics cause primordial follicle depletion through accelerated primordial follicle activation (Keating, 2009, 2011; Sobinoff et al., 2010, 2011). A study of VCD and MXC induced primordial follicle depletion has revealed a selective mechanism of pre-antral ovotoxicity involving small developing follicle atresia and primordial follicle activation both *in vitro* and *in vivo* (Sobinoff et al., 2010). Extracted neonatal mouse ovaries cultured in either VCD or MXC were immunopositive for the apoptotic markers caspase 2, caspase 3, and TUNEL in small developing follicles from the primary stage onward, but were absent in primordial follicles (Fig.5). In addition, the primordial follicles in VCD and MXC cultured ovaries expressed proliferating cell nuclear antigen (PCNA), a marker of primordial follicle activation (Picut et al., 2008; Tománek and Chronowska, 2006). VCD and MXC exposure also induces primordial follicle activation and developing follicle atresia *in vivo* as evidenced by increased primordial follicle PCNA expression and histomorphological analysis (Sobinoff et al., 2010). Microarray analysis confirmed via qPCR also showed VCD and MXC up-regulated PI3K/Akt and mTOR signalling, two synergistic pathways intimately associated with primordial follicle activation (Reddy et al., 2010). Further evidence for PI3K/Akt signalling in VCD induced primordial follicle activation comes from a study conducted by Hoyer et al (2009), in which LY294002, an inhibitor of PI3K, prevented primordial follicle depletion in cultured rat ovaries (Vlahos et al., 1994).

The polycyclic aromatic hydrocarbon DMBA, which was previously thought to cause indiscriminate follicular destruction, has also been shown to cause pre-antral ovotoxicity through selective immature follicle destruction and primordial follicle activation (Mattison and Schulman, 1980; Sobinoff et al., 2011). In addition to showing signs of maturing follicle atresia (caspase 2, caspase 3, TUNEL) and primordial follicle activation (PCNA), DMBA induced Akt1 phosphorylation, mTOR activation, and decreased FOXO3a expression in DMBA cultured primordial oocytes. All of these events occur downstream of the PI3K/Akt and mTOR signalling pathways, providing evidence for these pathways involvement in xenobiotic induced primordial follicle depletion (Reddy et al., 2010). Unlike VCD however, PI3K/Akt inhibitor studies utilising LY294002 in DMBA cultured rat ovaries caused accelerated primordial follicle depletion (Keating, 2009). In addition to its role in primordial follicle activation, PI3k/Akt signalling is also responsible for augmenting cellular survival by inhibiting the activation of proapoptotic proteins and transcription factors (Blume-Jensen et al., 1998; Testa and Bellacosa, 2001). Therefore, in addition to acting synergistically with mTOR signalling to cause primordial follicle activation, PI3k/Akt signalling may help preserve the primordial follicle pool in times of cytotoxic stress. Interestingly however, mTOR signalling does not require PI3k/Akt signalling to induce primordial follicle activation, and in fact may be the sole driver of DMBA induced primordial follicle activation (Adhikari et al., 2010).

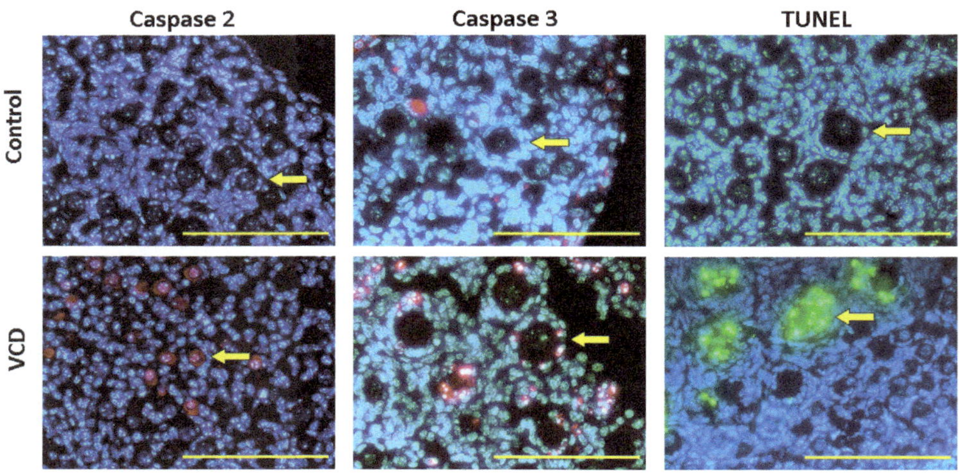

Fig. 5. Immunohistological staining of apoptotic markers in VCD exposed neonatal mouse ovaries. Blue staining (DAPI) represents nuclear staining; red staining (Cy-5) represents specific staining for the protein of interest; green staining (Fluorescein) represents specific staining for degraded DNA (TUNEL). Thin arrow=developing follicle; scale bar is equal to 50μm.

As xenobiotic induced primordial follicle activation is reportedly accompanied by small pre-antral follicular destruction, it has been hypothesised that xenobiotic induced primordial follicle depletion is the result of a homeostatic mechanism of follicular replacement (Keating, 2009; Sobinoff et al., 2010). In this hypothesis, the ovotoxic xenobiotic targets and destroys developing follicles, leading to increased primordial follicle recruitment to maintain the developing pool (Fig. 6). Although the developing pool may be maintained for some time, eventually the rate of developing follicle destruction will exceed the dwindling primordial follicle pools rate of replacement, resulting in POF. Indeed, it is well known that rapidly dividing cells, such as the granulosa cells of developing follicles, are highly susceptible to the action of cytotoxic xenobiotics (Blumenfeld and Haim, 1997; Hirshfield, 1991). Therefore, if the xenobiotic targeted these proliferating granulosa cells for destruction, the entire follicular structure would demise (Hughes and Gorospe, 1991). Even given the vulnerable nature of the primordial follicle explained earlier in this review, the primordial follicles quiescent nature may reduce their susceptibility to certain xenobiotics, and are only destroyed once a commitment to activation/recruitment has been made.

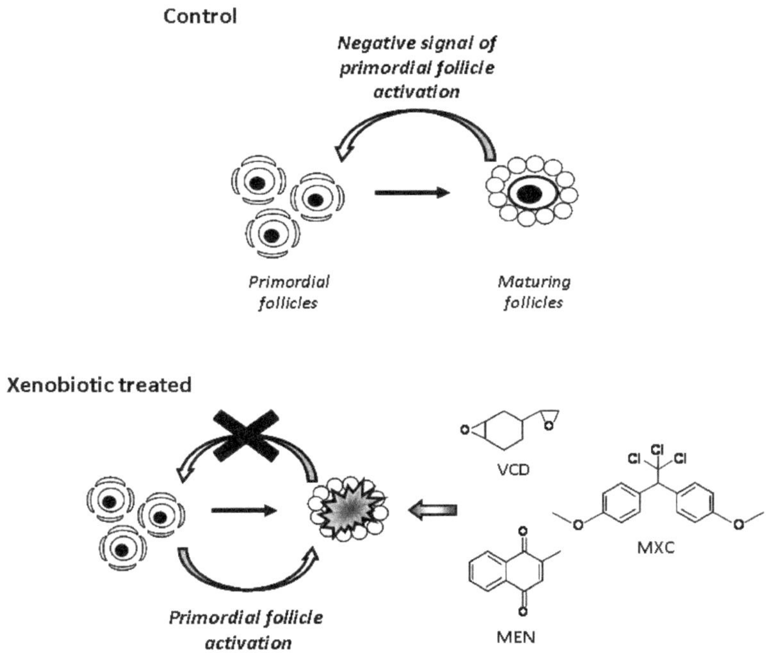

Fig. 6. Homeostatic mechanism of follicular replacement hypothesis. Under control conditions, premature primordial follicle activation is prevented by negative cytokine signals excreted from the developing pool of follicles. Xenobiotic exposure results in the destruction of this developing pool, removing these negative signals causing primordial follicle activation.

Another proposed mechanism of xenobiotic induced primordial follicle activation may involve perturbed signal transduction caused by oxidative stress. As described previously, ROS play a physiological role in regulating signal transduction by selectively oxidising cysteine residues on proteins resulting in a variety of reversible molecular interactions (Wells et al., 2009). It is therefore conceivable that increased levels of xenobiotic induced ROS could lead to abnormal cysteine oxidation and consequently dysregulated signal transduction. For example, the PI3K/Akt signalling pathway has been shown to be up-regulated by increased levels of ROS through the H_2O_2 oxidation of phosphatases which negatively regulate the pathway (Kim et al., 2005; Naughton et al., 2009). Given the PI3K/Akt pathway's role in the regulation of primordial follicle recruitment, increased ROS production could potentially cause primordial follicle activation in xenobiotic treated ovaries. Indeed, all three xenobiotics which have thus far been reported to induce primordial follicle activation also cause oxidative stress and induce the expression/activation of members of the PI3K/Akt signalling pathway in the ovary (Sobinoff et al., 2010, 2011; Tsai-Turton et al., 2007).

Xenobiotic induced primordial follicle activation may also be the result of abnormal cross-talk between signalling pathways. For example, DMBA exposure was shown to induce *Dnajb6* expression, a heat shock protein whose expression is normally induced by Nrf2 and Hsf1 in response to oxidative stress (Sobinoff et al., 2011; Thimmulappa et al., 2002; Wang, K. et al., 2009). Dnajb6 responds to stress by inhibiting nuclear factor of activated T cells (NFAT) transcriptional activity through the recruitment of class II histone deacetylase (Dai et al., 2005). In turn NFAT positively regulates PTEN expression, a known inhibitor of Akt1 phosphorylation. Therefore DMBA induced Dnajb6 expression may inhibit NFAT transcriptional activity, reducing PTEN expression and stimulating Akt1 phosphorylation, resulting in primordial follicle activation (Baksh et al., 2002; Reddy et al., 2010; Wang, Q. et al.).

4.5 Xenobiotic induced cell death

Ovarian follicles undergo physiological cell death via the apoptotic process of atresia, which is thought to select dysfunctional follicles and thus reserving the healthiest follicles for ovulation (Tilly et al., 1991). A number of studies have concluded that ovotoxic xenobiotics which target primordial follicles for destruction do so by inducing premature follicular atresia (Hu et al., 2001; Matikainen et al., 2001; Tilly and Robles, 1999). In this review we have already discussed the mechanisms by which ovotoxic xenobiotics may induce follicular atresia in primordial follicles (Ahr activation, Bioactivation, and ROS generation). However, other forms of cell death have been reportedly induced by xenobiotic exposure. Cell death by necrosis usually occurs in response to tissue injury, and elicits an inflammatory response in the surrounding tissue. Necrosis can be distinguished from apoptosis via histomorphological and ultrastructural analysis (Gobe and Harmon, 2001). In a study by Mattison (1980), the three PAHs BaP, 3-MC, and DMBA were shown to cause morphological changes in mouse primordial follicle oocytes which were consistent with necrosis (Mattison, 1980). The alkylating chemotherapeutic agent cyclophosphamide was also shown to cause necrotic damage in mouse primordial follicle oocytes three days after a single i.p injection (Plowchalk and Mattison, 1992). However, lower doses of cyclophosphamide produced atretic changes in primordial follicle oocytes, suggesting the type of cell death (apoptosis/necrosis) caused by xenobiotic exposure depends upon the dose given, and the duration of exposure. Therefore, concentrations of xenobiotic which cause mild cellular damage may result in active cell death, or apoptosis, while concentration which result in severe damage will result in passive cell death, or necrosis (Raffray and Gerald, 1997).

Autophagy or "self eating" is another possible non-apoptotic mechanism of cell death which may result in primordial follicle depletion. This conserved catabolic process involves the lysosomal-dependant turnover of cytoplasmic organelles and proteins during times of starvation or nutrient deficiency, allowing the regeneration of metabolic precursor molecules to ensure survival (Levine and Klionsky, 2004). Increased incidences of autophagy have also been observed in response to other environmental stresses, including hypoxia, oxidative stress, and xenobiotic exposure (Kiffin et al., 2006; Kondo et al., 2005). Under these conditions autophagy may renew damaged or dysfunctional organelles, thereby maintaining a healthy cell population. Although the activation of autophagy in response to cell stress may be a cellular adaptation to promote survival, excessive activation beyond a key threshold may result in cellular collapse and atrophy, a process known as autophagic cell death (Galluzzi et al., 2008). While debatable whether autophagic cell death is independent from apoptosis, it has been almost universally accepted that excess autophagy can induce apoptosis (Levine and Yuan, 2005; Maiuri et al., 2007). Recent studies have suggested autophagy as an alternate form of programmed cell death in the ovary, with evidence indicating it is the main mechanism by which oogonia are lost prior to primordial follicle formation (Duerrschmidt et al., 2006; Lobascio et al., 2007; Rodrigues et al., 2009). Thus prolonged xenobiotic exposure resulting in organelle damage may induce autophagic cell death in primordial follicles, resulting in depletion. Indeed, proteins responsible for regulating apoptosis, such as members of the Bcl2 family, have also been found to regulate autophagy (Maiuri et al., 2007; Shimizu et al., 2004). Therefore, gene expression studies in which these pathways have been thought to induce xenobiotic atresia could be inducing primordial follicle destruction by apoptotic independent or dependent autophagy (Flaws et al., 2006).

5. Ovotoxic xenobiotics as agents for wildlife fertility control

Population control of native and exotic pest species is necessary to prevent environmental degradation, competition and predation of native wildlife, the spread of pathogenic diseases, and conflicts with humans over food production. Traditionally, population control has involved the elimination of the target species through poisoning, trapping and shooting (McAlpine et al., 2007). Although effective immediately, these methods are seen as inhuman, unsustainable, and ineffective over the long term. Manipulating the reproductive rate, particularly in females, instead of increasing the mortality rate is potentially more humane, species specific, and effective at curtailing populations (Kirkpatrick, 2007). The use of ovotoxic xenobiotics as agents of contraception/sterilisation represents a novel approach to fertility control. Of particular interest are xenobiotics which have been shown to cause POF by specifically targeting the primordial follicle population for degradation (Hoyer and Devine, 2001; Sobinoff et al., 2010), thus causing permanent sterility.

To achieve widespread efficacy ovotoxic xenobiotics in fertility control must be delivered via single or minimal oral administration. To be successful an oral agent must also have permanent or very long lasting effects, be specific for the target pest species and be humane/environmentally safe (Castle and Dean, 1996). Rodents such as the rice-field rat represent a serious pest in cereal agriculture, accounting for an average annual loss of between 5-10% of rice crops in Asia, 17% of rice crops in Indonesia, between 15-100% of maize in Africa, and between 5-90% of total crop production in South America (Geddes, 1992; Mwanjabe and Leirs, 1997; Rodríguez and Jaime, 1993; Singleton, 2003; Taylor, 1968).

VCD represents an ideal fertility control agent due to its ability to induce rapid small follicle depletion resulting in POF in rodents at concentrations which do not cause widespread cytotoxicity (Springer et al., 1996). Additionally, VCD metabolism in the liver and hepatic tissue of rodents results in the production and excretion of the inert compound from the body, potentially reducing its effects on predators and its bioaccumulation in the environment (Flaws et al., 1994; Keating et al., 2010; Rajapaksa et al., 2007a). However, VCD does have disadvantages which make it fall short of the ideal fertility control agent. As described previously, VCD requires multiple doses to cause complete infertility in the rodent model (Springer et al., 1996). In addition, a VCD containing bait would need to be both attractive and palatable to the pest species, but not palatable or accessible to non-pest species. Currently, VCD is being trialled as an oral fertility control agent in the rice-field rat *Rattus argentiventer*. Registered by SenesTech Inc. as ContraPest®, the company website suggests the formulated bait is palatable, causes complete sterility within one month's ingestion, and does not adversely affect the animal's health and well being (http://www.senestech.com/). The use of other ovotoxicants as oral fertility control agents has been less successful. In a study by Sanders et al (2011) ERL-4221, a less toxic diepoxide, cycloaliphatic epoxide resin, which recently replaced VCD in industry, was investigated as a possible fertility control agent for pigs. A 20 day treatment period using palatable bait containing 16.0 mg ERL-4221 kg−1 bodyweight failed to produce any difference in follicular composition compared to control treated animals (Sanders et al., 2011). In summary, ovotoxicants represent potential fertility control agents, provided the xenobiotic delivers significant follicle depletion without side effects, and does not adversely affect the environment or food chain.

6. Conclusions

Ovotoxic xenobiotics cause primordial follicle depletion via several mechanisms which ultimately lead to their destruction or activation. These chemicals are rarely ovotoxic by themselves, and require hepatic or ovarian metabolism to exert their destructive effects on reproduction. This type of ovotoxicity is insidious in its nature, and is not usually detected until the primordial follicle pool has become severely depleted, resulting in premature reproductive senescence. Besides a loss in fertility, reproductive senescence is also associated with an increased incidence of a variety of health problems. Despite the negatives associated with ovotoxic xenobiotics, there is potential to use their destructive nature for wildlife control and agricultural gain. It is a form of poetic justice that ovotoxic xenobiotics which prevent women from conceiving may be used to combat one of the biggest causes of death in the third world, starvation. Future research should be aimed at further elaborating the specific mechanisms of primordial follicle ovotoxicity, improving our ability to predict/detect human risk from environmental exposure, and investigating the possibility of using these ovotoxicants for the environmental control of pest species.

7. Acknowledgments

The authors gratefully acknowledge the financial assistance to EAM by the Australian Research Council, Hunter Medical Research Institute and the Newcastle Permanent Building Society Charitable Trust. APS is the recipient of an Australian Postgraduate Award PhD scholarship. This work was supported by National Health and Medical Research Council (Project grant #510735) to EAM.

8. References

Adhikari, D., W. Zheng, Y. Shen, N. Gorre, T. Hämäläinen, A. J. Cooney, I. Huhtaniemi, Z. J. Lan and K. Liu (2010). "Tsc/mTORC1 signaling in oocytes governs the quiescence and activation of primordial follicles." *Human molecular genetics* 19(3): pp. 397-410, ISSN 0964-6906.

Baksh, S., H. R. Widlund, A. A. Frazer-Abel, J. Du, S. Fosmire, D. E. Fisher, J. A. DeCaprio, J. F. Modiano and S. J. Burakoff (2002). "NFATc2-mediated repression of cyclin-dependent kinase 4 expression." *Molecular cell* 10(5): pp. 1071-1081, 1097-2765.

Benedict, J. C., T. M. Lin, I. K. Loeffler, R. E. Peterson and J. A. Flaws (2000). "Physiological role of the aryl hydrocarbon receptor in mouse ovary development." *Toxicological Sciences* 56(2): pp. 382-388, ISSN 1096-6080.

Benedict, J. C., K. P. Miller, T. M. Lin, C. Greenfeld, J. K. Babus, R. E. Peterson and J. A. Flaws (2003). "Aryl hydrocarbon receptor regulates growth, but not atresia, of mouse preantral and antral follicles." *Biology of reproduction* 68(5): pp. 1511-1517, ISSN 0006-3363.

Bengtsson, M., J. Montelius, L. Mankowitz and J. Rydström (1983). "Metabolism of polycyclic aromatic hydrocarbons in the rat ovary comparison with metabolism in adrenal and liver tissues." *Biochemical Pharmacology* 32(1): pp. 129-136, ISSN 0006-2952.

Bhattacharya, P. and A. F. Keating (2011). "Ovarian metabolism of xenobiotics." *Experimental Biology and Medicine* 236(7): pp. 765-771, ISSN 1535-3702.

Blume-Jensen, P., R. Janknecht and T. Hunter (1998). "The kit receptor promotes cell survival via activation of PI 3-kinase and subsequent Akt-mediated phosphorylation of Bad on Ser136." *Current biology* 8(13): pp. 779-785, ISSN 0960-9822.

Blumenfeld, Z. and N. Haim (1997). "Prevention of gonadal damage during cytotoxic therapy." *Annals of medicine* 29(3): pp. 199-206, ISSN 0785-3890.

Bolton, J. L., M. A. Trush, M. Trevor, G. Dryhurst and T. J. Monks (2000). "Role of quinones in toxicology." *Chemical Research in Toxicology* 13(3): pp. 135-160, ISSN 0893-228X.

Bondy, S. C. and S. Naderi (1994). "Contribution of hepatic cytochrome P450 systems to the generation of reactive oxygen species." *Biochemical pharmacology* 48(1): pp. 155-159, ISSN 0006-2952.

Borman, S. M., P. J. Christian, I. G. Sipes and P. B. Hoyer (2000). "Ovotoxicity in female Fischer rats and B6 mice induced by low-dose exposure to three polycyclic aromatic hydrocarbons: comparison through calculation of an ovotoxic index." *Toxicology and applied pharmacology* 167(3): pp. 191-198, ISSN 0041-008X.

Cadenas, E. and K. J. A. Davies (2000). "Mitochondrial free radical generation, oxidative stress, and aging." *Free Radical Biology and Medicine* 29(3-4): pp. 222-230, ISSN 0891-5849.

Calvin, H. I., K. Grosshans and E. J. Blake (1986). "Estimation and manipulation of glutathione levels in prepuberal mouse ovaries and ova: relevance to sperm nucleus transformation in the fertilized egg." *Gamete research* 14(3): pp. 265-275, ISSN 1554-3919.

Carlson, D. B. and G. H. Perdew (2002). "A dynamic role for the Ah receptor in cell signaling? Insights from a diverse group of Ah receptor interacting proteins." *Journal of biochemical and molecular toxicology* 16(6): pp. 317-325, ISSN 1099-0461.

Caro, A. A. and A. I. Cederbaum (2004). "Oxidative Stress, Toxicology, and Pharmacology of CYP2E1*." *Annu. Rev. Pharmacol. Toxicol.* 44: pp. 27-42, ISSN 0066-4170.

Castle, P. E. and J. Dean (1996). "Molecular genetics of the zona pellucida: implications for immunocontraceptive strategies." *Journal of reproduction and fertility. Supplement* 50: pp. 1-8, ISSN 0449-3087.

Cenci, S., G. Toraldo, M. N. Weitzmann, C. Roggia, Y. Gao, W. P. Qian, O. Sierra and R. Pacifici (2003). "Estrogen deficiency induces bone loss by increasing T cell proliferation and lifespan through IFN- -induced class II transactivator." *Proceedings of the National Academy of Sciences of the United States of America* 100(18): pp. 10405-10410, ISSN 0027-8424.

Clague, N., M. Sevcik, G. Stuart, M. Brännström, P. O. Janson and J. F. Jarrell (1992). "The effect of estrous cycle and buthionine sulfoximine on glutathione release from the in vitro perfused rat ovary." *Reproductive Toxicology* 6(6): pp. 533-539, ISSN 0890-6238.

Conney, A. H. (1982). "Induction of microsomal enzymes by foreign chemicals and carcinogenesis by polycyclic aromatic hydrocarbons: GHA Clowes Memorial Lecture." *Cancer research* 42(12): pp. 4875-4917, ISSN 0008-5472.

Coulam, C. B., S. C. Adamson and J. F. Annegers (1986). "Incidence of premature ovarian failure." *Obstetrics & Gynecology* 67(4): pp. 604-606, ISSN 0029-7844.

Dai, Y. S., J. Xu and J. D. Molkentin (2005). "The DnaJ-related factor Mrj interacts with nuclear factor of activated T cells c3 and mediates transcriptional repression through class II histone deacetylase recruitment." *Molecular and cellular biology* 25(22): pp. 9936, 0270-7306.

Danielson, P. B. (2002). "The cytochrome P450 superfamily: biochemistry, evolution and drug metabolism in humans." *Current Drug Metabolism* 3(6): pp. 561-597, ISSN 1389-2002.

Dekant, W. (2009). "The role of biotransformation and bioactivation in toxicity." *Molecular, Clinical and Environmental Toxicology* 99: pp. 57-86, ISSN 1023-294X.

Denison, M. S. and S. Heath-Pagliuso (1998). "The Ah receptor: a regulator of the biochemical and toxicological actions of structurally diverse chemicals." *Bulletin of environmental contamination and toxicology* 61(5): pp. 557-568, ISSN 0007-4861.

Devine, P. J., I. G. Sipes and P. B. Hoyer (2004). "Initiation of delayed ovotoxicity by in vitro and in vivo exposures of rat ovaries to 4-vinylcyclohexene diepoxide." *Reproductive Toxicology* 19(1): pp. 71-77, ISSN 0890-6238.

Duerrschmidt, N., O. Zabirnyk, M. Nowicki, A. Ricken, F. A. Hmeidan, V. Blumenauer, J. Borlak and K. Spanel-Borowski (2006). "Lectin-like oxidized low-density lipoprotein receptor-1-mediated autophagy in human granulosa cells as an alternative of programmed cell death." *Endocrinology* 147(8): pp. 3851-3860, ISSN 0013-7227.

Edson, M. A., A. K. Nagaraja and M. M. Matzuk (2009). "The mammalian ovary from genesis to revelation." *Endocrine reviews* 30(6): pp. 624-712, ISSN 1945-7189.

Ekstrom, G. and M. Ingelman-Sundberg (1989). "Rat liver microsomal NADPH-supported oxidase activity and lipid peroxidation dependent on ethanol-inducible cytochrome P-450 (P-450IIE1)." *Biochemical pharmacology* 38(8): pp. 1313-1319, ISSN 0006-2952.

Flaws, J. A., S. L. Marion, K. P. Miller, P. J. Christian, J. K. Babus and P. B. Hoyer (2006). "Effect of bcl-2 overexpression in mice on ovotoxicity caused by 4-vinylcyclohexene." *Toxicology and applied pharmacology* 215(1): pp. 51-56, ISSN 0041-008X.

Flaws, J. A., K. L. Salyers, I. G. Sipes and P. B. Hoyer (1994). "Reduced ability of rat preantral ovarian follicles to metabolize 4-vinyl-1-cyclohexene diepoxide in vitro." *Toxicology and applied pharmacology* 126(2): pp. 286-294, ISSN 0041-008X.

Galluzzi, L., J. M. Vicencio, O. Kepp, E. Tasdemir, M. C. Maiuri and G. Kroemer (2008). "To die or not to die: that is the autophagic question." *Curr Mol Med* 8(2): pp. 78-91, ISSN 1566-5240.

Geddes, A. M. W. (1992). The relative importance of pre-harvest crop pests in Indonesia, *Natural Resources Institute Bulletin (United Kingdom)*, A. M. W. Geddes, pp. 1-70, Natural Resources Institute, ISBN 0-85954-295-5. Chatham, UK.

Gobe, G. and B. Harmon (2001). Apoptosis: Morphological Criteria and Other Assays, *Encyclopedia of Life Sciences*, Y. Zheng, <http://dx.doi.org/10.1002/9780470015902.a0002569.pub3>

Guraya, S. S. (1985). Primordial follicle, *Biology of Ovarian Follicles in Mammals*, pp. 3-14, Springer-Verlag, ISBN 9783540150220. Berlin, Germany.

Harrigan, J. A., C. M. Vezina, B. P. McGarrigle, N. Ersing, H. C. Box, A. E. Maccubbin and J. R. Olson (2004). "DNA adduct formation in precision-cut rat liver and lung slices exposed to benzo [a] pyrene." *Toxicological Sciences* 77(2): pp. 307-314, ISSN 1096-6080.

Hirshfield, A. N. (1991). "Development of follicles in the mammalian ovary." *International review of cytology* 124: pp. 43-101, ISSN 0074-7696.

Hooser, S. B., D. P. Douds, D. G. DeMerell, P. B. Hoyer and I. G. Sipes (1994). "Long-term ovarian and gonadotropin changes in mice exposed to 4-vinylcyclohexene." *Reproductive Toxicology* 8(4): pp. 315-323, ISSN 0890-6238.

Hou, Y. T., W. Xia, J. E. Pawlowski and T. M. Penning (1994). "Rat dihydrodiol dehydrogenase: complexity of gene structure and tissue-specific and sexually dimorphic gene expression." *Cancer research* 54(1): pp. 247-255, ISSN 0008-5472.

Hoyer, P. B. and P. J. Devine (2001). "Ovarian toxicity of 4-vinylcyclohexene diepoxide: a mechanistic model." *Toxicologic pathology* 29(1): pp. 91-99, ISSN 0192-6233.

Hoyer, P. B. and I. G. Sipes (1996). "Assessment of follicle destruction in chemical-induced ovarian toxicity." *Annual review of pharmacology and toxicology* 36(1): pp. 307-331, ISSN 0362-1642.

Hu, X., P. J. Christian, K. E. Thompson, I. Glenn Sipes and P. B. Hoyer (2001). "Apoptosis induced in rats by 4-vinylcyclohexene diepoxide is associated with activation of the caspase cascades." *Biology of reproduction* 65(1): pp. 87-93, ISSN 0006-3363.

Hughes, F. M. and W. C. Gorospe (1991). "Biochemical identification of apoptosis (programmed cell death) in granulosa cells: evidence for a potential mechanism underlying follicular atresia." *Endocrinology* 129(5): pp. 2415-2422, ISSN 0013-7227.

Igawa, Y., A. F. Keating, K. S. Rajapaksa, I. G. Sipes and P. B. Hoyer (2009). "Evaluation of ovotoxicity induced by 7, 12-dimethylbenz [a] anthracene and its 3, 4-diol metabolite utilizing a rat in vitro ovarian culture system." *Toxicology and applied pharmacology* 234(3): pp. 361-369, ISSN 0041-008X.

Jancovaa, P., P. Anzenbacherb and E. Anzenbacherovaa (2010). "Phase II drug metabolizing enzymes." *Biomed Pap Med Fac Univ Palacky Olomouc Czech Repub* 154(2): pp. 103-116, ISSN 1213-8118.

Janssen-Heininger, Y. M. W., B. T. Mossman, N. H. Heintz, H. J. Forman, B. Kalyanaraman, T. Finkel, J. S. Stamler, S. G. Rhee and A. van der Vliet (2008). "Redox-based regulation of signal transduction: principles, pitfalls, and promises." *Free Radical Biology and Medicine* 45(1): pp. 1-17, ISSN 0891-5849.

Jarrell, J. F., L. Bodo, E. V. Younglai, R. D. Barr and G. J. O'Connell (1991). "The short-term reproductive toxicity of cyclophosphamide in the female rat." *Reproductive Toxicology* 5(6): pp. 481-485, ISSN 0890-6238.

Keating, A. F. (2009). "Effect of phosphatidylinositol-3 kinase inhibition on ovotoxicity caused by 4-vinylcyclohexene diepoxide and 7, 12-dimethylbenz [a] anthracene in neonatal rat ovaries." *Toxicology and applied pharmacology* 241(2): pp. 127-134, ISSN 0041-008X.

Keating, A. F., S. M. Fernandez, C. J. Mark-Kappeler, N. Sen, I. G. Sipes and P. B. Hoyer (2011). "Inhibition of PIK3 Signaling Pathway Members by the Ovotoxicant 4-Vinylcyclohexene Diepoxide in Rats." *Biology of reproduction* 84(4): pp. 743-751, ISSN 0006-3363.

Keating, A. F., N. Sen, I. G. Sipes and P. B. Hoyer (2010). "Dual protective role for Glutathione S-transferase class pi against VCD-induced ovotoxicity in the rat ovary." *Toxicology and applied pharmacology* 247(2): pp. 71-75, ISSN 0041-008X.

Kidd, P. M. (1997). "Glutathione: systemic protectant against oxidative and free radical damage." *Alternative Medicine Review* 2(3): pp. 155-176, ISSN 1089-5159.

Kiffin, R., U. Bandyopadhyay and A. M. Cuervo (2006). "Oxidative stress and autophagy." *Antioxidants & redox signaling* 8(1-2): pp. 152-162, ISSN 1523-0864.

Kirkpatrick, J. F. (2007). "Measuring the effects of wildlife contraception: the argument for comparing apples with oranges." *Reproduction, Fertility and Development* 19(4): pp. 548-552, ISSN 1448-5990.

Klaunig, J. E. and L. M. Kamendulis (2004). "The role of oxidative stress in carcinogenesis." *Annu. Rev. Pharmacol. Toxicol.* 44: pp. 239-267, ISSN 0066-4170.

Kohalmy, K. and R. Vrzal (2011). "Regulation of Phase II Biotransformation Enzymes by Steroid Hormones." *Current Drug Metabolism* 12(2): pp. 104-123, ISSN 1389-2002.

Kondo, Y., T. Kanzawa, R. Sawaya and S. Kondo (2005). "Role of autophagy in cancer development and response to therapy." *Nature Reviews Cancer* 5(9): pp. 726-734, ISSN 1474-175X.

Levine, B. and D. J. Klionsky (2004). "Development by Self-Digestion:: Molecular Mechanisms and Biological Functions of Autophagy." *Developmental cell* 6(4): pp. 463-477, ISSN 1534-5807.

Levine, B. and J. Yuan (2005). "Autophagy in cell death: an innocent convict?" *Journal of Clinical Investigation* 115(10): pp. 2679-2688, ISSN 0021-9738.

Lobascio, A. M., F. G. Klinger, M. L. Scaldaferri, D. Farini and M. De Felici (2007). "Analysis of programmed cell death in mouse fetal oocytes." *Reproduction* 134(2): pp. 241-252, ISSN 1470-1626.

Luderer, U., T. J. Kavanagh, C. C. White and E. M. Faustman (2001). "Gonadotropin regulation of glutathione synthesis in the rat ovary." *Reproductive Toxicology* 15(5): pp. 495-504, ISSN 0890-6238.

Maiuri, M. C., E. Zalckvar, A. Kimchi and G. Kroemer (2007). "Self-eating and self-killing: crosstalk between autophagy and apoptosis." *Nature Reviews Molecular Cell Biology* 8(9): pp. 741-752, ISSN 1471-0072.

Martin, J. A., B. E. Hamilton, P. D. Sutton, S. J. Ventura, F. Menacker and M. L. Munson (2003). "Births: final data for 2002." *National vital statistics reports* 52(10): pp. 1-113, ISSN 1551-8930.

Matikainen, T., G. I. Perez, A. Jurisicova, J. K. Pru, J. J. Schlezinger, H. Y. Ryu, J. Laine, T. Sakai, S. J. Korsmeyer and R. F. Casper (2001). "Aromatic hydrocarbon receptor-driven Bax gene expression is required for premature ovarian failure caused by biohazardous environmental chemicals." *Nature Genetics* 28(4): pp. 355-360, ISSN 1061-4036.

Mattison, D. R. (1980). "Morphology of oocyte and follicle destruction by polycyclic aromatic hydrocarbons in mice." *Toxicology and Applied Pharmacology* 53(2): pp. 249-259, ISSN 0041-008X.

Mattison, D. R. and J. D. Schulman (1980). "How xenobiotic chemicals can destroy oocytes." *Contemp Obstet Gynecol* 15: pp. 157-169, ISSN 1471-8359.

Mattison, D. R., K. Shiromizu and M. S. Nightingale (1983a). "Oocyte destruction by polycyclic aromatic hydrocarbons." *American journal of industrial medicine* 4(1-2): pp. 191-202, ISSN 0271-3586.

Mattison, D. R., K. Shiromizu, J. A. Pendergrass and S. S. Thorgeirsson (1983b). "Ontogeny of ovarian glutathione and sensitivity to primordial oocyte destruction by cyclophosphamide." *Pediatric pharmacology (New York, NY)* 3(1): pp. 49-55, ISSN 0270-322X.

Mattison, D. R. and S. S. Thorgeirsson (1979). "Ovarian aryl hydrocarbon hydroxylase activity and primordial oocyte toxicity of polycyclic aromatic hydrocarbons in mice." *Cancer Research* 39(9): pp. 3471-3475, ISSN 0008-5472.

McAlpine, C. A., S. Heyenga, B. Taylor, A. N. N. Peterson and G. McDonald (2007). "Regional planning in Queensland's rangelands: challenges and prospects for biodiversity conservation." *Geographical Research* 45(1): pp. 27-42, ISSN 1745-5871.

McGee, E. A. and A. J. W. Hsueh (2000). "Initial and cyclic recruitment of ovarian follicles." *Endocrine reviews* 21(2): pp. 200-214, ISSN 1945-7189.

McNatty, K. P., A. E. Fidler, J. L. Juengel, L. D. Quirke, P. R. Smith, D. A. Heath, T. Lundy, A. O'Connell and D. J. Tisdall (2000). "Growth and paracrine factors regulating follicular formation and cellular function." *Molecular and Cellular Endocrinology* 163(1-2): pp. 11-20, ISSN 0303-7207.

Meirow, D., J. Dor, B. Kaufman, A. Shrim, J. Rabinovici, E. Schiff, H. Raanani, J. Levron and E. Fridman (2007). "Cortical fibrosis and blood-vessels damage in human ovaries exposed to chemotherapy. Potential mechanisms of ovarian injury." *Human reproduction* 22(6): pp. 1626-1633, ISSN 0268-1161.

Melendez-Colon, V. J., A. Luch, A. Seidel and W. M. Baird (1999). "Cancer initiation by polycyclic aromatic hydrocarbons results from formation of stable DNA adducts rather than apurinic sites." *Carcinogenesis* 20(10): pp. 1885-1891, ISSN 0143-3334.

Meunier, B., S. P. De Visser and S. Shaik (2004). "Mechanism of oxidation reactions catalyzed by cytochrome P450 enzymes." *Chemical reviews* 104(9): pp. 3947-3980, ISSN 0009-2665.

Mwanjabe, P. S. and H. Leirs (1997). "An early warning system for IPM-based rodent control in smallholder farming systems in Tanzania." *Belgian Journal of Zoology* 127: pp. 49-58, ISSN 0777-6276.

Nebert, D. W., D. D. Brown, D. W. Towne and H. J. Eisen (1984). "Association of fertility, fitness and longevity with the murine Ah locus among (C57BL/6N)(C3H/HeN) recombinant inbred lines." *Biology of reproduction* 30(2): pp. 363-373, ISSN 0006-3363.

Nebert, D. W., A. L. Roe, M. Z. Dieter, W. A. Solis, Y. I. Yang and T. P. Dalton (2000). "Role of the aromatic hydrocarbon receptor and [Ah] gene battery in the oxidative stress response, cell cycle control, and apoptosis." *Biochemical pharmacology* 59(1): pp. 65-85, ISSN 0006-2952.

Nguyen, L. P. and C. A. Bradfield (2007). "The search for endogenous activators of the aryl hydrocarbon receptor." *Chemical research in toxicology* 21(1): pp. 102-116, ISSN 0893-228X.

Park, J. H., S. Gopishetty, L. M. Szewczuk, A. B. Troxel, R. G. Harvey and M. Trevor (2005). "Formation of 8-oxo-7, 8-dihydro-2'-deoxyguanosine (8-oxo-dGuo) by PAH o-quinones: involvement of reactive oxygen species and copper (II)/copper (I) redox cycling." *Chemical research in toxicology* 18(6): pp. 1026-1037, ISSN 0893-228X.

Pavek, P. and Z. Dvorak (2008). "Xenobiotic-induced transcriptional regulation of xenobiotic metabolizing enzymes of the cytochrome P450 superfamily in human extrahepatic tissues." *Current drug metabolism* 9(2): pp. 129-143, ISSN 1389-2002.

Petrulis, J. R. and G. H. Perdew (2002). "The role of chaperone proteins in the aryl hydrocarbon receptor core complex." *Chemico-biological interactions* 141(1-2): pp. 25-40, ISSN 0009-2797.

Picut, C. A., C. L. Swanson, K. L. Scully, V. C. Roseman, R. F. Parker and A. K. Remick (2008). "Ovarian follicle counts using proliferating cell nuclear antigen (PCNA) and semi-automated image analysis in rats." *Toxicologic pathology* 36(5): pp. 674-679, ISSN 0192-6233.

Plowchalk, D. R. and D. R. Mattison (1992). "Reproductive toxicity of cyclophosphamide in the C57BL/6N mouse: 1. Effects on ovarian structure and function." *Reproductive Toxicology* 6(5): pp. 411-421, ISSN 0890-6238.

Pollenz, R. S., C. A. Sattler and A. Poland (1994). "The aryl hydrocarbon receptor and aryl hydrocarbon receptor nuclear translocator protein show distinct subcellular localizations in Hepa 1c1c7 cells by immunofluorescence microscopy." *Molecular pharmacology* 45(3): pp. 428-438, ISSN 0026-895X.

Raffray, M. and M. C. Gerald (1997). "Apoptosis and necrosis in toxicology: a continuum or distinct modes of cell death?" *Pharmacology & therapeutics* 75(3): pp. 153-177, ISSN 0163-7258.

Rajapaksa, K. S., E. A. Cannady, I. G. Sipes and P. B. Hoyer (2007a). "Involvement of CYP 2E1 enzyme in ovotoxicity caused by 4-vinylcyclohexene and its metabolites." *Toxicology and applied pharmacology* 221(2): pp. 215-221, ISSN 0041-008X.

Rajapaksa, K. S., I. G. Sipes and P. B. Hoyer (2007b). "Involvement of microsomal epoxide hydrolase enzyme in ovotoxicity caused by 7, 12-dimethylbenz [a] anthracene." *Toxicological sciences* 96(2): pp. 327-334, ISSN 1096-6080.

Ramesh, A., A. E. Archibong and M. S. Niaz (2010). "Ovarian Susceptibility to Benzo [a] Pyrene: Tissue Burden of Metabolites and DNA Adducts in F-344 Rats." *Journal of toxicology and environmental health. Part A* 73(23): pp. 1611-1625, ISSN 1528-7394.

Reddy, P., L. Liu, D. Adhikari, K. Jagarlamudi, S. Rajareddy, Y. Shen, C. Du, W. Tang, T. Hämäläinen and S. L. Peng (2008). "Oocyte-specific deletion of Pten causes premature activation of the primordial follicle pool." *Science* 319(5863): pp. 611-613, ISSN 0036-8075.

Reddy, P., W. Zheng and K. Liu (2010). "Mechanisms maintaining the dormancy and survival of mammalian primordial follicles." *Trends in Endocrinology & Metabolism* 21(2): pp. 96-103, ISSN 1043-2760.

Reyes, H., S. Reisz-Porszasz and O. Hankinson (1992). "Identification of the Ah receptor nuclear translocator protein (Arnt) as a component of the DNA binding form of the Ah receptor." *Science* 256(5060): pp. 1193-1195, ISSN 0036-8075.

Robles, R., Y. Morita, K. K. Mann, G. I. Perez, S. Yang, T. Matikainen, D. H. Sherr and J. L. Tilly (2000). "The aryl hydrocarbon receptor, a basic helix-loop-helix transcription factor of the PAS gene family, is required for normal ovarian germ cell dynamics in the mouse." *Endocrinology* 141(1): pp. 450-453, ISSN 0013-7227.

Rodrigues, P., D. Limback, L. K. McGinnis, C. E. Plancha and D. F. Albertini (2009). "Multiple mechanisms of germ cell loss in the perinatal mouse ovary." *Reproduction* 137(4): pp. 709-720, ISSN 1470-1626.

Rodríguez, M. and A. Jaime (1993). Rodent plague: a continuing problem in Latin America and the Caribbean. *Latin American Conference on Rodent Biology and Control of Plague,* ISBN 9718709053, Santa Cruz, Bolivia, October 1991.

Sanders, D. L., F. Xie, R. E. Mauldin, J. C. Hurley, L. A. Miller, M. R. Garcia, R. W. DeYoung, D. B. Long and T. A. Campbell (2011). "Efficacy of ERL-4221 as an ovotoxin for feral pigs (Sus scrofa)." *Wildlife Research* 38(2): pp. 168-172, ISSN 1448-5494.

Schafer, F. Q. and G. R. Buettner (2001). "Redox environment of the cell as viewed through the redox state of the glutathione disulfide/glutathione couple." *Free Radical Biology and Medicine* 30(11): pp. 1191-1212, ISSN 0891-5849.

Schmidt, J. V., G. H. Su, J. K. Reddy, M. C. Simon and C. A. Bradfield (1996). "Characterization of a murine Ahr null allele: involvement of the Ah receptor in hepatic growth and development." *Proceedings of the National Academy of Sciences* 93(13): pp. 6731-6736, ISSN 0027-8424.

Schroer, K., M. Kittelmann and S. Lütz (2010). "Recombinant human cytochrome P450 monooxygenases for drug metabolite synthesis." *Biotechnology and bioengineering* 106(5): pp. 699-706, ISSN 1097-0290.

Schwarz, D., P. Kisselev, I. Cascorbi, W. H. Schunck and I. Roots (2001). "Differential metabolism of benzo [a] pyrene and benzo [a] pyrene-7, 8-dihydrodiol by human CYP1A1 variants." *Carcinogenesis* 22(3): pp. 453-459, ISSN 0143-3334.

Shimada, T. and Y. Fujii Kuriyama (2004). "Metabolic activation of polycyclic aromatic hydrocarbons to carcinogens by cytochromes P450 1A1 and1B1." *Cancer science* 95(1): pp. 1-6, ISSN 1349-7006.

Shimada, T., Y. Oda, E. M. J. Gillam, F. P. Guengerich and K. Inoue (2001). "Metabolic activation of polycyclic aromatic hydrocarbons and other procarcinogens by cytochromes P450 1A1 and P450 1B1 allelic variants and other human cytochromes P450 in Salmonella typhimurium NM2009." *Drug metabolism and disposition* 29(9): pp. 1176-1182, ISSN 0090-9556.

Shimada, T., A. Sugie, M. Shindo, T. Nakajima, E. Azuma, M. Hashimoto and K. Inoue (2003). "Tissue-specific induction of cytochromes P450 1A1 and 1B1 by polycyclic aromatic hydrocarbons and polychlorinated biphenyls in engineered C57BL/6J mice of arylhydrocarbon receptor gene." *Toxicology and applied pharmacology* 187(1): pp. 1-10, ISSN 0041-008X.

Shimizu, S., T. Kanaseki, N. Mizushima, T. Mizuta, S. Arakawa-Kobayashi, C. B. Thompson and Y. Tsujimoto (2004). "Role of Bcl-2 family proteins in a non-apoptotic programmed cell death dependent on autophagy genes." *Nature cell biology* 6(12): pp. 1221-1228, ISSN 1465-7392.

Shiromizu, K. and D. R. Mattison (1985). "Murine oocyte destruction following intraovarian treatment with 3 methylcholanthrene or 7, 12 dimethylbenz (a) anthracene: Protection by alpha naphthoflavone." *Teratogenesis, carcinogenesis, and mutagenesis* 5(6): pp. 463-472, ISSN 1520-6866.

Shuster, L. T., D. J. Rhodes, B. S. Gostout, B. R. Grossardt and W. A. Rocca (2010). "Premature menopause or early menopause: long-term health consequences." *Maturitas* 65(2): pp. 161-166, ISSN 0378-5122.

Singleton, G. R. (2003). "Impacts of rodents on rice production in Asia." *IRRI Discuss. Pap. Ser* 45: pp. 1-30, ISSN 0115-3862.

Smith, B. J., D. R. Mattison and I. G. Sipes (1990). "The role of epoxidation in 4-vinylcyclohexene-induced ovarian toxicity." *Toxicology and applied pharmacology* 105(3): pp. 372-381, ISSN 0041-008X.

Sobinoff, A. P., M. Mahony, B. Nixon, S. D. Roman and E. A. McLaughlin (2011). "Understanding the villain: DMBA induced pre-antral ovotoxicity involves selective follicular destruction and primordial follicle activation through PI3K/Akt and mTOR signalling." *Toxicological Sciences*: pp. kfr195v1-kfr195, ISSN 1096-6080.

Sobinoff, A. P., V. Pye, B. Nixon, S. D. Roman and E. A. McLaughlin (2010). "Adding Insult to Injury: Effects of Xenobiotic-Induced Preantral Ovotoxicity on Ovarian Development and Oocyte Fusibility." *Toxicological Sciences* 118(2): pp. 653-666, ISSN 1096-6080.

Springer, L. N., M. E. McAsey, J. A. Flaws, J. L. Tilly, I. G. Sipes and P. B. Hoyer (1996). "Involvement of apoptosis in 4-vinylcyclohexene diepoxide-induced ovotoxicity in rats." *Toxicology and applied pharmacology* 139(2): pp. 394-401, ISSN 0041-008X.

Stapleton, H. M. and J. E. Baker (2003). "Comparing polybrominated diphenyl ether and polychlorinated biphenyl bioaccumulation in a food web in Grand Traverse Bay, Lake Michigan." *Archives of environmental contamination and toxicology* 45(2): pp. 227-234, ISSN 0090-4341.

Tarin, J. J. (1996). "Potential effects of age-associated oxidative stress on mammalian oocytes/embryos." *Molecular human reproduction* 2(10): pp. 717-724, ISSN 1360-9947.

Taylor, K. D. (1968). "An outbreak of rats in agricultural areas of Kenya in 1962." *East African Agricultural and Forestry Journal* 34(1): pp. 66-77, ISSN 0012-8325.

Terman, A., B. Gustafsson and U. T. Brunk (2006). "The lysosomal-mitochondrial axis theory of postmitotic aging and cell death." *Chemico-biological interactions* 163(1-2): pp. 29-37, ISSN 0009-2797.

Testa, J. R. and A. Bellacosa (2001). "AKT plays a central role in tumorigenesis." *Proceedings of the National Academy of Sciences* 98(20): pp. 10983-10985, ISSN 0027-8424.

Thimmulappa, R. K., K. H. Mai, S. Srisuma, T. W. Kensler, M. Yamamoto and S. Biswal (2002). "Identification of Nrf2-regulated genes induced by the chemopreventive agent sulforaphane by oligonucleotide microarray." *Cancer research* 62(18): pp. 5196-5203, ISSN 0008-5472.

Thor, H., M. T. Smith, P. Hartzell, G. Bellomo, S. A. Jewell and S. Orrenius (1982). "The metabolism of menadione (2-methyl-1, 4-naphthoquinone) by isolated hepatocytes. A study of the implications of oxidative stress in intact cells." *Journal of Biological Chemistry* 257(20): pp. 12419-12425, ISSN 0021-9258.

Tilly, J. L., K. I. Kowalski, A. L. Johnson and A. J. W. Hsueh (1991). "Involvement of apoptosis in ovarian follicular atresia and postovulatory regression." *Endocrinology* 129(5): pp. 2799-2801, ISSN 0013-7227.

Tilly, J. L. and R. Robles (1999). Apoptosis and its impact in clinical reproductive medicine, *Molecular biology in reproductive medicine*, B. C. J. M. Fauser, A. J. Rutherford, J. F. Strauss and A. Van Steirteghem, pp. 79–101, ISBN 1850709947. New York, USA.

Tománek, M. and E. Chronowska (2006). "Immunohistochemical localization of proliferating cell nuclear antigen (PCNA) in the pig ovary." *Folia histochemica et Cytobiologica* 44(4): pp. 269-274, ISSN 0239-8508.

Trevor, M., S. T. Ohnishi, T. Ohnishi and R. G. Harvey (1996). "Generation of reactive oxygen species during the enzymatic oxidation of polycyclic aromatic hydrocarbon trans-dihydrodiols catalyzed by dihydrodiol dehydrogenase." *Chemical research in toxicology* 9(1): pp. 84-92, ISSN 0893-228X.

Tsai-Turton, M., B. N. Nakamura and U. Luderer (2007). "Induction of apoptosis by 9, 10-dimethyl-1, 2-benzanthracene in cultured preovulatory rat follicles is preceded by a rise in reactive oxygen species and is prevented by glutathione." *Biology of Reproduction* 77(3): pp. 442-451, ISSN 0006-3363.

Valko, M., D. Leibfritz, J. Moncol, M. T. D. Cronin, M. Mazur and J. Telser (2007). "Free radicals and antioxidants in normal physiological functions and human disease." *The international journal of biochemistry & cell biology* 39(1): pp. 44-84, ISSN 1357-2725.

Valko, M., C. J. Rhodes, J. Moncol, M. Izakovic and M. Mazur (2006). "Free radicals, metals and antioxidants in oxidative stress-induced cancer." *Chemico-biological interactions* 160(1): pp. 1-40, ISSN 0009-2797.

van Wezel, I. L. and R. J. Rodgers (1996). "Morphological characterization of bovine primordial follicles and their environment in vivo." *Biology of reproduction* 55(5): pp. 1003-1011, ISSN 0006-3363.

Vlahos, C. J., W. F. Matter, K. Y. Hui and R. F. Brown (1994). "A specific inhibitor of phosphatidylinositol 3-kinase, 2-(4-morpholinyl)-8-phenyl-4H-1-benzopyran-4-one (LY294002)." *Journal of Biological Chemistry* 269(7): pp. 5241-5248, ISSN 0021-9258.

Wang, K., H. Fang, D. Xiao, X. Zhu, M. He, X. Pan, J. Shi, H. Zhang, X. Jia and Y. Du (2009). "Converting redox signaling to apoptotic activities by stress-responsive regulators

HSF1 and NRF2 in fenretinide treated cancer cells." *PLoS One* 4(10): pp. e7538, ISSN 1932-6203.

Wang, Q., Y. Zhou, L. N. Jackson, S. M. Johnson, C. W. Chow and B. M. Evers "Nuclear factor of activated T cells (NFAT) signaling regulates PTEN expression and intestinal cell differentiation." *Molecular Biology of the Cell* 22(3): pp. 412,

Wells, P. G., G. P. McCallum, C. S. Chen, J. T. Henderson, C. J. J. Lee, J. Perstin, T. J. Preston, M. J. Wiley and A. W. Wong (2009). "Oxidative stress in developmental origins of disease: teratogenesis, neurodevelopmental deficits, and cancer." *Toxicological sciences* 108(1): pp. 4-18, ISSN 1096-6080.

Wu, K. L. and T. Berger (2008). "Reduction in rat oocyte fertilizability mediated by S-(1, 2-dichlorovinyl)-L-cysteine: a trichloroethylene metabolite produced by the glutathione conjugation pathway." *Bulletin of environmental contamination and toxicology* 81(5): pp. 490-493, ISSN 0007-4861.

Permissions

The contributors of this book come from diverse backgrounds, making this book a truly international effort. This book will bring forth new frontiers with its revolutionizing research information and detailed analysis of the nascent developments around the world.

We would like to thank Dr. Tetsuji Nagata, for lending his expertise to make the book truly unique. He has played a crucial role in the development of this book. Without his invaluable contribution this book wouldn't have been possible. He has made vital efforts to compile up to date information on the varied aspects of this subject to make this book a valuable addition to the collection of many professionals and students.

This book was conceptualized with the vision of imparting up-to-date information and advanced data in this field. To ensure the same, a matchless editorial board was set up. Every individual on the board went through rigorous rounds of assessment to prove their worth. After which they invested a large part of their time researching and compiling the most relevant data for our readers. Conferences and sessions were held from time to time between the editorial board and the contributing authors to present the data in the most comprehensible form. The editorial team has worked tirelessly to provide valuable and valid information to help people across the globe.

Every chapter published in this book has been scrutinized by our experts. Their significance has been extensively debated. The topics covered herein carry significant findings which will fuel the growth of the discipline. They may even be implemented as practical applications or may be referred to as a beginning point for another development.

The editorial board has been involved in producing this book since its inception. They have spent rigorous hours researching and exploring the diverse topics which have resulted in the successful publishing of this book. They have passed on their knowledge of decades through this book. To expedite this challenging task, the publisher supported the team at every step. A small team of assistant editors was also appointed to further simplify the editing procedure and attain best results for the readers.

Our editorial team has been hand-picked from every corner of the world. Their multi-ethnicity adds dynamic inputs to the discussions which result in innovative outcomes. These outcomes are then further discussed with the researchers and contributors who give their valuable feedback and opinion regarding the same. The feedback is then collaborated with the researches and they are edited in a comprehensive manner to aid the understanding of the subject.

Apart from the editorial board, the designing team has also invested a significant amount of their time in understanding the subject and creating the most relevant covers. They scrutinized every image to scout for the most suitable representation of the subject and create an appropriate cover for the book.

The publishing team has been involved in this book since its early stages. They were actively engaged in every process, be it collecting the data, connecting with the contributors or procuring relevant information. The team has been an ardent support to the editorial, designing and production team. Their endless efforts to recruit the best for this project, has resulted in the accomplishment of this book. They are a veteran in the field of academics and their pool of knowledge is as vast as their experience in printing. Their expertise and guidance has proved useful at every step. Their uncompromising quality standards have made this book an exceptional effort. Their encouragement from time to time has been an inspiration for everyone.

The publisher and the editorial board hope that this book will prove to be a valuable piece of knowledge for researchers, students, practitioners and scholars across the globe.

List of Contributors

Thiago Monaco, Daniel Silvestre and Paulo S.P. Silveira
Department of Pathology, School of Medicine at the University of Sao Paulo, Sao Paulo, Brazil

Jaba Tkemaladze
Georgian Systemic Research Center, Tbilisi, Georgia

Alexander Tavartkiladze
Grigol Robakidze University's Oncology Department, Tbilisi, Georgia

Konstantin Chichinadze
Life Science Research Center, Tbilisi, Georgia
I. Javakhishvili Tbilisi State University, Tbilisi, Georgia

Mikael Altun, Max Grönholdt-Klein and Brun Ulfhake
Department of Neuroscience, Karolinska Institutet, Stockholm, Sweden

Lingzhan Wang
Department of Neuroscience, Karolinska Institutet, Stockholm, Sweden
Department of Human Anatomy, School of Medicine, Inner Mongolia University for Nationalities, Tongliao City, P. R. China

Nadine Hein and Katherine M. Hannan
Division of Cancer Research, Peter MacCallum Cancer Centre, Melbourne, Australia

Elaine Sanij
Division of Cancer Research, Peter MacCallum Cancer Centre, Melbourne, Australia
Department of Pathology, University of Melbourne, Melbourne, Australia

Jaclyn Quin
Division of Cancer Research, Peter MacCallum Cancer Centre, Melbourne, Australia
Department of Biochemistry and Molecular Biology, Australia

Austen Ganley
Institute of Natural Sciences, Massey University, Auckland, New Zealand

Ross D. Hannan
Division of Cancer Research, Peter MacCallum Cancer Centre, Melbourne, Australia
Department of Biochemistry and Molecular Biology, Australia
Department of Biochemistry and Molecular Biology, Monash University, Melbourne, Australia
School of Biomedical Sciences, The University of Queensland, Melbourne, Australia

Paula Medone, Jorge Rabinovich, Eliana Nieves and Soledad Ceccarelli
1Parasitological and Vector Studies Center (CEPAVE), National University of La Plata, La Plata, Argentina

Delmi Canale and Raúl L. Stariolo
Vectors Reference Center, National Coordination of Vectors Control, Direction of Transmitted Vectors Diseases, National Ministry of Health, Córdoba, Argentina

Frédéric Menu
Laboratory of Biometry and Biology Evolutionary, CNRS, UMR 5558, Lyon University F-69000 Lyon, Claude Bernard University, Lyon 1 F-69622, Villeurbanne, France

Tetsuji Nagata
Department of Anatomy and Cell Biology, Shinshu University School of Medicine, Matsumot, Japan
Shinshu Institute of Alternative Medicine and Welfare, Nagano, Japan

Keith Wheaton
Department of Biology, York University, Toronto, Canada

Diego Julio Arenas-Aranda and Fabio Salamanca-Gómez
Unit of Medical Research in Human Genetics, Medical National Center 21st Century,Mexican Institute of Social Insurance, Mexico

Elena Hernández-Caballero
Section of Posgrado's Studies and Investigation, High School of Medicine, Technical National Institute. Mexico City, Mexico

Milena Atanasova
Department of Biology, Medical University of Pleven, Pleven, Bulgaria

Aneliya Dimitrova
Department of Pathophysiology, Medical University of Pleven, Pleven, Bulgaria

Boryana Ruseva
Department of Physiology, Medical University of Pleven, Pleven, Bulgaria

Angelina Stoyanova
Department of Chemistry, Medical University of Pleven, Pleven, Bulgaria

Miglena Georgieva
Medical Center "Clinical Institute for Reproductive Medicine", Pleven, Bulgaria

Emiliana Konova
Medical Center "Clinical Institute for Reproductive Medicine", Pleven, Bulgaria
Center for Reproductive Health; Medical University of Pleven, Pleven, Bulgaria

Francisco J. Blanco and Carmelo Bernabéu
Centro de Investigaciones Biológicas, CSIC, Centro de Investigación Biomédica en Red de Enfermedades Raras, Spain

John N. Buchholz
Department of Physiology and Pharmcology, Loma Linda University, School of Medicine, Loma Linda, California, USA

William J. Pottorf
Department of Medical Affairs, Aztra Zeneca, Wilmington, Deleware, USA

Conwin K. Vanterpool
Department of Biological Sciences, Alabama A&M, Normal, Alabama, USA

Erik J. Behringer
Department of Medical Pharmacology and Physiology, University of Columbia, Columbia, Missouri, USA

Sue P. Duckles
Department of Pharmacology, College of Medicine, University of California, Irvine, California, USA

Alexander P. Sobinoff, Ilana R. Bernstein and Eileen A. McLaughlin
Priority Research Centre in Chemical Biology, University of Newcastle, Australia